Lecture Notes in Computational Science and Engineering

Michael Griebel • Marc Alexander Schweitzer
Editors

Meshfree Methods for Partial Differential Equations IV

With 157 Figures and 17 Tables

Michael Griebel
Marc Alexander Schweitzer
Institut für Numerische Simulation
Universität Bonn
Wegelerstrasse 6
53115 Bonn
Germany
griebel@ins.uni-bonn.de
schweitzer@ins.uni-bonn.de

ISBN 978-3-540-79993-1 e-ISBN 978-3-540-79994-8

Lecture Notes in Computational Science and Engineering ISSN 1439-7358

Library of Congress Control Number: 2008933852

Mathematics Subject Classification (2000): 65N99, 64M99, 65M12, 65Y99

Cover design: WMX Design GmbH, Heidelberg

Printed on acid-free paper

9 8 7 6 5 4 3 2 1

spinger.com

Preface

The Fourth International Workshop on *Meshfree Methods for Partial Differential Equations* was held from September 17 to September 20, 2007 in Bonn, Germany. One of the major goals of this workshop series is to bring together European, American and Asian researchers working in this exciting field of interdisciplinary research on a regular basis. To this end Ivo Babuška, Ted Belytschko, Michael Griebel, Antonio Huerta, Wing Kam Liu, and Harry Yserentant invited scientist from twelve countries to Bonn to strengthen the mathematical understanding and analysis of meshfree discretizations but also to promote the exchange of ideas on their implementation and application.

The workshop was again hosted by the Institut für Numerische Simulation at the Rheinische Friedrich-Wilhelms-Universität Bonn with the financial support of the Sonderforschungsbereich 611 *Singular Phenomena and Scaling in Mathematical Models.*

This volume of LNCSE now comprises selected contributions of attendees of the workshop. Their content ranges from applied mathematics to physics and engineering. This is also an indication that meshfree methods for the numerical solution of partial differential equations are becoming more and more mainstream in many areas of applications due to their flexiblity and wide applicability.

Bonn,
May, 2008

Michael Griebel
Marc Alexander Schweitzer

Contents

Circumventing Curse of Dimensionality in the Solution of Highly Multidimensional Models Encountered in Quantum Mechanics Using Meshfree Finite Sums Decomposition

Amine Ammar[1] and Francisco Chinesta[2]

[1] Laboratoire de Rhéologie, UJF - INPG - CNRS (UMR 5520)
1301 Rue de la Piscine,
BP 53 Domaine Universitaire, F-38041 Grenoble cedex 9, France
Amine.Ammar@ujf-grenoble.fr

[2] Laboratoire de Mécanique des Systèmes et des Procédés,
ENSAM - CNRS (UMR 8106)
151 Boulevard de l'Hôpital, F-75013 Paris, France
francisco.chinesta@paris.ensam.fr

Summary. The fine description of the mechanics and structure of materials at nanometric scale introduces some specific challenges related to the impressive number of degrees of freedom required due to the highly dimensional spaces in which those models are defined. This is the case of quantum mechanics models, in which the wavefunction is defined in a space of dimension $3 \times N_p$, being N_p the number of particles involved, that leads to the terrific curse of dimensionality. Despite the fact that spectacular progresses have been accomplished in the context of computational mechanics in the last decade, the treatment of those models, as we describe in the present work, needs further developments.

Key words: Quantum Mechanics, Curse of dimensionality, Schrödinger equation, Separated representations

1 Introduction

The brut force approach cannot be envisaged for solving highly dimensional models. Some recent approaches allowed computing models defined in moderate multidimensional spaces, as was the case of the sparse grid techniques [4]. However, when the space dimension increases significantly, its treatment becomes delicate. Some specialists as the Nobel Prize R.G. Laughlin, affirmed that no computer existing, or that will ever exist, can break the barriers found in quantum mechanics because it is a catastrophe of dimension [9].

We can understand the catastrophe of dimension by assuming a model defined in a hyper-cube in a space of dimension D, $\Omega =]-L, L[^D$. Now, if we define a grid to discretize the model, as it is usually performed in the vast majority of numerical methods (finite differences, finite elements, finite volumes, spectral methods etc.), consisting of N nodes on each direction, the total number of nodes will be N^D. If we assume that for example $N \approx 10$ (an extremely coarse description) and $D \approx 80$ (much lower than the usual dimensions required in quantum or statistical mechanics), the number of nodes involved in the discrete model reaches the astronomical value of 10^{80} that represents the presumed number of elementary particles in the universe!. We shall come back to the analysis of these systems later.

Thus, progresses on this field need further developments on the physical modelling as well as the introduction of new ideas and methods in the context of computational physics. In this work we present a recent solution procedure based on the use of a finite sums decomposition, that leads to a separated representation of the involved unknown fields. This solution strategy was successfully applied in [1] [2] for solving different highly dimensional models (involving hundreds of dimensions) encountered in the kinetic theory modelling of complex fluids. In this paper we analyze the suitability of its extension for treating some simple quantum systems. We will prove that the main difficulty related to the solution of the Schrödinger equation for fermions is more in the antisymmetry constraint that the Pauli's principle implies, than in its highly dimensional character.

1.1 Quantum systems

The quantum state of a given electronic distribution could be determined by solving the Schrödinger equation. This equation has been for longtime considered as one of the finest descriptions of the world. However, before focusing on the challenges of its numerical solution, we would like to recall that this equation is not relativistic and then it fails when it is applied to describe heavy atoms. Moreover the Pauli's principle constraint was introduced in the Schrödinger formalism in an "ad hoc" way, being the reason of the main numerical difficulties. In our opinion, the best description lies in the solution of the relativistic Dirac's equation (in which the Pauli's principle is implicitly taken into account) within the framework of the quantum field theory. However, the number of works addressing the solution of that equation is nowadays quite reduced.

Some simplificative hypotheses are usually introduced, as for example the Born-Oppenheimer that states that the nuclei can be in first approximation assumed as classical point-like particles, that the state of electrons only depends on the nuclei positions and that the electronic ground state corresponds to the one that minimizes the electronic energy for a nuclear configuration. This equation defines a multidimentionnal problem whose dimension increases linearly with the number of the electrons in the system.

Thus, the knowledge of a quantum system reduces to the determination of the wavefunction $\Psi(\mathbf{x}_1, \mathbf{x}_2, \cdots, \mathbf{x}_N, t; \mathbf{X}_1, \cdots, \mathbf{X}_M)$ (that establishes that the electronic wavefunction depends parametrically on the nuclei positions) whose evolution is governed by the Schrödinger equation:

$$i\hbar\frac{\partial\Psi}{\partial t} = -\frac{\hbar^2}{2m_e}\sum_{e=1}^{e=N}\nabla_e^2\Psi + \sum_{e=1}^{e=N-1}\sum_{e'=e+1}^{e'=N}V_{ee'}\Psi + \sum_{e=1}^{e=N}\sum_{n=1}^{n=M}V_{en}\Psi \tag{1.1}$$

where N is the number of electrons and M the number of nuclei, the last ones assumed located and fixed at positions $\mathbf{X}_j$. Each electron is defined in the whole physical space $\mathbf{x}_j \in \mathbb{R}^3$, $i = \sqrt{-1}$, $\hbar$ represents the Planck's constant divided by 2π and m_e is the electron mass.

The differential operator ∇_e^2 is defined in the conformation space of each particle, i.e.: $\nabla_e^2 \equiv \partial^2/\partial x_e^2 + \partial^2/\partial y_e^2 + \partial^2/\partial z_e^2$. The Coulomb's potentials accounting for the electron-electron and electron-nuclei interactions writes:

$$V_{ee'} = \frac{(q_e)^2}{\|\mathbf{x}_e - \mathbf{x}_{e'}\|} \tag{1.2}$$

$$V_{en} = -\frac{q_n q_e}{\|\mathbf{x}_e - \mathbf{X}_n\|} \tag{1.3}$$

where the electron charge is represented by q_e and the nuclei charge by $q_n = |q_e| \times Z$ (Z being the atomic number).

The time independent Schrödinger equation (from which one could determine the ground state, perform quantum static computations or accomplishing separated representations of the time-dependent solution) writes:

$$-\frac{\hbar^2}{2m_c}\sum_{e=1}^{e=N}\nabla_e^2\Psi + \sum_{e=1}^{e=N-1}\sum_{e'=e+1}^{e'=N}V_{ee'}\Psi + \sum_{e=1}^{e=N}\sum_{n=1}^{n=M}V_{en}\Psi = E\,\Psi \tag{1.4}$$

where the ground state corresponds to the eigenfunction Ψ_0 associated with the most negative eigenvalue E_0.

Several techniques have been proposed for solving this equation. Some of them lie in the direct solution of the (time-independent or time-dependent) Schrödinger equation. Due to the curse of dimensionality its solution is only possible for very reduced populations of electrons.

Other solution strategy is based on the Hartree-Fock (HF) approach and its derived approaches (post-Hartree-Fock methods). The main assumption of this approach lies in the approximation of the joint electronic wavefunction (related to the N electrons) as a product of N $3D$-functions (the molecular orbitals) verifying the antisymmetry restriction derived from the Pauli's principle. Thus, the original HF approach consists of writing the joint wavefunction from a single Slater's determinant. The Schödinger equation allows computing the N molecular orbitals after solving the resulting strongly non-linear problem. This technique has been extensively used in quantum chemistry to analyze the structure and behavior of molecules involving a moderate number of electrons. Of course, the HF assumption represents sometimes a too crude approximation which invalidate the derived results.

To circumvent this crude approximation different multi-determinant approaches have been proposed. Interested readers can refer the excellent overview of Cancès et al. [5] as well as the different chapters of the handbook on computational chemistry [8]. The simplest possibility consists in writing the solution as a linear combination of some Slater determinants built by combining n molecular orbitals, with $n > N$. These molecular orbitals are assumed known (e.g. the orbitals related to the hydrogen atom) and the weights are searched to minimize the electronic energy. When the molecular orbitals are built from the Hartree-Fock solution (by employing the ground state and some excited eigenfunctions) the technique is known as Configuration Interaction method (CI). A more sophisticated technique consists in writing this many-determinants approximation of the solution by using a number of molecular orbitals n (with $n > N$) assumed unknown. Thus, the minimization of the electronic energy leads to compute simultaneously the molecular orbitals as well as the associated coefficients of this many-determinants expansion. Obviously, each one of these unknown molecular orbitals are expressed in an appropriate functional basis (e.g. gaussian functions, ...). This strategy is known as Multi-Configuration Self-Consistent Field (MCSCF).

All the just mentioned strategies (and others like the coupled cluster or the Moller-Plesset perturbation methods) belong to the family of the wave-function based methods. In any case all these methods can be only used to solve quantum systems composed of a moderate number of electrons. As we confirm later the main difficulty is not in the dimensionality of the space, but in the use of the Slater determinants (needed to take into account the Pauli's principle) whose complexity scales on the factorial of the number of electrons, i.e. in $N!$.

The second family of approximation methods, widely used in quantum systems composed of hundreds, thousands and even millions of electrons, are based on the density functional theory (DFT). These models, more than looking for the expression of the wavefunction (with the associated multi-dimensional issue) look for the electronic distribution $\rho(\mathbf{x})$ itself. The main difficulties of this approach are related to the expressions of both the kinetic energy of electrons and the inter-electronic repulsion energy. The second term is usually modelled from the electrostatic self-interaction energy of a charge distribution $\rho(\mathbf{x})$. On the other hand the kinetic energy term is also evaluated in an approximate manner (from the electronic distribution itself in the Thomas-Fermi and related orbital-free DFT models or from a system of N non-interacting electrons –Kohn-Sham models–). Obviously, due to the just referred approximations introduced in the kinetic and inter-electronic interaction energies, a correction term is needed, the so-called exchange-correlation-residual-kinetic energy. However, no exact expression of this correction term exists and then different approximate expressions have been proposed and used. Thus, the validity and accuracy of the computed results will depend on the accuracy of the the exchange-correlation term that must be fitted for each system.

The models related to the Thomas-Fermi, less accurate in the practice because the too phenomenological expression of the kinetic energy coming from the reference system of an uniform non-interacting electron gas, allows to consider large multi-electronic systems. In a recent work, Gavini et al. [7] performed multi-million atom simulations by employing the Thomas-Fermi-Weizsacker family of orbital-free kinetic energy functionals. On the other hand, the Kohn-Sham based models are a priori more accurate, but they need the computation of the N eigenfunctions related to the N lowest eigenvalues of a non-physical atom composed of N non-interacting electrons. In [6] this last approach was considered, and enhanced numerical strategies based on the partition of unity paradigm were introduced.

Transient solutions are very common in the context of quantum gas dynamics (physics of plasma) but are more infrequent in material science when the structure and properties of molecules or crystals are concerned. For this reason, in what follows, we are focusing on the solution of the time-independent Schrödinger equation which leads to the solution of the associated multidimensional eigenproblem, whose eigenfunction related to the most negative eigenvalue constitutes the ground state of the system.

Quantum chemistry calculations performed in the Born-Oppenheimer setting consist either (i) in solving the geometry optimization problem, that is, to compute the equilibrium molecular configuration (nuclei distribution) that minimizes the energy of the system, finding the most stable molecular configuration that determines numerous properties like for instance infrared spectrum or elastic constants; or (ii) in performing an "*ab initio*" molecular dynamics simulation, that is, to simulate the time evolution of the molecular structure according to the Newton law of classical mechanics. Molecular dynamics simulations allow to compute various transport properties (thermal conductivity, viscosity, ...) as well as some others non-equilibrium properties.

1.2 From "ab initio" to molecular dynamics

Depending on the choice of the method, on the accuracy required, and on the computer facility available, the *ab initio* methods allow today for the simulations of systems up to ten, one hundred or some million atoms. In time dependent simulations, they are only convenient for small-time simulations, say not more than a picosecond. However, some times larger systems are concerned, and for this purpose one must focus on faster approaches, obviously less accurate. Two possibilities exist: the semi-empirical and the empirical approaches. The semi-empirical approaches speed up the *ab initio* methods by profiting of the information coming from experiments or previous simulations.

Empirical methods go on by considering explicitly only the nuclei, by introducing "empirical" potentials leading to the forces acting on the nuclei. Thus, in the stationary setting only the stable configuration is searched, and for this a geometrical optimization (to computed the nuclei equilibrium distribution) is addressed leading to the so-called molecular mechanics. The transient setting results in the classical molecular dynamics but now the computation is speed up of many orders of magnitude with respect to the molecular dynamics where the potentials are computed at the *ab initio* level.

Thus, if we assume a population of M nuclei (of mass m_n) and a two-body potential (many-body potentials are also available), now the Newton's law writes for a generic nuclei n:

$$m_n \frac{d^2 \mathbf{X}_n}{dt^2} = \sum_{k=1, k \neq n} \mathbf{F}_k^n, \qquad \forall n \in [1, \cdots, M] \tag{1.5}$$

where F_k^n denotes the force acting on nucleus n originated by the presence of nucleus k. Obviously these forces can be computed from the gradient of the assumed inter-particles potentials.

2 Solving the time independent Schrödinger equation

2.1 Dimensionless form

The dimensionless form of the Schrödinger equation is performed considering the characteristic time $\frac{\hbar^3}{q_e^4 m_e}$ and the characteristic length $\frac{\hbar^2}{q_e^2 m_e}$. The nuclei charge q_n becomes dimensionless by using the electron charge q_e. Thus, for the nuclei, the dimensionless charge results $Z = q_n / |q_e|$. For a system consisting of N electrons and M nuclei, the eigenproblem associated with the dimensionless time-independent Schrödinger equation writes:

$$(\mathcal{H} - E)\Psi = 0 \tag{2.6}$$

or

$$(V(\mathbf{x}_1, \mathbf{x}_2, \cdots, \mathbf{x}_{N_e}) - E)\Psi - \sum_{e=1}^{e=N} \frac{\nabla_e^2 \Psi}{2} = 0 \tag{2.7}$$

where the Hamiltonian writes

$$\mathcal{H}(\Psi) = -\sum_{e=1}^{e=N} \frac{\nabla_e^2 \Psi}{2} + \underbrace{\sum_{e=1}^{e=N} \sum_{n=1}^{n=M} V_{en} \Psi + \sum_{e=1}^{e=N-1} \sum_{e'=e+1}^{e'=N} V_{ee'} \Psi}_{V} \tag{2.8}$$

where

$$V_{ee'} = \frac{1}{\|\mathbf{x}_e - \mathbf{x}_{e'}\|} \tag{2.9}$$

and

$$V_{en} = -\frac{Z}{\|\mathbf{x}_e - \mathbf{X}_n\|} \tag{2.10}$$

2.2 Electronic density and Pauli's exclusion principle

As soon as the ground state wavefunction is known the electron density associated to the ground state can be computed by applying:

$$\rho_e(\mathbf{x}) = \int_{\mathbb{R}^{3\times(N-1)}} |\Psi_0|^2 \, d\mathbf{x}_1 \cdots d\mathbf{x}_{e-1} d\mathbf{x}_{e+1} \cdots d\mathbf{x}_N \tag{2.11}$$

that allows computing the electronic density at each point of the space according to:

$$\rho(\mathbf{x}) = \sum_{e=1}^{e=N} \rho_e(\mathbf{x}) \tag{2.12}$$

Probably, the most important difficulty in solving the time-independent Schrödinger equation lies in the fact that the electrons are indistinguishable, that is, they are not labelled. Thus, the many-electrons wavefunction must reflect this fact. If we use $\mathbf{x}_1$ and $\mathbf{x}_2$ to describe the coordinates of two different electrons, then:

$$|\Psi(\mathbf{x}_1, \mathbf{x}_2, \mathbf{x}_3, \cdots, \mathbf{x}_N)|^2 = |\Psi(\mathbf{x}_2, \mathbf{x}_1, \mathbf{x}_3, \cdots, \mathbf{x}_N)|^2 \tag{2.13}$$

Thus, if Π is any of the $N!$ permutations of the N electronic coordinates, then:

$$\Pi \, |\Psi(\mathbf{x}_1, \mathbf{x}_2, \mathbf{x}_3, \cdots, \mathbf{x}_N)|^2 = |\Psi(\mathbf{x}_1, \mathbf{x}_2, \mathbf{x}_3, \cdots, \mathbf{x}_N)|^2 \tag{2.14}$$

that implies just two possibilities:

$$\Pi\Psi(\mathbf{x}_1, \mathbf{x}_2, \mathbf{x}_3, \cdots, \mathbf{x}_N) = \Psi(\mathbf{x}_1, \mathbf{x}_2, \mathbf{x}_3, \cdots, \mathbf{x}_N) \tag{2.15}$$

or

$$\Pi\Psi(\mathbf{x}_1, \mathbf{x}_2, \mathbf{x}_3, \cdots, \mathbf{x}_N) = -\Psi(\mathbf{x}_1, \mathbf{x}_2, \mathbf{x}_3, \cdots, \mathbf{x}_N) \tag{2.16}$$

The most general statement of the Pauli's exclusion principle for electrons (fermions in the most general case) establishes that an acceptable many-electrons wavefunction must be antisymmetric with respect to the exchange of the coordinates of any two electrons.

This antisymmetry condition is usually expressed from the Slater's determinant containing all the possible permutations. Thus, if one consider that $F_1(\mathbf{x}_1)$, $F_2(\mathbf{x}_2)$, ..., $F_N(\mathbf{x}_N)$ are a set of N functions, each one defined in the space of the associated electron, then an antisymmetric form in the $\mathcal{R}^{3\times N}$ space is obtained by permuting these functions according to:

$$\mathcal{A}\left(\prod_{i=1}^{N} F_i(\mathbf{x}_i)\right) = \frac{1}{N!} \begin{vmatrix} F_1(\mathbf{x}_1) & F_1(\mathbf{x}_2) & \cdots & F_1(\mathbf{x}_N) \\ F_2(\mathbf{x}_1) & F_2(\mathbf{x}_2) & \cdots & F_2(\mathbf{x}_N) \\ \vdots & \vdots & \ddots & \vdots \\ F_N(\mathbf{x}_1) & F_N(\mathbf{x}_2) & \cdots & F_N(\mathbf{x}_N) \end{vmatrix} \tag{2.17}$$

In practice, this antisymmetry also affects the spin. For this reason we define the generalized electronic coordinates that involve the physical and the spin coordinates according to:

$$(\mathbf{x}_i)^{\mathrm{T}} = (x_i, y_i, z_i, s_i) \tag{2.18}$$

where s_i represents the spin of electron i with two possible values: $s_i = \pm s$.

2.3 Numerical issues related to the treatment of the Coulomb potential

In order to circumvent the difficulty related to the integration of the inverse of the distance function that appears in the Coulomb inter-electronic potential term in the variational formulation of the Schrödinger equation, two simple alternatives exist: (i) the first one consists in performing an integration of this function in the $6D$ space in which the integral is defined; and (ii) the second alternative lies in performing a smoothing of the inverse of the distance function using a smoothing parameter ε

$$\frac{1}{\|\mathbf{x}_e - \mathbf{x}_{e'}\|_\varepsilon} = \frac{1}{\sqrt{\varepsilon + \|\mathbf{x}_e - \mathbf{x}_{e'}\|^2}} \tag{2.19}$$

We verify numerically that for $\varepsilon < 0.01$ the computed solution of the dimensionless Schrödinger equation (that involves a unit characteristic length) were quite similar (with differences lower than one percent). The advantage of this second alternative is that, as described later, that smoothed function can be approximated by a finite sums decomposition (separated representation).

3 Solution of the multidimensionnal ground state problem

Firstly, we are describing the numerical procedure without addressing the antisymmetry constraint, even if the spin is explicitly considered. Thus, $\Omega = \mathcal{R}^3 \times \{-s, s\}$, the whole multidimensional domain being represented by Ω^N $\left(\Omega^N = \left(\mathcal{R}^3 \times \{-s, s\}\right)^N\right)$.

We introduce the following notation: N is the number of electrons; M is the number of nuclei; Q is the number of finite sums present in the decomposition of the inverse of the distance function; n is the number of finite sums approximating the wavefunction and m the number of terms representing the different Coulomb potentials $m = N \times (N-1) \times Q + M \times N$.

Now, the problem to be solved writes:

$$(V(\mathbf{x}_1, ..., \mathbf{x}_N) - E)\,\Psi - \frac{1}{2}\frac{\partial^2 \Psi}{\partial \partial \mathbf{x}^2} = 0 \tag{3.20}$$

with

$$\int_{\Omega^N} \Psi^2(\mathbf{x}) d\Omega^N = 1 \tag{3.21}$$

where V is the inter-electronic and electron-nucleus potential. These potentials contain two contributions:

$$\frac{1}{\|\mathbf{x}_i - \mathbf{x}_j\|_\varepsilon} = \sum_{l=1}^{Q} B_l^1(\mathbf{x}_i)\ B_l^2(\mathbf{x}_j) \prod_{k=1,\ k\neq i,j}^{N} 1(\mathbf{x}_k) \tag{3.22}$$

and

$$\frac{-Z_j}{\|\mathbf{x}_i - \mathbf{X}_j\|_\varepsilon} = \frac{-Z_j}{\|\mathbf{x}_i - \mathbf{X}_j\|_\varepsilon} \prod_{k=1,\ k\neq i}^{N} 1(\mathbf{x}_k) \tag{3.23}$$

where the nuclei positions $\mathbf{X}_j$ $(j = 1, \cdots, M)$, are assumed fixed and known, and $1(\mathbf{x})$ is the unit function. Thus, the potential can be written in the general form:

$$V(\mathbf{x}_1, ..., \mathbf{x}_N) = \sum_{h=1}^{m} \prod_{k=1}^{N} A_k^h(\mathbf{x}_k) \tag{3.24}$$

The variational formulation of the problem writes:

$$\int_{\Omega^N} \Psi^* \left(V(\mathbf{x}_1, ..., \mathbf{x}_N) - E\right) \Psi + \frac{1}{2} \frac{\partial \Psi^*}{\partial \mathbf{x}} \frac{\partial \Psi}{\partial \mathbf{x}} d\Omega^N = 0 \tag{3.25}$$

that can be rewritten in the compact form:

$$\int_{\Omega^N} \Psi^* \left(\mathcal{H} - E\right) \Psi \ d\Omega^N = 0 \tag{3.26}$$

where we assumed that on the boundary of Ω^N the wavefunction normal derivative vanishes.

For solving this problem we firstly apply a fixed point strategy, assuming E known. The solution after n iterations is assumed in the form:

$$\Psi(\mathbf{x}_1, \mathbf{x}_2, \cdots, \mathbf{x}_N) = \sum_{j=1}^{n} \alpha_j \prod_{k=1}^{N} F_{kj}(\mathbf{x}_k) \tag{3.27}$$

Now, we proceed in two steps: a projection and an enrichment stages, both performed in the finite elements framework.

1. **Projection stage:**
   Assuming known the functions $F_{kj}(\mathbf{x}_k)$ $\forall j = 1, .., n, \forall k = 1, .., N$, we look for the best coefficients α_j of the approximation $\Psi(\mathbf{x}_1, \cdots, \mathbf{x}_N) = \sum_{j=1}^{n} \alpha_j \prod_{k=1}^{N} F_{kj}(\mathbf{x}_k)$ by enforcing the variational formulation.

2. **Enrichment stage:**
Now, with the just computed coefficients we look for a new term of the finite sums decomposition, $\prod_{k=1}^{N} R_k(\mathbf{x}_k)$, of the wavefunction $\Psi(\mathbf{x}_1, \cdots, \mathbf{x}_N)$, i.e. $\Psi(\mathbf{x}_1, \cdots, \mathbf{x}_N) = \sum_{j=1}^{n} \alpha_j \prod_{k=1}^{N} F_{kj}(\mathbf{x}_k) + \prod_{k=1}^{N} R_k(\mathbf{x}_k)$, by enforcing again the variational formulation.

To approximate each one of the functions defined in the domain Ω in which each electron is defined, we introduce the vector $\mathbf{N}$ containing the finite element shape functions defined in Ω. The shape functions derivatives are grouped in $\mathbf{dN}$. The vectors containing the nodal values of functions F and R will be noted by $\mathbf{F}$ and $\mathbf{R}$ respectively.

Before to detail both algorithm steps, we are introducing the matrix form of the integrals involved in the variational formulation:

$$\mathbb{N} = \int_{\Omega} \mathbf{N}\,\mathbf{N}^T d\Omega, \qquad \mathbb{D} = \int_{\Omega} \mathbf{dN}\,\mathbf{dN}^T d\Omega, \qquad \mathbb{A}_k^h = \int_{\Omega} A_k^h(x)\mathbf{N}\,\mathbf{N}^T d\Omega \tag{3.28}$$

3.1 Projection stage

At this step, the unknowns are the approximation basis function coefficients. The unknown field (the wavefunction) at the present iteration is approximated as:

$$\Psi(\mathbf{x}_1, \cdots, \mathbf{x}_N) = \left[\prod_{k=1}^{N} \mathbf{N}^T \mathbf{F}_{k1}, \;\cdots\;, \prod_{k=1}^{N} \mathbf{N}^T \mathbf{F}_{kn} \right] \begin{bmatrix} \alpha_1 \\ \vdots \\ \alpha_n \end{bmatrix} \tag{3.29}$$

and the associated test field as:

$$\Psi^*(\mathbf{x}_1, \cdots, \mathbf{x}_N) = [\alpha_1^*, \cdots, \alpha_n^*] \begin{bmatrix} \prod_{k=1}^{N} \mathbf{F}_{k1}^T \mathbf{N} \\ \vdots \\ \prod_{k=1}^{N} \mathbf{F}_{kn}^T \mathbf{N} \end{bmatrix} \tag{3.30}$$

Thus, the variational formulation of the eigenproblem writes:

$$\alpha^{*T} \left(\mathbb{K}_H - E\mathbb{K}_L \right) \alpha = 0 \tag{3.31}$$

where

$$(\mathbb{K}_H)_{i,j} = \sum_{h=1}^{m} \prod_{k=1}^{N} \mathbf{F}_{ki}^T \mathbb{A}_k^h \mathbf{F}_{kj} + \frac{1}{2} \sum_{k=1}^{N} \left(\mathbf{F}_{ki}^T \mathbb{D} \mathbf{F}_{kj} \prod_{l=1,\; l \neq k}^{N} \mathbf{F}_{li}^T \mathbb{N} \mathbf{F}_{lj} \right) \tag{3.32}$$

$$(\mathbb{K}_L)_{i,j} = \prod_{k=1}^{N} \mathbf{F}_{ki}^T \mathbb{N} \mathbf{F}_{kj} \tag{3.33}$$

whose solution must be searched under the normality constraint:

$$\alpha^T \mathbb{K}_L \alpha = 1 \tag{3.34}$$

that is enforced within an iteration fixed point strategy. Then, a correction of the eigenvalue is performed according to:

$$E = \frac{\int_{\Omega^N} \Psi \mathcal{H} \Psi d\Omega^N}{\int_{\Omega^N} \Psi \Psi d\Omega^N} = \frac{\alpha^T \mathbb{K}_H \alpha}{\alpha^T \mathbb{K}_L \alpha} = \alpha^T \mathbb{K}_H \alpha \tag{3.35}$$

3.2 Basis enrichment stage

Now, a new optimal product $\prod_{k=1}^N R_k(\mathbf{x}_k)$ of functions is searched by enforcing the variational formulation:

$$\Psi(\mathbf{x}_1, \cdots, \mathbf{x}_N) = \sum_{j=1}^{n} \alpha_j \prod_{k=1}^{N} F_{kj}(\mathbf{x}_k) + \prod_{k=1}^{N} R_k(\mathbf{x}_k) \tag{3.36}$$

The fixed point strategy is used again for computing each function $R_d(\mathbf{x}_d)$ involved in the product, by assuming known the remaining $N-1$ functions. Thus, the test functions are expressed as:

$$\Psi^*(\mathbf{x}_1, \cdots, \mathbf{x}_N) = R_d^*(\mathbf{x}_d) \prod_{k=1,\ k\neq d}^{N} R_k(\mathbf{x}_k) \tag{3.37}$$

that introduced in the vartiational formulation leads to the linear system:

$$\mathbf{R}_d^*\ \mathbf{V}(\mathbf{R}_1, \cdots, \mathbf{R}_N) + \mathbf{R}_d^*\ \mathbb{K}(\mathbf{R}_1, \cdots, \mathbf{R}_N)\, \mathbf{R}_d = \mathbf{0} \tag{3.38}$$

where

$$\begin{aligned} \mathbb{K} = \left(-E\mathbb{N} + \frac{1}{2}\mathbb{D}\right) \prod_{k=1,\ k\neq d}^{N} \mathbf{R}_k^T \mathbb{N} \mathbf{R}_k + \sum_{h=1}^{m} \mathbb{A}_d^h \prod_{k=1,\ k\neq d}^{N} \mathbf{R}_k^T \mathbb{A}_k^h \mathbf{R}_k + \\ + \frac{1}{2} \sum_{l=1,\ l\neq d}^{N} \mathbb{N}\left(\mathbf{R}_l^T \mathbb{D} \mathbf{R}_l\right) \prod_{k=1,\ k\neq d,l}^{N} \mathbf{R}_k^T \mathbb{N} \mathbf{R}_k \end{aligned} \tag{3.39}$$

and

$$\begin{aligned} \mathbf{V} = \sum_{j=1}^{n} \alpha_j \Bigg(\left(-E\mathbb{N}\mathbf{F}_{dj} + \frac{1}{2}\mathbb{D}\mathbf{F}_{dj}\right) \prod_{k=1, k\neq d}^{N} \mathbf{R}_k^T \mathbb{N} \mathbf{F}_{kj} + \\ + \sum_{h=1}^{m} \mathbb{A}_d^h \mathbf{F}_{dj} \prod_{k=1, k\neq d}^{N} \mathbf{R}_k^T \mathbb{A}_k^h \mathbf{F}_{kj} + \end{aligned}$$

$$+\frac{1}{2}\sum_{l=1,\, l\neq d}^{N} \mathbb{N}\mathbf{F}_{dj}\left(\mathbf{R}_l^T \mathbb{D}\mathbf{F}_{lj}\right) \prod_{k=1,\, k\neq d,l}^{N} \mathbf{R}_k^T \mathbb{N}\mathbf{F}_{kj}\Bigg) \tag{3.40}$$

solved again under the non-linear normality constraint:

$$\mathbf{R}_d^T \mathbb{N}\mathbf{R}_d = 1 \tag{3.41}$$

that is enforced within an iteration fixed point scheme that update at each iteration the eigenvalue.

3.3 Solution algorithm

The solution algorithm can be summarized as follows:

- Let $E = E_0$ be the first trial eigenvalue.
- Compute until convergence the following steps:
  1. Proceed with the enrichment step to compute $\mathbf{R}_k$, $\forall k$. Note that at the first iteration $\mathbf{V} = 0$.
  2. Update the approximation basis with the just computed functions.
  3. Proceed with the projection step.
  4. Updated the eigenvalue E.

3.4 Computing the electronic density

For each electron k we can compute its spatial distribution from:

$$\rho_k(\mathbf{x}) = \sum_{i=1}^{n}\sum_{j=1}^{n} \alpha_i \alpha_j F_{ki}(\mathbf{x}) F_{kj}(\mathbf{x}) \prod_{l=1,\, l\neq k}^{N} \mathbf{F}_{li}^T \mathbb{N}\mathbf{F}_{lj} \tag{3.42}$$

The total electronic density is then obtained by adding the spatial distribution related to each electron.

3.5 Taking into account the antisymmetry

In order to compute an antisymmetric solution we could apply the just proposed algorithm applying an antisymmmetrizer after each enrichment step. This results equivalent to enrich with the $N!$ functions derived from the product of functions just computed in the enrichment step, via the Slater's determinant. This strategy converges slowly and requires the storage of numerous functions.

An alternative procedure proposed by Beylkin and Mohlenkamp [3] allows alleviating the storage cost. Since the operator $(\mathcal{H} - E)$ is purely symmetric it could commute with the antisymmetrizer $\mathcal{A}$. Thus, we can defer the application of the antisymmetrizer and the operator. The idea is to incorporate the antisymmetrizer in the algorithm by evaluating its effect. Of coarse the pseudowave functions obtained are not the solution of the schrodinger equation and need to be antisymmetrized to obtain the exact solution.

In this work the antisymmetrizer operator will apply on the test functions. Thus, the variational formulation writes:

$$\int_{\Omega^N} (\mathcal{A}\Psi^*)\,(\mathcal{H}-E)\,\Psi\; d\Omega^N = 0 \tag{3.43}$$

that requires the modification of the test function approximation at the projection step:

$$\mathcal{A}\Psi^*(\mathbf{x}_1,\cdots,\mathbf{x}_N) = [\alpha_1^*,\ \cdots\ ,\alpha_n^*] \begin{bmatrix} \mathcal{A}\prod_{k=1}^N \mathbf{F}_{k1}^T\mathbf{N} \\ \vdots \\ \mathcal{A}\prod_{k=1}^N \mathbf{F}_{kn}^T\mathbf{N} \end{bmatrix} \tag{3.44}$$

as well as at the enrichment step:

$$\mathcal{A}\Psi^*(\mathbf{x}_1,\cdots,\mathbf{x}_N) = \mathcal{A}\left[R_d^*(x_d) \prod_{k=1,k\neq d}^{N} R_k(\mathbf{x}_k)\right] \tag{3.45}$$

Thus, the algorithm remains basically unchanged, but an additional condition at the enrichment step must be introduced. In fact, if the N functions F_{kj} $(j=1,..,n)$ are linearly dependent, then the determinant operator vanishes. Thus, an orthogonalization is performed at each enrichment step according to:

$$< R_d(\mathbf{x}_d), R_k(\mathbf{x}_k) >= R_d^T(\mathbf{x}_d)\,\mathbb{N}\,R_k(\mathbf{x}_k) = 0 \tag{3.46}$$
$$\forall d = 1,\cdots,N;\quad \forall k = 1,\cdots,d-1,d+1,\cdots,N$$

4 Some preliminary numerical results

To illustrate the solution procedure we start solving systems composed of a single nucleus ($Z = 3$) and different number of electrons (from 1 to 5). Figure 4.1 depicts, assuming a one-dimensional physical space, the electronic distributions as well as the differences between each couple of consecutive electronic distributions. This simulation was carried out by assuming a large enough one-dimensional domain such that both the electronic distribution and its derivative vanish on its boundary. Obviously, the numerical model could be improved by using larger domains and non-uniform one-dimensional nodal distributions, but in this first attempt we considered the simplest strategy. The length of the computational domain was set to 10 dimensionless units as depicted in figure 4.1.

We can notice that the first two electrons are occupying a s-type orbital. When an additional electron is introduced, and due to the Pauli's exclusion

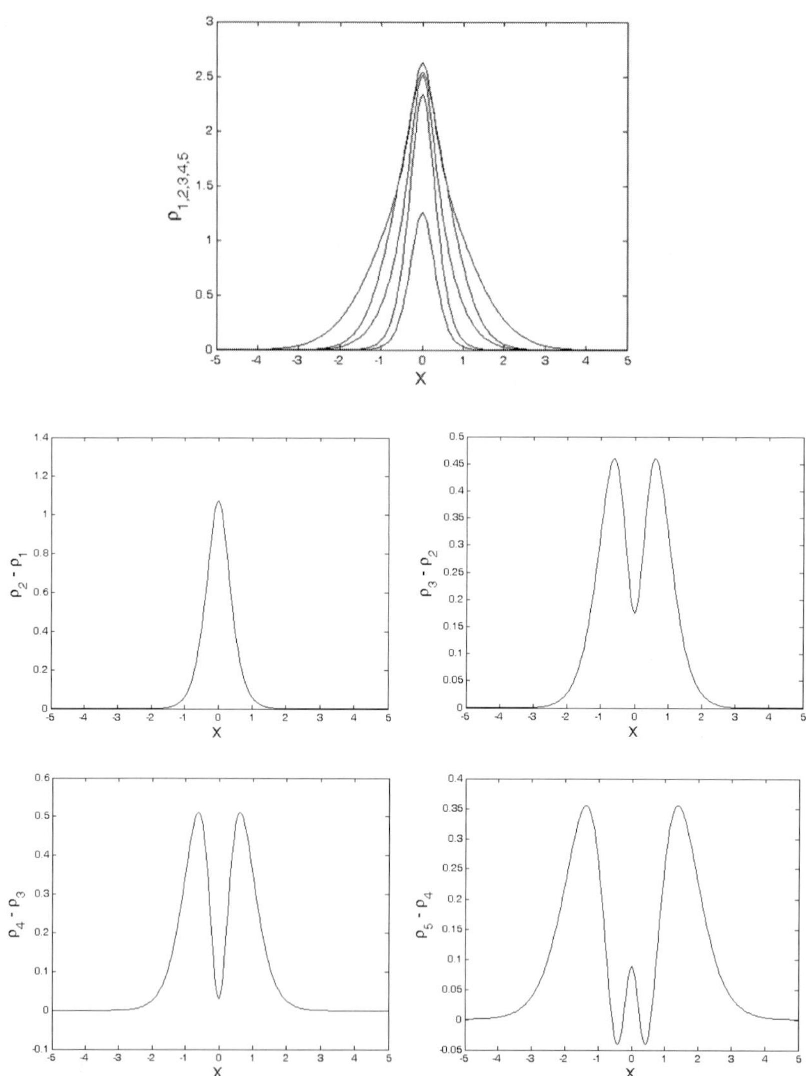

Figure 4.1. Evolution of the electronic distribution of a system composed of a single nucleus ($Z = 3$) and an increasing number of electrons (one-dimensional physical space).

principle, the three electrons cannot occupy the same orbital. Thus, a kind of *p*-orbital is encountered. This behavior is also noticed when $2D$ physical spaces are considered, as figure 4.2 illustrates.

Now, we are considering the hypothetical one-dimensional molecules of helium (He_2) and of LiH. For this purpose we consider the system composed

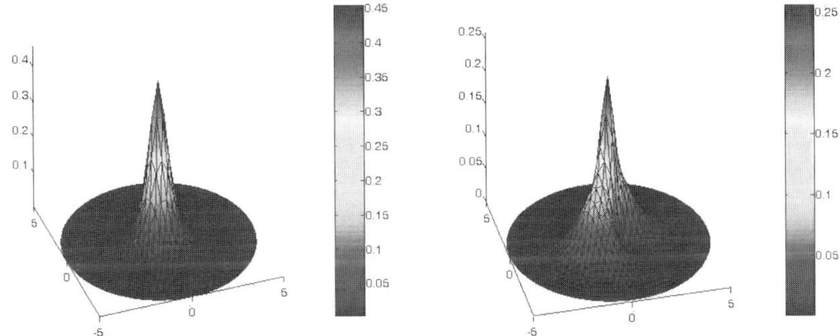

Figure 4.2. Evolution of the electronic distribution of a system composed of a single nucleus ($Z = 3$) and an increasing number of electrons (two-dimensional physical space). (*See also* Color Plate on page 359)

of 4 electrons and two nuclei. Now, the total energy (which is the sum of the ground state energy plus the one associated to the inter-nuclei Coulomb's potential) is obtained as a function of the distance between both nuclei. The computational domain consists of 16 dimensionless units (that seems to be large enough to ensure the nullity of both the electronic distribution and its derivative on the domain boundary for all the relative positions between the nuclei later considered).

The computed evolution is depicted in 4.3(a) that explain the higher stability of the LiH molecule. In the same figure we depict the electronic distribution when the inter-nuclei distance takes the dimensionless values of 0.5, 1, 2, 3 and 4. When the distance increases we can notice that the electronic distribution for the He_2 remains symmetric. On the contrary, an asymmetric charge distribution is noticed in the case of the LiH molecule.

In all the simulations reported in this section we considered the usual 3D Coulomb potential that was simply and crudely restricted to 1D or 2D, even if this reduction has not any physical meaning. The smoothing parameter ϵ in Eq. (2.19) was set to $\epsilon = 0.01$. Different separated representations of the inverse of the distance function (according to Eq. (3.22)) were performed by increasing the approximation accuracy. Thus, the number of finite sums used in the approximation (3.22) ranged in the interval $Q \in [100, 300]$. Finally, in all the simulations that were performed the accurate representation of wavefunctions needed for the use of around ten finite sums, i.e. in Eq. (3.27) n was $n \approx 10$.

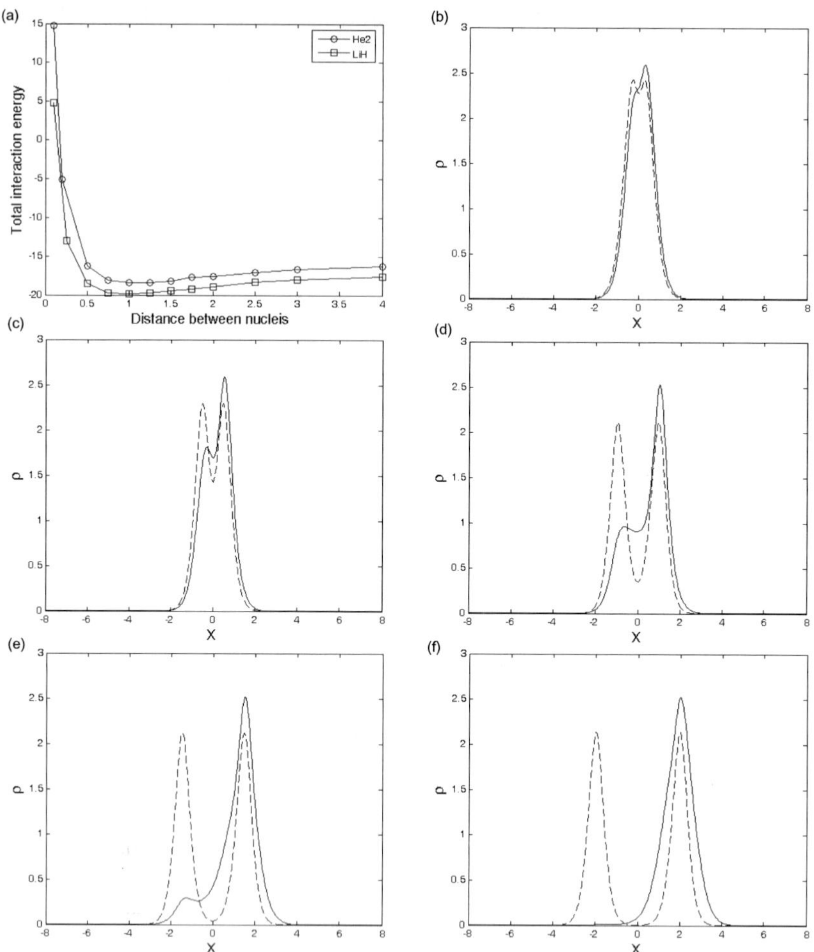

Figure 4.3. Analysis of the hypothetical one-dimensional LiH and He_2 molecules.

5 Conclusions

This paper analyzed the suitability of a finite sums decomposition based on a separated representation, to address highly dimensional models, as the ones encountered in quantum mechanics when the solution of the Schrödinger equation is envisaged.

Based in our former experience on the solution of models defined in highly dimensional spaces (involving hundreds of dimensions) we have extended the numerical technique that we proposed in [1] [2] to the direct solution of the Schrödinger equation.

The main conclusion of this analysis was that the main difficulty related to the solution of the Schrödinger equation for fermions is more the antisymmetry constraint that the Pauli's principle implies, than its highly dimensional character. The curse of dimensionality can be circumvented efficiently using finite sums decompositions based on separated representations as we proved in [1] [2]. However an efficient treatment of the antisymmetry constraint needs for further developments, if one want to address the direct solution of the Schrödinger equation. One possibility is to focus on the improvement of approximated approaches (the ones derived from the DFT or the Hartree-Fock approaches). Others alternatives need further developments.

References

1. A. Ammar, B. Mokdad, F. Chinesta, R. Keunings, A new family of solvers for some classes of multidimensional partial differential equations encountered in kinetic theory modeling of complex fluids, Journal of Non-Newtonian Fluid Mechanics, 139, 2006, 153-176.
2. A. Ammar, B. Mokdad, F. Chinesta, R. Keunings, A new family of solvers for some classes of multidimensional partial differential equations encountered in kinetic theory modeling of complex fluids. Part II: transient simulation using space-time separated representation, Journal of Non-Newtonian Fluid Mechanics, 144, 2007, 98-121.
3. G. Beylkin, M.J. Mohlenkamp, Algorithm for numerical analysis in high dimensions, SIAM J. Sci. Comput., 26/6, 2005, 2133-2159.
4. H.J. Bungartz, M. Griebel, Sparse grids, Acta Numer., 13, 2004, pp. 1-123.
5. E. Cancès, M. Defranceschi, W. Kutzelnigg, C. Le Bris, Y. Maday, Computational Quantum Chemistry: a primer, Handbook of Numerical Analysis, Vol. X, Elsevier, 2003, pp. 3-270.
6. J.S. Chen, W. Hu, M. Puso, Orbital HP-clouds for solving Schrdinger equation in quantum mechanics, Comput. Methods Appl. Mech. Engrg., 196/37-40, 2007, pp. 3693-3705.
7. V. Gavini, K. Bhattacharya, M. Ortiz, Quasi-continuum orbital-free density-functional theory: A route to multi-million atom non-periodic DFT calculation, Journal of the Mechanics and Physics of Solids, 55/4, 2007, pp. 697-718.
8. Handbook of Numerical Analysis, Vol. X: Computational Chemistry, C. Le Bris editor, Elsevier, 2003.
9. R.B. Laughlin, The theory of everything, Proceeding of the U.S.A. National Academy of Sciences, 2000.

A Pressure Correction Approach Coupled with the MLPG Method for the Solution of the Navier-Stokes Equations

Rubén Avila and Apolinar Pérez

Departamento de Termoenergía, Facultad de Ingeniería, Universidad Nacional Autónoma de México, México, D.F. C.P. 04510, ravila@servidor.unam.mx

Summary. We present a pressure-velocity correction approach for the solution of the Navier-Stokes equations. The Meshless Local Petrov Galerkin (MLPG) method is used to solve the two dimensional, incompressible and steady state viscous fluid flow equations. The weak form of these equations, which are formulated in the Cartesian coordinate system, are integrated in a local standard domain by the Gauss-Lobatto-Legendre quadrature rule. The Moving Least Square (MLS) scheme is used to generate the interpolation shape functions. The pressure-velocity correction approach (segregated solution procedure) follows an iterative process, in which the momentum equations are solved sequentially to obtain the velocities v_1^{**} and v_2^{**} from initial guessed values for the velocity (v_1^* and v_2^*) and pressure (p^*) fields. Using the corrected velocities $v_i = v_i^{**} + v_i'$ and pressure $p = p^* + p'$ in the weak form of the continuity and momentum equations, we generate a system of three equations with three unknown variables (a fully implicit method): the velocity corrections (v_1' and v_2') and the pressure correction (p'). Using the correction values the pressure is updated and the velocities are corrected to satisfy the continuity equation. The updated values are taken as the new guessed values, and the iterative process continues until convergence. We apply the method for the solution of four (low Rayleigh number and low Reynolds number) fluid flow problems. We conclude that the MLPG method coupled with an implicit procedure to calculate the corrections of pressure and velocities can be used as a reliable methodology for the solution of the Navier-Stokes equations.

Key words: Meshfree, incompressible flow, MLPG, Navier-Stokes

1 Introduction

The solution of fluid flow problems in science and engineering has been traditionally based on the use of numerical schemes that solve the fluid equations in an Eulerian frame of reference. The widely used mesh based methods such as the finite difference, finite element, finite volume, spectral element, etc., have provided satisfactory results, which have been useful for engineering design purposes as well as for basic fluid dynamics research. However, due to the mesh constraints and the required computer time to generate three-dimensional meshes, the optimal use of the mesh based methods for the solution of problems involving moving boundaries (phase change, free-surface fluid-flexible structure interaction, etc.) and fluid flow in three-dimensional systems with complex geometries, is still under investigation. A family of methods known as meshless (particle or nodes) methods has been developed since the last two decades to overcome those difficulties inherent in the mesh based methods such as: (1) connectivity between the cells, (2) aspect ratio of the cells, (3) interpolation functions generated upon the mesh, (4) very fine mesh in problems with high gradients, (5) adaptive re-meshing and (6) mapping of the dependent variables from the old mesh to the new mesh [10]. The mathematical basis and the general characteristics and advantages of the up to now developed meshfree methods may be consulted in the open literature [10], [5], [2], [11]. Among the main advantages of the meshless methods we can mention the following: (1) there are no mesh constraints, (2) they allow an accurate representation of complex geometry domains, (3) the connectivity between the particles is generated as part of the computational process and (4) the local domains surrounding each particle can intersect each other and overlap. Even though the meshfree methods have been applied successfully in fields such as solid mechanics and fluid dynamics, most of them are still under development.

In this paper we use the Meshless Local Petrov Galerkin (MLPG) method coupled with a fully implicit pressure-velocity correction approach to solve the Navier-Stokes (N-S) equations formulated in terms of primitive variables (velocity and pressure). Using the MLPG method the integration of the weak form of the fluid equations is carried out in a local domain, which is defined by the set of particles surrounding each particle located in the computational domain. We have formulated the weak form of the governing equations in a Cartesian coordinate system. The shape functions needed to approximate the flow variables (velocities, pressure and temperature) in a local domain have been generated by using the Moving Least Square (MLS) scheme. To integrate the fluid equations and to differentiate the dependent variables in systems with complex geometry, we introduce a one-to-one iso-parametric mapping between the physical and computational domains. The integration of the equations in the local computational domain is performed by using the Gauss-Lobatto-Legendre quadrature rule. The weight function used in the MLS scheme and in the weak formulation of the equations, is a compact support fourth order spline.

To the knowledge of the authors with the exception of the paper by Lin and Atluri [3] no additional work has been carried out to solve the N-S equations (formulated in terms of primitive variables) through the use of the MLPG method. Most of the research has been focused to solve either potential flow problems, the convection-diffusion equation or the two dimensional stream function-vorticity formulation of the N-S equations. Lin and Atluri [4] solved the convection-diffusion equation and introduced the upwind schemes (by modifying the shape and location of the weight function used in the MLS and weighted residuals method) to stabilize the numerical approximation for high Peclet number flow conditions. Lin and Atluri [3] solved the two-dimensional steady state incompressible fluid flow equations formulated in terms of primitive variables. To incorporate the incompressibility constraint they used a standard mixed formulation, and in an attempt to circumvent the governing stability conditions, and to avoid the presence of spurious pressure values, they modified the mixed formulation by adding perturbation terms to the local integral form of the continuity equation (previously weighted by a test function). Arefmanesh et al. [7] modified the MLPG method and developed the Meshless Control Volume method (selecting a weight function equal to unity in the weak formulation) to solve the transient heat conduction problem, the potential flow problem and the advection-diffusion equation. Wu et al. [6] used the MLPG method to solve the steady state, two dimensional, incompressible fluid flow equations formulated in derived variables (vorticity-stream function). Ma [8] also used the MLPG method to solve a two-dimensional, non steady, non-viscous, nonlinear water wave problem, however the transient solution only involves the integration of a Poisson equation for the pressure at each time step. In this investigation we have introduced the incompressibility constraint by using a segregated solution procedure and not by the use of a mixed formulation [3]. In the segregated procedure, the discretized momentum equations are solved sequentially to obtain the velocities u_i^{**} (for the two-dimensional case $i = 1, 2$) with a guessed pressure field p^*, and guessed velocity values u_i^* [9]. The initial guessed pressure field does not guarantee that the obtained velocities u_i^{**} satisfy the incompressibility constraint. Consequently the velocities must be improved $u_i = u_i^{**} + u_i'$ and the pressure field must also be corrected $p = p^* + p'$. However a governing equation for the pressure field can not be generated. In order to overcome this difficulty an iterative pressure-correction approach has been introduced. Using the iterative process the improved velocities (calculated by a corrected pressure field) gradually must satisfy the set of the fluid equations (mass conservation and momentum). In the literature it is possible to find algorithms to improve the velocities and pressures, most of them are based on the SIMPLE method (Semi-Implicit Method for Pressure-Linked Equations) [9]. The major approximation made in the SIMPLE-like algorithms is that in the discretized equations for the velocity corrections u_i', the effects of the velocity corrections of the neighboring cells are neglected (semi-implicit approach), leading to a discretized equation for the pressure corrections p' which is a function of the residual of mass and the pressure corrections of the neighboring cells. In this investigation, we propose to use a fully implicit pressure correction approach. In this methodology the velocity and pressure corrections of the particles surrounding each particle are taken into account. The correction fields are obtained by simultaneously solving the weak form of the equations for the corrections of the velocity (u_i')

and pressure (p'). After the calculation of the corrections, the velocities and pressure are updated ($u_i^* = u_i$ and $p^* = p$) and the iterative process continues until convergence. We have to mention that our results have been obtained for low Rayleigh number and low Reynolds number flow conditions. Hence no criteria has been made to increase the stability of the flow in convective dominated flow conditions. Additional effort should be made to introduce in the MLPG method the upwind schemes originally proposed by Lin and Atluri [4].

In the second section of this paper we present the weak formulation of the fluid equations. In the third section we describe the segregated solution strategy based on the pressure-velocity correction method. In the fourth section we describe the MLPG method and the MLS methodology. In the fifth section we present numerical results for low Reynolds number and low Rayleigh number flow conditions. Finally we present our conclusions.

2 Weak formulation of the fluid equations

The weak form of the steady state, two dimensional, incompressible fluid flow equations is written as [3]

Continuity Equation

$$-\int_{\Omega}\left(v_1\frac{\partial W}{\partial x_1}+v_2\frac{\partial W}{\partial x_2}\right)d\Omega+\int_{\Gamma}\left(v_1\eta_1 W+v_2\eta_2 W\right)dS=0 \tag{2.1}$$

x_1-Momentum Equation

$$\begin{aligned}
&\int_{\Omega}W\left(v_1\frac{\partial v_1}{\partial x_1}+v_2\frac{\partial v_1}{\partial x_2}\right)d\Omega-\int_{\Omega}Wg_1 d\Omega-\int_{\Omega}\frac{1}{\rho}p\frac{\partial W}{\partial x_1}d\Omega+\\
&\int_{\Omega}\nu\left(\frac{\partial v_1}{\partial x_1}\frac{\partial W}{\partial x_1}+\frac{\partial v_1}{\partial x_2}\frac{\partial W}{\partial x_2}\right)d\Omega+\int_{\Omega}\nu\left(\frac{\partial v_1}{\partial x_1}\frac{\partial W}{\partial x_1}+\frac{\partial v_2}{\partial x_1}\frac{\partial W}{\partial x_2}\right)d\Omega-\\
&\int_{\Gamma}\nu\left(\frac{\partial v_1}{\partial x_1}W\eta_1+\frac{\partial v_1}{\partial x_2}W\eta_2\right)dS-\int_{\Gamma}\nu\left(\frac{\partial v_1}{\partial x_1}W\eta_1+\frac{\partial v_2}{\partial x_1}W\eta_2\right)dS+\\
&\int_{\Gamma}\frac{1}{\rho}pW\eta_1 dS=0
\end{aligned} \tag{2.2}$$

x_2-Momentum Equation

$$\begin{aligned}
&\int_{\Omega}W\left(v_1\frac{\partial v_2}{\partial x_1}+v_2\frac{\partial v_2}{\partial x_2}\right)d\Omega-\int_{\Omega}Wg_2 d\Omega-\int_{\Omega}\frac{1}{\rho}p\frac{\partial W}{\partial x_2}d\Omega+\\
&\int_{\Omega}\nu\left(\frac{\partial v_2}{\partial x_1}\frac{\partial W}{\partial x_1}+\frac{\partial v_2}{\partial x_2}\frac{\partial W}{\partial x_2}\right)d\Omega+\int_{\Omega}\nu\left(\frac{\partial v_1}{\partial x_2}\frac{\partial W}{\partial x_1}+\frac{\partial v_2}{\partial x_2}\frac{\partial W}{\partial x_2}\right)d\Omega-\\
&\int_{\Gamma}\nu\left(\frac{\partial v_2}{\partial x_1}W\eta_1+\frac{\partial v_2}{\partial x_2}W\eta_2\right)dS-\int_{\Gamma}\nu\left(\frac{\partial v_1}{\partial x_2}W\eta_1+\frac{\partial v_2}{\partial x_2}W\eta_2\right)dS+\\
&\int_{\Gamma}\frac{1}{\rho}pW\eta_2 dS=0
\end{aligned} \tag{2.3}$$

Energy Equation

$$\int_{\Omega} W \left(v_1 \frac{\partial T}{\partial x_1} + v_2 \frac{\partial T}{\partial x_2} \right) d\Omega + \int_{\Omega} \alpha \left(\frac{\partial T}{\partial x_1} \frac{\partial W}{\partial x_1} + \frac{\partial T}{\partial x_2} \frac{\partial W}{\partial x_2} \right) d\Omega - \int_{\Gamma} \alpha W \left(\frac{\partial T}{\partial x_1} \eta_1 + \frac{\partial T}{\partial x_2} \eta_2 \right) dS \quad (2.4)$$

where ρ is the density, v_1 and v_2 are the fluid velocities along the x_1 and x_2 directions respectively, p is the pressure, g_1 and g_2 are the components of the gravity vector along the x_1 and x_2 directions respectively, ν is the kinematic viscosity, T is the temperature, α is the thermal diffusivity, η_1 and η_2 are the cosine directors of the outward normal unit vector on the surface Γ and W is a weight function.

3 A fully implicit pressure-correction method for the solution of the N-S equations

A segregated solution methodology has been used to solve (2.1)-(2.3) in a sequential process. To obtain a converged solution where the velocities v_1 and v_2 satisfy the incompressibility constraint and the momentum equations, we follow an iterative implicit pressure-correction approach. The steps of the iterative process are the following:

1).-Propose initial guessed values for the velocities (v_1^* and v_2^*) and the pressure p^*. Solve the x_1-momentum equation (see (2.2)).

2).-Now we have an updated field for the v_1 velocity, i.e. v_1^{**}. Using this updated velocity we solve the x_2-momentum equation (see (2.3)).

3).-Now we have an updated field for the v_2 velocity, i.e. v_2^{**}. Substituting the updated velocity values v_1^{**} and v_2^{**} into the continuity equation (2.1) we generate an equation (EQMass1) with a mass residual $\Delta\dot{m}$.

4).-The values for the correction of velocities (v_1' and v_2') and pressure p' are introduced as

$$v_1 = v_1^{**} + v_1' \ , \qquad v_2 = v_2^{**} + v_2' \quad \text{and} \quad p = p^* + p' \quad (3.5)$$

We assume that the fields given by (3.5), satisfy the continuity equation. Hence by substituting (3.5) into the continuity equation (2.1) we generate an equation (EQMass2) with residual equal to zero. Subtracting (EQMass2) from (EQMass1) we have

$$\int_{\Omega} v_1' \frac{\partial W}{\partial x_1} d\Omega + \int_{\Omega} v_2' \frac{\partial W}{\partial x_2} d\Omega - \int_{\Gamma} v_1' \eta_1 W dS - \int_{\Gamma} v_2' \eta_2 W dS = \Delta\dot{m} \quad (3.6)$$

5).-Substituting the previously calculated velocity v_1^{**} into the x_1-momentum equation, we generate an equation (EQ1) whose residual equals zero. Substituting the corrected values $v_1 = v_1^{**} + v_1'$ and $p = p^* + p'$ again into the

x_1-momentum equation we generate a second equation (EQ2) whose residual is also zero. Subtracting (EQ2) from (EQ1) we obtain an equation for the correction of velocity v_1' and for the correction of pressure p', which is written as;

$$-\int_{\Omega}\rho W v_1^*\frac{\partial v_1'}{\partial x_1}d\Omega-\int_{\Omega}\rho W v_2^*\frac{\partial v_1'}{\partial x_2}d\Omega+\int_{\Omega}p'\frac{\partial W}{\partial x_1}d\Omega-\int_{\Omega}\mu\frac{\partial v_1'}{\partial x_1}\frac{\partial W}{\partial x_1}d\Omega-$$
$$\int_{\Omega}\mu\frac{\partial v_1'}{\partial x_2}\frac{\partial W}{\partial x_2}d\Omega-\int_{\Omega}\mu\frac{\partial v_1'}{\partial x_1}\frac{\partial W}{\partial x_1}d\Omega+\int_{\Gamma}\mu\frac{\partial v_1'}{\partial x_1}W\eta_1 dS+\int_{\Gamma}\mu\frac{\partial v_1'}{\partial x_2}W\eta_2 dS+$$
$$\int_{\Gamma}\mu\frac{\partial v_1'}{\partial x_1}W\eta_1 dS-\int_{\Gamma}p'W\eta_1 dS=0 \tag{3.7}$$

6).-Substituting the previously calculated velocity v_2^{**} into the x_2-momentum equation, we generate an equation (EQ3) whose residual equals zero. Substituting the corrected values $v_2 = v_2^{**} + v_2'$ and $p = p^* + p'$ again into the x_2-momentum equation we generate a second equation (EQ4) whose residual is also zero. Subtracting (EQ4) from (EQ3) we obtain an equation for the correction of the velocity v_2' and for the correction of pressure p', which is written as;

$$-\int_{\Omega}\rho W v_1^{**}\frac{\partial v_2'}{\partial x_1}d\Omega-\int_{\Omega}\rho W v_2^*\frac{\partial v_2'}{\partial x_2}d\Omega+\int_{\Omega}p'\frac{\partial W}{\partial x_2}d\Omega-\int_{\Omega}\mu\frac{\partial v_2'}{\partial x_1}\frac{\partial W}{\partial x_1}d\Omega-$$
$$\int_{\Omega}\mu\frac{\partial v_2'}{\partial x_2}\frac{\partial W}{\partial x_2}d\Omega-\int_{\Omega}\mu\frac{\partial v_2'}{\partial x_2}\frac{\partial W}{\partial x_2}d\Omega+\int_{\Gamma}\mu\frac{\partial v_2'}{\partial x_1}W\eta_1 dS+\int_{\Gamma}\mu\frac{\partial v_2'}{\partial x_2}W\eta_2 dS+$$
$$\int_{\Gamma}\mu\frac{\partial v_2'}{\partial x_2}W\eta_2 dS-\int_{\Gamma}p'W\eta_2 dS=0 \tag{3.8}$$

7).-Solve the system of three equations (3.6), (3.7) and (3.8) with three unknown variables v_1', v_2' and p'.
8).- Update the velocities v_1, v_2 and pressure p as

$$v_1 = v_1^*+\alpha_1\left[(v_1^{**}+v_1')-v_1^*\right], \qquad v_2 = v_2^*+\alpha_2\left[(v_2^{**}+v_2')-v_2^*\right], \qquad p = p^*+\alpha_p'$$

where $\alpha_1 = \alpha_2 = 0.5$ and $\alpha_p = 0.1$ are under-relaxation factors.
9).-The updated values are defined as the new guessed values for the next iteration: $v_1^* = v_1$, $v_2^* = v_2$ and $p^* = p$.
10).-Go to the first step until convergence.

4 The MLPG method and the MLS technique

The MLPG method is based on a local weak formulation of the fluid equations. To discretize the governing equations a set of particles (nodes) is placed in the

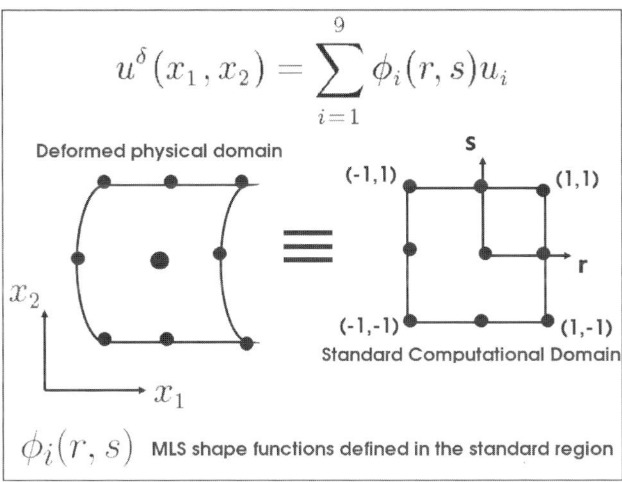

Figure 3.1. Deformed physical domain mapped to a local standard computationl domain. The MLS shape functions are digitally generated within the local domain (r,s) by using 9 neighbours located in the near vicinity of each particle. (*See also* Color Plate on page 360)

physical domain. The integration of the equations is performed in a local sub-domain surrounding each particle. We have solved fluid flow problems with complex geometry, hence the local sub-domain can be a four sides deformed region defined by nine particles (see Figure 3). Each of the deformed regions is mapped to a standard computational region ($-1 \leq r \leq 1$ and $-1 \leq s \leq 1$) defined in a local coordinate system (r, s). The integration of the fluid equations (2.2)-(2.4) is carried out by a Gauss-Lobatto-Legendre quadrature rule, whereas the system of equations for the correction of velocities and pressure (3.6), (3.7) and (3.8) has been integrated by using a Gauss-Legendre quadrature rule. We have approximated the flow field variables by a polynomial expansion of the form

$$u^\delta(x_1, x_2) = \sum_{i=1}^{N=9} \phi_i(r,s)u_i(x_1, x_2)$$

where x_1 and x_2 are the global Cartesian coordinates and r and s are the local coordinates defined in the standard region. The summation is carried out over the local region defined by the number of neighbours ($N = 9$) surrounding each particle. The shape functions $\phi_i(r, s)$ defined in the standard region are digitally generated by using the MLS numerical technique. The MLS approximation $u^\delta(\mathbf{x})$ of the function $u(\mathbf{x})$ is written as

$$u^\delta(x_1, x_2) = \mathbf{P}^T(r,s) \cdot \mathbf{a} \tag{4.9}$$

where

$$\mathbf{P}^T(r,s) = \begin{bmatrix} 1 \; r \; s \end{bmatrix}$$

is a monomial basis of order three and

$$\mathbf{a}(\mathbf{x}) = \begin{bmatrix} a_1 \\ a_2 \\ a_3 \end{bmatrix}$$

is a vector of unknown coefficients. In the MLS method the coefficients a_j are determined by minimizing a weighted discrete L_2-norm

$$J(r,s) = \sum_{i=1}^{N=9} \hat{W}_i(r,s) \left[\mathbf{P}^T(r,s) \cdot \mathbf{a} - u_i\right]^2 \tag{4.10}$$

where $\hat{W}_i(r,s)$ is the weight function associated with the ith particle in the local standard domain. By minimizing (4.10) we obtain the expression for the coefficients a_j;

$$\mathbf{a} = \mathbf{M}^{-1}\mathbf{B}\mathbf{u} \tag{4.11}$$

Using equation (4.11) in equation (4.9), we find the expression for the MLS shape functions

$$\phi^T(r,s) = \mathbf{P}^T(r,s) \cdot \mathbf{M}^{-1}\mathbf{B} \tag{4.12}$$

the matrices $\mathbf{M}^{-1}$ and $\mathbf{B}$ are also defined in the standard domain (r,s).

We have used a fourth order weight function of compact support, defined as

$$\hat{W}_i(r,s) = \begin{cases} 1 - 6q^2 + 8q^3 - 3q^4 & \text{for } q = \frac{|\vec{\xi} - \vec{\xi}_i|}{\rho_m} \leq 1 \\ 0 & \text{for } q = \frac{|\vec{\xi} - \vec{\xi}_i|}{\rho_m} > 1 \end{cases}$$

where ρ_m is a length scale directly related with the size of the local sub-domain [2], and $| \vec{\xi} - \vec{\xi}_i |$ is the distance between the ith neighbour (defined by its position vector $\vec{\xi}_i$) and the point of estimate defined by its position vector $\vec{\xi}$. The derivative of the MLS shape functions with respect to the local coordinates r and s is written as (see (4.12)):

$$\frac{\partial \phi^T(r,s)}{\partial r} = \frac{\partial}{\partial r}\left(\mathbf{P}^T\mathbf{M}^{-1}\mathbf{B}\right) = \left(\frac{\partial \mathbf{P}^T}{\partial r}\mathbf{M}^{-1}\mathbf{B} + \mathbf{P}^T\frac{\partial \mathbf{M}^{-1}}{\partial r}\mathbf{B} + \mathbf{P}^T\mathbf{M}^{-1}\frac{\partial \mathbf{B}}{\partial r}\right)$$

and

$$\frac{\partial \phi^T(r,s)}{\partial s} = \frac{\partial}{\partial s}\left(\mathbf{P}^T\mathbf{M}^{-1}\mathbf{B}\right) = \left(\frac{\partial \mathbf{P}^T}{\partial s}\mathbf{M}^{-1}\mathbf{B} + \mathbf{P}^T\frac{\partial \mathbf{M}^{-1}}{\partial s}\mathbf{B} + \mathbf{P}^T\mathbf{M}^{-1}\frac{\partial \mathbf{B}}{\partial s}\right)$$

where the derivative of the matrix $\mathbf{M}^{-1}$ can be obtained as

$$\frac{\partial \mathbf{M}^{-1}}{\partial r} = -\mathbf{M}^{-1}\frac{\partial \mathbf{M}}{\partial r}\mathbf{M}^{-1}$$

and

$$\frac{\partial \mathbf{M}^{-1}}{\partial s} = -\mathbf{M}^{-1}\frac{\partial \mathbf{M}}{\partial s}\mathbf{M}^{-1}$$

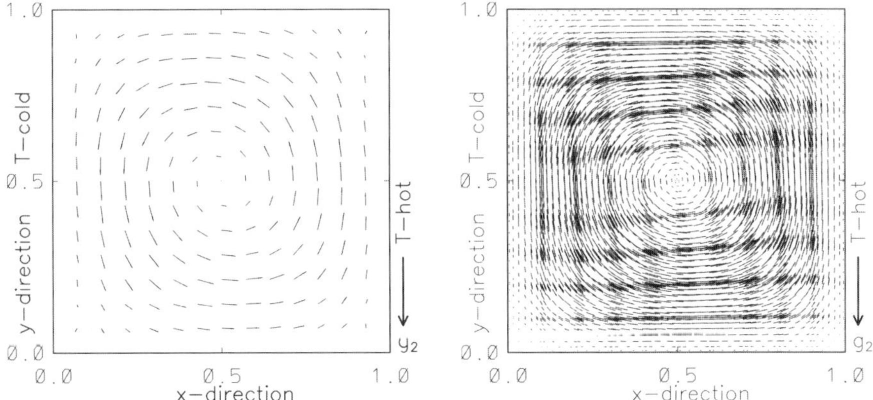

Figure 4.2. Natural convection in a square cavity ($Ra = 1520$). Velocity field. The two vertical walls have a Dirichlet boundary condition. The two horizontal walls have a Neumann boundary condition (adiabatic walls). *Left*: MLPG solution. *Right*: SEM results.

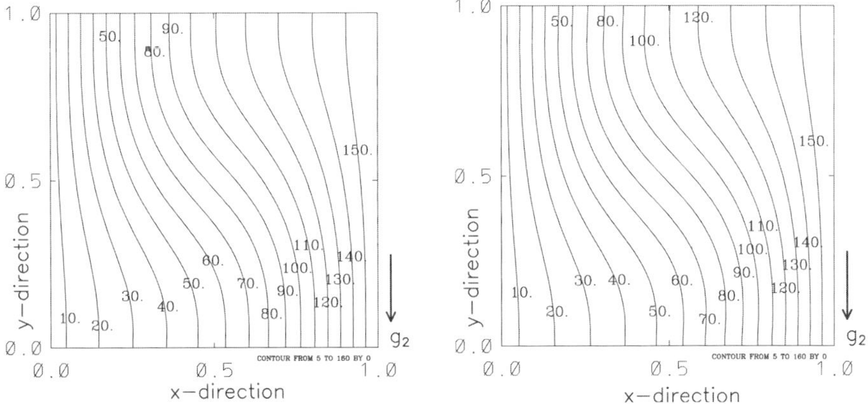

Figure 4.3. Natural convection in a square cavity ($Ra = 1520$). Temperature field. Comparison between the MLPG solution with 15×15 particles (*left*) and the SEM results with 10×10 macro elements and a polynomial expansion $p = 9$ (*right*).

The expression for the derivatives of the weight function $\hat{W}_i(r, s)$ with respect to the local variables (r, s) has been obtained by previously defining the two-dimensional weight function as the tensor product of the weight functions defined along each direction of the local coordinate system. To perform the numerical quadrature and to differentiate a function over the deformed local physical domain which is a function of (x_1,x_2), we have transformed this region into the standard region defined in terms of (r,s) [1]. To impose the boundary conditions required to solve the fluid equations, it is well known that the MLS interpolation functions do not satisfy the Kronecker delta function criterion. Hence in order to enforce the Dirichlet boundary conditions we have used an interpolation technique [2].

5 Results

In this section we present the numerical simulation of four two-dimensional, steady state, incompressible fluid flow problems. The numerical results obtained by the MLPG method for the cases 1-3, have been compared with the numerical solution provided by a standard *h-p* Spectral Element Method (SEM) [1]. The MLPG results for the third case have also been compared with an analytical solution.

5.1 Natural convection in a cavity

In this section we present the numerical simulation of the natural convection of a fluid confined in a square cavity using the MLPG technique and the SEM. The results have been obtained for low Rayleigh number $Ra = g\beta \Delta T L^3/(\nu\alpha)$ flow conditions. Figure 4.2 shows the physical problem and the velocity field for $Ra = 1520$ ($\Delta T = 155$ K). In the cavity all the boundaries have a non-slip dynamic boundary condition. Regarding the thermal conditions, the vertical walls have a Dirichlet boundary condition, where the left wall is the cold boundary and the right wall is the hot boundary. The two horizontal walls have a Neumann boundary condition (adiabatic walls). The gravity force is acting along the y direction (negative sense). The MLPG solution was obtained by using a uniform distribution of particles (15×15 particles). The SEM solution was obtained by using 10×10 macro elements and a polynomial expansion $p = 9$. A comparison of the temperature field (isothermal lines) calculated by the MLPG method and the SEM is shown in figure 4.3. Notice that the imposed natural boundary condition (adiabatic condition) at the horizontal surfaces is satisfied by the MLPG method. The temperature profile along the y line located at $x = 0.5$ m and the temperature profile along the x line located at $y = 0.5$ m are shown in figure 5.4. We can observe that the MLPG solution is in agreement with the SEM calculation.

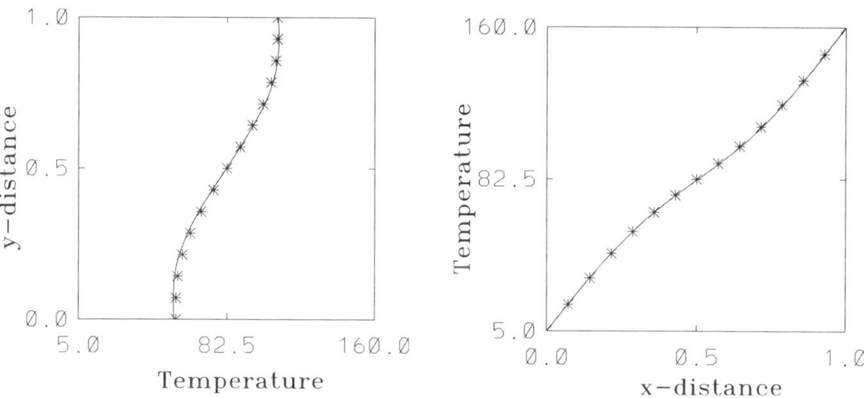

Figure 5.4. Natural convection in a square cavity ($Ra = 1520$). Temperature profile along the y line located at $x = 0.5$ m (*left*). Temperature profile along the x line located at $y = 0.5$ m (*right*). Continuous line: SEM solution. Symbols: MLPG solution.

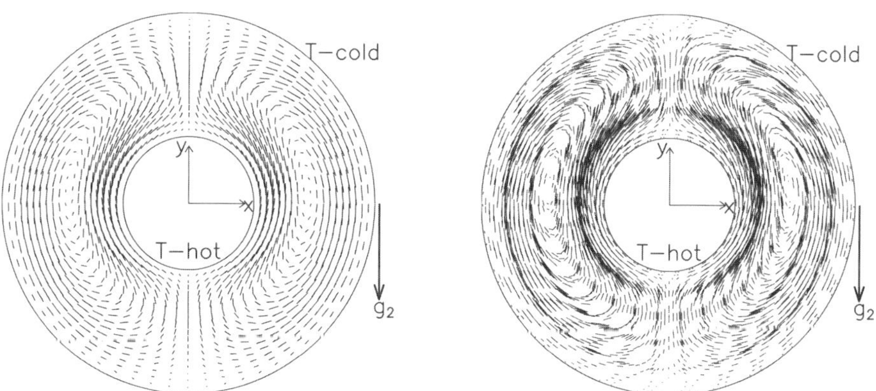

Figure 5.5. Natural convection in concentric cylinders ($Ra = 942$). Velocity field. The temperature of the internal cylinder is higher than the temperature of the external cylinder. *Left*: MLPG solution. *Right*: SEM results.

5.2 Natural convection in concentric cylinders

We present the numerical simulation of the natural convection of a fluid confined between two concentric infinite cylinders. The results have been obtained for a low Rayleigh number ($Ra = g\beta\Delta T d^3/(\nu\alpha) = 942$, where d is the size of the gap between the cylinders) flow condition. Figure 5.5 shows the physical problem and the velocity field obtained by the MLPG method and by the SEM. The origin of the Cartesian coordinate system (x_1,x_2) or (x,y) is located at the center of the concentric cylinders. The aspect ratio is $r = R_{int}/R_{ext} = 0.35$. The gravity acceleration is acting along the x_2- (or y)

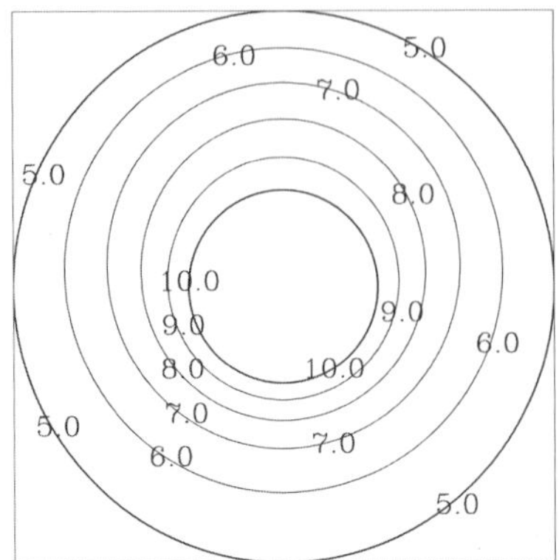

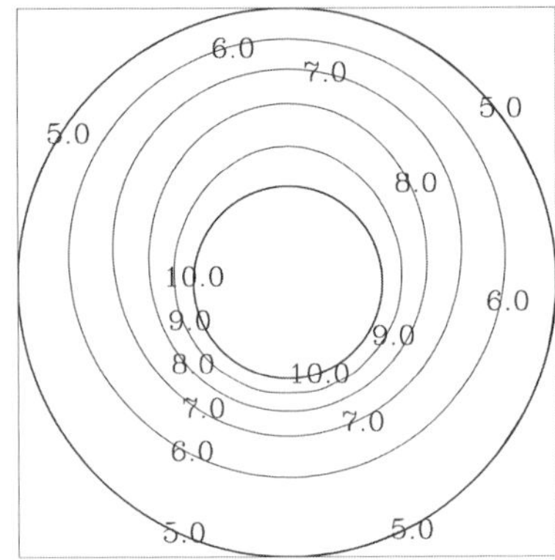

Figure 5.6. Natural convection in concentric cylinders ($Ra = 942$). Isothermal lines. *Left*: MLPG solution. *Right*: SEM results.

axis (with negative sense). In the MLPG simulation we used 60 particles along the angular direction and 20 particles along the radial direction (1200 particles). The particles have been uniformly distributed. The SEM simulation was carried out by using 150 macro elements and a polynomial expansion $p = 9$. The temperature of the internal cylinder is higher than the temperature of the external cylinder. Figure 5.6 shows the isothermal lines obtained by the two numerical methods. It can be observed that due to the buoyancy force, the isothermal lines are not concentric. The isothermal curves obtained by the SEM have more eccentricity, particularly those whose value is close to the internal temperature. The average Nusselt number at the internal and at the external cylinders was also computed. By the use of the MLPG method we obtain $\overline{Nu_{int}} = 1.75$ and $\overline{Nu_{ext}} = 0.65$. Using the SEM we have $\overline{Nu_{int}} = 1.9$ and $\overline{Nu_{ext}} = 0.65$.

5.3 Couette flow in concentric cylinders

The shear flow of a fluid confined between two concentric cylinders is presented in this section. The shear flow is induced by the rotation of the internal cylinder or by the rotation of the external cylinder. We carried out numerical calculations for both cases: (i) rotation of the internal cylinder (fixed external cylinder) with Reynolds number $Re = \rho \Omega_{int} R_{int}(R_{ext} - R_{int})/\mu = 23$ and (ii) rotation of the external cylinder (fixed internal cylinder) with Reynolds number $Re = \rho \Omega_{ext} R_{ext}(R_{ext} - R_{int})/\mu = 6.8$. The geometrical characteristics of the concentric cylinders for this case are the same as for the previous case (see section 5.2). For the internal cylinder rotation case the angular velocity is $\Omega_{int} = -10^{-2}$ rad/s. For the external cylinder rotation case the angular velocity is $\Omega_{ext} = -10^{-3}$ rad/s. The analytical solution for the circumferential component of the velocity $\hat{v}(r)$ as a function of the radial direction r in a steady state, two-dimensional Taylor-Couette flow is

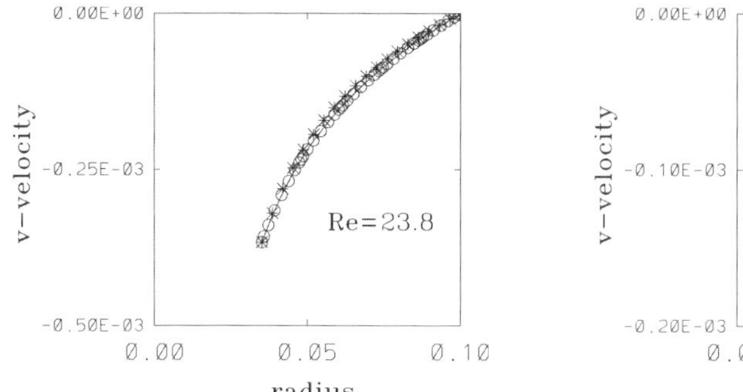

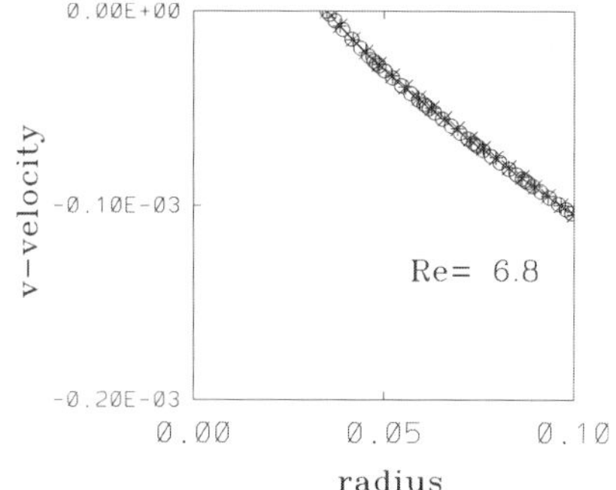

Figure 5.7. Couette flow in concentric cylinders. v velocity profile along the x-axis at $y = 0$ (along the radius). Left: Internal cylinder rotation ($Re = 23.8$). Right: External cylinder rotation ($Re = 6.8$). Circles: SEM solution. Asterisks: MLPG solution. Continuous line: analytical solution (5.13).

$$\hat{v}(r) = R_{int}\Omega_{int}\frac{R_{ext}/r - r/R_{ext}}{R_{ext}/R_{int} - R_{int}/R_{ext}} + R_{ext}\Omega_{ext}\frac{r/R_{int} - R_{int}/r}{R_{ext}/R_{int} - R_{int}/R_{ext}} \tag{5.13}$$

For the MLPG simulation we used 60 particles along the angular direction and 20 particles along the radial direction (1200 particles), again the particles have been uniformly distributed. The SEM simulation was carried out by using 150 macro elements and a polynomial expansion $p = 9$. Figure 5.7 shows the v velocity profiles along the x-axis at $y = 0$ (hence along the radius) for both cases. The curve on the left shows the results for the internal cylinder rotation case. The curve on the right shows the results for the external cylinder rotation case. It can be observed in figure 5.7 that the results provided by the MLPG method are in agreement with the results obtained by the SEM and the circumferential component of the velocity $\hat{v}(r)$ given by the analytical solution (5.13).

5.4 Flow around an infinite circular cylinder

The fluid flow around an infinite cylinder for low Reynolds number conditions is presented in this section. The Reynolds number is defined as a function of the diameter of the cylinder $Re_D = \rho DU/\mu$. Figure 5.8 shows the physical domain and the distribution of the MLPG particles. We have used a non-uniform distribution of particles (567 particles). Figure 5.8 also shows the stream lines for $Re_D = 40$. Its possible to observe the recirculation region (see the two symmetric recirculation zones formed downstream of the cylinder) and the point of separation, this last takes place around $\theta = 110°$ from the leading point which is in agreement with theoretical and experimental results [12]. We have calculated the skin friction drag coefficient C_{Df} due to viscous shear forces produced at the cylinder surface. It has been reported in the literature that the expression to calculate the friction drag coefficient for low Reynolds

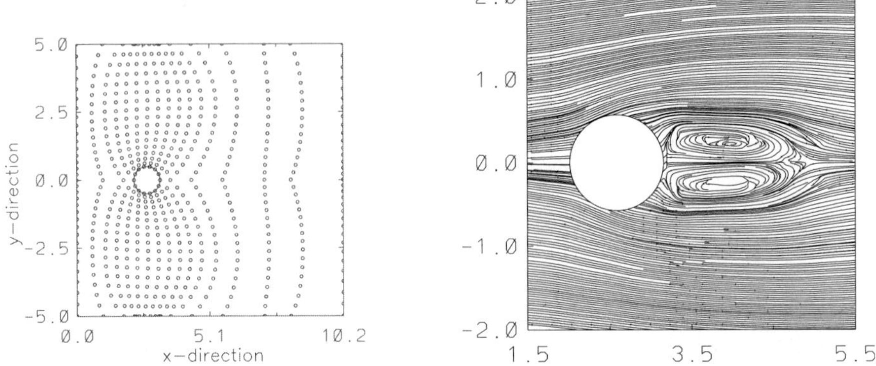

Figure 5.8. Flow around an infinite cylinder. Domain definition and particle distribution. Free stream U velocity from left to right. Top, bottom and left boundaries: Dirichlet boundary condition. Right boundary: Neumann boundary condition (outlet).

numbers is $C_{Df} = 5.786/\sqrt{Re_D}$ [12]. Using the MLPG method for $Re_D = 20$ we obtain $C_{Df} = 1.54$ and for $Re_D = 40$ we have $C_{Df} = 0.89$. Using the Khan et al. [12] expression for low Reynolds number flow we have: for $Re_D = 20$, $C_{Df} = 1.29$, and for $Re_D = 40$, $C_{Df} = 0.91$ these values are in agreement with the MLPG results.

6 Conclusions

The MLPG method coupled with a fully implicit velocity-pressure correction algorithm has been proposed to solve the two-dimensional, steady state incompressible fluid flow equations. One of the advantages of the proposed methodology is that the pressure and the velocity corrections of the particles located in the neighborhood of each node (particle) are fully taken into account, leading to an increase of the rate of convergence of the iterative procedure. The main disadvantage of the proposed methodology is that the algebraic system of equations which is generated to calculate the corrections of velocity (v_1' and v_2') and pressure (p'), requires a great amount of memory. The numerical results obtained by the MLPG method are in agreement with the results provided by a mesh based method: the Spectral Element Method. In order to overcome the governing stability conditions and to increase the capabilities of our code to compute high Reynolds number and high Rayleigh number fluid flow problems, we have to include additional modelling such as: (1) staggered particles (used in Finite Volume methods), (2) different number of neighbours for the solution of each equation (mixed order interpolation used in Finite Element methods) and (3) the upwind algorithms [4].

Acknowledgement

The authors acknowledge the financial support provided by DGAPA-UNAM (PAPIIT project IN102506-3). Most of the computations were carried out in the supercomputers of DGSCA-UNAM.

References

1. G. E. Karniadakis and S. J. Sherwin, *Spectral/hp Element Methods for CFD*, Oxford University Press, 1999.
2. G. R. Liu, *Mesh Free Methods, Moving Beyond the Finite Element Method*, CRC Press, 2003.
3. H. Lin and S. N. Atluri, *The Meshless Local Petrov-Galerkin (MLPG) Method for Solving Incompressible Navier-Stokes Equations*, CMES **2** (2) (2001), 117–142.
4. H. Lin and S. N. Atluri, *The Meshless Local Petrov-Galerkin (MLPG) Method for convection-diffusion problems*, CMES **1** (2) (2000), 45–60.
5. T. Belytschko, Y. Krongauz, D. Morgan, M. Glemming and P. Krysl, *Meshless methods: An overview and recent developments*, Comput. Methods in Appl. Mech. and Engrg. **39** (1996), 3–139.
6. Y.L. Wu, G.R. Liu and Y. T. Gu, *Application of Meshless Local Petrov Galerkin (MLPG) Approach to Simulation of Incompressible Flow*, Numerical Heat Transfer, Part B **48** (2005), 459–475.
7. A. Arefmanesh, M. Najafi and H. Abdi, *A Meshless Local Petrov-Galerkin Method for Fluid Dynamics and Heat Transfer Applications*, Journal of Fluids Engineering **127** (2005), 647–654.
8. Q.W. Ma, *Meshless local Petrov-Galerkin method for two-dimensional nonlinear water wave problems*, Journal of Computational Physics **205** (2005), 611–625.
9. W. Q. Tao, Z. G. Qu and Y. L. He, *A novel segregated algorithm for incompressible fluid flow and heat transfer problems-CLEAR (Coupled and Linked Equations Algorithm Revised) Part I: Mathematical formulation and solution procedure*, Numerical Heat Transfer, Part B **45** (2004), 1–17.
10. S. Li. and W. K. Liu, *Meshfree and particle methods and their applications*, App. Mech. Rev. **55** (1) (2004), 1–34.
11. S. N. Atluri and S. Shen, *The Meshless Local Petrov-Galerkin (MLPG) Method*, Tech. Science Press, 2002.
12. W. A. Khan and J. R. Culham and M. M. Yovanovich, *Fluid flow around and heat transfer from an infinite circular cylinder*, Journal of Heat Transfer **127** (2005), 785–790.

Large Scale, Multiresolution Flow Simulations Using Remeshed Particle Methods

Philippe Chatelain, Michael Bergdorf, and Petros Koumoutsakos

Computational Science, ETH Zurich, CH-8092, Switzerland
pchatela@ethz.ch, bergdorf@ethz.ch, petros@ethz.ch

Summary. Particle methods are a robust and versatile computational tool for the simulation of continuous and discrete physical systems ranging from Fluid Mechanics to Biology and Social Sciences. In advection dominated problems particle methods can be considered as the method of choice due to their inherent robustness, stability and Lagrangian adaptivity. At the same time however, smooth particle methods encounter major difficulties in simulating the equations they set out to discretize when their computational elements fail to overlap, a condition necessary for their convergence [2]. A number of ad-hoc parameters and artificial dissipation techniques are often introduced in techniques such as Smoothed Particle Hydrodynamics (SPH) [15, 19] in order to remedy these difficulties.

In the present paper we demonstrate that the convergence of smooth particle methods can be ensured by a periodic remeshing of the particles using high-order interpolation kernels. This procedure retains the Lagrangian character and stability of particle methods and enables the control of their accuracy [5, 9, 16, 17] while introducing numerical dissipation at levels well below those introduced by temporal discretizations.

In addition, remeshing enables two major improvements over grid-free particle methods : First by exploiting the regularity of the remeshed particles, it reduces by at least an order of magnitude their computational cost [6, 10] and facilitates their massively parallel implementation. Second, remeshing enables the development of consistent multiresolution techniques such as wavelet-particle methods [4]. This approach has been implemented efficiently in massively parallel computer architectures allowing for unprecedented vortex dynamics simulations using billions of particles.

Key words: particle methods, vortex methods, smoothed particle hydrodynamics, wavelets, multiresolution

1 Introduction

Particle methods are distinguished by their robustness and adaptivity in simulations of convection dominated flows. The Lagrangian formulation allows for automatic adaptivity and permits significantly larger time steps than the ones imposed by stability requirements by the the corresponding Eulerian discretizations. At the same time smooth particle methods become inaccurate, computationally inefficient and fail to provide consistent and convergent simulations of flows that lead to particle distortion.

In the recent years remeshed particle techniques (see [16, 17, 24] and references therein) have been proposed where a mesh is used along with the particles in order to develop efficient and accurate computations of vortical flows. The mesh is used in order to reinitialize the distorted particle locations [5, 9, 13, 16, 24] thus ensuring the convergence of the method. In addition the mesh enables the efficient computation of differential operators, and the use of fast elliptic solvers for the computation of the field equations. The particles and the mesh exchange field quantities and particle strengths via moment conserving interpolations.

Vortex methods exemplify the computational advantages and challenges of particle methods in simulations of incompressible vortical flows. These simulations are based on the discretization of the vorticity-velocity formulation of the Navier-Stokes equations in a Lagrangian form. The methodological advances outlined above introduce a number of challenges for the massively parallel implementation of vortex methods that have hindered in the past large scale Direct Numerical Simulations. In Section 3.1 we report on efficient domain decompositions and optimized data mappings that rely on the Message Passing Interface (MPI). The algorithm is implemented for the distributed-memory architecture of the IBM BlueGene/L using up to 16K CPUs and involving up to six billion particles; it enables unprecedented resolutions for the DNS of long wavelength instabilities.

The reinitialization of particle locations detracts from automatic adaptivity though. In a Particle-Mesh method adaptivity needs to be reintroduced and controlled. Because of the role of the mesh in the present hybrid approach, several multi-resolution tools can be adopted [3]. We present in Section 2.2 such a tool: the Particle-Wavelet Method, and assess its capabilities on interface growth problems.

2 Particle-Mesh Methods

Particle methods are well-suited for the solution of conservation laws in $\mathbb{R}^d$ with advection terms

$$\frac{\partial q}{\partial t} + \nabla \cdot (\mathbf{u}\, q) = \mathcal{L}(q, \mathbf{x}, t) \tag{2.1}$$

where $\mathbf{u}$ is the velocity field and the operator $\mathcal{L}$ collects the non-advective terms, e.g. diffusive fluxes, sinks and sources. The transported quantity q is discretized on particle locations and is approximated as

$$q(\boldsymbol{x},t) \approx \sum_p Q_p(t)\zeta^h\left(\mathbf{x}-\mathbf{x}_p(t)\right) , \tag{2.2}$$

where h denotes the mesh spacing and $Q_p = v_p q(\mathbf{x}_p)$ are the particle strengths, and v_p the particle volumes. The accuracy of Eq. 2.2 depends on the smoothness and moment-conservation properties of the kernel $\zeta^h(\mathbf{x}) = h^{-d}\zeta(\mathbf{x}/h)$, see [9] for a detailed analysis.

The discretization of the Lagrangian form in Eq. 2.1 using the particle approximation of Eq. 2.2, yields a set of Ordinary Differential Equations (ODEs) [17] for the particle strengths, positions, and volumes which have to be integrated numerically

$$\begin{aligned} \frac{dQ_p}{dt} &= v_p\left(\mathcal{L}^h q\right)(\boldsymbol{x}_p,t) , \\ \frac{d\boldsymbol{x}_p}{dt} &= \boldsymbol{u}(\boldsymbol{x}_p,t) , \\ \frac{dv_p}{dt} &= v_p\left(\nabla^h\cdot\boldsymbol{u}\right)(\boldsymbol{x}_p,t) , \end{aligned} \tag{2.3}$$

for particles $p = 1,\ldots,N$.

In the present approach, we use a mesh conjointly with the particles [3, 7, 17]. We evaluate the function q on regular grid locations, compute the differential operators on the grid, and then interpolate back onto the particle locations. In terms of operations, one Euler step for the system of Eq. 2.3 is completed by

- (P $\rightarrow$ M) Interpolate particle quantities on a lattice by evaluating Eq. 2.2 on grid locations

$$q(\mathbf{x}_{ij\ldots}) = \sum_p Q_p\zeta^h\left(\mathbf{x}_{ij\ldots}-\mathbf{x}_p\right) \tag{2.4}$$

  where $\mathbf{x}_{ij\ldots}$ is a grid node and $ij\ldots$ are node indices
- (M $\rightarrow$ M) Perform operations on the grid, *e.g.* evaluate right-hand sides of the system of Eq. 2.3
- (M $\rightarrow$ P) Interpolate quantities, right-hand sides, respectively back onto the particles,

$$\mathcal{L}(\mathbf{x}_p) = \sum_i\sum_j\sum_{\ldots} h^{-d}\mathcal{L}(\mathbf{x}_{ij\ldots})\zeta^h\left(\mathbf{x}_p-\mathbf{x}_{ij\ldots}\right) \tag{2.5}$$

  and advance the quantities and locations.

Like any particle method, this hybrid approach remains linearly stable for advection. Non-linear stability, however, requires that particle trajectories do not cross. This Lagrangian CFL constraint reads

$$\mathrm{LCFL} = \delta t\,\|\nabla\mathbf{u}\|_\infty < C\,. \tag{2.6}$$

The automatic adaptivity of particle methods, synonymous with uncontrolled Lagrangian distortion, comes at the expense of their convergence and accuracy [1, 2]. An irregular distribution of particles leads to a loss of convergence, as the function approximation (Eq. 2.2) ceases to be well sampled. It is therefore necessary to periodically regularize the particle locations by "remeshing" them onto regular positions [16]. This redistribution is readily carried out through a P $\rightarrow$ M interpolation of the particle weights and generating a new set of particles with positions at the grid locations.

2.1 Efficient implementation and massively parallel simulations

The Particle Mesh methodology described above not only ensures the accuracy of the simulations but also allows the use of fast and scalable techniques for the computation of differential operators, e.g. finite differences, and the handling of elliptic problems with multigrid or Fourier solvers.

In contrast, in the Smoothed Particle Hydrodynamics (SPH) method [15, 19] or the Particle Strength Exchange (PSE) scheme [11], one computes the differential operators on the right-hand sides of Eq. 2.3 directly at the particle locations. This requires the summation of contributions from particles within the kernel cut-off and thus finding these neighboring particles. This process involves additional data structures, accessing non-contiguous data, more memory cache misses, and in the end, efficiency degradation.

The scalability of this Particle-Mesh approach has allowed large scale simulations on massively parallel architectures [7]. The parallel implementation of this hybrid methodology is provided by the open source Parallel Particle Mesh (PPM) library [21]. PPM provides a general-purpose framework that can handle the simulation of particle-only, mesh-only or particle-mesh systems. The library can define topologies, i.e. space decompositions and the assignment of sub-domains to processors, which achieve particle- and mesh-based load balancing. The data communication is organized in mappings which can be applied to either mesh points or particles. As an example, a local neighbor-to-neighbor mapping re-assigns particles which are advected from a sub-domain into a neighboring one. PPMis written in Fortran 90 on top of the Message Passing Interface (MPI)[1]. In Section 3.1, we report on the performance of this framework on the massively parallel architecture of IBM BlueGene/L and its application to aircraft wakes.

2.2 Multiresolution Particle-Mesh methods with Wavelet-based adaptation

The "remeshing" procedure introduced by the Particle Mesh technique ensures particle overlap and preserves the accuracy of the method [17]. As a

[1] PPM is available at www.cse-lab.ethz.ch

consequence, it distances itself from classical particle methods with an automatic adaptivity that is synonym of degraded accuracy. Adaptation can be reintroduced in Particle-Mesh techniques though, in a more controlled fashion, through adaptive mesh refinement techniques [3] or the Wavelet-based [4] approach presented here.

Wavelet-based adaptation of remeshed particle quantities

In the present framework we implement tensor-product wavelets $\psi^{l,\mu}$ and scaling functions φ^l on a sequence of $L+1$ dyadically refined grids with mesh spacings $\{h_l\}_{l=0}^L = \{h_0\,2^{-l}\}_{l=0}^L$ and grid points $\boldsymbol{k} \in \{\mathcal{K}^l\}_{l=0}^L$. The scaling functions and wavelets are related as:

$$\varphi_{\boldsymbol{j}}^l = \sum_{\boldsymbol{k}} H_{\boldsymbol{j},\boldsymbol{k}}^l\, \varphi_{\boldsymbol{k}}^{l+1}\,, \qquad \psi_{\boldsymbol{j}}^{l,\mu} = \sum_{\boldsymbol{k}} G_{\boldsymbol{j},\boldsymbol{k}}^{l,\mu}\, \varphi_{\boldsymbol{k}}^{l+1}\,, \tag{2.7}$$

where $\mu = 1, \ldots, 2^d - 1$ for d-dimensional wavelets. The discrete filters $H_{\boldsymbol{j},\boldsymbol{k}}^l$ and $G_{\boldsymbol{j},\boldsymbol{k}}^{l,\mu}$ depend on the specific choice of wavelets employed. Using these bases the function $q(\boldsymbol{x})$ is expressed as

$$q(\boldsymbol{x}) = \sum_{\boldsymbol{k}\in\mathcal{K}^0} c_{\boldsymbol{k}}^0\, \varphi_{\boldsymbol{k}}^0(\boldsymbol{x}) + \sum_{l=0}^{L-1} \sum_{\boldsymbol{k}\in\mathcal{K}^l} \sum_{\mu=1}^{2^d-1} d_{\boldsymbol{k}}^{l,\mu}\, \psi_{\boldsymbol{k}}^{l,\mu}(\boldsymbol{x})\,. \tag{2.8}$$

The scaling coefficients $c_{\boldsymbol{k}}^l$ and detail coefficients $d_{\boldsymbol{k}}^{l,\mu}$ can be efficiently computed using a Fast Wavelet Transform. In areas where the function $q(\boldsymbol{x})$ is smooth the detail coefficients of fine levels l will tend to be small, and a compressed representation of $q(\boldsymbol{x})$ is obtained by discarding detail coefficients for which $|d_{\boldsymbol{k}}^{l,\mu}| < \varepsilon$. The error introduced by this compression is bounded by

$$\|q(\boldsymbol{x}) - q_{\geq}(\boldsymbol{x})\|_\infty \leq C_1\, \varepsilon \leq C_2\, \mathcal{N}^{P/d}\,, \tag{2.9}$$

where $q_{\geq}$ is the compressed q, P is the order of the wavelets and $\mathcal{N}$ is the number of active coefficients.

It is important to note that each detail coefficient is associated with a grid point on the next finer grid, as illustrated in Fig. 2.1. Hence, the compressed representation $q_{\geq}$ is inherently linked with a adapted grid, composed only of the grid points whose detail coefficients are significant, *i.e.* $|d_{\boldsymbol{k}}^{l,\mu}| \geq \varepsilon$.

Particle-Mesh interpolation and Wavelet MRA

The wavelet-based MRA of the remeshed particle properties enables an enhanced multiresolution particle function representation. In order to allow for the emergence of small scales between two remeshing steps we follow the conservative approach of Liandrat and Tchamitchian [18] and additionally activate all children of the active grid points.

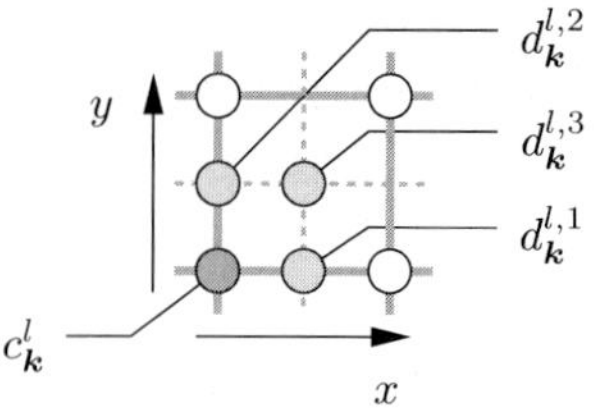

Figure 2.1. Each detail coefficient $d_{\boldsymbol{k}}^{l,\mu}$, with $\mu = 1, \ldots, 2^d - 1$ corresponds to a specific grid point on the next finer level.

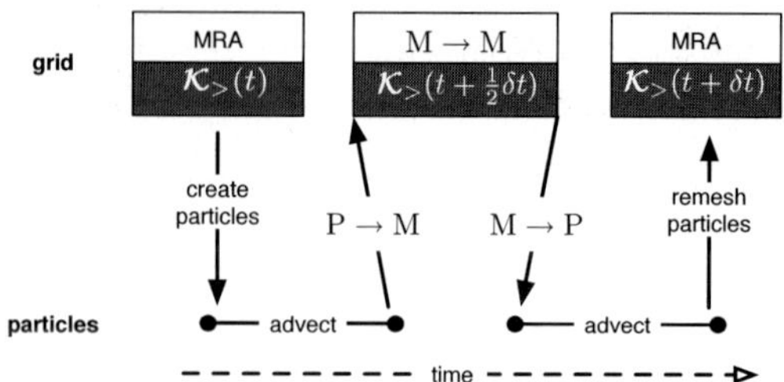

Figure 2.2. Particles are created on the adapted grid $\mathcal{K}_>(t)$ and advected. In the context of a two-step ODE integration scheme, the particle function representation is evaluated (P → M) on an intermediate grid $\mathcal{K}_>(t + \frac{1}{2}\delta t)$ and the right-hand sides that are evaluated on this grid (M → M) are interpolated back onto the particles (M → P). At the end of the time step the particles are remeshed onto a mesh $\mathcal{K}_>(t + \delta t)$ on which the next MRA is performed.

The computational steps (P → M, M → M, M → P) are outlined in Fig. 2.2 for the case of a two-step ODE integration scheme), and are based on level-wise operations. For a detailed description of multiresolution P → M and M → P interpolations we refer to [4].

3 Results

3.1 Massively parallel simulation of aircraft wakes

Vortex Particle Mesh Method

We apply the Particle-Mesh methodology to the Navier-Stokes equations in their velocity($\mathbf{u}$)-vorticity($\boldsymbol{\omega} = \nabla \times \mathbf{u}$) form

$$\frac{\partial \boldsymbol{\omega}}{\partial t} + \nabla \cdot (\mathbf{u}\,\boldsymbol{\omega}) = (\boldsymbol{\omega} \cdot \nabla)\,\mathbf{u} + \nu \nabla^2 \boldsymbol{\omega} \tag{3.10}$$

$$\nabla \cdot \mathbf{u} = 0 \tag{3.11}$$

where ν is the kinematic viscosity. In this case, particles carry circulations $\mathbf{Q}_p = \int_{v_p} \boldsymbol{\omega} d\mathbf{x}$. The computation of the right-hand sides of the particle ODEs involves derivatives of the vorticity and velocity fields. We note that velocity is the solution of the Poisson problem

$$\nabla^2 \mathbf{u} = -\nabla \times \boldsymbol{\omega} \, . \tag{3.12}$$

The domain is a periodic box in $\mathbb{R}^3$, we compute derivatives on the mesh with fourth order accurate finite differences and solve Eq. 3.12 in Fourier space.

Scalability

The parallel efficiency was assessed on a test problem taken from [25] for $512 \leq N_{\text{CPU}} \leq 16384$ of IBM BlueGene/L (BG/L). We measure the strong efficiency as

$$\eta_{\text{strong}} = \frac{N_{\text{CPUS}}^{\text{ref}} \, T(N_{\text{CPUS}}^{\text{ref}})}{N_{\text{CPUS}} \, T(N_{\text{CPUS}})} \tag{3.13}$$

where T is the average computation time of one time step. In order to test the code up to the large sizes allowed by BG/L, we used $N_{\text{CPUS}}^{\text{ref}} = 2048$ and a problem size of $768 \times 1024 \times 2048$ or 1.6 billion particles. This brings the per-processor problem size from 786432 down to 98304 when we run on the maximum number of processors. The curve (Fig. 3(b)) displays a plateau up to $N_{\text{CPUS}} = 4096$, with the per-processor problem size becoming progressively smaller and communication overhead overwhelming the computing cycles.

From this result, we base our weak scalability study on a constant per-processor number of particles of $M_{\text{per CPU}} \simeq 4\,10^5$. We used the following measure

$$\eta_{\text{weak}} = \frac{T(N_{\text{CPUS}}^{\text{ref}}, M^{\text{ref}})}{T(N_{\text{CPUS}}, \frac{N_{\text{CPUS}}}{N_{\text{CPUS}}^{\text{ref}}} M^{\text{ref}})} \, . \tag{3.14}$$

where we took $N_{\text{CPUS}}^{\text{ref}} = 512$. The code displays (Fig. 3(a)) excellent scalability up to $N_{\text{CPUS}} = 4096$. Eq. 3.14 assumes linear complexity for the problem at hand. There is however an $O(N \log N)$ component to the overall complexity of the present problem as we are solving the Poisson equation for the convection velocity. The two curves (with and without the cost for the solution of the Poisson equation) are shown in (Fig. 3(a)). We note that using 16K processors enables unprecedented simulations using $O(10^{10})$ vortex particles.

Aircraft trailing vortices

The evolution and eventual destruction of aircraft trailing vortices is affected by several types of instabilities, usually classified according to their wavelength. A rapidly growing, medium-wavelength instability has been the focus

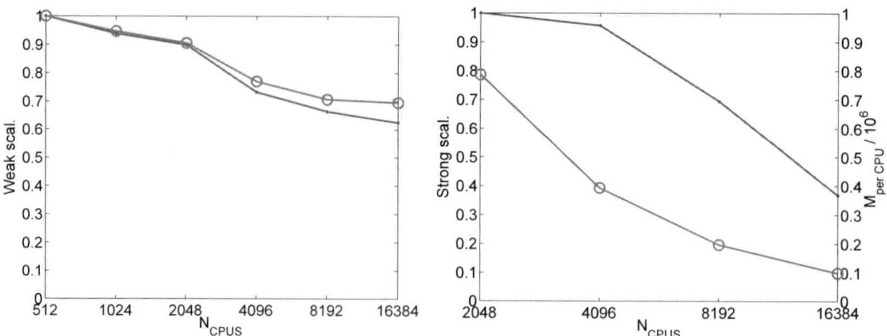

(a) Weak scalability for a per-processor problem size $= 4 \cdot 10^5$; full problem (solid dots) and excluding the Poisson solver (circles)

(b) Strong scalability (solid dots) and per-processor size (circles)

Figure 3.3. Medium-wavelength instability of counter-rotating vortices: parallel efficiencies on IBM BlueGene/L. (*See also* Color Plate on page 360)

of recent experimental [12,20,22] and numerical studies [8,23,25]. This instability occurs in the presence of a secondary vortex pair that is counter-rotating relative to the main pair. These secondary vortices are generated by a sufficient negative load on the horizontal tail or the inboard edge of outboard flaps. Being weaker, they eventually wrap around the primary ones in so-called Ω-loops, leading to the reconnection of vortices of unequal circulations. This in turn triggers an accelerated vortex destruction.

We carry out the Direct Numerical Simulation of the onset of this instability at $Re = \Gamma_1/\nu = 6000$ where Γ_1 is the circulation of the main vortex pair. We use a long domain and a noisy initial condition to allow the growth of several modes. The present simulation is afforded thanks to a mesh resolution of $2048 \times 1024 \times 768$ and 1.6 billion particles. It is run on 4096 CPUs; the wall-clock computation time was 39s on average per time step. Figure 3.4 shows the evolution of this system and how it picks up the medium wavelength mode. We refer the reader to [7] for additional numerical details and discussion of this simulation.

3.2 PMW simulation of interface evolution

We apply the Particle-Mesh Wavelet method to the dynamics of interfaces driven by complex physics. We consider the growth of dendrites in a supercooled liquid according to the sharp-interface model. The governing equations are given by

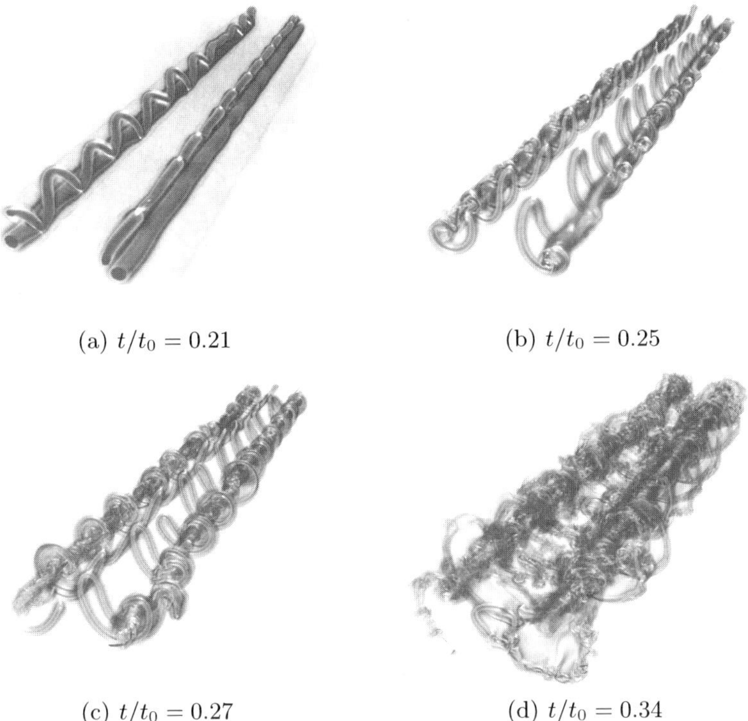

(a) $t/t_0 = 0.21$ (b) $t/t_0 = 0.25$

(c) $t/t_0 = 0.27$ (d) $t/t_0 = 0.34$

Figure 3.4. Trailing vortices, instability initiation by ambient noise: visualization of the vorticity structures by volume rendering. High vorticity norm regions correspond to red and opaque; low vorticity ones are blue and transparent. (*See also* Color Plate on page 361)

$$
\begin{aligned}
\frac{\partial T}{\partial t} &- \nabla \cdot (k \nabla T)\,, \\
T|_{\partial\Omega} &= T_{\partial\Omega}\,, \\
T|_{\Gamma} &= T_{\Gamma}\,, \\
\boldsymbol{u}|_{\Gamma} &= -\boldsymbol{n}\,[k\,\nabla T \cdot \boldsymbol{n}]_{\Gamma}\,, \\
\frac{\partial \varphi}{\partial t} + \boldsymbol{u} \cdot \nabla \varphi &= 0\,, \\
\Gamma &= \{\, x \mid \varphi(x) = 0 \,\}\,,
\end{aligned}
\tag{3.15}
$$

where T is the temperature, φ the level set function, Ω is the computational domain, $\Gamma = \partial\Omega_r$ is the phase boundary, $\boldsymbol{n}$ is the outward normal on Γ, k is the thermal conductivity. The temperature at the interface is given by the Gibbs-Thomson relation

$$
T_{\Gamma} = -\epsilon_c \kappa - \epsilon_v\, u_n\,, \tag{3.16}
$$

where κ denotes the curvature on the interface, ϵ_c is the surface tension coefficient and ϵ_v the molecular kinetic coefficient. We focus on a mathematically unstable example from [14] that gives rise to small-scale features and demonstrates the regularizing properties of the particle level set formulation: $\Omega = [-1.5, 1.5]^2$, $k = 1.0$, $\epsilon_c = \epsilon_v = 0$, and the initial conditions for T and the interface are given by

$$\Gamma = [-0.1, 0.1] \times [-0.1, 0.1]$$
$$T(\boldsymbol{x}, t = 0) = \begin{cases} -0.5\,, & \boldsymbol{x} \in \Omega \backslash \Omega_r \\ 0.0\,, & \text{else}\,. \end{cases} \tag{3.17}$$

For this simulation $h_0 = 3/32$ and $N_{\text{lev}} = 5$. We compare two cases: in one case $\varepsilon = 10^{-4}$ and in the other $\varepsilon = 10^{-7}$. Fig. 3.5 depicts the propagation of the phase boundary for the two cases. It is evident that with bigger ε the 45°- symmetry, respectively the anisotropy, is more pronounced. This expected symmetry break is caused by the MRA and other grid effects. Fig. 3.6 illustrates the enhanced resolution at the phase boundary at the beginning of the simulation and once the dendritic fingering has evolved.

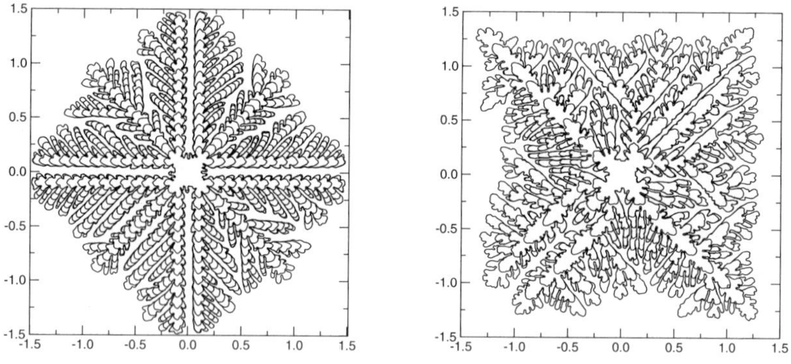

Figure 3.5. Dendritic growth in a pure melt. *Left:* $\varepsilon = 10^{-7}$; *right:* $\varepsilon = 10^{-4}$. Contours represent the temporal evolution of the phase boundary.

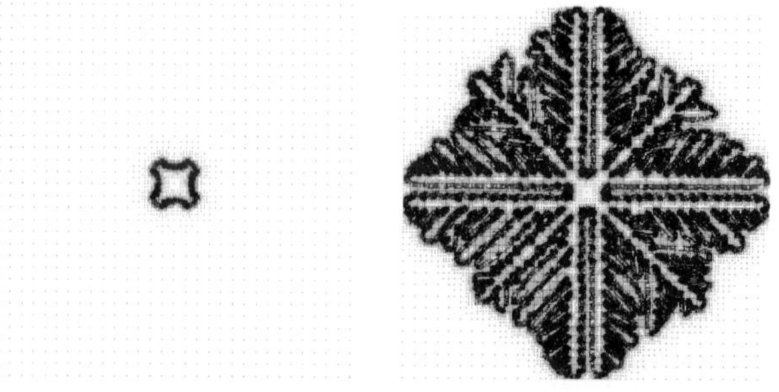

Figure 3.6. Active grid points/particles at two different times of the simulation of dendritic growth.

4 Conclusions

This paper outlines advances in remeshed particle methods. The proposed Particle-Mesh coupling ensures computational efficiency and preserves the accuracy of particle methods when their locations are distorted. In addition to ensuring convergence, and maintaining the large time steps afforded by the Lagrangian convection of particles, the present use of a mesh enables fast calculations of differential operators and the introduction of a novel wavelet-based, multiresolution particle method. The method is implemented efficiently in massivley parallel computer architectures and its capabilities are demonstrated in simulations of aircraft wakes.

References

1. J. T. Beale, *On the accuracy of vortex methods at large times*, in Proc. Workshop on Comput. Fluid Dyn. and React. Gas Flows, IMA, Univ. of Minnesota, 1986, B. E. et al., ed., Springer-Verlag, New York, 1988, p. 19.
2. J. T. Beale and A. Majda, *Vortex methods I: convergence in 3 dimensions*, Mathematics of Computation, 39 (1982), pp. 1–27.
3. M. Bergdorf, G.-H. Cottet, and P. Koumoutsakos, *Multilevel adaptive particle methods for convection-diffusion equations*, Multiscale Model. Simul., 4 (2005), pp. 328–357.
4. M. Bergdorf and P. Koumoutsakos, *A Lagrangian Particle-Wavelet method*, Multiscale Modeling & Simulation: A SIAM Interdisciplinary Journal, 5 (2006), pp. 980–995.
5. A. Chaniotis, D. Poulikakos, and P. Koumoutsakos, *Remeshed smoothed particle hydrodynamics for the simulation of viscous and heat conducting flows*, J. Comput. Phys., 182 (2002), pp. 67–90.
6. P. Chatelain, G.-H. Cottet, and P. Koumoutsakos, *PMH: Particle Mesh Hydrodynamics*, International Journal of Modern Physics C, 18 (2007), pp. 610–618.
7. P. Chatelain, A. Curioni, M. Bergdorf, D. Rossinelli, W. Andreoni, and P. Koumoutsakos, *Billion vortex particle direct numerical simulations of aircraft wakes*, Computer Methods in Applied Mechanics and Engineering, 197 (2008), pp. 1296–1304.
8. R. Cocle, L. Dufresne, and G. Winckelmans, *Investigation of multiscale subgrid models for les of instabilities and turbulence in wake vortex systems*, Lecture Notes in Computational Science and Engineering, 56 (2007), pp. 141–159.
9. G.-H. Cottet and P. Koumoutsakos, *Vortex Methods, Theory and Practice*, Cambridge University Press, 2000.
10. G.-H. Cottet and L. Weynans, *Particle methods revisited: a class of high-order finite-difference schemes*, C. R. Acad. Sci. Paris, Sér. I, 343 (2006), pp. 51–56.
11. P. Degond and S. Mas-Gallic, *The weighted particle method for convection-diffusion equations. part 1: The case of an isotropic viscosity*, Mathematics of Computation, 53 (1989), pp. 485–507.

12. D. A. Durston, S. M. Walker, D. M. Driver, S. C. Smith, and Ö. Savas, *Wake vortex alleviation flow field studies*, J. Aircraft, 42 (2005), pp. 894–907.
13. J. D. Eldredge, T. Colonius, and A. Leonard, *A vortex particle method for two dimensional compressible flow*, J. Comp. Phys., 179 (2002), pp. 371–399.
14. F. Gibou, R. Fedkiw, R. Caflisch, and S. Osher, *A level set approach for the numerical simulation of dendritic growth*, SIAM J. Sci. Comput., 19 (2003), pp. 183–199.
15. R. Gingold and J. Monaghan, *Smoothed particle hydrodynamics: theory and application to non-spherical stars*, Mon. Not. Roy. Astron. Soc., 181 (1977), p. 375.
16. P. Koumoutsakos, *Inviscid axisymmetrization of an elliptical vortex*, J. Comput. Phys., 138 (1997), pp. 821–857.
17. ———, *Multiscale flow simulations using particles*, Annu. Rev. Fluid Mech., 37 (2005), pp. 457–487.
18. J. Liandrat and P. Tchamitchian, *Resolution of the 1D regularized burgers equation using a spatial wavelet approximation*, ICASE Report 90-83, NASA Langley Research Center, 1990.
19. J. J. Monaghan, *Smoothed particle hydrodynamics*, Reports on Progress in Physics, 68 (2005), pp. 1703–1759.
20. J. M. Ortega and Ö. Savas, *Rapidly growing instability mode in trailing multiple-vortex wakes*, AIAA Journal, 39 (2001), pp. 750–754.
21. I. F. Sbalzarini, J. H. Walther, M. Bergdorf, S. E. Hieber, E. M. Kotsalis, and P. Koumoutsakos, *PPM a highly efficient parallel particle mesh library for the simulation of continuum systems*, J. Comput. Phys., 215 (2006), pp. 566–588.
22. R. Stuff, *The near-far relationship of vortices shed from transport aircraft*, in AIAA Applied Aerodynamics Conference, 19th, Anaheim, CA, AIAA, ed., AIAA, 2001, pp. AIAA–2001–2429.
23. E. Stumpf, *Study of four-vortex aircraft wakes and layout of corresponding aircraft configurations*, J. Aircraft, 42 (2005), pp. 722–730.
24. G. Winckelmans, *Vortex methods*, in Encyclopedia of Computational Mechanics, E. Stein, R. De Borst, and T. J. Hughes, eds., vol. 3, John Wiley and Sons, 2004.
25. G. Winckelmans, R. Cocle, L. Dufresne, and R. Capart, *Vortex methods and their application to trailing wake vortex simulations*, C. R. Phys., 6 (2005), pp. 467–486.

On the Stabilization of Stress-Point Integration in the Element Free Galerkin Method

Qinglin Duan[1] and Ted Belytschko[2]

[1] Department of Mechanical Engineering, Northwestern University, 2145 Sheridan Road, Evanston, IL 60208, U.S.A. q-duan@northwestern.edu

[2] Department of Mechanical Engineering, Northwestern University, 2145 Sheridan Road, Evanston, IL 60208, U.S.A. tedbelytschko@northwestern.edu

Summary. Stabilized stress-point integration schemes based on Least-Squares Stabilization (LSS), Taylor series Expansion Based Stabilization (TEBS) and Finite Increment Gradient (FIG) are compared for linear elastostaticity problems and some relations between them are described. Particular emphasis is placed on stress-point integration procedures with stabilization. The convergence and stability properties of stabilized methods in the framework of the element free Galerkin (EFG) method with stress-point integration are studied by numerical examples. It is shown that stabilized stress-point integration consumes much less computational time than full integration and exhibits higher accuracy and much better convergence and stability than unstabilized stress-point integration and stabilized nodal integration.

Key words: Stress-point integration; Nodal integration; EFG; meshfree; stabilization

1 Introduction

In recent years, many efforts have been devoted to improving the speed and efficiency of meshfree methods. Development of simple and efficient domain integration methods for Galerkin based meshfree methods, such as the EFG method [1] or the RKPM method [2], is still an open topic.

In the context of meshfree Galerkin methods, background Gauss integration is commonly used in the integration of the Galerkin weak form. However, due to the non-polynomial character of the moving least-squares (MLS) approximants employed for most meshfree methods, higher-order Gauss quadrature is required than in the finite element method, which impairs the computational efficiency of meshfree Galerkin methods. These methods have been studied and improved by Dolbow and Belytschko [3] and Griebel and Schweitzer [4]. However, additional gains are needed to match the efficiency of FEM.

Nodal integration schemes, i.e. evaluating the integrals of the weak form only at the nodes, are very attractive due to their high computational efficiency and their absence of the background cell structure or background mesh employed for Gauss quadrature. However, they tend to suffer from spurious singular modes due to the underintegration of the weak form. Beissel and Belytschko [5] first addressed this problem and proposed a stabilized nodal integration procedure by adding a residual of the equilibrium equation to the potential energy functional in the EFG framework. Based on a strain smoothing procedure, Chen et al. developed a stabilized conforming nodal integration (SCNI) method [6,7] for Galerkin meshless methods and also extended it to the natural-element method [8]. Other improvements for nodal integration have also been reported by Bonet et al. in his CSPH method, see [9] for details.

Domain integration of the weak form can also be performed by an intermediate scheme often called stress-point integration. In this scheme, additional slave particles called stress points are added between the original particles and serve as additional quadrature points. This method was first proposed by Dyka et al. [10, 11] for tensile instabilities in SPH and was later extended to higher dimensions by Randles and Libersky [12]. In their implementations, only the stress points are used for quadrature. Belytschko et al. [13] pointed out that stress points do not circumvent the tensile instabilities, but instead, in many cases they restore the positive definiteness of the linear equations, i.e. they correct rank deficiency. In [14], Rabczuk and Belytschko et al. developed an alternative form of stress-point integration for multi-dimensions that employs both the particles and stress points as quadrature points and it shows better stability properties than nodal integration. Fries and Belytschko [15] have studied the influence of stress-point integration on the convergence and stability properties in the Laplace and Poisson equations. They concluded that for regular nodal arrangements, good rates of convergence can be achieved, however, for non-uniform nodal arrangements, stress-point integration is associated with a mild instability which can result in poor convergence. Thus, stabilization is needed in these cases to improve the convergence of stress-point integration.

The purpose of this paper is to examine several stabilized stress-point integration schemes for elastostatic problems and to examine their relationships. The stabilization schemes studied are least-square stabilization (LSS) [5, 15], Taylor series expansion based stabilization (TEBS) [16–18] and the finite increment gradient (FIG) stabilization [19]. The relationships between these stabilization techniques are clearly brought out by our formulations. Numerical results show that LSS and TEBS successfully improve the convergence and stability properties of stress-point integration while the FIG fails. Moreover, the superiority of stabilized stress-point integrations over the stabilized nodal integrations is also clearly demonstrated by our numerical examples.

The outline of this paper is as follows. The MLS approximation with the computation of its second order derivatives is first reviewed in section 2. The equations and the corresponding Galerkin weak form of elastostatic problems are briefly described in section 3. Stress-point integration scheme with the placement of stress points and the determination of the integration weights are discussed in section 4. Several stabilization strategies including LSS, TEBS and FIG are described and their relations are discovered in section 5. Section 6 demonstrates the effectiveness of the stabilizations for stress-point integration and its superiority over stabilized nodal integrations by numerical examples. Conclusions are discussed in section 7.

2 Moving Least Squares (MLS) approximation

Given a set of particles $\mathbf{x}_I$ in the domain $\Omega \subset \mathbb{R}^n$ (n is the dimension number of the space), meshfree approximation for a scalar function $u(\mathbf{x})$ can be written in a form similar to those used in FEM

$$u^h(\mathbf{x}) = \sum_I N_I(\mathbf{x}) u_I \tag{2.1}$$

where $u^h(\mathbf{x})$ is the approximation of $u(\mathbf{x})$, u_I the nodal parameters, in most cases, $u_I \neq u^h(\mathbf{x})$. $N_I(\mathbf{x})$ is the meshfree shape function and usually constructed by the moving least squares (MLS) or equivalently by the reproducing kernel approximation procedures, in which $N_I(\mathbf{x})$ can be written as

$$N_I(\mathbf{x}) = \mathbf{p}^T(\mathbf{x}_I) w_I(\mathbf{x}) \boldsymbol{\alpha}(\mathbf{x}) \tag{2.2}$$

where $w_I(\mathbf{x}) = w(\mathbf{x} - \mathbf{x}_I)$ is a weight function (positive, symmetric and with a compact support). In this paper, we consider the following weight as

$$w_I(\mathbf{x}) = w_I(\bar{s}) = \begin{cases} 1.0 - 15\bar{s}^4 + 24\bar{s}^5 - 10\bar{s}^6 & \text{for } \bar{s} \leq 1 \\ 0 & \text{for } \bar{s} > 1 \end{cases} \tag{2.3}$$

where $\bar{s} = s/r$, r is the radius of the domain of influence measuring the size of the support, $s = |\mathbf{x} - \mathbf{x}_I|$ the distance from a sampling point $\mathbf{x}$ to the node $\mathbf{x}_I$. Note that weight function given in Eq.(2.3) is C^2 with respect to x or y which is required in the following stabilization procedures. The quartic spline weight function [20], which is widely used in meshfree method, is not smooth enough for this purpose, because its second derivative with respect to x or y is singular at the node itself, i.e. at $\bar{s} = 0$. $\mathbf{p}(\mathbf{x})$ in Eq.(2.2) is the vector of interpolation base functions which is usually constructed to constitute a complete basis of the subspace of polynomials.

The unknown vector $\boldsymbol{\alpha}(\mathbf{x})$ is determined by the so-called reproducibility or consistency approximation which implies that meshfree approximation defined in Eq.(2.1) can reproduce exactly the polynomials basis vector $\mathbf{p}(\mathbf{x})$, i.e.

$$\mathbf{p}(\mathbf{x}) = \sum_I \mathbf{p}(\mathbf{x}_I) N_I(\mathbf{x}) \tag{2.4}$$

Substitution of Eq.(2.2) into the above equation gives

$$\mathbf{A}(\mathbf{x})\boldsymbol{\alpha}(\mathbf{x}) = \mathbf{p}(\mathbf{x}) \tag{2.5}$$

where

$$\mathbf{A}(\mathbf{x}) = \sum_I \mathbf{p}(\mathbf{x}_I)\mathbf{p}^T(\mathbf{x}_I)w_I(\mathbf{x}) \tag{2.6}$$

is equivalent to the Gram matrix **A** in the standard MLS procedure [1]. The process to construct the MLS shape functions can be summarized as below

1. Construct matrix **A** by Eq.(2.6);
2. Solve Eq.(2.5) to determine $\boldsymbol{\alpha}(\mathbf{x})$;
3. Substitute $\boldsymbol{\alpha}(\mathbf{x})$ in Eq.(2.2) to obtain the MLS shape functions $N_I(\mathbf{x})$.

Derivatives of the shape function $N_{I,i}(\mathbf{x})$ may be calculated by applying the chain rule to Eq.(2.2)

$$N_{I,i} = \mathbf{p}^T(w_{I,i}\boldsymbol{\alpha} + w_I\boldsymbol{\alpha}_{,i}) \tag{2.7}$$

where $\boldsymbol{\alpha}_{,i}$ can be obtained by taking the derivative of Eq.(2.5)

$$\mathbf{A}\boldsymbol{\alpha}_{,i} = \mathbf{p}_{,i} - \mathbf{A}_{,i}\boldsymbol{\alpha} \tag{2.8}$$

with

$$\mathbf{A}_{,i} = \sum_I \mathbf{p}(\mathbf{x}_I)\mathbf{p}^T(\mathbf{x}_I)w_{I,i} \tag{2.9}$$

Similarly, by differentiating Eq.(2.7), the second derivative may be calculated as

$$N_{I,ij} = \mathbf{p}^T(w_{I,ij}\boldsymbol{\alpha} + w_{I,i}\boldsymbol{\alpha}_{,j} + w_{I,j}\boldsymbol{\alpha}_{,i} + w_I\boldsymbol{\alpha}_{,ij}) \tag{2.10}$$

where $\boldsymbol{\alpha}_{,ij}$ may be determined by differentiating Eq.(2.8)

$$\mathbf{A}\boldsymbol{\alpha}_{,ij} = \mathbf{p}_{,ij} - \mathbf{A}_{,ij}\boldsymbol{\alpha} - \mathbf{A}_{,i}\boldsymbol{\alpha}_{,j} - \mathbf{A}_{,j}\boldsymbol{\alpha}_{,i} \tag{2.11}$$

with

$$\mathbf{A}_{,ij} = \sum_I \mathbf{p}(\mathbf{x}_I)\mathbf{p}^T(\mathbf{x}_I)w_{I,ij} \tag{2.12}$$

Note that Eq.(2.8) and Eq.(2.11) have the same matrix **A**. If **A** is LU decomposed with the decomposition stored, then, solving Eq.(2.8) and Eq.(2.11) are just back substitutions, see [21, 22] for details.

3 Governing equations and spatial discretization

Consider a elastostatic boundary value problem on a two-dimensional domain Ω bounded by Γ. The equation of equilibrium is

$$\nabla \cdot \boldsymbol{\sigma}(\mathbf{u}) + \mathbf{b} = 0 \tag{3.13}$$

where $\boldsymbol{\sigma}(\mathbf{u})$ is the stress tensor which corresponds to the displacement field $\mathbf{u}$, and $\mathbf{b}$ is a body force vector. The boundary conditions are

$$\boldsymbol{\sigma} \cdot \mathbf{n} = \bar{\mathbf{t}} \text{ on } \Gamma_t \tag{3.14}$$

$$\mathbf{u} = \bar{\mathbf{u}} \text{ on } \Gamma_u \tag{3.15}$$

where the superposed bar denotes prescribed boundary values, and $\mathbf{n}$ is the unit normal to the boundary.

The variational (or weak) form of the equilibrium equation can be written as

$$\int_{\Omega} \delta\boldsymbol{\epsilon}^T \boldsymbol{\sigma} d\Omega - \int_{\Omega} \delta\mathbf{u}^T \mathbf{b} d\Omega - \int_{\Gamma_t} \delta\mathbf{u}^T \bar{\mathbf{t}} d\Gamma + \beta \int_{\Gamma_u} \delta\mathbf{u}^T (\mathbf{u} - \bar{\mathbf{u}}) d\Gamma = 0 \tag{3.16}$$

where

$$\mathbf{u} = \{\, u_x \; u_y \,\}^T \; \delta\mathbf{u} = \{\, \delta u_x \; \delta u_y \,\}^T \tag{3.17}$$

$$\delta\boldsymbol{\epsilon} = \{\, \delta\epsilon_{xx} \; \delta\epsilon_{yy} \; 2\delta\epsilon_{xy} \,\}^T \; \delta\boldsymbol{\sigma} = \{\, \delta\sigma_{xx} \; \delta\sigma_{yy} \; \delta\sigma_{xy} \,\}^T \tag{3.18}$$

Note that the last term of the left side of Eq.(3.16) is a penalty term to enforce the essential boundary condition Eq.(3.15) and β is the penalty parameter. For linear elasticity, the stress-strain law can be written as

$$\boldsymbol{\sigma} = \mathbf{D}\boldsymbol{\epsilon} \tag{3.19}$$

with

$$\boldsymbol{\epsilon} = \{\, \epsilon_{xx} \; \epsilon_{yy} \; 2\epsilon_{xy} \,\}^T \tag{3.20}$$

$$\mathbf{D} = \frac{\bar{E}}{1 - \bar{\nu}^2} \begin{bmatrix} 1 \; \bar{\nu} & 0 \\ \bar{\nu} \; 1 & 0 \\ 0 \; 0 & \dfrac{(1 - \bar{\nu})}{2} \end{bmatrix} \tag{3.21}$$

where $\bar{E} = E$, $\bar{\nu} = \nu$ for plane stress and $\bar{E} = \frac{E}{1-\nu^2}$, $\bar{\nu} = \frac{\nu}{1-\nu}$ for plane strain with E and ν are Young's modulus and Poisson's ratio respectively. The strain-displacement relation can be written as

$$\boldsymbol{\epsilon} = \{\, \epsilon_{xx} \; \epsilon_{yy} \; 2\epsilon_{xy} \,\}^T = \mathbf{L}\mathbf{u} \tag{3.22}$$

where $\mathbf{L}$ is the strain operator matrix

$$\mathbf{L} = \begin{bmatrix} \partial/\partial x & 0 \\ 0 & \partial/\partial y \\ \partial/\partial y & \partial/\partial x \end{bmatrix} \tag{3.23}$$

To spatially discretize above equation, the trial and test function in Eq.(3.17) are approximated using MLS shape functions as

$$\mathbf{u}^h = \begin{Bmatrix} u_x^h \\ u_y^h \end{Bmatrix} = \mathbf{N}(\mathbf{x})\mathbf{d} = \mathbf{N}_I\mathbf{d}_I \; \delta\mathbf{u}^h = \begin{Bmatrix} \delta u_x^h \\ \delta u_y^h \end{Bmatrix} = \mathbf{N}(\mathbf{x})\delta\mathbf{d} = \mathbf{N}_I\delta\mathbf{d}_I \quad (3.24)$$

The matrix of shape funcition is

$$\mathbf{N}(\mathbf{x}) = \begin{bmatrix} \mathbf{N}_1 \; \mathbf{N}_2 \cdots \; \mathbf{N}_n \end{bmatrix} \quad (3.25)$$

where $\mathbf{N}_I = N_I\mathbf{I}_2$, n is the total number of particles. Substitutiing Eqs.(3.17-3.25) into Eq.(3.16), the following equation results

$$(\mathbf{K} + \beta\mathbf{K}^p)\mathbf{d} = \mathbf{f} + \beta\mathbf{f}^p \quad (3.26)$$

where

$$\mathbf{K} = \int_\Omega \mathbf{B}^T\mathbf{D}\mathbf{B}d\Omega \quad (3.27)$$

$$\mathbf{f} = \int_\Omega \mathbf{N}^T\mathbf{b}d\Omega + \int_{\Gamma_t} \mathbf{N}^T\bar{\mathbf{t}}d\Gamma \quad (3.28)$$

$$\mathbf{K}^p = \int_{\Gamma_u} \mathbf{N}^T\mathbf{N}d\Gamma \quad (3.29)$$

$$\mathbf{f}^p = \int_{\Gamma_u} \mathbf{N}^T\bar{\mathbf{u}}d\Gamma \quad (3.30)$$

where $\mathbf{B} = \mathbf{L}\mathbf{N}$.

4 Stress-point integration scheme

Stress-point integration, introduced in [10–12] and [13–15], introduces additional integration points between the nodes. In this paper, the stress points are added as shown in Fig. 4.1a, i.e. the original nodes are connected to construct a triangle mesh, then the stress points are placed in the centers of the elements. There are different methods to determine the integration weights for all integration points (i.e. original nodes and stress points), such as using a Voronoi diagram [14,15]. In this paper, we compute the integration weights in the following way which is similar to that described in [7,18] for nodal integration. First, a Delaunay triangulation is constructed based on all integration points; then, by joining the centroids of the triangles and the mid-edge points as shown in Fig. 4.1b, a polygon Ω_J surrounding the integration point J can be created and the integration weight V_J is set to the area of this polygon.

Using the stress-point integration scheme, the domain integration of Eq.(3.27) can be written as

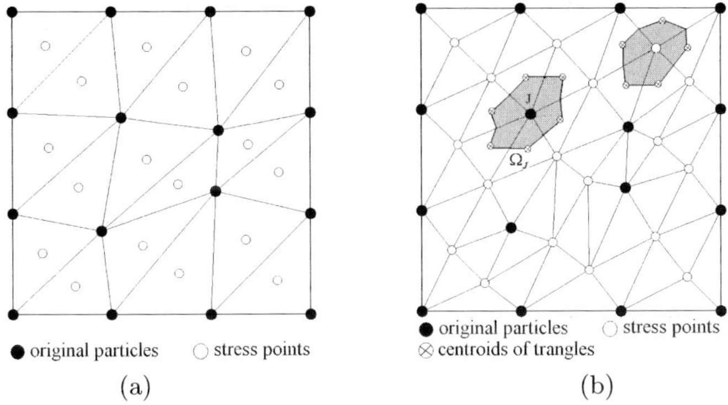

Figure 4.1. Stress-point integration: (a) Placement of stress-points (b) Determination of integration weights

$$\mathbf{K} = \sum_{J=1}^{n+m} V_J \mathbf{B}(\mathbf{x}_J)^T \mathbf{D} \mathbf{B}(\mathbf{x}_J) \tag{4.31}$$

where n and m are the numbers of original particles and stress points respectively. Note that the method described above can also be applied to nodal integration, in which the Delaunay triangulation is constructed only based on the original nodes. For the boundary integrals in Eqs.(3.28-3.30), one dimensional Gaussian integration is used in this paper.

5 Stabilizations for stress-point integration

5.1 Least-Squares Stabilization (LSS)

In least-squares stabilization [5, 15], the weak form of the governing equation (3.13) can be written as

$$\int_\Omega \delta u \cdot (\nabla \cdot \boldsymbol{\sigma} + \mathbf{b}) d\Omega - \int_\Omega \gamma (\nabla \cdot \delta\boldsymbol{\sigma})(\nabla \cdot \boldsymbol{\sigma} + \mathbf{b}) d\Omega = 0 \tag{5.32}$$

In the above equation, the first term on the left side is the usual Galerkin part, and the second part is the stabilization term which is a weighted residual of the governing equation and γ is the stabilization parameter. Follow the same derivation from Eq.(3.16) to Eq.(3.26), the following equation results

$$(\mathbf{K} + \mathbf{K}^\gamma + \beta \mathbf{K}^p)\mathbf{d} = \mathbf{f} + \mathbf{f}^\gamma + \beta \mathbf{f}^p \tag{5.33}$$

where

$$\mathbf{K}^\gamma = \int_\Omega \gamma (\mathbf{L}^T \mathbf{D} \mathbf{L} \mathbf{N})^T (\mathbf{L}^T \mathbf{D} \mathbf{L} \mathbf{N}) d\Omega \tag{5.34}$$

$$\mathbf{f}^{\gamma} = -\int_{\Omega} \gamma (\mathbf{L}^T\mathbf{DLN})^T \mathbf{b} d\Omega \tag{5.35}$$

Other matrix and vector are the same as Eqs.(3.27–3.30). In [5], the stabilization parameter γ is determined by

$$\gamma = \frac{2\alpha_s l_c^2}{E} \tag{5.36}$$

where α_s is the dimensionless stabilization parameter; l_c, a characteristic length scale of the discretization (or nodal arrangement); and E, the Young's modulus. In this way, γ is a constant in the whole computational domain and it can move out of the integration operator in Eqs.(5.34–5.35). In section 5.3, we will give another formulation to determine γ in which γ can vary in the computational domain.

5.2 Taylor series Expansion Based Stabilization (TEBS)

Several stabilization methods are based on Taylor series expansions [16–18]. The integrand in Eq.(3.27) can be approximated in the vicinity of a integration point $\mathbf{x}_J$ by a Taylor series as follows

$$\begin{aligned}\mathbf{B}(\mathbf{x})^T\mathbf{DB}(\mathbf{x}) = \mathbf{B}(\mathbf{x}_J)^T\mathbf{DB}(\mathbf{x}_J) + (\mathbf{x}-\mathbf{x}_J)\cdot\nabla\left[\mathbf{B}(\mathbf{x})^T\mathbf{DB}(\mathbf{x})\right]_{\mathbf{x}=\mathbf{x}_J} \\ +\frac{1}{2}(\mathbf{x}-\mathbf{x}_J)\cdot\nabla^2\left[\mathbf{B}(\mathbf{x})^T\mathbf{DB}(\mathbf{x})\right]_{\mathbf{x}=\mathbf{x}_J}\cdot(\mathbf{x}-\mathbf{x}_J)\end{aligned} \tag{5.37}$$

Substitution of above equation into Eq.(3.27), and after using the chain rule of differentiation and neglecting the terms containing the third order of shape functions, Eq.(3.27) can be written as

$$\begin{aligned}\mathbf{K} = \int_{\Omega}\mathbf{B}^T\mathbf{DB}d\Omega = \sum_{J=1}^{ni}\int_{\Omega_J}\mathbf{B}^T\mathbf{DB}d\Omega = \sum_{J=1}^{ni}\left[\mathbf{B}^T\mathbf{DB}\right]_{\mathbf{x}_J}V_J \\ +\sum_{J=1}^{ni}\left[\frac{\partial\mathbf{B}^T}{\partial x}\mathbf{DB}+\mathbf{B}^T\mathbf{D}\frac{\partial\mathbf{B}}{\partial x}\right]_{\mathbf{x}_J}\int_{\Omega_J}(x-x_J)d\Omega \\ +\sum_{J=1}^{ni}\left[\frac{\partial\mathbf{B}^T}{\partial y}\mathbf{DB}+\mathbf{B}^T\mathbf{D}\frac{\partial\mathbf{B}}{\partial y}\right]_{\mathbf{x}_J}\int_{\Omega_J}(y-y_J)d\Omega \\ +\sum_{J=1}^{ni}\left[\frac{\partial\mathbf{B}^T}{\partial x}\mathbf{D}\frac{\partial\mathbf{B}}{\partial x}\right]_{\mathbf{x}_J}\int_{\Omega_J}(x-x_J)^2d\Omega \\ +\sum_{J=1}^{ni}\left[\frac{\partial\mathbf{B}^T}{\partial y}\mathbf{D}\frac{\partial\mathbf{B}}{\partial y}\right]_{\mathbf{x}_J}\int_{\Omega_J}(y-y_J)^2d\Omega \\ +\sum_{J=1}^{ni}\left[\frac{\partial\mathbf{B}^T}{\partial x}\mathbf{D}\frac{\partial\mathbf{B}}{\partial y}+\frac{\partial\mathbf{B}^T}{\partial y}\mathbf{D}\frac{\partial\mathbf{B}}{\partial x}\right]_{\mathbf{x}_J}\int_{\Omega_J}(x-x_J)(y-y_J)d\Omega\end{aligned} \tag{5.38}$$

where V_J is the area of the polygon Ω_J associated to integration point $\mathbf{x}_J$, and

$$M_J^x = \int_{\Omega_J}(x - x_J)d\Omega \; M_J^y = \int_{\Omega_J}(y - y_J)d\Omega \tag{5.39}$$

$$\begin{aligned} M_J^{xx} &= \int_{\Omega_J}(x - x_J)^2 d\Omega \qquad & M_J^{yy} = \int_{\Omega_J}(y - y_J)^2 d\Omega \\ M_J^{xy} &= \int_{\Omega_J}(x - x_J)(y - y_J)d\Omega \end{aligned} \tag{5.40}$$

are respectively the first-order and the second-order area moments of Ω_J. These moments are required for implementing this stabilization method which can be evaluated in a theoretic way or a numerical way such as Gaussian quadrature.

5.3 Approximated TEBS and determination of stabilization parameter of LSS

To eliminate the burden of the evaluation of the area moments, it is assumed that Ω_J is a square (or a circle) with the same area V_J and the integration point $\mathbf{x}_J$ is located in the center of the square (or the circle). In this way, we have

$$M_J^x = 0 \quad M_J^y = 0 \quad M_J^{xy} = 0 \tag{5.41}$$

$$M_J^{xx} = M_J^{yy} = \frac{V_J^2}{12} \text{for square} \quad M_J^{xx} = M_J^{yy} = \frac{V_J^2}{4\pi} \text{for circle} \tag{5.42}$$

Then the evaluation of stiffness matrix, i.e. Eq.(5.38), can be simplified as (assume Ω_J is a square)

$$\mathbf{K} = \sum_{J=1}^{ni} \left[\mathbf{B}^T\mathbf{D}\mathbf{B}\right]_{\mathbf{x}_J} \mathbf{V}_J + \sum_{J=1}^{ni} \mathbf{K}^S(\mathbf{x}_J)V_J \tag{5.43}$$

where ni is the total number of integration points, and

$$\mathbf{K}^S(\mathbf{x}_J) = \frac{V_J^2}{12}\left[\frac{\partial \mathbf{B}(\mathbf{x_J})^T}{\partial x}\mathbf{D}\frac{\partial \mathbf{B}(\mathbf{x_J})}{\partial x} + \frac{\partial \mathbf{B}(\mathbf{x_J})^T}{\partial y}\mathbf{D}\frac{\partial \mathbf{B}(\mathbf{x_J})}{\partial y}\right] \tag{5.44}$$

is the stabilization matrix introduced by Taylor series expansion. This method can be called approximated TEBS (ATEBS).

In the least squares stabilization method described above, the stabilization matrix can be evaluated at integration point $\mathbf{x}_J$ according to Eq.(5.34) as

$$\mathbf{K}^\gamma = \int_\Omega \gamma(\mathbf{L}^T\mathbf{D}\mathbf{L}\mathbf{N})^T(\mathbf{L}^T\mathbf{D}\mathbf{L}\mathbf{N})d\Omega = \sum_{J=1}^{ni} \mathbf{K}^\gamma(\mathbf{x}_J)V_J \tag{5.45}$$

where

$$\mathbf{K}^{\gamma}(\mathbf{x}_J) = \gamma_J \left[\mathbf{L}^T\mathbf{D}\mathbf{B}(\mathbf{x}_J)\right]^T \left[\mathbf{L}^T\mathbf{D}\mathbf{B}(\mathbf{x}_J)\right] \tag{5.46}$$

Note that $\mathbf{K}^{\gamma}(\mathbf{x}_J)$ is similar to $\mathbf{K}^S(\mathbf{x}_J)$ in the sense that both of them only contain the second order derivatives of the shape functions. Also note that $\mathbf{K}^{\gamma}(\mathbf{x}_J)$ contains two $\mathbf{D}$ matrices, while $\mathbf{K}^S(\mathbf{x}_J)$ only has one. With this in mind, it is not difficult to obtain

$$\gamma_J = \eta \frac{1}{E}\frac{V_J}{12} = \eta \frac{1}{E}\frac{h_J^2}{12} \tag{5.47}$$

where η is a dimensionless stabilization parameter and $\alpha_s = \frac{\eta}{24}$; h_J is the length of the side of the square Ω_J, i.e. $V_J = h_J^2$. Note that Eq.(5.47) is just Eq.(5.36) with $l_c = h_J$, i.e. using the characteristic length scale of the local discretization (near integration point $\mathbf{x}_J$) h_J instead of the global one. Now, it is clear that why the stabilization parameter γ is chosen as Eq.(5.36) in [5].

5.4 Finite increment gradient stabilization

The following is our interpretation of the method given in [19]. By using a Taylor series expansion, the displacement $u^h(\mathbf{x}+\mathbf{h})$ can be expressed in terms of $u^h(\mathbf{x})$ as

$$u^h(\mathbf{x}+\mathbf{h}) = u^h(\mathbf{x}) + \mathbf{h}\cdot\nabla u^h(\mathbf{x}) \tag{5.48}$$

or with one order higher as

$$u^h(\mathbf{x}+\mathbf{h}) = u^h(\mathbf{x}) + \mathbf{h}\cdot\nabla u^h(\mathbf{x}) + \frac{1}{2}\mathbf{h}\cdot\nabla(\nabla u^h(\mathbf{x}))\cdot\mathbf{h} = u^h(\mathbf{x}) + \mathbf{h}\cdot\widetilde{\nabla} u^h(\mathbf{x}) \tag{5.49}$$

where

$$\widetilde{\nabla} u^h(\mathbf{x}) = \nabla u^h(\mathbf{x}) + \nabla(\nabla u^h(\mathbf{x}))\cdot\frac{\mathbf{h}}{2} \tag{5.50}$$

is called the finite increment gradient [19] of $u^h(\mathbf{x})$ which contains the second order derivatives of $u^h(\mathbf{x})$. Applying the spatial approximation Eq.(3.24) to Eq.(5.50), the following modified derivatives of shape functions are obtained

$$\frac{\partial \widetilde{N}_I}{\partial x} = \frac{\partial N_I}{\partial x} + \frac{h_x}{2}\frac{\partial^2 N_I}{\partial x^2} + \frac{h_y}{2}\frac{\partial^2 N_I}{\partial x \partial y} \tag{5.51}$$

$$\frac{\partial \widetilde{N}_I}{\partial y} = \frac{\partial N_I}{\partial y} + \frac{h_x}{2}\frac{\partial^2 N_I}{\partial x \partial y} + \frac{h_y}{2}\frac{\partial^2 N_I}{\partial y^2} \tag{5.52}$$

In the vicinity of integration point $\mathbf{x}_J$, we choose

$$h_x = h_y = h_J \tag{5.53}$$

and $h_J = \sqrt{V_J}$ as defined in section 5.3. Then the matrix $\mathbf{B}(\mathbf{x}_J)$ can be constructed using these modified derivatives instead of the original ones as

$$\widetilde{\mathbf{B}}(\mathbf{x}_J) = \left[\widetilde{\mathbf{B}}_1(\mathbf{x}_J)\; \widetilde{\mathbf{B}}_2(\mathbf{x}_J) \cdots\; \widetilde{\mathbf{B}}_n(\mathbf{x}_J) \right] \tag{5.54}$$

with

$$\widetilde{\mathbf{B}}_I(\mathbf{x}_J) = \begin{bmatrix} \frac{\partial \widetilde{N}_I(\mathbf{x}_J)}{\partial x} & 0 \\ 0 & \frac{\partial \widetilde{N}_I(\mathbf{x}_J)}{\partial y} \\ \frac{\partial \widetilde{N}_I(\mathbf{x}_J)}{\partial y} & \frac{\partial \widetilde{N}_I(\mathbf{x}_J)}{\partial x} \end{bmatrix} \tag{5.55}$$

Finally the stiffness matrix can be evaluated as

$$\mathbf{K} = \sum_{J=1}^{n+m} V_J \widetilde{\mathbf{B}}(\mathbf{x}_J)^T \mathbf{D} \widetilde{\mathbf{B}}(\mathbf{x}_J) \tag{5.56}$$

In fact, finite increment gradient stabilization is closely related to the TEBS given above, which can be clearly discovered if we let

$$M_J^x = M_J^y = \frac{h_J}{2} V_J \tag{5.57}$$

$$M_J^{xx} = M_J^{yy} = M_J^{xy} = \frac{h_J^2}{4} V_J \tag{5.58}$$

Substitution of Eqs.(5.57-5.58) into Eq.(5.38) yields

$$\mathbf{K} = \sum_{J=1}^{n+m} \widehat{\mathbf{B}}(\mathbf{x}_J)^T \mathbf{D} \widehat{\mathbf{B}}(\mathbf{x}_J) V_J \tag{5.59}$$

where

$$\widehat{\mathbf{B}}(\mathbf{x}_J) = \mathbf{B}(\mathbf{x}_J) + \frac{h_J}{2} \frac{\partial \mathbf{B}(\mathbf{x}_J)}{\partial x} + \frac{h_J}{2} \frac{\partial \mathbf{B}(\mathbf{x}_J)}{\partial y} \tag{5.60}$$

By substitution of Eqs.(5.51–5.52) and Eq.(5.55) into Eq.(5.54), we can easily get

$$\widehat{\mathbf{B}}(\mathbf{x}_J) = \widetilde{\mathbf{B}}(\mathbf{x}_J) \tag{5.61}$$

In this sense, finite increment gradient stabilization can also be classified as an approximated TEBS with the assumption of Eqs.(5.57–5.58).

6 Numerical examples

Two numerical examples are studied in this section. Plane stress conditions were assumed for the two-dimensional calculations. The material parameters used in the following examples are $E = 10^7 Pa$ and $\nu = 0.3$. The L_2 norms of the displacement and energy errors are defined as follows:

$$L_2^U = \left(\frac{\sum_{i=1}^n (u_i^h - u_i^e)^2}{\sum_{i=1}^n (u_i^e)^2} \right)^{\frac{1}{2}} \tag{6.62}$$

$$L_2^E = \frac{1}{A} \left(\frac{\int_\Omega (\epsilon^e - \epsilon^h)^T \mathbf{D} (\epsilon^e - \epsilon^h) d\Omega}{\int_\Omega \epsilon^{e^T} \mathbf{D} \epsilon^e d\Omega} \right)^{\frac{1}{2}} \tag{6.63}$$

where the superscript e and h denote analytic and numerical solutions respectively and n is the total number of the particles and A the area of problem domain. To construct the MLS approximation, the quadratic basis $[\,1\ x\ y\ x^2\ xy\ y^2\,]$ is used and the radius of influence domain for node I is defined as

$$r_I = c h_I^{max} \tag{6.64}$$

where

$$h_I^{\max} = \max_{J \in S_I} \|\mathbf{x}_I - \mathbf{x}_J\| \tag{6.65}$$

where S_I is the set of neighbouring points of $\mathbf{x}_I$ which construct a polygon surrounding point $\mathbf{x}_I$ and $c = 2.5$ for quadratic basis. To enforce essential boundary conditions, we choose the penalty method with a penalty parameter $\beta = 10^5 \bar{E}$.

6.1 Pressure-loaded half plane

As the first example, a pressure-loaded half plane problem is examined. The exact solution, expressed in terms of the configuration of Fig. 6.2a, is given by

$$\begin{aligned}
\sigma_{xx} &= \frac{p}{2\pi} \{2(\theta_1 - \theta_2) - \sin 2\theta_2 + \sin 2\theta_1\} \\
\sigma_{yy} &= \frac{p}{2\pi} \{2(\theta_1 - \theta_2) + \sin 2\theta_2 - \sin 2\theta_1\} \\
\sigma_{xy} &= \frac{p}{2\pi} \{\cos 2\theta_1 - \cos 2\theta_2\}
\end{aligned} \tag{6.66}$$

and

$$\begin{aligned}
u_x &= \frac{p}{4\pi\mu} \left\{ (\kappa - 1)[(x - a)\theta_2 - (x + a)\theta_1] - (\kappa - 1) y \ln(\frac{r_2}{r_1}) \right\} \\
u_y &= \frac{p}{4\pi\mu} \left\{ (\kappa - 1) y (\theta_2 - \theta_1) + (\kappa + 1)[x \ln(\frac{r_2}{r_1}) - a \ln(\frac{r_2}{2a}) - a \ln(\frac{r_1}{2a})] \right\}
\end{aligned} \tag{6.67}$$

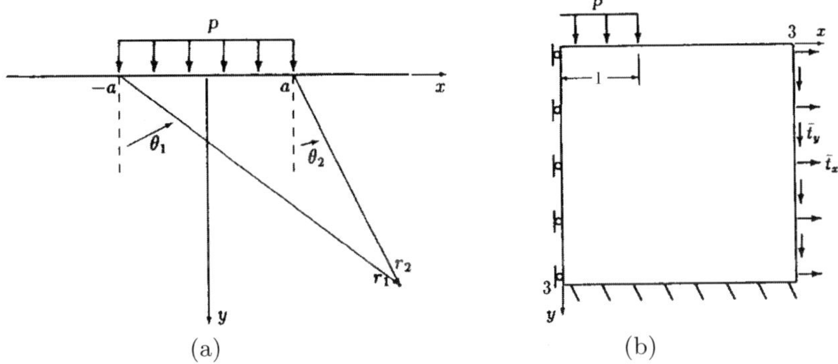

Figure 6.2. Problem of pressure-loaded half plane: (a) Schematic diagram (b) Region of numerical solution

where

$$\mu = \frac{\bar{E}}{2(1+\mu)} \qquad \kappa = \frac{3-\bar{\mu}}{1+\bar{\mu}} \tag{6.68}$$

and $2a$ is the width of the pressure load p.

Due to symmetry, the numerical model for this problem is a $3a$ by $3a$ region with $a = 1$, as shown in Fig. 6.2b. The displacement along the line of symmetry, $x = 0$, is fixed in the x-direction and free in y-direction. Along $y = 3a$ the displacements are prescribed, and along $x = 3a$ the tractions are prescribed, in accordance with the exact solutions given above.

The four uniform grids shown in Fig. 6.3 are employed for convergence study and the results are plotted in Fig. 6.4. For such uniform grids, stress-point integration displays excellent convergence in both displacement and energy which are almost comparable to those of full integration. This conforms with the conclusions of the convergence properties of stress-point integration in uniform grid given in [15]. Note that even the nodal integration in such cases also display less consistent convergence albeit less than those of stress-point integration. This was also observed in [5].

However, this conclusion does not hold for non-uniform grids. To show this, we make the node distribution of these four grids irregular as shown in Fig. 6.5 and the convergence properties of different methods are plotted in Fig. 6.6. Nodal integration does not display convergence for such irregular node distributions. In contrast, stress-point integration still shows reasonable convergence when the grids are coarse. However, the curve tails up and the convergence is lost when the grids become dense. On the contrary, three stabilized stress-point integrations, i.e. SI-LSS, SI-TEBS and SI-ATEBS, display convergence for the whole range that we tested. Moreover, their convergence rates are higher and their errors are lower than those of stress-point integra-

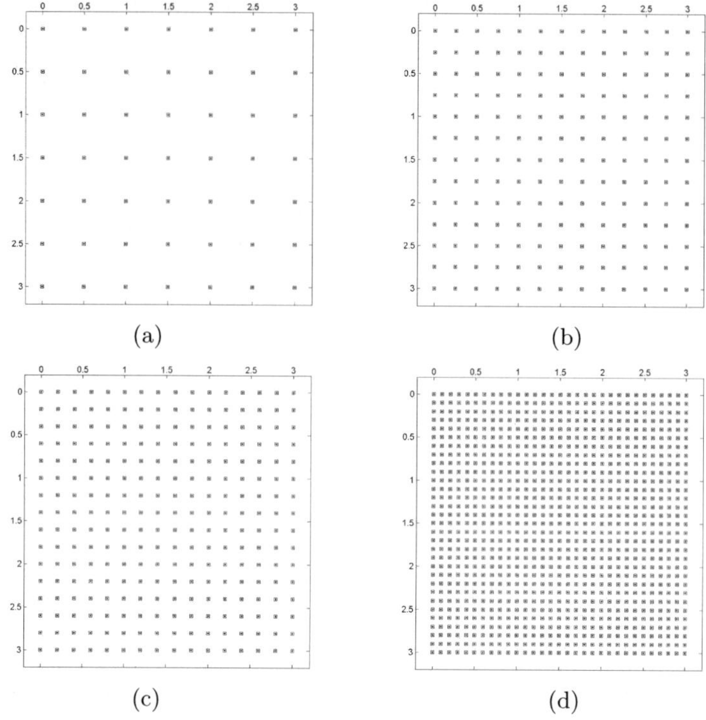

Figure 6.3. Regular nodal arrangements for the pressure-loaded half plane problem: (a) 49-node grid; (b) 169-node grid; (c) 256-node grid; (d) 961-node grid

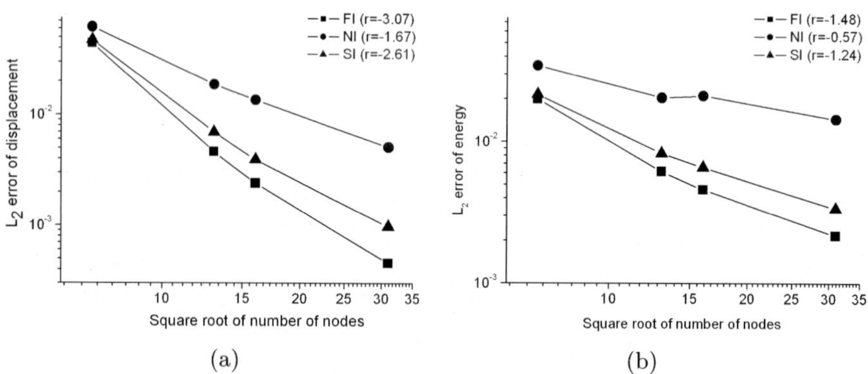

Figure 6.4. Convergence properties of different integration methods for the pressure-loaded half plane problem using uniform grids: (a) displacement; (b) energy

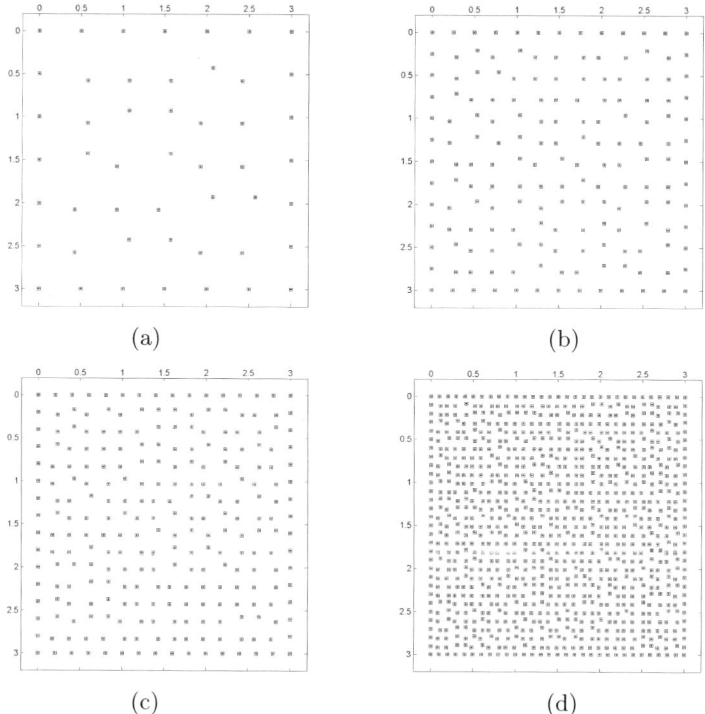

Figure 6.5. Irregular nodal arrangements for the pressure-loaded half plane problem: (a) 49-node grid; (b) 169-node grid; (c) 256-node grid; (d) 961-node grid.

tion. The improved convergence and the increased accuracy of these three stabilized solutions over the unstabilized solution are evidence of the effectiveness of their stabilizations. However, the accuracy of stabilized stress-point integration is lower than that of full integration. This is the price of the increased speed, which is clearly demonstrated by the comparison of CPU times as shown in Fig. 6.7. Obviously, full integration consumed the most CPU time and its computational efficiency is not acceptable, especially in the case of dense grids. However, the increased CPU time of stabilized stress-point integration over nodal integration is acceptable. Note that, for this example, the stabilization based on modifying derivative of shape functions, i.e. the SI-FIG, shows high solution errors, albeit it displays some convergence.

The improved performance of SI-LSS, SI-TEBS and SI-ATEBS over the direct stress-point integration can be further demonstrated by the comparison of the profiles of σ_{yy} along the line $y = 1$, as shown in Fig. 6.8. Stress-point integration shows some small oscillations which are removed or at least are alleviated by the three stabilized methods.

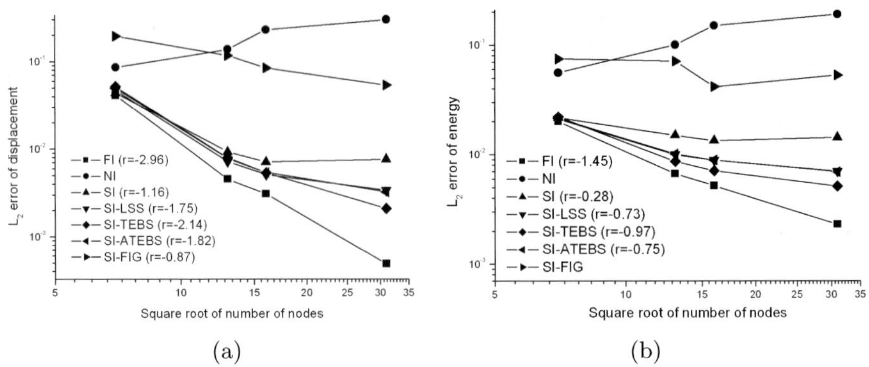

Figure 6.6. Convergence properties of different integration methods for the pressure-loaded half plane problem using non-uniform grids: (a) displacement; (b) energy.

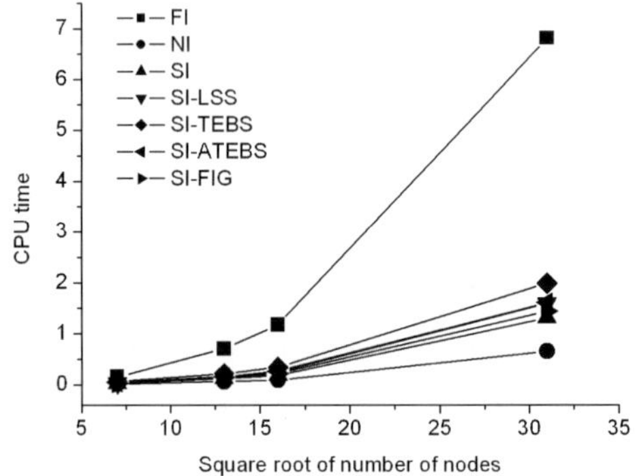

Figure 6.7. Comparison of CPU time consumed by different integration methods for the pressure-loaded half plane problem using four different grids.

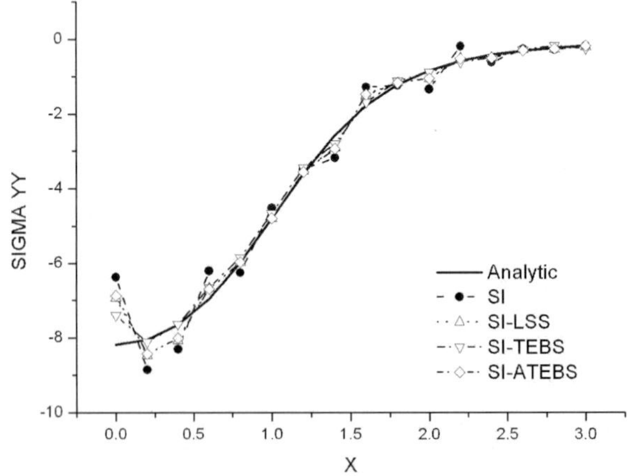

Figure 6.8. Profile of σ_{yy} along y = 1 for the pressure-loaded half plane problem.

6.2 Plate with a hole

The problem of a plate with a hole shown in Fig. 6.9a is next examined. The exact solution of this problem is

$$
\begin{aligned}
\sigma_{xx} &= T_x \left[1 - \frac{a^2}{r^2} \left(\frac{3}{2} \cos 2\theta + \cos 4\theta \right) + \frac{3}{2} \frac{a^4}{r^4} \cos 4\theta \right] \\
\sigma_{yy} &= -T_x \left[\frac{a^2}{r^2} \left(\frac{1}{2} \cos 2\theta - \cos 4\theta \right) + \frac{3}{2} \frac{a^4}{r^4} \cos 4\theta \right] \\
\sigma_{xy} &= -T_x \left[\frac{a^2}{r^2} \left(\frac{1}{2} \sin 2\theta + \sin 4\theta \right) - \frac{3}{2} \frac{a^4}{r^4} \sin 4\theta \right]
\end{aligned}
\tag{6.69}
$$

and

$$
\begin{aligned}
u_x &= \frac{1+\bar{\nu}}{\bar{E}} T_x \left(\frac{1}{1+\bar{\nu}} r \cos\theta + \frac{2}{1+\bar{\nu}} \frac{a^2}{r} \cos\theta + \frac{1}{2} \frac{a^2}{r} \cos 3\theta - \frac{1}{2} \frac{a^4}{r^3} \cos 3\theta \right) \\
u_y &= \frac{1+\bar{\nu}}{\bar{E}} T_x \left(\frac{-\bar{\nu}}{1+\bar{\nu}} r \sin\theta - \frac{1-\bar{\nu}}{1+\bar{\nu}} \frac{a^2}{r} \sin\theta + \frac{1}{2} \frac{a^2}{r} \sin 3\theta - \frac{1}{2} \frac{a^4}{r^3} \sin 3\theta \right)
\end{aligned}
\tag{6.70}
$$

where a is the radius of the hole and T_x is the traction in the x direction loaded at infinity as indicated by Fig. 6.9a.

Due to two-fold symmetry, only the first quadrant is modeled as shown in Fig. 6.9b. Along the line $x = 5$ and $y = 5$, natural boundary conditions are applied in accordance with the exact solutions Eq.(6.69). The essential boundary conditions $u_x = 0$ is applied along the line $x = 0$ and $u_y = 0$ is applied along the line $y = 0$. Six grids as shown in Fig. 6.10 are employed for convergence study and the results are plot in Fig. 6.11. Again, full integration is clearly more accurate and displays a higher rate of convergence than other methods.

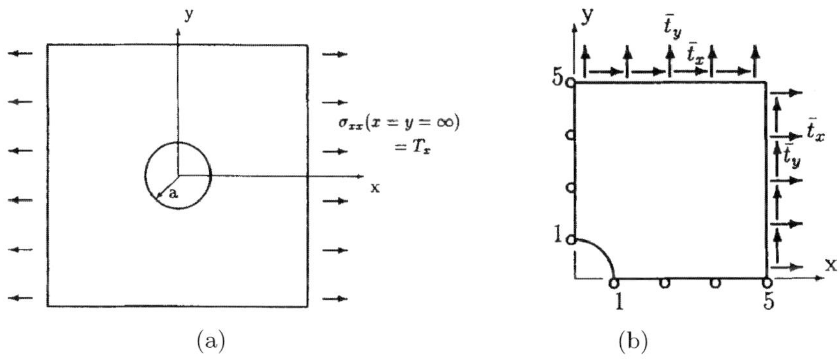

Figure 6.9. Problem of infinite plate with a hole loaded at infinity by $\sigma_{xx} = T_x$: (a) Schematic diagram (b) Region of numerical solution.

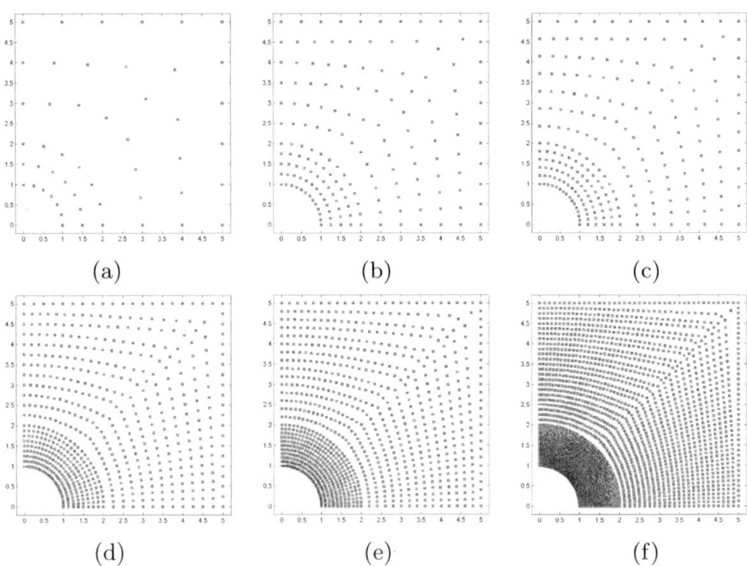

Figure 6.10. Nodal arrangements for the plate with a hole: (a) 50-node grid; (b) 168-node grid; (c) 238-node grid; (d) 630-node grid; (e) 960-node grid; (f) 2403-node grid.

However, its computational time increase rapidly as the number of nodes increases, which is indicated by Fig. 6.12. This drawback limits its application to large scale industrial analysis. Stress-point integration for this problem shows a low rate convergence in displacement and its convergence in energy is lost when the grid becomes very dense as indicated by Fig. 6.11. In contrast, the three stabilized stress-point integration methods, i.e. SI-LSS, SI-TEBS and SI-ATEBS, show better convergence rate both in displacement and in energy over those of the unstabilized one and at the same time they consume much less computational time than that of full integration as shown in Fig. 6.12. This demonstrates the superiority of stabilized stress-point integrations over full integration and unstabilized SI. Note that, in both examples given above, SI-TEBS gives more accurate results and better convergence rates than those of SI-LSS and SI-ATEBS, the latter two methods give almost identical performance. However, it consumes more CPU time than SI-LSS and SI-ATEBS, and it also needs to compute the area moments defined in Eq.(5.39–5.40) which are completely avoided by the other two methods.

We also applied these stabilization strategies to nodal integration, thus set up the methods NI-LSS, NI-TEBS, NI-ATEBS and NI-FIG. Their convergence properties are also examined by this example and the results are plot in Fig. 6.13. Clearly, NI-FIG is not convergent. The other three methods show some convergence when the grids are coarse, however, their convergence rates almost vanish when the grids become very dense. In contrast, SI-TEBS method show

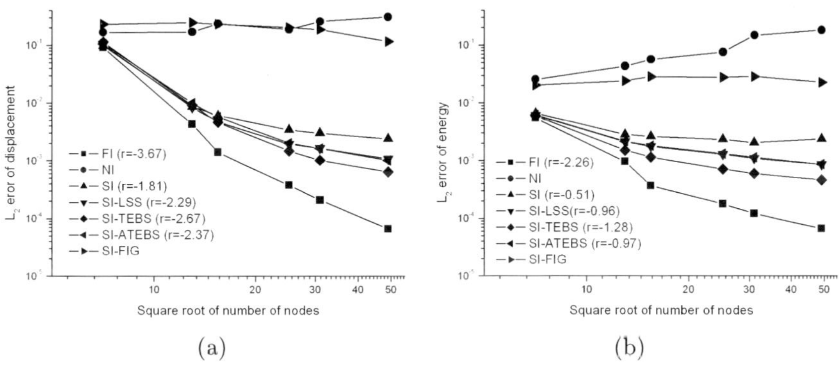

(a) (b)

Figure 6.11. Convergence properties of different integration methods for the plate with a hole problem: (a) displacement; (b) energy.

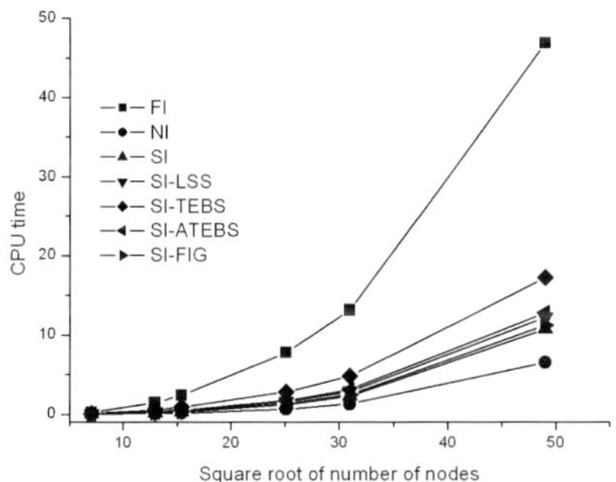

Figure 6.12. Comparison of CPU time consumed by different integration methods for the plate with a hole problem using six different grids.

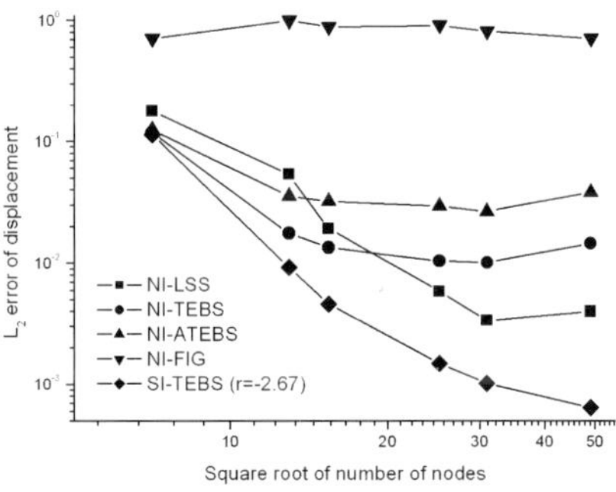

Figure 6.13. Comparison of convergence in displacement obtained by different stabilized nodal integration methods and the stress-point integration with TEBS.

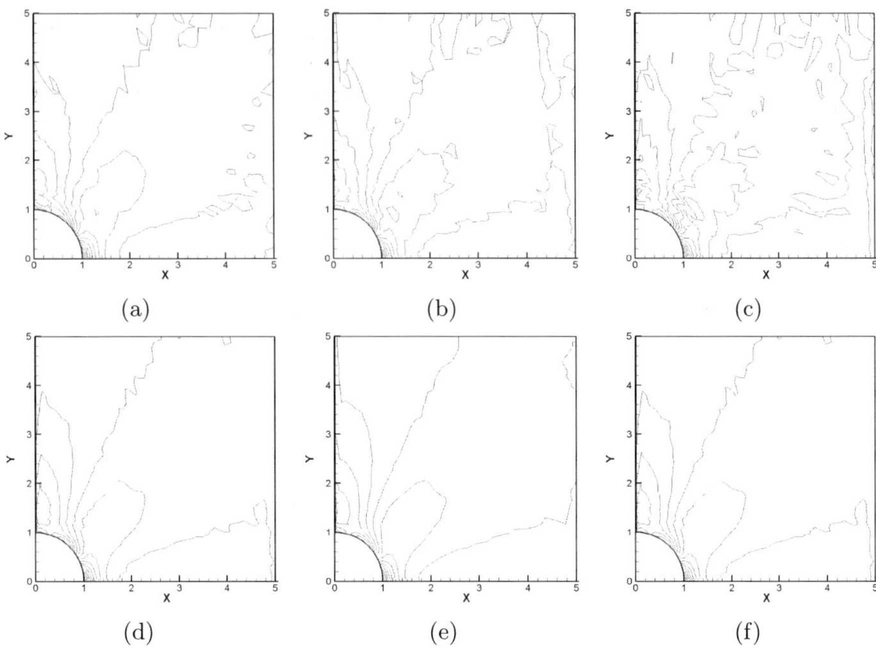

Figure 6.14. σ_{yy} Contours for the plate with a hole problem obtained by different integration methods: (a) NI-LSS; (b) NI-TEBS; (c) NI-ATEBS; (d) SI-LSS; (e) SI-TEBS; (f) SI-ATEBS.

more accurate results and better convergence rate than these stabilized nodal integration methods as indicated in Fig. 6.13.

The superiority of stabilized stress-point integrations over stabilized nodal integrations can be further demonstrated by the comparison of the σ_{yy} contours obtained by these two kinds of methods as shown in Fig. 6.14. Stabilized nodal integrations still exhibit severe oscillations, while stabilized stress-point integrations remove most of such oscillations and the results become very smooth.

7 Conclusions

In this work, stabilized stress-point integrations based respectively on LSS, TEBS and FIG were studied for linear elastostatic problems and their relations are investigated. Numerical results show that LSS and TEBS greatly improve the accuracy and convergence properties of standard stress-point integration while FIG performs poorly. In comparison with full integration, SI-LSS, SI-TEBS and SI-ATEBS require much less CPU times. On the other hand, in comparison with nodal integration or standard stress-point integration, they exhibit higher accuracy and better convergence rates and at the same time the increase of CPU time is still acceptable. Another observation made in this study that is worth noting is that the stabilized stress-point integrations show much better performance in accuracy, convergence and stability than the stabilized nodal integration, which further demonstrates that the insertion of stress points is preferable.

Acknowledgement

The support of the Office of Naval Research under grant N00014-06-1-0380 is gratefully acknowledged.

References

1. T. Belytschko, Y. Y. Lu, and L. Gu, *Element-free Galerkin methods*, Int. J. Numer. Meth. Engrg. **37** (1994), 229–256.
2. W. K. Liu, S. Ju, and Y. F. Zhang, *Reproducing kernel particle methods*, Int. J. Numer. Meth. Fluids **20** (1995), 1081–1106.
3. J. Dolbow, T. Belytschko, *Numerical integration of the Galerkin weak form in meshfree methods*, Computational Mechanics **23** (1999), 219–230.
4. M. Griebel, M. A. Schweitzer, *A particle-partition of unity method-part II: Efficient cover construction and reliable integration*, SIAM Journal on Scientific Computing **23** (2002), 1655–1682.
5. S. Beissel, T. Belytschko, *Nodal integration of the element-free Galerkin method*, Computer Methods in Applied Mechanics and Engineering **139** (1996), 49–74.
6. J. S. Chen, C. T. Wu, S. Yoon, and Y. You, *A stabilized conforming nodal integration for Galerkin mesh-free methods*, Int. J. Numer. Meth. Engrg. **50** (2001), 435–466.
7. J. S. Chen, S. Yoon, C. T. Wu, *Non-linear version of stabilized conforming nodal integration for Galerkin mesh-free methods*, Int. J. Numer. Meth. Engrg. **53** (2002), 2587–2615.
8. J. W. Yoo, B. Moran, J. S. Chen, *Stabilized conforming nodal integration in the natural-element method*, Int. J. Numer. Meth. Engrg. **60** (2004), 861–890.
9. J. Bonet, S. Kulasegaram, *Correction and stabilization of smooth particle hydrodynamics methods with applications in metal forming simulations*, Int. J. Numer. Meth. Engrg. **47** (2000), 1189–1214.
10. C. T. Dyka, R. P. Ingel, *An approach for tension instability in smoothed particle hydrodynamics*, Computers and Structures **57** (1995), 573–580.
11. C. T. Dyka, P. W. Randles, R. P. Ingel, *Stress points for tension instability in SPH*, Int. J. Numer. Meth. Engrg. **40** (1997), 2325–2341.
12. P. W. Randles, L. D. Libersky, *Normalized sph with stress points*, Int. J. Numer. Meth. Engrg. **48** (2000), 1445–1462.
13. T. Belytschko, Y. Guo, W. K. Liu and S. P. Xiao, *A unified stability analysis of meshless particle methods*, Int. J. Numer. Meth. Engrg. **48** (2000), 1359–1400.
14. T. Rabczuk, T. Belytschko, S. P. Xiao, *Stable particle methods based on Lagrangian kernels*, Computer Methods in Applied Mechanics and Engineering **193** (2004), 1035–1063.

15. T. P. Fries, T. Belytschko, *Convergence and stabilization of stress-point integration in mesh-free and particle methods*, Int. J. Numer. Meth. Engrg. Published online.
16. W. K. Liu, J. S. Ong, R. A. Uras, *Finite element stabilization matrices-a unification approach*, Computer Methods in Applied Mechanics and Engineering **53** (1985), 13–46.
17. T. Nagashima, *Node-by-node meshless approach and its applications to structural analysis*, Int. J. Numer. Meth. Engrg. **46** (1999), 341–385.
18. G. R. Liu, G. Y. Zhang, Y. Y. Wang, etc, *A nodal integration technique for meshfree radial point interpolation method (NI-RPIM)*, Int. J. Solids and Structures **44** (2007), 3840–3860.
19. J. Bonet, S. Kulasegaram, *Finite increment gradient stabilization of point integrated meshless methods for elliptic equations*, Communications in Numerical Methods in Engineering **16** (2000), 475–483.
20. T. Belytschko, Y. Krongauz, D. Organ, M. Fleming, P. Krysl, *Meshless methods: An overview and recent developments*, Computer Methods in Applied Mechanics and Engineering **139** (1996), 3–47.
21. T. Belytschko, M. Fleming, *Smoothing, enrichment and contact in the element-free Galerkin method*, Computers and Structures **71** (1999), 173–195.
22. T. Belytschko, Y. Krongauz, M. Fleming, D. Organ, W. K. Liu, *Smoothing and accelerated computations in the element free Galerkin method*, Journal of Computational and Applied Mathematics **74** (1996), 111–126.

The Partition of Unity Meshfree Method for Solving Transport-Reaction Equations on Complex Domains: Implementation and Applications in the Life Sciences

Martin Eigel et al.

[1] Mathematics Institute, University of Warwick, Coventry, UK
eigel@maths.warwick.ac.uk
[2] Mathematics Institute, University of Warwick, Coventry, UK
E.George@warwick.ac.uk
[3] Mathematics Institute, University of Warwick, Coventry, UK
mak@maths.warwick.ac.uk

Summary. There is a wide range of highly significant scientific problems which on appropriate physical scales can be formulated as partial differential equations defined on so-called complex domains. Such complex domains often occur when material is transported through an environment of high geometrical complexity, for example porous media, domains with many obstacles, or membrane systems that are folded in a topologically complex configuration. The latter often occurs in cell biology, where the biological membranes inside the cell are strikingly topologically complex. In addition the medium in which, for example, proteins diffuse in the cell nucleus, is a complex porous media type of environment as many macro-molecules and protein-DNA complexes like the chromatin form a highly irregular structure in which many bio-molecular interactions occur. The distribution of biomolecules inside cells and tissues, their over-abundance or absence in metabolism, signalling etc., is the cause of many human diseases, therefore numerical simulations will be essential for future diagnostic abilities. Under appropriate assumptions the resulting molecular transport can be formulated as a PDE (Partial Differential Equation). The first challenge for any numerical discretisation is the generation of a cover for the underlying computational domain. Here, the meshfree Partition of Unity Method (PUM) offers a number of new degrees of freedom, as patches can be shifted, their size increased or diminished, with no need to create a non-overlapping cover at all times as is characteristic for traditional Finite Element and Finite Volume discretisations. Further advances in cover creation algorithms as discussed in this paper will allow the routine simulation of problems on domains with more complex geometries than have been treatable before.

Key words: Complex Domains, PUM, Meshfree Discretisation, Cover Construction, Cell Biology

1 Introduction

Many and perhaps the most important problems in Cell Biology can be mathematically formulated as transport processes, and most of them must be defined on complex domains. The complexity of the domain is due to the immense richness of cellular sub-structures, like membranes folding in a particular way, or macro-molecular assemblies like the cytoskeleton or the chromatin changing constantly their shapes. Traditionally the focus in Cell Biology has been on other scales, for example looking at the molecular design (Structural Biology, for a current review see [14]) of proteins, protein complexes or other macro-molecules. One of the most important general new research questions in Cell Biology is posed on a more macroscopic scale, but takes these microscopic molecular properties into account: *How abundant and with what kind of distributions are certain types of bio-molecules present in the cell? On what spatial and temporal scales do molecule distributions change in the cell?* If these questions could be answered in more detail the most urgent problems in Cell Biology could be understood and solved. Most questions on a systems level about the function and variability of regulatory and metabolic networks could be answered as they ultimately depend on the interaction of bio-molecules, which itself is dependent on the spatial and temporal variation of different molecular species. Even the memory of the system, for example encoded in the DNA, is depending on such molecular interaction.

The experimental basis allowing to pose questions on cellular molecular distributions is relatively new and rapidly evolving. This has been triggered by new microscopy techniques covering a wide range of spatial scales, the development of photon emitting molecular markers allowing life cell imaging of dynamical processes (such as the most common GFP, Green Fluorescent Protein), and techniques to manipulate photon emission or link the emission to molecular transport or interaction events. Here we mention as examples FRET (Fluorescence Resonance Energy Transfer), FLIP (Fluorescence Loss In Photobleaching) or FRAP (Fluorescence Recovery After Photobleaching). Some references where such techniques are explained and used successfully are for example [10], [11] and [6]. It is tempting to make such microscopy techniques more quantitative. For a recent research paper in this direction see [13]. Using such methods the complexity of the macro-molecular cellular structure can sometimes be shown to exhibit anomalous diffusion, see [19]. In combination with cloning, antibody production and modern techniques of image analysis to post-process the captured images there is a chance to understand how cellular (and tissue) geometry in combination with molecular properties influence changes in molecular distribution and composition.

As cellular form and function has/is such a highly complex organisation/process there is a need to develop new numerical techniques that are able to simulate spatial-temporal processes on domains which can be retrieved by image analysis tools from microscopy images. This is in no way a simple task. Many even sophisticated grid generating algorithms working in two spatial dimensions (on which we restrict ourselves in this paper) have severe problems to generate automatically meshes delivering the necessary computational grid used by general purpose Finite Element (FEM) or Finite Volume Methods (FVM). Here the recent exploration of a discretisation based on a Partition of Unity Method (PUM) using overlapping patches to create a cover has proven to be a very promising alternative ([1, 3, 4, 12, 25, 27]). In this paper we will discuss the basic ideas of the method, introduce new cover generation algorithms, and demonstrate the feasibility of the whole discretisation with the help of different complex test domains. We also show that we can introduce successfully cover adaptivity for PUM when considering temporal processes, i.e. solving parabolic problems.

Similar complex domain problems are of interest in other areas of applications. We would like to mention engineering, where certain fine structures have to be resolved explicitly, for example in designing optimal catalysators. On an appropriate scale also porous media, for example relevant in geology and hydrology should be mentioned. This holds primarily in cases where the homogenised equation is not used for all the medium and some geometrical structures in the soil are resolved explicitly. All these and areas with similar problems will benefit from the increased flexibility PUM has to offer, especially in the area of cover creation.

This paper will mainly focus on the two-dimensional geometrical aspects of transport-reaction equations modelling typical problems in Cell Biology, with emphasis on simple transport behaviour, i.e. diffusion processes. We will not go into detail with discussing the reaction part of this class of equations. The reaction part essentially is determined from the structure of the metabolic and regulatory networks present in the cell. In most cases such reaction networks incorporating a large number of interacting molecular species are considered without any reference to the geometrical properties of cells. In other words such complex network models covering biochemical interactions rely on the basic hyptheses that the cell is a *homogeneously mixed reactor*, perhaps with some structure when simple cellular compartments are included. As there are typically many reactions even under such simplifying homogeneity assumption the analysis of such reaction networks (for example their qualitative behaviour) is not simple. Recent advances can be found in [7], a recent review in [8]. It will be an additional future challenge to consider the high-dimensional transport-reaction equations that simulate problems modelling spatially explicit reaction networks. The second challenge is surely the consideration of three-dimensional physical space in combination with this problem class.

Finally some comments on multi-scale analysis of cellular processes. The equations used to model transport equations are of continuum type, formulated as Partial Differential Equations (PDE). This can be justified mathematically when considering the typical resolution of a confocal microscope and the abundance of molecules in typical experiments based on fluorescence markers. Nevertheless all PDE in this context must be interpreted as limit equations of underlying stochastic processes describing the movement of particles, or/and derived by averaging over geometrical fine structures (finer here should mean finer relative to the scale chosen, i.e. the resolution of the microscope). The latter is usually called a homogenisation process inside the mathematical literature (see for example [22]). Nevertheless much insight will be gained by deriving such limit equations from microscopic rules, both about relevant properties of macro-molecules (such as enzymes) and about statistics of the cellular media in which these macro-molecules are moving. For recent advances on one part of the problem, a multi-scale analysis of macro-molecules (with the help of describing microscopic properties by Markov chains), see [15], [16] and [17]. As a next step such methods need to be given a spatial extension. Such a procedure should deliver meaningful refined models of transport-reaction processes with high explanatory value that can eventually be solved with a PUM as described in this paper.

2 The Partition of Unity Method and its Implementation

2.1 Overview

In the following we develop the version of the PUM that has been taken as the basis of our PDE discretisation. Let H be an appropriate Hilbert space ($H = H^1$ for 2nd order problems). We can obtain the variational form of the PDE using a continuous bilinear form $a : H \times H \to R$ and a linear form $l \in H'$ along with appropriate boundary conditions. The final problem we seek to solve may be summarised as

$$\text{Find} \quad u \in H \quad \text{s.t.} \quad a(u,v) = l(v) \qquad \forall v \in H. \tag{2.1}$$

The basic method of discretisation in the PUM framework is then given by the following steps:

- Given a domain Ω on which a linear scalar PDE is defined, open sets called *patches* are used to form a cover of the domain. ($\Omega_N := \{\omega_i\}_{i=1}^N$, with $\overline{\Omega} \subset \bigcup_i \omega_i$).
- A partition of unity $\{\varphi_i\}_{i=1}^N$ subordinate to the cover is constructed.
- The local function space on patch ω_i, $1 \le i \le N$, is given by $\mathcal{V}_i := \text{span}\{\psi_i^k\}_{k=1}^{p_i}$, with $\{\psi_i^k\}_{k=1}^{p_i}$ being a set of base functions defining the approximation space for each patch. The global approximation space, also called the trial or the *PUM space*, is defined by $V_{\text{PU}} := \text{span}\{\varphi_i \psi_i^k\}_{i,k}$. Replacing H by the finite dimensional subspace[4] V_{PU}, a global approxi-

[4] note that V_{PU} is conforming for the Neumann problems we are concerned with in this article

mation u_h to the unknown solution u of the PDE is defined as a (weighted) sum of local approximation functions on the patches:

$$u_h(x) = \sum_{i=1}^{N} \varphi_i(x) \left(\sum_k \xi_i^k \psi_i^k(x) \right).$$

- The unknown coefficients ξ_i^k are determined by substituting the above approximation into the PDE and using the method of weighted residuals to derive an algebraic system of equations

$$A\xi = b. \tag{2.2}$$

More detail on the PUM, including a description of its approximation properties and how to construct the PUM space, may be found in, for example, [3, 9, 23]. We have implemented the PUM in a C++ code called the *Generic Discretisation Framework* (*GDF*), and in this paper we focus on the aspects of our implementation of the method which are unique.

An arbitrary cover of patches can lead to a set of local basis functions whose union is not linearly independent on some patches. While this might not be a major problem as long as only a few function spaces exhibit linear dependence, the solution process may become more difficult. A recent investigation of this topic can be found in [18].

By the cover construction algorithm described in Sect. 2.2 we completely abolish this difficulty by ensuring linear independent function spaces. In [26] Schweitzer defined the so called *flat top property* which is a sufficient (but not necessary) condition for this. In a partition of unity exhibiting this property all patches ω_i have a subset $\tilde{\omega_i}$ larger than a null-set in the Lebesgue sense with no other patches overlapping. See [26] for a more formal definition.

2.2 Cover Construction

Introduction

A key first step in the discretisation of a set of partial differential equations in the mesh-free partition of unity framework, is the formation of a finite open cover $\Omega_N = \{\omega_i\}_{i=1}^N$ of the domain of interest, $\Omega \subset \bigcup_i \omega_i$. Here N represents the number of patches in the cover, and we typically associate one point or node in a patch with that patch leading to N nodes. Recall from Sec. 2 that a global approximation space $\mathcal{V}_{\mathrm{PU}}$ is created by blending together local approximation spaces $\mathcal{V}$ with $i = 1, \ldots, N$. The support of each $\mathcal{V}_i$ is the closure of a unique cover patch, $\overline{\omega}_i$. The cover greatly influences the efficiency of the mesh-free method, with, for example, the neighbouring relationships $N_i = \{\omega_j \in \Omega_N \mid \omega_i \cap \omega_j \neq \emptyset\}$ determining the sparseness of the linear system resulting from the discretisation and determining the speed with which the approximate solution can be evaluated at a given point x. That is why we focus on this aspect of the discretisation, making the method very versatile when dealing with complex domains.

Core Cover Construction Algorithm

Cover construction in *GDF* is based on the algorithm described in [12], however our focus on subdomains and extremely complicated domains has lead to several unique extensions to that existing algorithm. These extensions are primarily encapsulated later in this section as well as in steps 6, 7, and 8 of our core cover construction algorithm summarised below in Algorithm 1. The following notation is used throughout this section:

- Ω : the domain of interest (with $\overline{\Omega}$ being its closure).
- N_0 : the initial number of points to be inserted in Ω to help form the cover. These points can be either regularly or irregularly distributed.
- N : the final number of patches in the cover.
- ω_i $(i = 1 \ldots N)$: the ith cover patch (with $\overline{\omega}_i$ being its closure). These patches are chosen to be simple $d-$dimensional boxes $(d = 1, 2, 3)$ in practical computations.
- Ω_N : a cover of Ω consisting of N patches $(\Omega_N = \{\omega_i\}_{i=1}^N)$.
- x_i : the point (or node) associated with patch ω_i. Typically, this point is in the centre of the patch.
- α_i : the *cover factor* of the ith patch = the number by which the length (and width and breadth) of patch i is multiplied in order to extend it so that patches overlap and form a proper cover of the domain (as opposed to simply partitioning the domain). Often, $\alpha_i = \alpha_j \;\; \forall \;\; i, \; j \in \{1, \ldots, N\}$ and we refer to the cover factor simply as α.

Algorithm 1

1. Form a (minimum) bounding box $B \supseteq \Omega$.
2. Insert the first point x_1 into the domain and let $\omega_1 = B$ initially.
3. For each new x_j, $j = 2, \ldots, N_0$,
   - insert x_j into Ω.
   - if $x_j \notin \omega_i$ for $i = 1, \ldots, j-1$, name the patch into which x_j was inserted ω_j.
   - else if $x_j \in \omega_i$ for some $i \in \{1, \ldots, j-1\}$ subdivide ω_i into 2 (1d) or 4 (2d) or 8 (3d) equally sized patches. Rename ω_i as the new smaller patch containing x_i.
   - repeat the preceding two sub-steps until x_i and x_j are in different patches.

   Note that if the spatial dimension is greater than 1*, many empty patches may be created by the preceding steps. For example, subdividing a 2d box only once to isolate two points into separate sub-patches creates two empty patches. Furthermore, non-empty patches obtained from these first three steps will likely not cover the entire domain (see Fig. 2.1 (a) and (b)).*

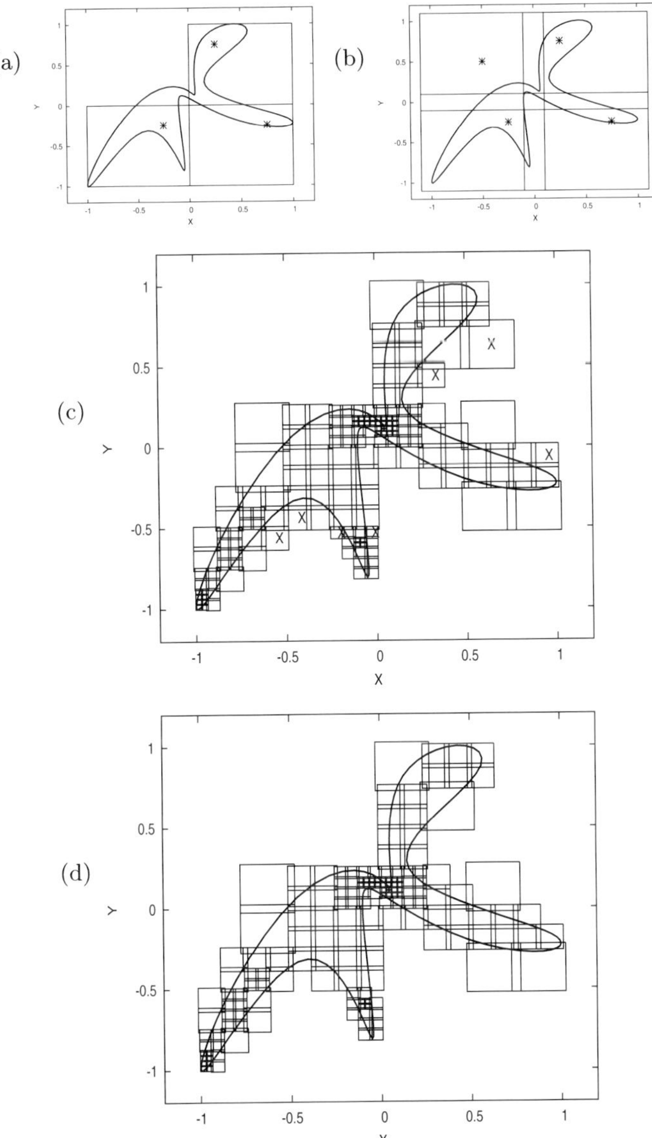

Figure 2.1. Key stages of cover construction. (a) Three points distributed randomly in a complex domain and the initial cover. (Steps 1–3 of Algorithm 1). (b) An increase in the number of patches so that the whole domain is covered. Patches have also been extended by $\alpha = 1.2$. (Steps 4 and 5). (c) Optional refinements of the cover. For clarity, the points are omitted and only the associated patch pictured. Seven patches whose intersection with the domain are subsets of another patch are labelled 'X' and will be removed in the final stage of basic cover construction. (Step 7). (d) The final cover of 159 patches with the seven patches from frame (c) removed. (Step 8).

4. For each empty patch created in step 3,
   - insert a point x_k, $k = N_0 + 1, \ldots, N$, into Ω and name that patch ω_k.

   This gives a set of patches $\{\omega_i\}_{i=1}^N$ whose intersection with the domain partition the interior of the domain. To get an *overlapping* cover:
5. For each ω_i, $i = 1, \ldots, N$
   - extend ω_i by a multiplicative *cover factor* α_i typically chosen to be $1 < \alpha_i < 2$.
6. If the domain contains one or more subdomains, check if any subdomain is a proper subset of a patch. If so, refine that patch until it intersects the boundary of the subdomain. Patches created during this refinement which are not disjoint from the domain are added to the cover while patches which are disjoint from the domain are removed, and the value of N is adjusted accordingly. This step guarantees that subdomain boundaries are always properly resolved by the cover.
7. The preceding steps give an overlapping cover of the domain as required by the PUM. One may optionally perform further refinements of this cover. There are two refinements which we perform to aid our computations:
   - To facilitate integration of the region(s) interior to the domain on a patch which is intersected by the boundary, we refine boundary patches if necessary until no boundary patch is intersected by the boundary more than twice.
   - To permit more efficient computation, we refine patches so that no two adjacent patches differ by more that k (for some integer $k \geq 0$) refinement levels. In practise, $k = 1$ is chosen. This ensures that on a simple d–dimensional rectangular domain and for any cover factor, $1 < \alpha < 2$, no patch will contain another patch as a proper subset.
8. Finally,

   For $i = 1, \ldots, N$
   - if $(\omega_i \cap \Omega) \subseteq (\omega_j \cap \Omega)$ for some $j = 1, \ldots, N,\ j \neq i,$, then
     - $\Omega_N \longrightarrow \Omega_N \setminus \omega_j$
     - $N \longrightarrow N - 1$

The data structures used for the subdivision process in the algorithm are binary trees in 1d, quadtrees in 2d, and octrees in 3d. By restricting the difference in size of adjacent patches to no more than one level of refinement and restricting the cover factor α to $1 < \alpha < 2$, we guarantee that the non-null intersection of a patch occurs only with immediately adjacent patches. A further restriction on α to $1 < \alpha < 1.5$ on rectangular domains guarantees that each patch contains a centrally-located subset on which all partition of unity functions are zero except the one associated with that patch. This is the *flat top property* outlined in Sec. 2, and it is a sufficient condition for the union of all local basis functions to be linearly independent. In the next subsection, we outline extensions to the above algorithm which guarantee the flat top property on arbitrary domains. Note however that step 8 of Algorithm

1 greatly reduces the likelihood of linear dependence in the union of basis functions and is all we typically need for practical computations.

Fig. 2.1 exhibits key stages in the construction of a 2d cover of a complex domain. Of particular interest to us are domains containing many holes, a typical situation encountered in cell biology. For such complex domains we again form covers as described above, treating the interior boundaries just like the exterior boundary of the domain. Fig. 2.2 illustrates one such case for a domain containing 8 holes and 208 points. If it is necessary to solve equations on a subdomain, a separate cover is created for that subdomain as described in Algorithm 1.

Optional Extensions to the Core Cover Construction Algorithm

Many extensions to Algorithm 1 are possible, forming an important area of research similar to mesh generation theory for mesh-based methods. We will introduce three such extensions in the following, with the first two geared towards ensuring that the resulting cover has the *flat top property*. The third extension discussed makes the flat top property likely but does not guarantee it, but it does have the advantage of reducing the overall number of patches used and also shows the flexibility of our method since in involves the use of patches of varying cover factors.

Firstly, a simple adjustment to Algorithm 1 guarantees that the cover has the *flat top property*. In step 7 of the algorithm, continue subdividing until the centre point of each patch which intersects the domain is in the domain (discarding patches which are disjoint from the domain). Since this extension can lead to a vast increase in the number of patches over which to integrate and since we find that our partition of unity method works well without it, we do not use this extension in practical computations.

Secondly, we can again guarantee the flat top property by modifying step 8 of Algorithm one as follows: eliminate each boundary patch whose intersection with the domain is fully contained within *the union of* neighbouring patches.

A third extension involves eliminating boundary patches whose intersection with the domain is small (relative to their sizes) and expanding adjacent patches appropriately to ensure that the domain remains covered. This also exhibits the flexibility of our approach since we use cover factors of *varying sizes* within the same cover. Importantly, this also leads to a reduction in the total number of cover patches and hence improves the efficiency of the method. This extension to Algorithm 1 is outlined in Algorithm 2. Fig. 2.3 illustrates the effects of the extended algorithm with a domain having a complex shape. A cover created by Algorithm 1 is modified to remove six patches and thirteen neighbouring patches are extended to ensure that the domain remains covered. In Fig. 2.3, the value of TOL from step 2 of Algorithm 2 is 0.25.

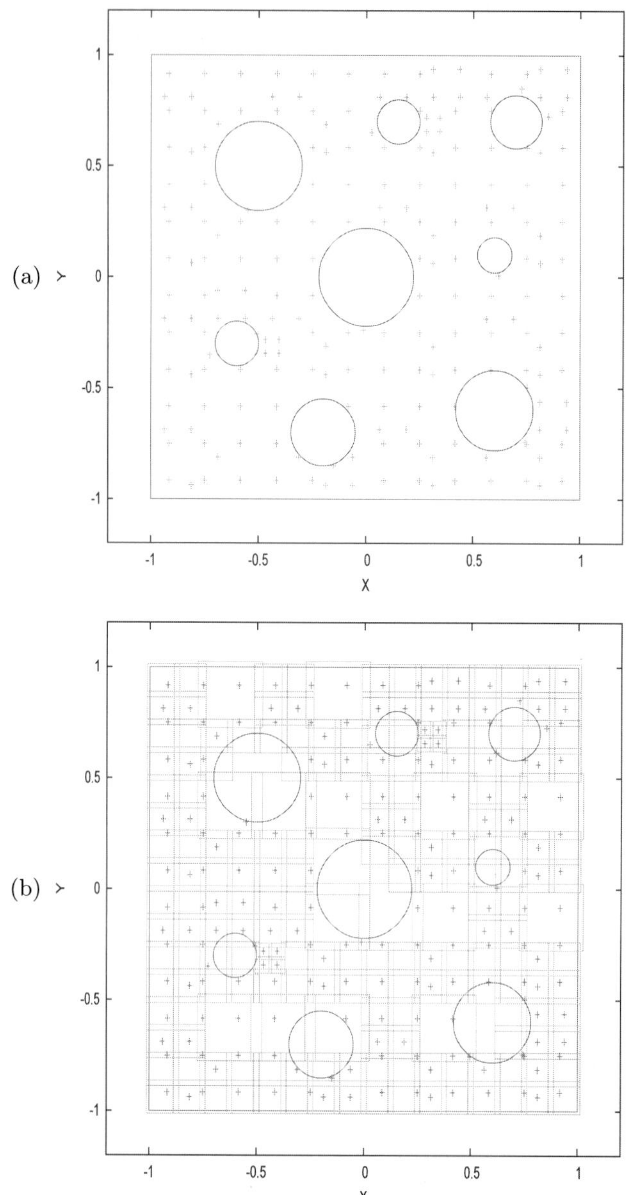

Figure 2.2. (a) View of a domain containing 208 points and eight holes of varying sizes, and (b) domain with cover, with a cover factor of 1.2. (*See also* Color Plate on page 362)

Algorithm 2

1. Perform Algorithm 1.

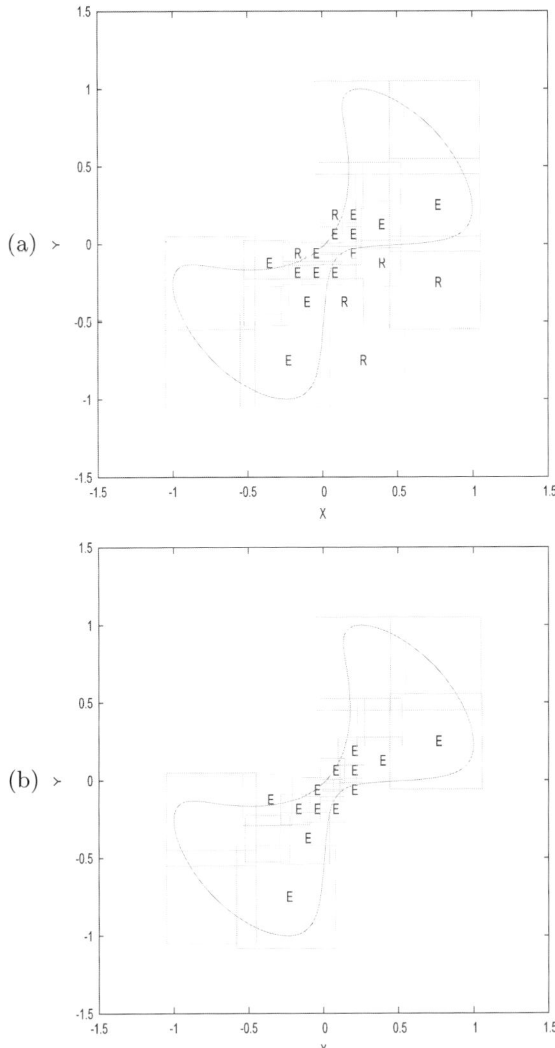

Figure 2.3. (a) A cover created by Algorithm 1, with patches to be removed labelled 'R' and patches which will be extended to maintain a complete cover labelled 'E'. (b) The cover generated at the end of Algorithm 2, with six patches from the original cover removed and thirteen patches (labelled 'E') extended. (*See also* Color Plate on page 363)

2. For each boundary patch, check what proportion of its size (area in 2d) lies within the domain. If this proportion is $<$ TOL, tag the patch for potential removal.
3. For each patch tagged for potential removal, do the following:

- Determine which of its neighbours could be expanded to take its place (notably, such a patch cannot be also tagged for removal), and determine the maximum amount by which these neighbours could be expanded without crossing the centre point of an adjacent patch.
- Gradually increase the size (*i.e.* cover factor) of all patches to be expanded until the union of these expanded patches covers the domain region of the patch to be removed *(success)* or until the cover factor of an expanded patch is within some preset tolerance of the maximum allowed cover factor *(failure)*.
- If the preceding step ends successfully and the expanded patches do not intersect the boundary of the domain more than twice, remove the patch tagged for removal and expand the neighbouring patches by the cover factors determined in the preceding step. Otherwise, do nothing to the patch tagged for removal and its neighbours.

Outlook

Cover construction is one of the most crucial parts of the PUM in *GDF* since it greatly impacts efficiency and even accuracy of the entire method. It is therefore under constant review for possible improvements. The key goal is to cover the domain with as few patches as possible while resolving all of the important features of the geometry. We are constantly exploring new ways of doing this, with features such parametric maps being considered as a potential tool to reduce the need for some refinement steps at boundary patches (see Sec. 2.3). This ongoing work will be discussed further as the development and applications of *GDF* progress.

2.3 Discretisation and Quadrature

In order to build the algebraic equation system (2.2) for the partition of unity method, the finite analogue of problem (2.1) has to be evaluated with respect to all patches ω_i and all basis functions $\{\varphi_i \psi_i^k\}_{k=1}^{p_i}$ defined on the respective patch. This is the most time consuming part of the method due to the (potentially) arbitrary distribution of nodes. It is necessary to find for each patch all of its neighbours in order to determine the integration areas. Fortunately, this costly procedure can be alleviated by reusing information about overlapping patches, as long as nodes are static. The numerical integration of boundary patches is another demanding task, depending on the shape of the domain. A simple approach would be to stick with regular integration cells (e.g. squares in 2d) and successively refine integration areas along boundaries to asymptotically approximate the real integration area. Another method using curvilinear patches with higher accuracy and fewer integration areas is mentioned next. For more details on the discretisation, see [9] or [23].

While most integration areas are simple in the sense that an affine map can be applied for transforming the reference cell $[-1,1]^d$ (where d = the

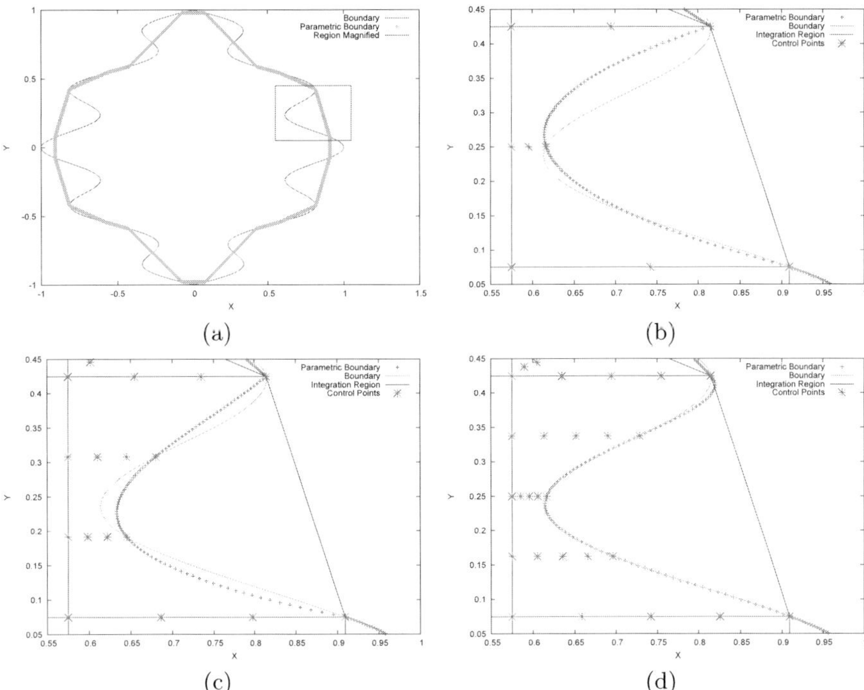

Figure 2.4. (a) A domain with a linear approximation to its boundary. The region enclosed in the small box is shown magnified in the remaining frames. (b) Close-up of a quadratic approximation to the boundary. (c) Close-up of a cubic approximation to the boundary. (d) Close-up of a quartic approximation to the boundary. In the last three frames, we also plot the control points [24] used to create the parametric maps. (*See also* Color Plate on page 364)

spatial dimension) onto the physical patch, at boundaries this approach can lead to large errors and degrade accuracy. One approach is to use higher order polynomial approximations for integration areas of boundary patches. Parametric transformations (of the same order as the approximation spaces, i.e. isoparametric) are common in finite element codes. In our simulations, we support arbitrary order maps in patches intersecting any interface. In Fig. 2.4, the effect of these parametric maps in approximating the real boundary of a domain is demonstrated. The fourth order parametric map almost perfectly recreates the original domain boundary, as demonstrated by the magnified view of one of the curved sections of that boundary. As further evidence of the merits of parametric maps, the following table demonstrate a decrease in the L^2 error with increasing parametric map order when solving

$$-\Delta u + u = f \qquad \text{in } \Omega, \tag{2.3}$$

$$\frac{\partial}{\partial \nu} u = g \qquad \text{on } \partial\Omega. \tag{2.4}$$

on the domain of Fig. 2.4 using *GDF*. The f and g are chosen to give a solution of $u(x, y) = x^2 y$ and all normal derivatives are computed on the real domain boundary.

Parametric Map Order	L^2 Error
1 (Linear)	0.0515333
2 (Quadratic)	0.0473768
3 (Cubic)	0.0453204
4 (Quartic)	0.0436432

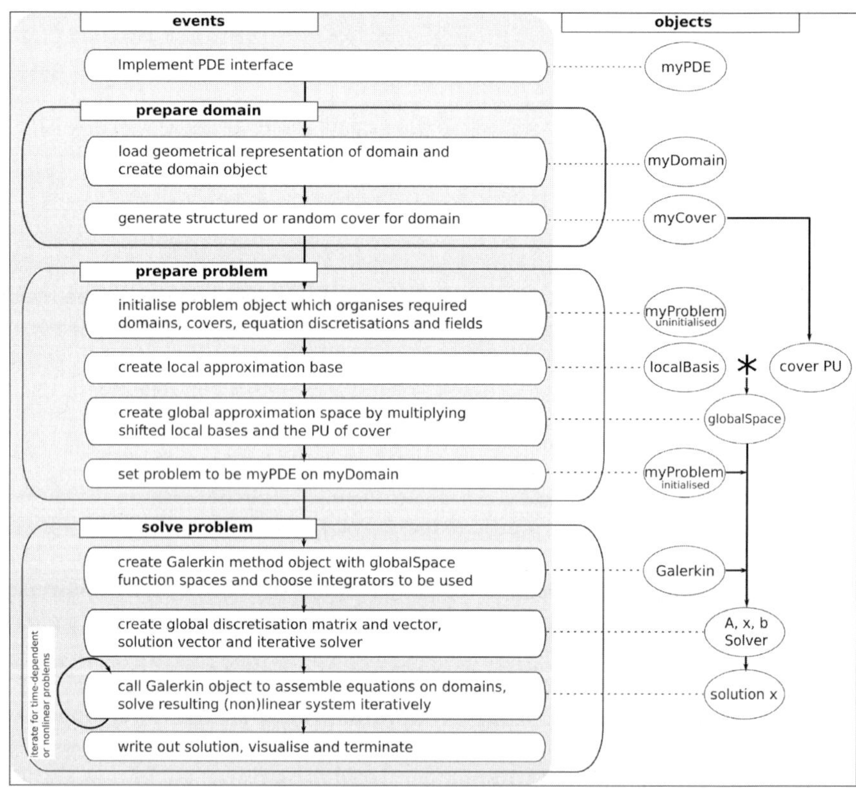

Figure 2.5. *GDF* flow chart

2.4 The Generic Discretisation Framework, GDF

The PUM is implemented by the authors in a C++ code called the *Generic Discretisation Framework* (*GDF*). Modern paradigms for software design such as reusability by object orientation and generic programming through templates are incorporated into the design of *GDF*. Fig. 2.5 shows the steps required for setting up a problem and solving it in the framework. In the left column the events the user manages are detailed while the right column shows the objects being created during that process.

3 Numerical Examples

Standard h and p convergence tests have been carried out on both simple and complex ($h-$refinement only) domains using *GDF*. The results indicate that even on complicated domains such as those containing many holes, optimal convergence rates are obtained [9].

Here we demonstrate the flexibility of *GDF* in handling computations on complex domains. We perform adaptive parabolic computations and then we simulate several biologically relevant scenarios.

3.1 Adaptive refinement

Two of the main advantages a cover has when compared to a mesh is the flexibility to resolve geometrical features (demonstrated in the next section) and to easily adapt to local approximation requirements. In this section, we use a simple explicit error indicator in order to steer a time-dependent adaptive refinement of our cover.

As usual (see e.g. [2,20]), we define boundary and interior residuals subject to an approximate solution $u_h \in V_{\mathrm{PU}}$

$$r := f + \Delta u_h - u_h \qquad \text{in } \Omega \tag{3.5}$$

$$R := g - \frac{\partial}{\partial \nu} u_h \qquad \text{on } \partial\Omega. \tag{3.6}$$

From the bilinear form $a(\cdot,\cdot)$ of the weak problem and the Galerkin orthogonality $a(e,v) = 0 \; \forall v \in V_{\mathrm{PU}}$ for the error $e = u - u_h$, we get

$$\begin{aligned} a(e,v) = l(v) - a(u_h,v) &= \int_\Omega fv + \int_{\partial\Omega} \partial_\nu u_h v + \int_\Omega (\Delta u_h - u)v \\ &= \int_\Omega rv + \int_{\partial\Omega} Rv = \sum_i \int_\Omega \varphi_i r(v - v_h) + \sum_i \int_{\partial\Omega} \varphi_i R(v - v_h). \end{aligned} \tag{3.7}$$

In this integration by parts argument we exploited the regularity of the PUM shape functions. In our simulations, we employ a PU of degree 2 (i.e. $\varphi_i \in$

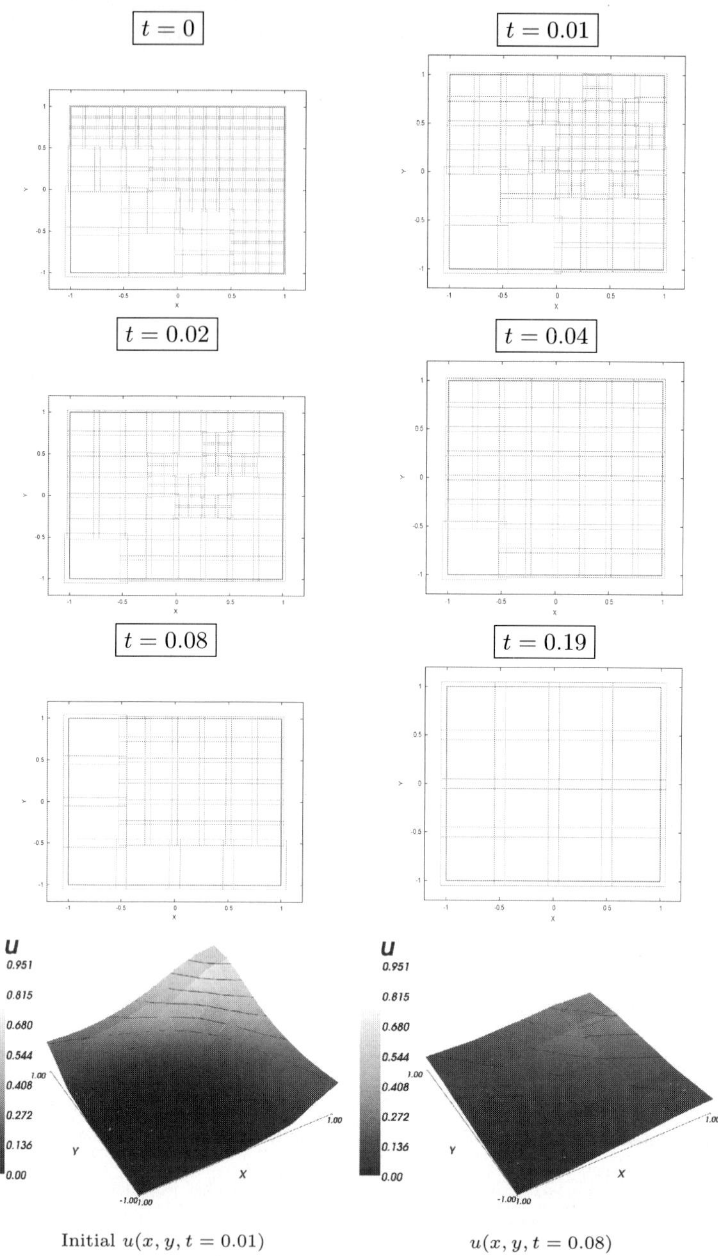

Figure 3.6. Successive refinement and coarsening of the cover in an adaptive parabolic problem at selected timesteps (TS). The initial and final (see TS = 19) cover consists of 16 equally-sized patches. The solution $u(x, y, t)$ after an initial time step and at time step 8 of the simulation are shown in the last two frames. (*See also* Color Plate on page 365)

$C^2(\omega_i)$) based on cubic splines. Hence, there are no additional boundary terms on the interior patch edges or within the patches that need to be taken into account. After a series of Cauchy-Schwarz estimates and with the standard approximation assumptions, one obtains, with unknown $C > 0$, the local explicit error indicator

$$\eta_i = C(h\|r\|_{L^2(\omega_i \cap \Omega)} + h^{1/2}\|R\|_{L^2(\partial(\omega_i \cap \Omega))}). \tag{3.8}$$

In the energy norm induced by the bilinear form there then holds the estimate

$$\|e\|_E^2 \leq C \left(\sum_i \eta_i^2 \right)^{1/2} . \tag{3.9}$$

Of course, the explicit error estimator used here exhibits the usual shortcomings of this approach. Mainly, the constants of the estimate are difficult to determine with reasonable accuracy. However, in this article, we like to put the emphasis on the adaptive cover rather than the actual error assessment. For this specific purpose, (3.8) can be used as an (heuristic) indicator with low computational overhead and satisfactory results.

In order to dynamically adjust the cover, a simple refinement process for evolutionary equations is defined. Here we denote by Ω_k the cover at the end of timestep k.

Algorithm 3

1. At time step k: if $k = 0$ or cover $\Omega_k \neq \Omega_{k-1}$, transfer previous solution u_h^{k-1} to u_h^k on current cover by an L^2 projection.
2. refinement loop: do while number of maximal refinements per time step is not reached and while previous refinement iteration has actually changed cover Ω_k
   - solve current time step
   - evaluate patch error indicators η_i
   - mark those patches for refinement whose indicator is larger than an upper tolerance τ
   - mark those patches for unrefinement whose indicator is smaller than a lower tolerance $\underline{\tau}$
   - refine marked patches and unrefine those patches which (a) have not been refined previously in this time step and (b) whose immediate neighbours, sharing the same parent patch, are also marked
   - ensure validity of cover and perform additional refinements if necessary
3. GOTO 1. to iterate time steps

The initial value of each time step has to be projected from the previous function space to the space of the refined cover. Similarly to the level transfer operators described in [23], there are several ways to accomplish this. On the one hand, one can employ a global approach which constructs L^2 projections of the global system. On the other hand, local projections exploiting the knowledge about which patches were refined or coarsened from the cover of the previous time step can be much more efficient.

We solve the heat equation for 19 timesteps on $\overline{\Omega} = [-1, 1]^2$ with an initial condition of $u(x, y, t = 0) = e^{-20(x-1)^2(y-1)^2}$, a function with height concentrated around the upper right hand corner of the domain. The initial cover consists of 4×4 equally sized patches. An initially large error is indicated in the corner of the domain where the initial function exhibits large values. In the multi-step refinement procedure described above, patches in the upper right corner are thus successively refined. As the problem we are solving does not include additional sources in the domain, the solution quickly becomes smooth and the refinement algorithm identifies regions whose average error is sufficiently small to unrefine sets of patches again. By the time the simulation has ended, the refinements and coarsening of the patches as the solution $u(x, y, t)$ smoothens, have lead back to the original coarse cover (see Fig. 3.6).

3.2 Complex domain

We solve Equations (2.3) and (2.4) using quadratic local approximation spaces on a complicated domain, Ω. The right-hand-side functions f and g are chosen to give a solution of $u(x, y) = sin(2\pi(x + y) - cos(2\pi y) + e^{-10(x^2+y^2)}$. The domain with its cover is exhibited in frame (a) of Fig. 3.7 while the corresponding L^2 errors, $||e||_{L^2}$, are plotted in frame (b). The L^2 and H^1 errors are summarised in the following table. In the table, in addition to the notation already explained, the following notation is used:

- p = the degree of the (polynomial) local approximation spaces.

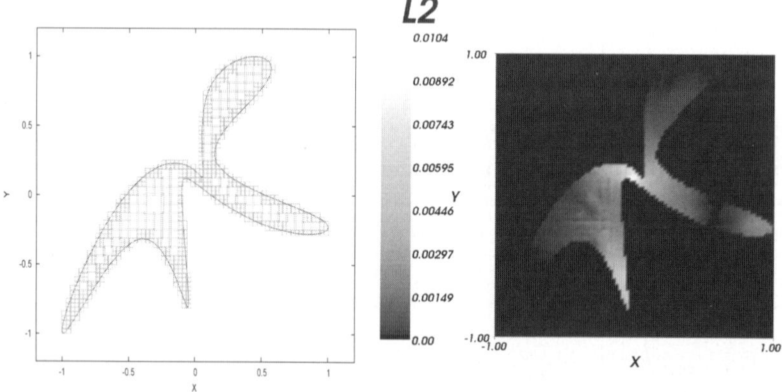

Figure 3.7. (a) The domain of Example 3.2 with a cover of 642 patches with cover factor 1.2. (b) The point-wise L^2 error. (*See also* Color Plate on page 366)

- dof = the degrees of freedom = $\sum_i \dim(V_i)$.
- $\|e\|_{H^1}$= The H^1 norm of the error.

N	p	dof	$\|e\|_{L^2}$	$\|e\|_{H^1}$
642	2	3852	0.0315751	0.467831

3.3 Signal transport from the cell membrane into the nucleus - an example with nested subdomains

We now illustrate a typical biological process, cellular signalling, that poses new and challenging problems to numerical mathematics. First the compartmentalisation of the cell requires that the computational domain can be split into nested sub-domains. Each of these sub-domains will have its own cover where the different cover constructions discussed earlier will be applied independently. GDF (see section 2.4) has been constructed to allow theoretically for an arbitrary hierarchy of such nested sub-domains. Between the sub-domains interface conditions have to be specified. Such interface242 conditons model for example the rate of protein translocation, or the rate with which molecules pass the nuclear pore complexes (NPC). Again the biological background induces a complex mixture of boundary and interface conditions acting between the nested sub-domains. The overall problem setting with respect to sub-domain strucure which we will discuss in the following can be seen in Figure 3.8. It is related to typical processes which need to be understood for cell signalling. The figure shows a typical mammalian fibroblast, a very flat cell type justifying the two-dimensional approach. One important experiment discussed in [6] is ERK1/2 shuttling between the cell nucleus and cytoplasm. The extracellular signal-regulated protein kinase ERK1/2 is a crucial effector linking extracellular stimuli to cellular responses: upon phosphorylation ERK [also known as mitogen-activated protein kinase P42/P44 (MAPK)] concentrates in the nucleus where it activates specific programmes of gene expression. Little is known about the time course and regulation of ERK exchange between nucleus and cytoplasm in living cells, and here simulations can add immensely to our understanding. In [6] the speed of ERK2 shuttling between nucleus and cytoplasm was measured quantitatively and it was determined that shuttling accelerated after ERK activation. Finally it was demonstrated that ERK2 did not diffuse freely in the nucleus and that diffusion was further impeded after phosphorylation, suggesting the formation of complexes of low mobility, or the binding to nuclear sub-structures. To understand this and similar processes better GDF was especially constructed to allow additional state variables on every boundary and interface, i.e. on lower dimensional manifolds, for example describing a concentration of bound molecules along the surface.

We can test, as one possible hypothesis, that the nucleoli create impenetrable obstacles $\Omega_{1\ldots4}$ which act as binding partners for the diffusing substrate u describing ERK1/2. Another idea would be that ERK1/2 can bind anywhere

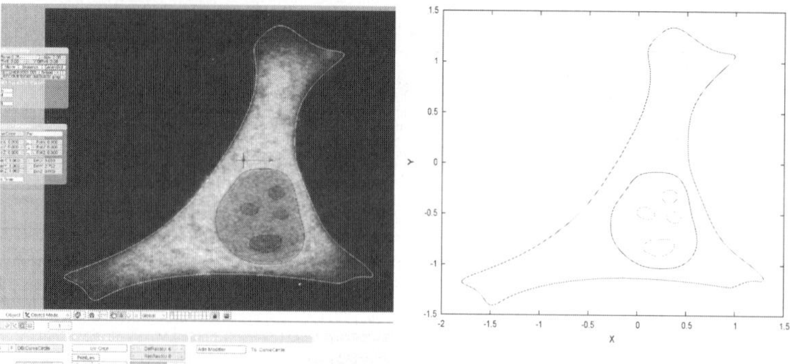

Figure 3.8. Domain with nested subdomains. Left the original fibroplast microscopy image inside a modelling programme to retrieve the (sub-)domain shapes, courtesy of Gimmi Ratto, see also [6]. Right: The main domain represents the cell body, the main sub-domain represents the nucleus, and the four inner sub-subdomains with interfaces $\partial\Omega_{1\ldots4}$ (green) represent nucleoli. (*See also* Color Plate on page 366)

at the chromatin structure. This would create a typical two-phase model which GDF is also able to handle. But here we first isolate the nucleus, assume a certain inflow of ERK1/2 from one side of the nuclear envelope, and include a binding/unbinding-term to the interfaces $\partial\Omega_{1\ldots4}$ (the nucleoli) with rates $\kappa_I = 1$ and $\kappa_B = 5$, i.e. the substrate dissociates with a higher rate. Note that both diffusion and binding parameters are non-dimensionalised and not yet adapted to a real experiment. Here we only demonstrate the principles. While bound to an interface, the substrate u_i on the interface $\Gamma_i \in \partial\Omega_i$ is stationary and does not diffuse any more, as has been suggested in [6]. Then the equations restricted to the nuclear dynamics (treating the nuclear envelope as a boundary) become (with Ω only describing the nucleus, not the cytoplasm):

$$\partial_t u - D\Delta u = 0 \qquad \text{in } \Omega \times (0,T] \tag{3.10a}$$

$$\partial_n u = g \qquad \text{on } \Gamma \times (0,T] \tag{3.10b}$$

$$\partial_n u = -(\kappa_I u - \kappa_B u_i) \qquad \text{on } \Gamma_i \times (0,T], \quad i = 1\ldots4 \tag{3.10c}$$

$$u_i = \kappa_I u - \kappa_B u_i \qquad \text{on } \Gamma_i \times (0,T], \quad i = 1\ldots4 \tag{3.10d}$$

and initial conditions

$$u = 0 \qquad \text{on } \Omega \times \{0\} \tag{3.10e}$$

$$u_i = 0 \qquad \text{on } \partial\Omega_i \times \{0\}, \quad i = 1\ldots4. \tag{3.10f}$$

The time dependent boundary function g is chosen such that the described influx during the first time steps is obtained. A typical solution of this process, where ERK1/2 is only allowed to flow in from one side of the nuclear

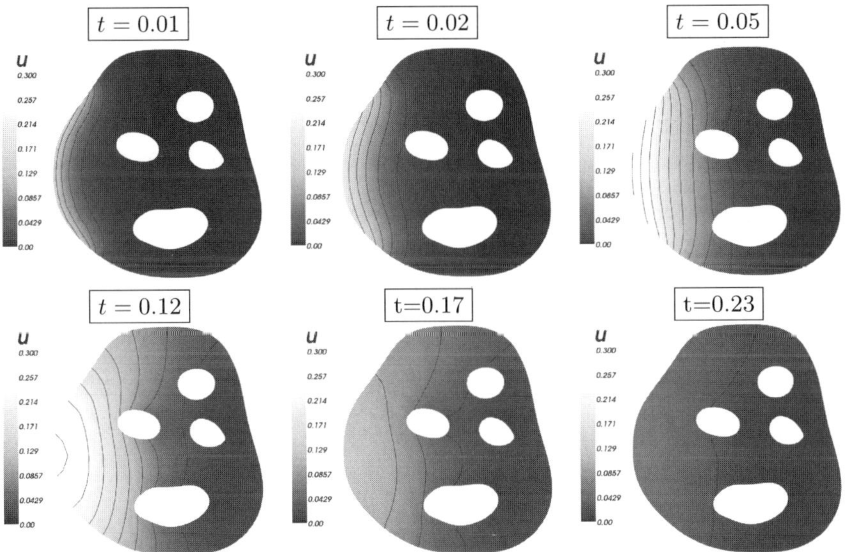

Figure 3.9. Simulation of hindered diffusion process with binding to obstacle interfaces, shown at different time steps (dt = 0.01).

envelope for a short time, can be seen in Figure 3.9. Figure 3.10 shows that the numerical scheme performs quite well with respect to mass conservation, where material is exchanged from the surfaces with the interior of the domain. Finally Figure 3.11 shows that we can scale up the problem to include the whole fibroblast as the computational domain. In this case the boundary condition for the nuclear envelope (described by $\Gamma := \partial\Omega$ in Eq. 3.10) had been changed into an interface condition, where molecules can translocate in and out of the nucleus through the NPC, where the translocation rate anywhere along Γ (i.e. the NPCs are assumed to have a 'concentration' in the nuclear envelope, they are not modelled as discrete entities) does depend on the concentration gradient.

3.4 Current Work

Here we give some general comments on our implementation. Our current work focuses on interface problems in biology, in which processes in a main domain affect and are affected by processes on an interface adjacent to the domain. In this work (currently in 2d) the codimension 1 boundaries of subdomains are discretised using finite element methods and diffusion equations are solved on that interface. Meanwhile, the surrounding main domain is discretised using the partition of unity meshfree method and interchange between the two domains (interior and meshfree) are handled by flux type interface conditions, which include the possibility of exploring different binding and

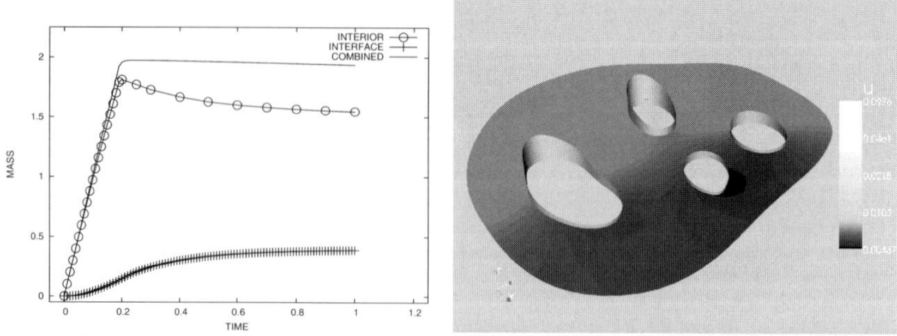

Figure 3.10. Left: Mass distribution of both the mobile and immobile fraction of molecules versus time. In this simulation the immobile fraction is modelled to get bound at the sub-domain boundaries. The total mass of the system is conserved as molecules cannot cross (artificially) the outer domain boundary representing the nuclear envelope. **Right:** Boundary surface concentrations can be visualised as heights in the z-direction. At particular time points the surfaces can act as a reservoir releasing particles into the volume (time step = 40, dt = 0.01).

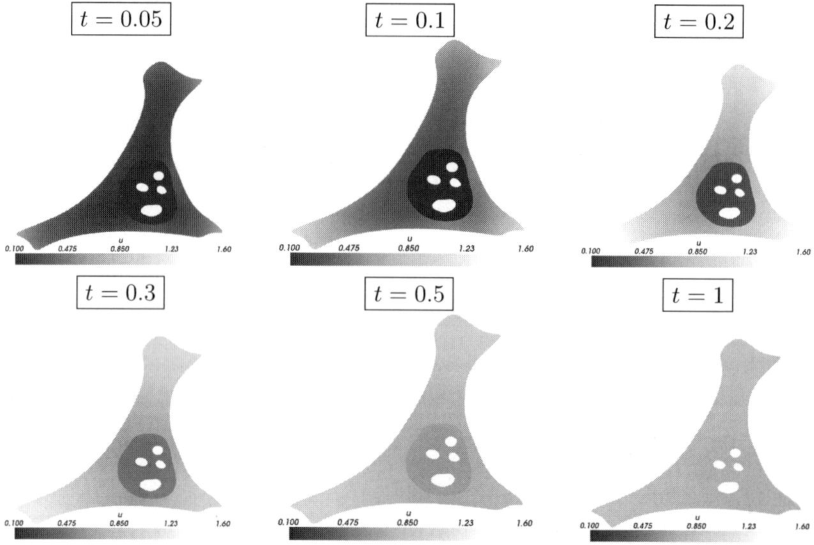

Figure 3.11. Concentration function $u(x, y)$ at selected times for a simulation on the entire two-dimensional fibroblast shape. In this simplified signalling problem a molecule is released (activated) uniformly at the outer cell membrane, diffuses inside the cytoplasm, can enter the nucleus (a sub-domain, i.e. a nested domain of level 1) following a simple linear relationship based on concentration differences (not modelling the NPCs in detail), and finally distributes itself inside the nucleus where it can get bound to the nucleoli modelled as sub-sub-domains (nested domains of level 2).

unbinding rates of a species to and from the interface. This work is explored in a forthcoming manuscript by the authors.

4 Conclusion and Outlook

It was demonstrated that cellular biology and new microscopy techniqes offer many new challenging problems for modelling and numerical analysis. We showed that the meshfree PUM can be extended to problems with complex domains. There is a need to investigate and test new cover creation algorithms in order to automise the work flow going from microscopy images to simulations detecting basic transport modes in cellular signalling or any other regulatory or metabolic process in the cell. Moreover we showed that the cover can be adaptively refined and coarsened for time-dependent problems.

Future challenging extensions of the methods presented would be automatic cover creation algorithms in three dimensions. Similarly adaptive cover generating algorithms will need to be analysed and tested for higher-dimensional systems of parabolic equations describing the simultaneous transport and interaction of different molecular species. Finally, as mentioned in the introduction, there is a need to derive a better understanding of molecular transport behaviour on the typical scales of the confocal microscope. This should be based generally on a multi-scale analysis of both the medium in which the molecules move, as well as on the microscopic properties of the many molecular machines involved in the process. As an important step a detailed multi-scale analysis of the nuclear pore complexes (NPCs) would immensely help our understanding of cellular signalling.

Acknowledgements

This paper is part of the research activities supported by UniNet contract 12990 funded by the European Commission in the context of the VI Framework Programme. Martin Eigel was partly funded by the EPSRC Life Science Interface via the IPCR project at the University of Warwick. Erwin George was funded by BBSRC, reference BB/C00437X/1.

References

1. I. Babuška, U. Banerjee, and J. E. Osborn, *Meshless and Generalized Finite Element Methods: A Survey of Some Major Results*, Meshfree Methods for Partial Differential Equations (M. Griebel and M. A. Schweitzer, eds.), Lecture Notes in Computational Science and Engineering, vol. 26, Springer, 2002, pp. 1–20.
2. Ainsworth, Mark and Oden, J. Tinsley (1993). A unified approach to a posteriori error estimation using element residual methods. Numerische Mathematik, 65(1)

3. Babuška and J. M. Melenk, *The partition of unity method*, Internat. J. Numer. Methods Engrg., **40**(4) (1997), 727–758.
4. Belytschko, T., Y. Y. Lu, and L. Gu, *Element-free Galerkin methods*, Int. J. Numer. Meth. Engrg. **37** (1994), 229–256.
5. Bouchekhima, A.-N., Frigerio, L., and Kirkilionis, M. A. (2007). Geometric Quantification of the Plant Endoplasmatic Reticulum. Warwick Preprint 11/2007. Submitted to Journal of Microscopy.
6. Costa, M., Marchi, M., Cardarelli, F., Roy, A, Beltram, F., Maffei, L. and Ratto, G.M., Dynamic regulation of ERK2 nuclear translocation and mobility in living cells, Journal of Cell Science 119, pp. 4952-4963, 2006.
7. Domijan, M. and Kirkilionis, M. (2007). Bistability and Oscillations in Chemical Reaction Networks. Warwick Preprint 04/2007. Submitted to Journal of Mathematical Biology.
8. Domijan, M. and Kirkilionis, M. (2007). Graph Theory and Qualitative Analysis of Reaction Networks. Warwick Preprint 13/2007. Accepted: Networks and Heterogeneous Media, 2008.
9. Eigel, M., E. George, and M. Kirkilionis, *A Meshfree Partition of Unity Method for Diffusion Equations on Complex Domains*, Warwick Preprint:10/2007. Submitted to IMA J. Num. Anal.
10. Gerlich, D. and Ellenberg, J., 4D imaging to assay complex dynamics in live specimens. Nat Cell Biol , Vol. Suppl, pp. S14-S19, 2003.
11. Goodwin, J. S. and Kenworthy, A. K, Photobleaching approaches to investigate diffusional mobility and trafficking of Ras in living cells.. Methods Cell Biol, Vol. 37, pp. 154-164, 2005.
12. Griebel, M. and Schweitzer, M.A., *A Particle-partition of unity method. II. Efficient cover construction and reliable integration*, SIAM J. Sci. Comput., **23(5)** (2002), 1655–1682 (electronic).
13. McNally, J. G., Quantitative FRAP in analysis of molecular binding dynamics in vivo. Methods Cell Biol , Vol. 85, pp. 329-351, 2008.
14. Robinson, C. V., Sali, A., and Baumeister, W. (2007). The molecular sociology of the cell. Nature, 450(7172), 973-982.
15. Sbano, L. and Kirkilionis, M. (2007). Molecular Systems with Infinite and Finite Degrees of Freedom. Part I: Continuum Approximation. Warwick Preprint 05/2007. Submitted to J. Math. Biol.
16. Sbano, L. and Kirkilionis, M. (2007). Molecular Systems with Infinite and Finite Degrees of Freedom. Part II: Deterministic Dynamics and Examples. Warwick Preprint 07/2007. Submitted to J. Math. Biol.
17. Sbano, L. and Kirkilionis, M. (2007). Multiscale Analysis of Reaction Networks. Warwick Preprint 12/2007.
18. Tian, R., G. Yagawa, H. Terasaka, *Linear dependence problems of partition of unity-based generalized FEMs*, Comput. Methods Appl. Mech. Engrg., **195** (2006), 4768–4782.
19. Weiss, M.; Hashimoto, H. and Nilsson, T. Anomalous protein diffusion in living cells as seen by fluorescence correlation spectroscopy. Biophys J, Vol. 84, pp. 4043-4052, 2003.
20. Ainsworth, M. and Tinsley, J., A Posteriori Error Estimation in Finite Element Analysis, John Wiley & Sons., 2000.
21. Braess, D., *Finite elements: Theory, fast solvers, and applications in solid mechanics*, Cambridge University Press, 2001.

22. Cioranescu, D. and Donato, P.. An Introduction to homogenization. Oxford University Press, 1999.
23. Schweitzer, M.A., *A parallel multilevel partition of unity method for elliptic partial differential equations*, Lecture Notes in Computational Science and Engineering, vol. 29, Springer, 2003.
24. Strang, G. and Fix, G. J., *An Analysis of the Finite Element Method*, Prentice-Hall, 1973.
25. Griebel, M. and Schweitzer, M.A., (eds.), *Meshfree Methods for Partial Differential Equations*, Lecture Notes in Computational Science and Engineering, vol. 26, Springer, 2002.
26. Griebel, M. and Schweitzer, M.A., *A Particle-Partition of Unity Method Part VII: Adaptivity*, Lecture Notes in Computational Science and Engineering, vol. 57, Springer, 2007.
27. Griebel, M. and Schweitzer, M.A., (eds.), *Meshfree Methods for Partial Differential Equations II*, Lecture Notes in Computational Science and Engineering, vol. 43, Springer, 2005.

Solving One Dimensional Scalar Conservation Laws by Particle Management

Yossi Farjoun and Benjamin Seibold

Department of Mathematics, Massachusetts Institute of Technology,
77 Massachusetts Avenue, Cambridge MA 02139, USA
{yfarjoun,seibold}@math.mit.edu

Summary. We present a meshfree numerical solver for scalar conservation laws in one space dimension. Points representing the solution are moved according to their characteristic velocities. Particle interaction is resolved by purely local particle management. Since no global remeshing is required, shocks stay sharp and propagate at the correct speed, while rarefaction waves are created where appropriate. The method is TVD, entropy decreasing, exactly conservative, and has no numerical dissipation. Difficulties involving transonic points do not occur, however inflection points of the flux function pose a slight challenge, which can be overcome by a special treatment. Away from shocks the method is second order accurate, while shocks are resolved with first order accuracy. A postprocessing step can recover the second order accuracy. The method is compared to CLAWPACK in test cases and is found to yield an increase in accuracy for comparable resolutions.

Key words: conservation law, meshfree, particle management

1 Introduction

Lagrangian particle methods approximate the solution of flow equations using a cloud of points which move with the flow. Examples are vortex methods [1], smoothed particle hydrodynamics (SPH) [5, 13], or generalized SPH methods [3]. The latter are typically based on generalized meshfree finite difference schemes [11]. An example is the finite pointset method (FPM) [9]. Moving the computational nodes with the flow velocity $\boldsymbol{v}$ allows the discretization of the governing equations in their more natural Lagrangian frame of reference. The material derivative $\frac{D}{Dt} = \frac{\partial}{\partial t} + \boldsymbol{v} \cdot \nabla$ becomes a simple time derivative. For a conservation law, the natural velocity is the characteristic velocity. In a frame of reference which is moving with this velocity, the equation states that the function value remains constant. Of course, this is only valid where the solution is smooth. In this case, characteristic particle methods are very simple solution methods for conservation laws.

In spite of the obvious advantages of particle methods, almost all numerical methods for conservation laws operate on a fixed Eulerian grid, even though significant work has to be invested to solve even a simple advection problem preserving sharp features and without creating oscillations. Leaving aspects of implementation complexity aside, two main reasons favor fixed grid methods: First, a fixed grid allows an easy generalization to higher space dimensions using dimensional splitting. Second, in particle methods one has to deal with the interaction of characteristics. While the former point remains admittedly present, the latter aspect is addressed in this contribution.

Most methods which use the characteristic nature of the conservation law circumvent the problem of crossing characteristics by remeshing. Before any particles can interact, the numerical solution is interpolated onto "nicely" distributed particles, for instance onto an equidistant grid – in which case the approach is essentially a fixed grid method. The CIR-method [2] is an example. Such approaches incur multiple drawbacks: First, the shortest interaction time determines the global time step. Second, the error due to the global interpolation may yield numerical dissipation and dispersion. Finally, such schemes are not conservative when shocks are present. In practice, finite volume approaches, such as Godunov methods with appropriate limiters [14], or ENO [6]/WENO [12] schemes are used to compute weak entropy solutions that show neither too much oscillations nor too much numerical dissipation.

With moving particles, two fundamental problems arise: On the one hand, neighboring particles may depart from each other, resulting in poorly resolved regions. On the other hand, a particle may (if left unchecked) overtake a neighbor, which results in a "breaking wave" solution. The former problem can be remedied by inserting particles. The latter has to be resolved by merging particles. When characteristic particles interact (i.e. one overtakes the other) one is dealing with a shock, thus particles must be merged.

In this contribution, we present a local and conservative particle management (inserting and merging particles) that yields no numerical dissipation (where solutions are smooth) and correct shock speeds (where they are not). The particle management is based on exact conservation properties between neighboring particles, which are derived in Sect. 2. In Sect. 3, we outline our numerical method. The heart of our method, the particle management, is derived in Sect. 4. There, we also show that the method is TVD. In Sect. 5, we prove that the numerical solutions satisfy the Kružkov entropy condition, thus showing that the solutions we find are entropy solutions for any convex entropy function. In Sect. 6, we apply the method to examples and compare it to traditional finite volume methods using CLAWPACK. In Sect. 7 we present how non-convex flux functions can be treated. Finally, in Sect. 8, we outline possible extensions and conclusions. These include applications of the 1D solver itself as well as possible extensions beyond the 1D scalar case.

2 Scalar Conservation Laws

Consider a one-dimensional scalar conservation law

$$u_t + (f(u))_x = 0, \quad u(x,0) = u_0(x) \tag{2.1}$$

with f' continuous. As long as the solution is smooth, it can be obtained by the method of characteristics [4]. The function $u(x(t),t)$ is constant along the characteristic curve

$$x(t) = x(0) + f'(u(x(0),0))\,t\,. \tag{2.2}$$

For nonlinear functions f the characteristic curves can "collide", resulting in a shock, whose speed is given by the Rankine-Hugoniot condition [4]. Discontinuities are shocks only if the characteristic curves run into them. Other discontinuities become rarefaction waves, i.e. continuous functions which attain every value between the left and the right states. If the flux function f is convex or concave between the left and right state of a discontinuity, then the solution is either a shock or a rarefaction. If f'' switches sign between between the two states, then a combination of a shock and a rarefaction occur. These physical solutions are defined by a weak formulation of (2.1) accompanied by an entropy condition.

2.1 Conservation Properties

Conservations laws conserve the total area under the solution

$$\frac{d}{dt}\int_{-\infty}^{\infty} u(x,t)\,\mathrm{d}x = 0\,. \tag{2.3}$$

The change of area between two *moving* points $b_1(t)$ and $b_2(t)$ is given by

$$\begin{aligned}\frac{d}{dt}\int_{b_1(t)}^{b_2(t)} u(x,t)\,\mathrm{d}x &= b_2'(t)\,u(b_2(t),t) - b_1'(t)\,u(b_1(t),t) + \int_{b_1(t)}^{b_2(t)} u_t(x,t)\,\mathrm{d}x \\ &= (b_2'(t)\,u(b_2(t),t) - b_1'(t)\,u(b_1(t),t)) - (f(u(b_2(t),t)) - f(u(b_1(t),t)))\,.\end{aligned}$$

If $x_1(t)$ and $x_2(t)$ are *characteristic* points, that is, points following the characteristics of a smooth solution as in equation (2.2), we have that $x_1'(t) = f'(u_1)$ and $x_2'(t) = f'(u_2)$. Therefore, the change of area between x_1 and x_2 is

$$(f'(u_2)u_2 - f'(u_1)u_1) - (f(u_2) - f(u_1)) = [f'(u)u - f(u)]_{u_1}^{u_2}\,, \tag{2.4}$$

where $[g(x)]_a^b = g(b) - g(a)$. Equation (2.4) implies that the change of area between two characteristic points does *not* depend on the positions of the points, only on the left state u_1 and right state u_2 and the flux function. Since the two states do not change as the points move, the area between the two points changes linearly, as does the distance between them:

$$\frac{d}{dt}(x_2(t) - x_1(t)) = x_2'(t) - x_1'(t) = f'(u_2) - f'(u_1) = [f'(u)]_{u_1}^{u_2} . \tag{2.5}$$

If the two points move at different speeds, then there is a time t_0 (which may be larger or smaller than t) at which they have the same position. Thus at time $t = t_0$ the distance between them, and the area between them equal zero. From (2.4) and (2.5) we have that

$$\begin{aligned} \int_{x_1(t)}^{x_2(t)} u(x,t)\,\mathrm{d}x &= (t - t_0) \cdot [f'(u)u - f(u)]_{u_1}^{u_2} , \\ x_2(t) - x_1(t) &= (t - t_0) \cdot [f'(u)]_{u_1}^{u_2} . \end{aligned}$$

In short, the area between two Lagrangian points can be written as

$$\int_{x_1(t)}^{x_2(t)} u(x,t)\,\mathrm{d}x = (x_2(t) - x_1(t))\, a(u_1, u_2) , \tag{2.6}$$

where $a(u_1, u_2)$ is the nonlinear average function

$$a(u_1, u_2) = \frac{[f'(u)u - f(u)]_{u_1}^{u_2}}{[f'(u)]_{u_1}^{u_2}} = \frac{\int_{u_1}^{u_2} f''(u)\,u\,\mathrm{d}u}{\int_{u_1}^{u_2} f''(u)\,\mathrm{d}u} . \tag{2.7}$$

The integral form shows that a is indeed an average of u, weighted by f''. This last observation needs one additional assumption: that the points x_1 and x_2 remain characteristic point between t and t_0. That is, that a shock does not develop between the two points before t_0. Our numerical method relies heavily on the nonlinear average $a(\cdot,\cdot)$.

Lemma 1. *Let f be strictly convex or concave in $[u_L, u_U]$, that is $f'' < 0$ or $f'' > 0$ in (u_L, u_U). Then for all $u_1, u_2 \in [u_L, u_U]$, the average (2.7) is...*

1. *the same for f and $-f$. Thus we assume $f'' > 0$ WLOG;*
2. *symmetric, $a(u_1, u_2) = a(u_2, u_1)$. Thus we assume $u_1 \leq u_2$ WLOG;*
3. *an average, i.e. $a(u_1, u_2) \in (u_1, u_2)$, for $u_1 \neq u_2$;*
4. *strictly increasing in both u_1 and u_2; and*
5. *continuous at $u_1 = u_2$, with $a(u_1, u_1) = u_1$.*

Proof. We only include here the proof of 4. We show that $a(u_1, u_2)$ is strictly increasing in the second argument. Let $u_1 < u_2 < u_3$, $u_i \in [u_L, u_U]$. Then

$$\begin{aligned} a(u_1, u_3) &= \frac{\int_{u_1}^{u_2} f''(u)u\,\mathrm{d}u + \int_{u_2}^{u_3} f''(u)u\,\mathrm{d}u}{\int_{u_1}^{u_3} f''(u)\,\mathrm{d}u} \\ &> \frac{a(u_1, u_2)\int_{u_1}^{u_2} f''(u)\,\mathrm{d}u + a(u_1, u_2)\int_{u_2}^{u_3} f''(u)\,\mathrm{d}u}{\int_{u_1}^{u_3} f''(u)\,\mathrm{d}u} = a(u_1, u_2) . \end{aligned}$$

Similar arguments show the result for the first argument. □

3 Description of the Particle Method

The first step is to approximate the initial function u_0 by a finite number of points $x_1, \ldots, x_m$ with function values $u_1, \ldots, u_m$. A straightforward strategy is to place $x_1, \ldots, x_m$ equidistantly on the interval of interest and assign $u_i = u_0(x_i)$. More efficient adaptive sampling strategies can be used, since our method does not impose any requirements on the point distribution. For instance, one can choose x_i and u_i to minimize the L^1 error, using the specific interpolation introduced in Sect. 4. This strategy is the topic of future work. The points are ordered so that $x_1 < \cdots < x_m$. The evolution of the solution is found by moving each point x_i with speed $f'(u_i)$. This is possible as long as there are no "collisions" between points. Two neighboring points $x_i(t)$ and $x_{i+1}(t)$ collide at time $t + \Delta t_i$, where

$$\Delta t_i = -\frac{x_{i+1} - x_i}{f'(u_{i+1}) - f'(u_i)} . \tag{3.8}$$

A positive Δt_i indicates that the two points will eventually collide. Thus, $t + \Delta t_\mathrm{s}$ is the time of the next particle collision[1], where

$$\Delta t_\mathrm{s} = \min_i \{\Delta t_i | \Delta t_i \geq 0\} .$$

For any time increment $\Delta t \leq \Delta t_\mathrm{s}$ the points can be moved directly to their new positions $x_i + f'(u_i)\Delta t$. Thus, we can step forward an amount Δt_s, and move all points accordingly. Then, at least one particle will share its position with another. To proceed further, we merge each such pair of particles. If the collision time Δt_i is negative, the points depart from each other. Although at each of the points the correct function value is preserved, after some time their distance may be unsatisfyingly large, as the amount of error introduced during a merge grows with the size of the gaps in the neighboring particles. To avoid this, we insert new points into large gaps between points *before* merging particles. In Sect. 4.1 we derive positions and values of the new particles that assure that the method is conservative, TVD, and entropy diminishing.

4 Interpolation and Particle Management

The movement of the particles is given by a fundamental property of the conservation law (2.1): its characteristic equation (2.2). We derive particle management to satisfy another fundamental property: the conservation of area (2.3). Using the conservation principles derived in Sect. 2, the function value of an inserted or merged particle is chosen, such that area is conserved exactly. A simple condition on the particles guarantees that the entropy does not increase. In addition, we define an interpolating function between two neighboring particles, so that the change of area satisfies relation (2.4). Furthermore, this interpolation is an analytical solution to the conservation law.

[1] If the set $\{i | \Delta t_i \geq 0\}$ is empty, then $\Delta t_\mathrm{s} = \infty$.

4.1 Conservative Particle Management

Consider four neighboring particles located at $x_1 < x_2 \le x_3 < x_4$ with associated function values u_1, u_2, u_3, u_4. Assume that the flux f is strictly convex or concave on the range of function values $[\min_i(u_i), \max_i(u_i)]$. If $u_2 \neq u_3$, the particles' velocities must differ $f'(u_2) \neq f'(u_3)$, which gives rise to two possible cases that require particle management:

- **Inserting:** The two particles deviate, i.e. $f'(u_2) < f'(u_3)$. If the distance $x_3 - x_2$ is larger than a predefined maximum distance d_{max}, we insert a new particle (x_{23}, u_{23}) with $x_2 < x_{23} < x_3$ and u_{23} chosen so that the area between x_2 and x_3 is preserved by the insertion:

$$(x_{23} - x_2)\, a(u_2, u_{23}) + (x_3 - x_{23})\, a(u_{23}, u_3) = (x_3 - x_2)\, a(u_2, u_3)\ . \quad (4.9)$$

  This condition defines a function, connecting (x_2, u_2) with (x_3, u_3), on which the new particle has to lie. This function is the interpolation defined in Sect. 4.2 and illustrated in Fig. 4.2.
- **Merging:** The two particles collide, i.e. $f'(u_2) > f'(u_3)$. If the distance $x_3 - x_2$ is smaller than a preset value d_{min} ($d_{\min} = 0$ is possible), we replace both with a new particle (x_{23}, u_{23}). The position of the new particle x_{23} is chosen with $x_2 < x_{23} < x_3$ and u_{23} is chosen so that the total area between x_1 and x_4 is unchanged:

$$\begin{aligned}&(x_{23} - x_1)\, a(u_1, u_{23}) + (x_4 - x_{23})\, a(u_{23}, u_4)\\ &\quad = (x_2 - x_1)\, a(u_1, u_2) + (x_3 - x_2)\, a(u_2, u_3) + (x_4 - x_3)\, a(u_3, u_4)\ . \quad (4.10)\end{aligned}$$

  Any particle (x_{23}, u_{23}) with $x_2 < x_{23} < x_3$ that satisfies (4.10) would be a valid choice. We choose $x_{23} = \frac{x_2 + x_3}{2}$, and then obtain u_{23} such that (4.10) is satisfied. Figure 4.1 illustrates the merging step.

Observe that inserting and merging are similar in nature. Conditions (4.9) and (4.10) for u_{23} are nonlinear (unless f is quadratic, see Remark 1). For most cases $u_{23} = \frac{u_2 + u_3}{2}$ is a good initial guess, and the correct value can be obtained by a few Newton iteration steps. The next few claims attest that we can find a unique value u_{23} that satisfies (4.9) and (4.10).

Lemma 1. *The function value u_{23} for the particle at x_{23} is unique.*

Proof. We show the case for merging. The argument for insertion is similar. From Lemma 1 we have that both $a(u_1, \cdot)$ and $a(\cdot, u_4)$ are strictly increasing. Thus, the LHS of (4.10) is strictly increasing, and cannot be the same for different values of u_{23}. □

Lemma 2. *If $x_2 = x_3 = x_{23}$, there exists $u_{23} \in [u_2, u_3]$ which satisfies* (4.10).

Proof. WLOG we assume that $u_2 \le u_3$. First, we define

$$\begin{aligned} A \quad &= (x_2 - x_1)\, a(u_1, u_2) + (x_4 - x_2)\, a(u_3, u_4), \quad \text{and} \\ B(u) &= (x_2 - x_1)\, a(u_1, u) + (x_4 - x_2)\, a(u, u_4)\,. \end{aligned}$$

Equation (4.10) is now simply $B(u_{23}) = A$. The monotonicity of a implies that

$$B(u_2) \leq A \leq B(u_3)\,. \tag{4.11}$$

Since a is continuous, so is B, and the existence of u_{23} follows the intermediate value theorem. □

Corollary 1. *If particles are merged only according to Lemma 2, then the total variation of the solution is either the same as before the merge, or smaller.*

Merging points only when $x_2 = x_3$ can be too restrictive. Fortunately, the following claim allows for a little more freedom.

Theorem 1. *Consider four consecutive particles* $(x_i, u_i)\ \forall i = 1, \ldots, 4$. *Merging particles 2 and 3 so that* $x_{23} = \frac{x_2+x_3}{2}$ *yields* $u_{23} \in [u_2, u_3]$ *if*

$$\frac{x_3 - x_2}{|u_3 - u_2|} \leq \frac{1}{16} \left(\frac{\min |f''|}{\max |f''|} \right)^6 \frac{\min\left(x_4 - x_2, x_3 - x_1\right)}{|\max(u_3, u_2) - \min(u_4, u_1)|}\,. \tag{4.12}$$

Here the min and max of f'' are taken over the maximum range of $u_1, \ldots, u_4$. This condition is naturally satisfied if $x_2 = x_3$.

Proof (outline). The full proof will be given in a future paper. The idea is to merge in two steps: First, we find a value $\tilde{u}$ such that setting $u_2 = u_3 = \tilde{u}$ (while leaving x_2 and x_3 unchanged) results in the same area. Then, we merge the two points to u_{23}. In the first step we bound $\tilde{u}$ *away* from u_2 and u_3 (but inside $[u_2, u_3]$), and in the second step we bound $|u_{23} - \tilde{u}|$ from above. □

Theorem 2. *The particle method can run to arbitrary times.*

Proof. Let $u_L = \min_i u_i$, $u_U = \max_i u_i$, and $C = \max_{[u_L, u_U]} |f''(u)| \cdot (u_U - u_L)$. For any two particles, one has $|f'(u_{i+1}) - f'(u_i)| \leq C$. Define $\Delta x_i = x_{i+1} - x_i$. After each particle management, the next time increment (as defined in Sect. 3) is at least $\Delta t_s \geq \frac{\min_i \Delta x_i}{C}$. If we do not insert particles, then in each merge one particle is removed. Hence, a time evolution beyond any given time is possible, since the increments Δt_s will increase eventually. When a particle is inserted (whenever two points are at a distance more than d_{max}), the created distances are at least $\frac{d_{max}}{2}$, preserving a lower bound on the following time increment. □

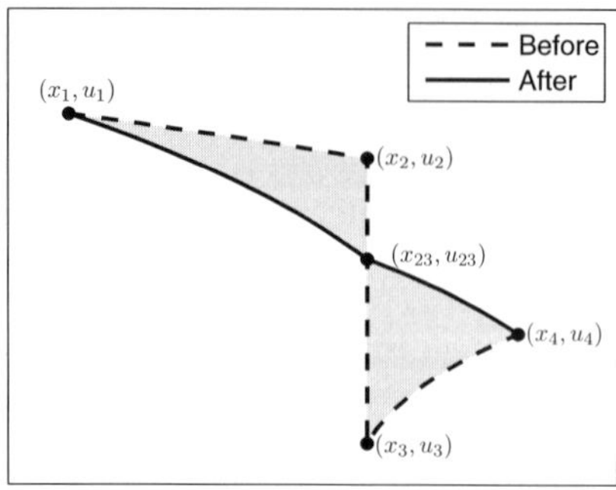

Figure 4.1. Merging two particles

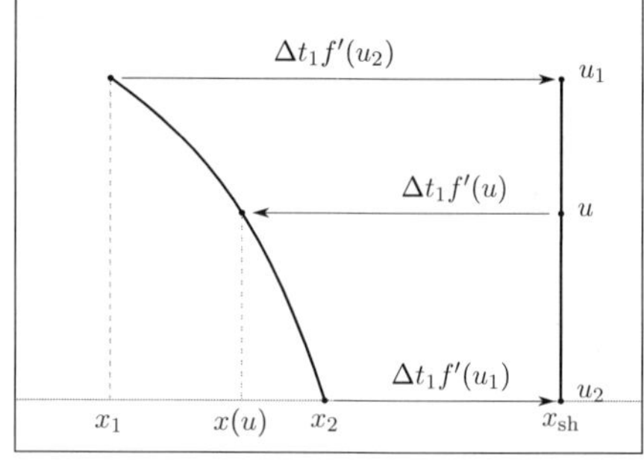

Figure 4.2. Definition of the interpolation

4.2 Conservative Interpolation

The particle management does not require an interpolation between points. As it stands, it complements the characteristic movement to yield a full particle method for the conservation law (2.1) that can run for arbitrarily long times. Yet, for plotting the solution and interpreting approximation properties, it is desirable to define an interpolation that is compatible with the conservation principles of the underlying partial differential equation. We define such an interpolation between each two neighboring points (x_1, u_1) and (x_2, u_2).

In the case $u_1 = u_2$, we define the interpolation to be a constant. In the following, we describe the case $u_1 \neq u_2$. Assume that f is strictly convex or concave in $[u_1, u_2]$. Therefore $f'(u_1) \neq f'(u_2)$. Hence, as derived in Sect. 2.1, the solution either came from a discontinuity (i.e. it is a rarefaction) or it will become a shock. The time Δt_1 until this discontinuity happens is given by (3.8). At time $t + \Delta t_1$ the points have the same position $x_1 = x_2 = x_{\text{sh}}$, as shown in Fig. 4.2. At this time the interpolation must be a straight line connecting the two points, representing a discontinuity at x_{sh}. We require any point of the interpolating function (x, u) to move with its characteristic velocity $f'(u)$ in the time between t and $t+\Delta t_1$. This defines the interpolation uniquely as

$$x(u) = x_1 - t_1 \left(f'(u) - f'(u_1)\right) = x_1 + \frac{f'(u) - f'(u_1)}{f'(u_2) - f'(u_1)}(x_2 - x_1) \,. \qquad (4.13)$$

Defining x as a function of u is in fact an advantage, since at a discontinuity characteristics for all intermediate values u are defined. Thus, rarefaction fans arise naturally if $f'(u_1) < f'(u_2)$. Since f'' has no inflection points between u_1 and u_2, the inverse function $u(x)$ exists. However, it is only required at a single point for particle management. For plotting purposes we can always plot $x(u)$ instead.

Lemma 3. *The interpolation* (4.13) *is a solution of the conservation law* (2.1).

Proof. Using that $\dot{x}_i(t) = f'(u_i)$ for $i = 1, 2$ one obtains

$$\begin{aligned}\frac{\partial x}{\partial t}(u,t) &= \dot{x}_1 + \frac{f'(u) - f'(u_1)}{f'(u_2) - f'(u_1)}(\dot{x}_2 - \dot{x}_1) \\ &= f'(u_1) + \frac{f'(u) - f'(u_1)}{f'(u_2) - f'(u_1)}(f'(u_2) - f'(u_1)) = f'(u)\,.\end{aligned}$$

Thus every point on the interpolation $u(x,t)$ satisfies the characteristic equation (2.2). □

Corollary 2 (exact solution property). *Consider characteristic particles with $x_1(t) < x_2(t) < \cdots < x_n(t)$ for $t \in [t_1, t_2]$. At any time consider the function defined by the interpolation* (4.13)*. This function is a classical (i.e. continuous) solution to the conservation law* (2.1)*. In particular, it satisfies the conservation properties given in Sect. 2.1.*

Theorem 3 (TVD). *With the assumptions of Theorem 1, the particle method is total variation diminishing.*

Proof. Due to Corollary 2, the characteristic movement yields an exact solution, thus the total variation is constant. Particle insertion simply refines the interpolation, thus preserves the total variation. Due to Theorem 1, merging of particles yields a new particle with a function value u_{23} between the values of the removed particles. Thus, the total variation is the same as before the merge or smaller. □

Remark 1. The method is particularly efficient for quadratic flux functions. In this case the interpolation (4.13) between two points is a straight line, since f' is linear. Furthermore, the average (2.7) is the arithmetic mean $a(u_1, u_2) = \frac{u_1+u_2}{2}$, since f'' is constant. Consequently, the function values for particle insertion and merging can be computed explicitly.

Remark 2. The method has some similarity to *front tracking* by Holden et al. [7], and some fundamental differences. In front tracking, one approximates the flux function by a piecewise linear, and the solution by a piecewise constant function. Shocks are moved according to the Rankine-Hugoniot condition. In comparison, our method uses the wave solutions. Hence, in front tracking everything is a shock; in the particle method, everything is a wave.

5 Entropy

We have argued in Sect. 4.2 that due to the constructed interpolation the particle method naturally distinguishes shocks from rarefaction fans. In this section, we show that the method in fact satisfies the entropy condition

$$\eta(u)_t + q(u) \leq 0 \tag{5.14}$$

if a technical assumption on the resolution of shocks is satisfied. We consider the Kružkov entropy pair $\eta_k(u) = |u-k|$, $q_k(u) = \operatorname{sign}(u-k)(f(u)-f(k))$. As argued by Holden et al. [8], if (5.14) is satisfied for η_k, then it is satisfied for any convex entropy function. Relation (5.14) implies that the total entropy $\int \eta_k(u(x))\,\mathrm{d}x$ does not increase in time for all values of k. Using the interpolation (4.13) we show that the numerical solution obtained by the particle method satisfies this condition.

Lemma 1 (entropy for merging). *Consider four particles located at $x_1 < x_2 = x_3 < x_4$, with the middle two to be merged. We consider the case $f'' > 0$, i.e. $u_2 > u_3$ WLOG.*[2] *If the resulting value u_{23} satisfies $u_1 \geq u_{23} \geq u_4$, then the Kružkov entropy does not increase due to the merge.*

Proof. We consider the segment $[x_1, x_4]$. Let $u(x)$ and $\tilde{u}(x)$ denote the interpolating function before resp. after the merge. The area under the function is preserved. We present the proof for $k \leq u_{23}$. For $k \geq u_{23}$ the proof is similar. The interpolating function u is monotone in the value of its endpoints, thus $u(x) \leq \tilde{u}(x)$ for $x \in [x_2, x_4]$. Since $|x| = x - 2\Theta(-x)$, where $\Theta(x)$ is the Heaviside step function, we can write

$$
\begin{aligned}
\int_{x_1}^{x_4} |u-k|\,\mathrm{d}x &= \int_{x_1}^{x_4} (u-k)\,\mathrm{d}x - 2\int_{x_1}^{x_4} (u-k)\Theta(k-u)\,\mathrm{d}x \\
&= \int_{x_1}^{x_4} (\tilde{u}-k)\,\mathrm{d}x - 2\int_{x_2}^{x_4} (u-k)\Theta(k-u)\,\mathrm{d}x \\
&\geq \int_{x_1}^{x_4} (\tilde{u}-k)\,\mathrm{d}x - 2\int_{x_2}^{x_4} (\tilde{u}-k)\Theta(k-u)\,\mathrm{d}x \geq \int_{x_1}^{x_4} |\tilde{u}-k|\,\mathrm{d}x\,.
\end{aligned}
$$

Thus, the entropy does not increase due to the merge. □

The assumption of Lemma 1 implies that shocks must be reasonably well resolved before the points defining it are merged. It is satisfied if left and right of a shock points are not too far away. In the method, it can be ensured by an *entropy fix*: A merge is rejected *a posteriori* if the resolution condition is not satisfied. Then, points are inserted near the shock, and the merge is re-attempted. It remains to show in future work that with this procedure Theorem 2 still holds.

Theorem 1. *The presented particle method yields entropy solutions.*

Proof. During the characteristic movement of the points the entropy is constant, since due to Corollary 2 the interpolation is a classical solution to the conservation law. Particle insertion does not change the interpolation, thus it does not change the entropy. Merging does not increase the entropy if the conditions of Lemma 1 are satisfied. □

[2] For the case $f'' < 0$, all following inequality signs must be reversed.

6 Numerical Results

The particle method is particularly well suited for initial conditions that are composed of similarity solutions. By construction, the movement of the particles yields the exact solution as long as the solution is smooth. General initial conditions can be approximated by the interpolation (4.13). Good strategies of sampling initial conditions shall be addressed in future work. Figure 6.3 shows a smooth initial function $u_0(x) = \exp\left(-x^2\right)\cos\left(\pi x\right)$ and its time evolution under the flux function $f(u) = \frac{1}{4}u^4$. The curved shape of the interpolation is due to the nonlinearity in f'. At time $t = 0.25$ the solution (obtained by CLAWPACK using 80000 points) is still smooth, and thus represented exactly on the particles. At time $t = 8$ shocks and rarefactions have occurred and interacted. Although the numerical solution uses only a few points, it represents the true solution well.

The accuracy of the particle method is measured numerically. We consider the flux function and initial conditions as used in Fig. 6.3. For a sequence of resolutions h, the initial data are sampled, and the particle method is applied ($d_{max} = 1.9h$). Figure 6.4 shows the L^1-error to the correct solution (obtained by a computation with much higher resolution, verified with CLAWPACK). While the solution is smooth ($t = 0.25$), the method is second

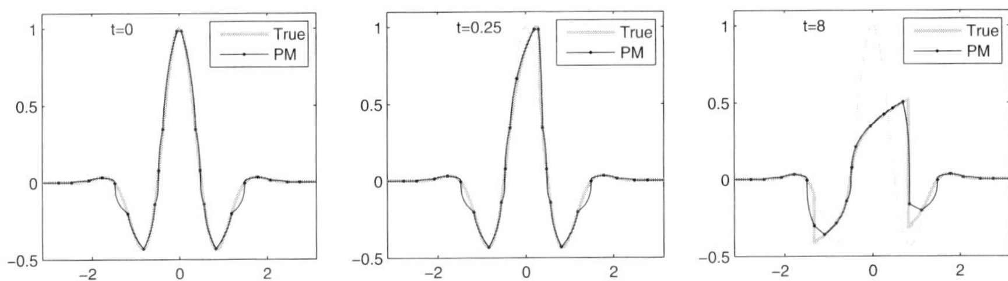

Figure 6.3. The particle method for $f(u) = \frac{1}{4}u^4$ before and after a shock.

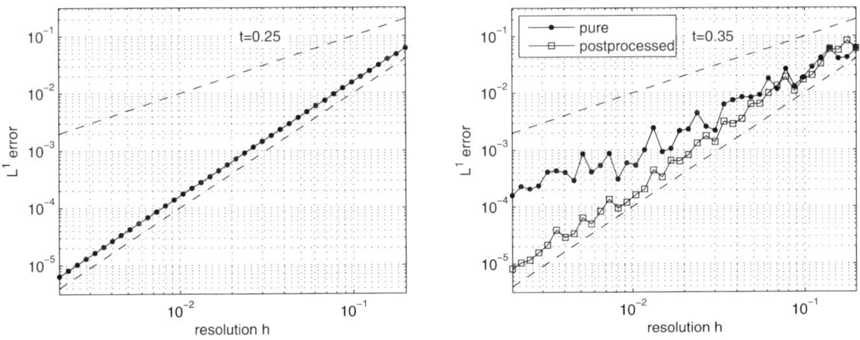

Figure 6.4. Error to the correct solution before and after a shock.

order accurate, as is sampling the initial data. After a shock has occurred ($t = 0.35$), the approximate solution (dots) becomes only first order accurate, since the shock has just been treated by particle management, thus an error of the order height×width of the shock is made. A postprocessing step (squares) can recover the second order accuracy: At merged particles, discontinuities are placed so that the total area is preserved.

7 Non-Convex Flux Functions

So far we have only considered flux functions with no inflection points (i.e. f'' has always the same sign) on the region of interest. In this section, we generalize to flux functions for which f'' has a finite number of zero crossings $u_1^* < \cdots < u_k^*$. Between two such points $u \in [u_i^*, u_{i+1}^*]$ the flux function is either convex or concave. We impose the following requirement for any set of particles: Between any two neighboring particles for which f'' has opposite sign, there must be an *inflection particle* (x, u_i^*). Thus, between two neighboring particles, f has never an inflection point, and most results from the previous sections apply. The interpolation between any two particles is uniquely defined by (4.13). It has infinite slope at the inflection points, but this is harmless. The characteristic movement of particles is the same as for flux functions without inflection points. The only complication is merging of particles when an inflection particle is involved: The standard approach, as presented in Sect. 4.1, removes two colliding points and replaces them with a point of a different function value. If an inflection particle is involved in a collision, we must merge points in a different way so that an inflection particle remains.

We present one such special merge for dealing with a single inflection point (we do not consider here the interaction of two inflection points). Also, we only consider a collision where the positions are exactly the same. Since the inflection particle must remain (although its position may change), we consider five neighboring particles and not four as before. Let $(x_i, u_i), i = 1, \ldots, 5$ be these particle so that $x_2 = x_3$, $f''(u_3) = 0$, and (WLOG) $f''' > 0$, i.e. the inflection particle is the slowest. The other cases are simple symmetries of this

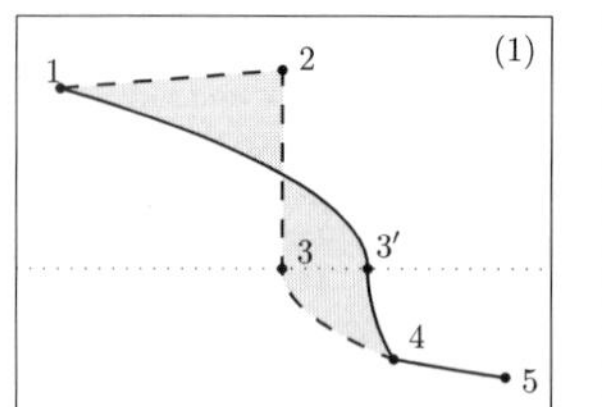

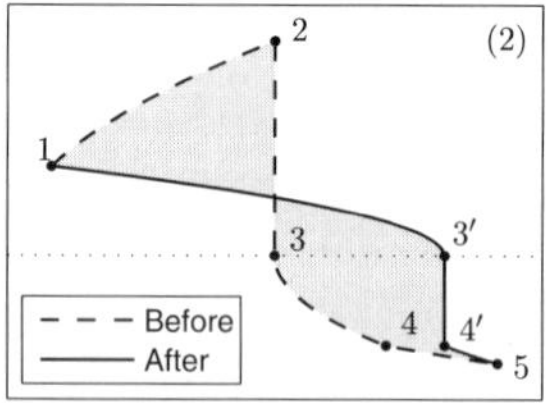

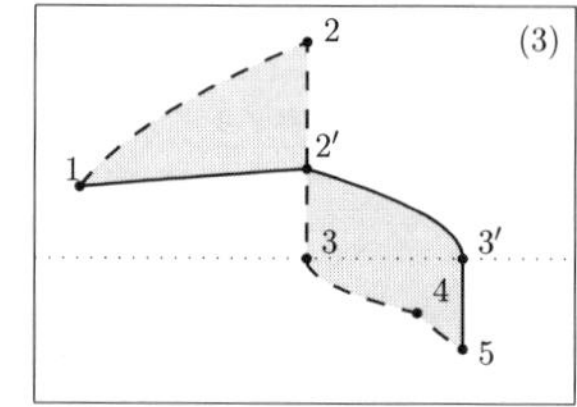

Figure 7.5. Particle management around an inflection particle ($f''(u_3) = 0$).

situation. We present three successive steps to finding the final configuration of the particles. Each next step is attempted if the previous one failed.

1. Remove particle 2 and increase x_3 to satisfy the area condition. This fails if x_3 needs to be increased beyond x_4.
2. Remove particle 2, set $x_3 = x_4$ and increase both to satisfy the area condition. This fails if x_3, x_4 need to be increased beyond x_5.
3. Remove particle 4, set $x_3 = x_5$ and find u_2 to satisfy the area condition. This cannot fail if the previous two have failed.

Formally, the resulting configuration could require another, immediate, merge (since $x_3 = x_4$ or $x_3 = x_5$). However, we need not merge these points as they move away from each other. The five point particle management guarantees that in each merging step one particle is removed, thus Theorem 2 holds.

As numerical evidence of the performance, we apply our method to the Buckley-Leverett equation (see LeVeque [10]), defined by the flux function $f(u) = \frac{u^2}{u^2+\frac{1}{2}(1-u)^2}$. It is a simple model for two-phase fluid flow in a porous medium. We consider piecewise constant initial data with a large downward jump crossing the inflection point, and a small upward jump. The large jump develops a shock at the bottom and a rarefaction at the top, the small jump is a pure rarefaction. Around $t = 0.2$, the two similarity solutions interact, thus lowering the separation point between shock and rarefaction. Figure 7.6 shows the reference solution (solid line, by CLAWPACK using 80000 points). The solution obtained by our particle method (dots) is compared to a second order CLAWPACK solution (circles) of about the same resolution. While the finite volume scheme loses the downward jump very quickly, the particle method captures the behavior of the solution almost precisely. Only directly near the shock inaccuracies are visible, which are due to the crude resolution. The solution away from the shock is nearly unaffected by the error at the shock. Note that although we impose a special treatment only at the inflection point, the switching point between shock and rarefaction is identified correctly.

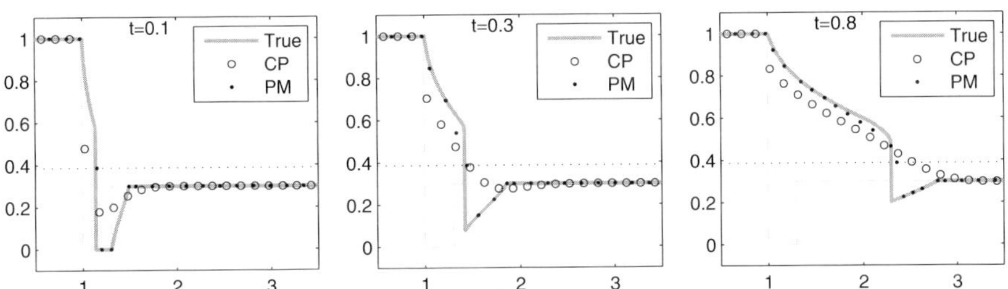

Figure 7.6. Numerical results for the Buckley-Leverett equation.

8 Conclusions and Outlook

We have presented a particle method for 1D scalar conservation laws, which is based on the characteristic equations. The associated interpolation yields an analytical solution wherever the solution is smooth. Particle management resolves the interaction of characteristics locally while conserving area. Thus, shocks are resolved without creating any numerical dissipation away from shocks. The method is TVD and entropy decreasing, and shows second-order accuracy. It deals well with non-convex flux functions, as the results for the Buckley-Leverett equation show. The particle method serves as a good alternative to fixed grid methods whenever 1D scalar conservation laws have to be solved with few degrees of freedom, but exact conservation and sharp shocks are desired. An application, which we plan to investigate in future work, is nonlinear flow in networks (e.g. traffic flow in road networks). For large networks, only a few number of unknowns can be devoted to the numerical solution on each edge. We regard the current work as a first step towards a more general particle method. Future work will focus on three main directions:

- **Source terms:** Source terms in an equation $u_t + (f(u))_x = g(x, u)$ could be handled using a fractional step method: In each time step, we would first move the particles according to $u_t + (f(u))_x = 0$ (including particle management), then change their values according to an integral formulation of $u_t = g(x, u)$. In the latter step, the constructed interpolation can be used.
- **Systems of conservation laws:** The particle method is based on similarity solutions of the conservation law. For simple systems, such as the shallow water equations in 1D, the analytical solutions to Riemann problems are known. Two complications arise in the generalization of the method:
  - To connect two general states in a hyperbolic system, intermediate states have to be included.
  - For a general system it is not clear at which velocity to move the particles.
- **Higher space dimensions:** Scalar conservation laws in higher space dimensions can be reduced to 1D problems to be solved by fractional steps. In principle, this dimensional splitting can be used with the particle method. However, remeshing would be required between the different spacial directions, thus the benefits of the meshfree approach would be lost. For the generalization to a fundamentally meshfree approach in higher space dimensions, the following problem has to be overcome: In 1D one is never truly meshfree, since the points have a natural ordering. The method uses this in the interpolation and to detect shocks. In 2D/3D shocks can occur without particles colliding, as they can move past each other. Other meshfree methods, such as FPM applied to the Euler equations, circumvent this issue by using the pressure to regulate shocks.

Acknowledgments

We would like to thank R. LeVeque for helpful comments and suggestions. The authors would like to acknowledge the support of the National Science Foundation. Y. Farjoun was partially supported by grant DMS–0703937.

References

1. A. J. Chorin, *Numerical study of slightly viscous flow*, J. Fluid Mech., 57 (1973), pp. 785–796.
2. R. Courant, E. Isaacson, and M. Rees, *On the solution of nonlinear hyperbolic differential equations by finite differences*, Comm. Pure Appl. Math., 5 (1952), pp. 243–255.
3. G. A. Dilts, *Moving least squares particles hydrodynamics I, consistency and stability*, Internat. J. Numer. Methods Engrg., 44 (1999), pp. 1115–1155.
4. L. C. Evans, *Partial differential equations*, vol. 19 of Graduate Studies in Mathematics, American Mathematical Society, 1998.
5. R. A. Gingold and J. J. Monaghan, *Smoothed particle hydrodynamics – Theory and application to nonspherical stars*, Mon. Not. R. Astron. Soc., 181 (1977), p. 375.
6. A. Harten, B. Engquist, S. Osher, and S. Chakravarthy, *Uniformly high order accurate essentially non-oscillatory schemes. III*, J. Comput. Phys., 71 (1987), pp. 231–303.
7. H. Holden, L. Holden, and R. Hegh-Krohn, *A numerical method for first order nonlinear scalar conservation laws in one dimension*, Comput. Math. Appl., 15 (1988), pp. 595–602.
8. H. Holden and N. H. Risebro, *Front Tracking for Hyperbolic Conservation Laws*, Springer, 2002.
9. J. Kuhnert and S. Tiwari, *A meshfree method for incompressible fluid flows with incorporated surface tension*, Revue Européenne des elements finis, 11 (2002), pp. 965–987.
10. R. J. Le Veque, *Finite volume methods for hyperbolic problems*, Cambridge University Press, first ed., 2002.
11. T. Liszka and J. Orkisz, *The finite difference method at arbitrary irregular grids and its application in applied mechanics*, Comput. & Structures, 11 (1980), pp. 83–95.
12. X.-D. Liu, S. Osher, and T. Chan, *Weighted essentially non-oscillatory schemes*, J. Computat. Phys., 115 (1994), pp. 200–212.
13. L. Lucy, *A numerical approach to the testing of the fission hypothesis*, Astronomical Journal, 82 (1977), pp. 1013–1024.
14. B. van Leer, *Towards the ultimate conservative difference scheme II. Monotonicity and conservation combined in a second order scheme*, J. Computat. Phys., 14 (1974), pp. 361–370.

Stability of Energy Transfer in the Weak Coupling Method

Konstantin Fackeldey[1], Rolf Krause[1], and Marc Alexander Schweitzer[1]

Institute for Numerical Simulation, University Bonn
{fackeldey, krause, schweitzer}@ins.uni-bonn.de

Summary. In this paper we are concerned with a weak coupling technique for the concurrent simulation of multiscale phenomena. In particular we focus on the construction of an initial embedding of discrete atomic data fields in an appropriate subspace $\mathcal{H}_N(\Omega) \subset L^2(\Omega)$ which provides the foundation for the proposed coupling technique in a function space setting. Thus, we must consider the regularity of the coupling information and the stability of the resulting basis.

Key words: multiscale, weak coupling, molecular dynamics, continuum mechanics

1 Introduction

In many physical phenomena, the macroscopic behavior of solids in structural dynamics is governed by effects on a microscopic scale. As a consequence an accurate representation of large scale behavior requires to capture the effects of all scales from micro to macro. On the macroscopic level a description by continuum mechanics can be used. Since those macroscopic models are usually based on a partial differential equation (PDE), they are –at least formally– incompatible with the discrete displacements on an atomistic level. On the fine scale, models involving detailed information about crystalline and defect structure, such as molecular dynamics yield satisfactory accuracy. Here the interactions are defined by inter atomic potentials. These microscopic mechanics are non-linear and strongly non-local. In contrast to continuum mechanics the description is based on a system of ODEs associated with discrete points in the Euclidean space. However, a complete fine scale description of the problem on the complete macroscopic domain is often computationally infeasible. Thus multiscale models must be employed, where different models are used simultaneously within a single simulation process. Due to the coupling of different effects on the different scales, the development of these methods is a demanding task.

The different multiscale methods vary not only in scope and the underlying assumptions, but also in their approach to broader questions such as a hierarchically and concurrent multiscale approach. In the first class, the computations are performed on each scale separately. Often, the scale coupling is done by transferring problem parameters, i.e., the results obtained on one scale determine the parameters for the computational model on another scale [1,5]. Thus for instance a continuum model can be derived from the atomic information [2]. Another approach is pursued in the concurrent coupling techniques. Here, computations on different scales are carried out simultaneously [4,16,17]. For a recent overview on multiscale techniques we refer to [14].

As afore mentioned, depending on the scale of interest, the relevant dynamics may require the use of quite different models on the respective scales.

Let us consider a displacement field $\nu \in \mathcal{H}_N(\Omega)$, where $\mathcal{H}_N(\Omega)$ is a linear subspace of $L^2(\Omega)$ and $\Omega \subset \mathbb{R}^d$, $d = 1,2,3$, is a domain. Here, the displacement field ν is determined by coarse as well as fine influences. In our multiscale context, we now aim at a decomposition of ν, which separates the high frequency components ν' of the total solution ν described by the atomic interactions from the smooth part $\bar{\nu}$ of the displacements. We therefore follow the approach of [9], where in the space $L^2(\Omega)$ the total displacement field $\nu \in \mathcal{H}_N(\Omega)$ is decomposed as

$$\nu = \bar{\nu} + \nu' \,. \tag{1.1}$$

Here, $\bar{\nu}$ is the coarse part of the total displacement and ν' refers to the fine scale displacements. Let us note that we have chosen a function space oriented setting, since this allows for more flexible decomposition approaches.

As a matter of fact, in case of molecular dynamics, the material behavior on the micro or fine scale is modeled by means of an system of isolated atoms or molecules. More precisely, on the micro scale the atoms x_α and their "discrete displacements" $\tilde{\nu}_\alpha$ form the scattered data set

$$\chi_N(\Omega) := \{(x_\alpha, \tilde{\nu}_\alpha) \,|\, \alpha = 1, ..., N,\, x_\alpha \in \Omega\,, \tilde{\nu}_\alpha \in \mathbb{R}^d\}\,. \tag{1.2}$$

We note that since $\tilde{\nu} \in \mathbb{R}^{dN}$, the atomic displacements on the fine scale cannot be interpreted as an element of $\mathcal{H}_N(\Omega)$. As a consequence, a direct sum decomposition of the underlying function space as in [9] is not possible. This is due to the fact, that the coarse scale values $\bar{\nu}$ are given as a displacement field in $L^2(\Omega)$, whereas the displacement field of the fine scale is given in terms of (1.2).

In the Bridging Scale Method [16] the decomposition given by (1.1) is the starting point for the multiscale simulation. Here, the decomposition is performed in a completely discrete setting. To do so an interpolation operator evaluating an interpolation function at the equilibrium position of the underlying atoms is used. The coarse scale is then the image of a projection from the fine scale displacement onto a displacement field.

In this paper, we pursue a different approach by performing the scale decomposition of the total displacement also for the atomic displacements in $L^2(\Omega)$. At a first glance this function space oriented decomposition seems impossible, since the displacement on the fine scale $\tilde{\nu}$ is given by (1.2). We therefore interpret the discrete displacements $\tilde{\nu} \in \mathbb{R}^{dN}$ as elements of the function space $\mathcal{H}_N(\Omega)$. This is done by means of the linear operator

$$\iota\colon \chi_N(\Omega) \to \mathcal{H}_N(\Omega) \subset L^2(\Omega)\,, \qquad \iota(\tilde{\nu}) = \nu. \tag{1.3}$$

This embedding can be chosen in a problem-dependent fashion and the properties of the resulting multiscale decomposition depend strongly on the choice of a basis $\{\phi_\alpha\}$ for $\mathcal{H}_N(\Omega)$.

For the discretization of the coarse scale we choose linear finite elements. Let $\mathcal{S}^h$ be a finite element space over Ω related to a mesh $\mathcal{T}_h$. In order to identify the coarse scale displacement $\bar{\nu}$, we employ an L^2-projection $\pi : \mathcal{H}_N(\Omega) \to \mathcal{S}^h$. This projection is designed to extract from the total displacement ν the coarse part, which is assumed to be in the finite element space $\mathcal{S}^h$. More precisely, for any total displacement $w \in \mathcal{H}_N(\Omega)$, the corresponding coarse scale displacement is given by $\pi(w) \in \mathcal{S}^h$ which satisfies

$$(\pi(w), \mu)_{L^2(\Omega)} = (w, \mu)_{L^2(\Omega)} \qquad \forall \mu \in M^h, \tag{1.4}$$

where M^h denotes a suitable multiplier space. In order to compute the algebraic representation of π in (1.4), we need to assemble two (generalized) mass matrices. For the first matrix, we need to evaluate integrals of the form $\int_\Omega \mu_p \lambda_q \, dx$, where μ_p are the basis functions spanning the multiplier space M^h, and λ_q are the basis functions of $\mathcal{S}^h$. Here, p, q are assumed to be in some index set $\mathcal{N}^h$ with $|\mathcal{N}_h| = \dim(\mathcal{S}^h)$. The computation of the resulting mass matrix can be done in a similar fashion as the assembly of the standard mass matrix. For the second matrix, we need to evaluate integrals of the form

$$\int_\Omega \mu_p \phi_\alpha \, dx \tag{1.5}$$

where the ϕ_α are the basis functions for the space $\mathcal{H}_N(\Omega)$, i.e. $\mathcal{H}_N(\Omega) = \operatorname{span}\langle\phi_\alpha\rangle$. In order to compute these integrals the intersection between the support of μ_p and the support of ϕ_α has to be computed. For details concerning an efficient way for computing these intersections, we refer to [7].

Summing up the weak coupling concept involves several steps: In the first step, an approximation of the fine scale displacement embeds the discrete values into a function space. In the second step we perform an L^2-like projection separating the coarse from the fine scale. In case the molecular dynamic simulation is restricted to a subset of Ω, in a third and final step, the resulting low-frequency contributions have to be extended to the whole computational domain Ω. For further details we refer to [6].

In the forthcoming we examine the first step of our transfer scheme, namely the operator ι. A possible choice for ι could be $\iota(\tilde{\nu}) = \sum_\alpha \tilde{\nu}_\alpha \phi_\alpha$. For this choice, we need to specify "suitable" basis functions ϕ_α. In the context of multiscale methods the term suitable refers to the kind of information which we are interested in. More precisely we can expect that for the transfer of gradient based information from the fine to the coarse scale a basis which can only reproduce constants exactly is insufficient. Thus, for the construction of these basis functions, we will employ techniques from scattered data approximation which allow for a flexible choice of the basis functions ϕ_α.

One class of shape function used for scattered data interpolation are the radial basis functions [10]. Another approach to construct shape functions ϕ_α from the collection of particles $\chi_N(\Omega)$ is based in the moving least squares technique (MLS) that we shortly summarize in the next section, see e.g. [8] for details.

2 Moving Least Squares Method

Here, we construct the operator ι from (1.3) on the basis of the moving least squares approach which originated in scattered data approximation. We suppose, that the scattered data set $\chi_N(\Omega)$ in (1.2) is given. Our aim is to find a function $u : \bar{\Omega} \to \mathbb{R}$, such that

$$u(x_\alpha) \approx \tilde{\nu}_\alpha \text{ for all } \alpha = 1, ..., N. \tag{2.6}$$

In order to construct a moving least squares (MLS) fit, we consider the approximation space being the space $\mathbb{P}_m$ of polynomials with the basis $\{P_i\}_{i=1}^N$ of degree m in d variables [1] and a set of non-negative weight functions

$$W_\alpha : \mathbb{R}^d \to \mathbb{R}_0^+ \text{ with } \operatorname{supp}(W_\alpha) =: \omega_\alpha,$$

and the dilatation parameter h_α of $W_\alpha(x) = W\left(\frac{x-x_\alpha}{h_\alpha}\right)$. Now, we minimize for each x the quadratic functional

$$J(\tau)(x) = \sum_{\alpha=1}^{N} W_\alpha(x)(\nu_\alpha - \tau(x_\alpha))^2 \tag{2.7}$$

over all $\tau \in \mathbb{P}_m$.

In order to minimize (2.7), we set the derivative of (2.7) equal to zero and obtain the system of equations

$$\sum_{\alpha=1}^{N} W_\alpha(x)\tilde{\nu}_\alpha P_j(x_\alpha) = \sum_{\alpha=1}^{N} W_\alpha(x) \sum_{i=1}^{n} P_i(x_\alpha)P_j(x_\alpha)c(x) \qquad j = 1, ..., n. \tag{2.8}$$

[1] The approximation space can be generalized to an abstract approximation space $V(\Omega)$. Note however, that we then obtain reproduction of $V(\Omega)$ by ϕ_α of (2.11).

With the definitions

$$\begin{aligned} P(x) &:= [P_1(x)\ P_2(x)\ \cdots\ P_N(x)]^T \\ W(x) &:= [W_1(x)\ W_2(x)\ \cdots\ W_N(x)]^T \\ B &:= (B_{ij})_{i,j=1,\dots,N},\ B_{ij} = W_i(x)P_j(x) \\ f &:= [\tilde{\nu}_1\ \tilde{\nu}_2\ \cdots\ \tilde{\nu}_N]^T \\ A(x) &:= (A_{ij})_{i,j=1,\dots,n},\ A_{ij} = \sum_{\alpha=1}^{N} P_i(x_\alpha)W_\alpha(x)P_j(x_\alpha) \\ c(x) &:= [c_1(x)\ c_2(x)\ \dots\ c_n(x)]^T, \end{aligned}$$

equation (2.8) can be written as

$$A(x)c(x) = B(x)f. \tag{2.9}$$

The above matrix $A(x)$ is also known as Gram's matrix. The minimizer $u(x)$ f (2.7) is given by the linear combination

$$u(x) = \sum_{\alpha=1}^{N} \tilde{\nu}_\alpha \phi_\alpha(x) \tag{2.10}$$

where the shape functions ϕ_α satisfy

$$\phi_\alpha(x) = P^T(x_\alpha)[A(x)]^{-1}W_\alpha(x)P(x_\alpha). \tag{2.11}$$

Properties of the Gram-Matrix

Note that (2.11) involves the inverse of the Gram matrix $A(x)$ for each point of evaluation. Thus, we must be concerned with the regularity of $A(x)$ for all $x \in \Omega$. Here, we attain the positive definiteness of $A(x)$ for all $x \in \Omega$ from the $\mathbb{P}_m$-unisolvence of the sets $\chi_N(\Omega) \cap \omega_\alpha$ for all α.

Weight Functions and Scaling

The size of the support of the weight functions W_α, i.e. of the shape functions ϕ_α can be determined by

$$\omega_\alpha = \{y \in \mathbb{R}^d \mid \|x_\alpha - y\| < h_\alpha\}$$

where the dilatation parameter h_α can in principle be chosen individually for each data site x_α. However, this choice is closely related to the accuracy and stability of the approximation and thus crucial for the stability of the projection operator π. Recall that the $\mathbb{P}_m$-unisolvence of $\chi_N(\Omega) \cap \omega_\alpha$ for all α must be ensured. Note also that the smoothness of the approximation depends on the smoothness of the weight function, i.e. if $W_\alpha \in C^r(\Omega)$ then $\phi_\alpha \in C^r(\Omega)$.

Reproduction Properties

From (2.10) with $\nu_\alpha = u(x_\alpha)$ for $u \in \mathbb{P}_m$ and (2.11) it is clear that $\mathbb{P}_m \subset \operatorname{span}\langle\phi_\alpha\rangle$, thus reproduction of polynomials of order m in MLS is guaranteed.

Partition of Unity and Shepard's Approach

We denote $\{\phi_\alpha\}$ as a partition of unity of order q if the reproducing property

$$\sum_{\alpha=1}^{N} \phi_\alpha(x) b(x_\alpha) = b(x)$$

and the derivative reproducing conditions

$$\sum_{\alpha=1}^{N} D^s \phi_\alpha(x) b(x_\alpha) = D^s b(x), \quad |s| \le q$$

hold for all $b \in \mathbb{P}_m$. In the case of $m = 0$, the approximation space is given by $\mathbb{P}_m = \{\mathbf{1}\}$ and the Gram matrix reduces to

$$A(x) = \sum_{\alpha=1}^{N} W_\alpha(x).$$

Thus the shape functions are given by

$$\phi_\beta(x) = \frac{W_\beta(x)}{\sum_{\alpha=1}^{N} W_\alpha(x)} = W(x) \cdot (A(x))^{-1}$$

which is also known as Shepard's method. One can thus easily verify, that

$$0 < \phi_\beta(x) < 1 \ \text{ and } \ \sum_{\beta=1}^{N} \phi_\beta(x) = 1 \ \forall x \in \Omega.$$

The Shepard partition of unity is an efficient method for the approximation of scattered data, since the Gram matrix reduces to a scalar, and thus an explicit form of ϕ_α is given. As a drawback, the type of information which we can transfer from a coarse to a fine scale is confined to displacements. For gradient based information a higher order MLS method has to be applied, which requires the implicit representation (2.11).

In general the MLS approximant is non interpolating, i.e. they do not satisfy the Kronecker delta property. However an interpolating approximant can be constructed by using singular weighting functions at all nodes [11].

For the deduction of the MLS shape function we can use different starting points like the minimization of a weighted least-squares functional, or a Taylor-Series expansion, or the direct imposition of the reproducing conditions. There are also other techniques, which produce a partition of unity like e.g. the Reproducing Kernel Particle Methods (RKPM) [3, 12, 13]. Even though that the RKPM and the MLS have different origins their equivalence can be shown. We want to employ the MLS functions as the basis for our space $\mathcal{H}_N(\Omega)$, i.e. for the range of the embedding. Our construction is essentially L^2 based and so Shepard's method should be sufficient to obtain at least first order in L^2. If we also need to bound the error in H^1 then MLS of first order should be employed.

3 Numerical Experiment

To confirm this assertion, we consider the idealized but representative reference scattered data approximation problem (2.6) via the minimization of (2.7) for the data $f_\alpha = u(x_\alpha)$ where $u(x) = x^2$. We compare the results obtained via the MLS approach for the point set $[-3, 3]$ with $h = 1$ using the approximation spaces $\mathbb{P}_m$ with $m = 0, 1$. Here, we anticipate to find an asymptotic convergence behavior of $O(h)$ in the L^2-norm for $m = 0$ and $O(h^2)$ for $m = 1$. Furthermore, the approximation error will stagnate with respect to the H^1-norm for the Shepard functions with $m = 0$ whereas the MLS shape functions with $m = 1$ will provide an $O(h)$ convergence also in H^1. This expected convergence behavior can be clearly observed from Figure 3.2.

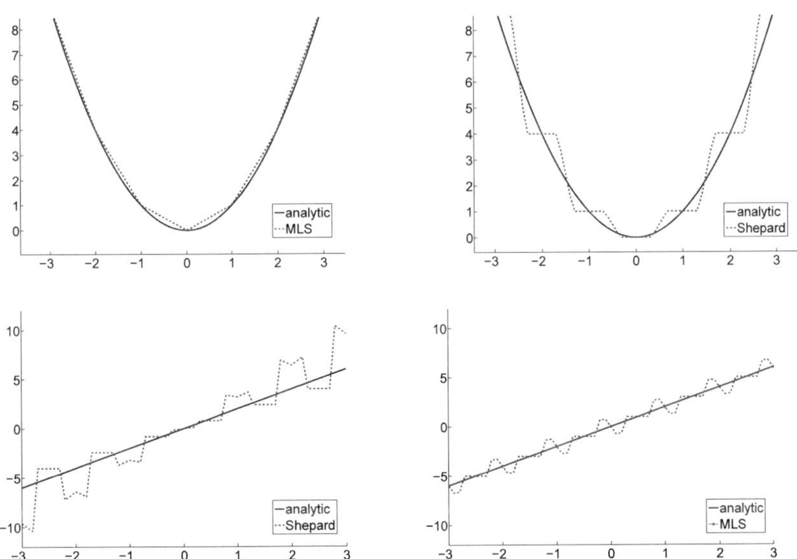

Figure 3.1. Approximation (top row) by Shepard's Method (left) and MLS (right) and the respective derivatives (bottom row). (*See also* Color Plate on page 367)

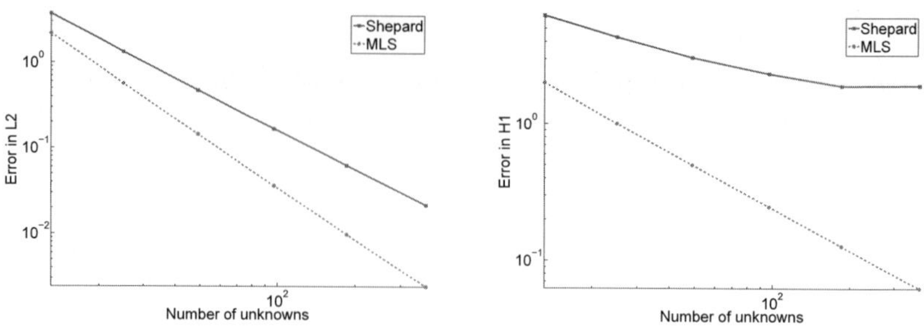

Figure 3.2. Error in the L^2-norm (left) and H^1-norm (right) of Shepard's method (solid) and the MLS (dashed). (*See also* Color Plate on page 367)

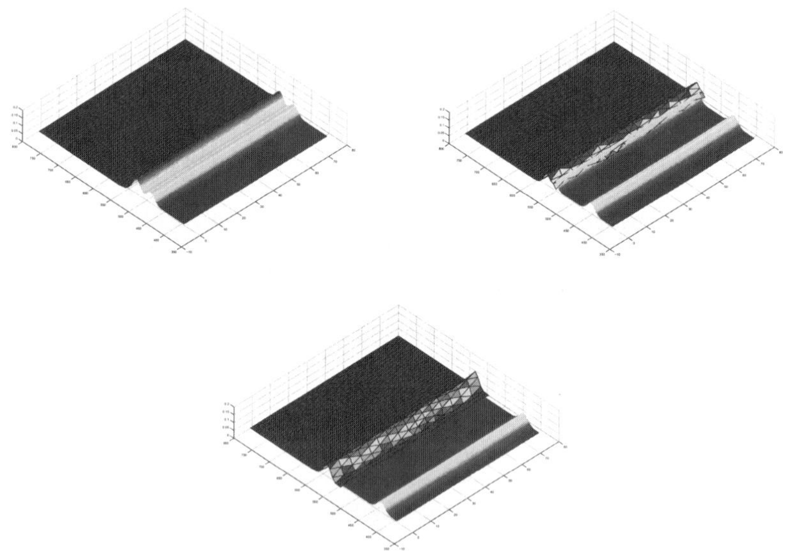

Figure 3.3. Weak scale transfer (2*d*) based on (1.4) using Shepard's approach ($m = 0$). (*See also* Color Plate on page 368)

Thus, the construction of a weak coupling operator aimed at transferring function values may be based on the Shepard functions (if the error bound of $O(h)$ where h is related to the maximal atomic distance is acceptable), compare Figure 3.1. However, if the transfer of gradient information is required (or the jump in the resolution between coarse function space $\mathcal{S}_h$ and $\mathcal{H}_N(\Omega)$ is too small) the use of higher order moving least squares functions is necessary.

References

1. M. Arndt and M. Griebel, Derivation of higher order gradient continuum models from atomistic models for crystaline solids, *Multiscale Mod. Sim.* **4(2)** (2005), 531-562.
2. X. Blanc, C. Le Bris and P.-L. Lions, From Molecular Models to Continuum Models, *Arch. Rat. Mech. Anal.*, **164 (4)** (2002), pp. 341–381.
3. J.S. Chen and C. Pan and C.I. Wu, Large Defomation Analysis of Rubber based on a Reproducing Kernel Particle Method, *Comp. Mech.*,**19** (1997), pp. 211 227
4. W. A. Curtin and R. E. Miller, Atomistic/Continuum Coupling in Computational Materials Science, *Mod. and Sim. Mat. Sc. and Engrg.*, **11** (2003), pp. R33–R68.
5. W. E and B. Engquist, Multiscale modeling and computation, *Notices Am. Math. Soc.* **50(9)** (2003), 1062-1070.
6. K. Fackeldey and R. Krause, Multiscale Coupling in Function Space - Weak Coupling between Molecular Dynamics and Continuum Mechanics, *Sonderforschungsbereich 611, Preprint* no. 350, 2007.
7. K. Fackeldey and D. Krause and R. Krause, Implementational Aspects of the Weak Coupling Method, *in preparation.*
8. G. Fasshauer, Meshfree approximation methods with Matlab, *Interdisciplinary Mathematical Sciences 6. Hackensack, NJ: World Scientific, 2007.*
9. T.J.R. Hughes, G. Feijoo, L. Mazzei and J.-B. Quincy, The variational multiscale method: a paradigm for computational mechanics, *Comput. Meth. Appl. Mech. Engrg.*, **166** (1998), pp. 3-24.
10. A. Iske, Multiresolution Methods in Scattered Data Modelling *Lecture Notes in Computational Science and Engineering, Vol. 37, Springer, 2004.*
11. P. Lancaster, K. Salkauskas, Surfaces generated by Moving Least Squares Methods, *Math. Comp.*, **37** (1981), pp. 141-158
12. W.K. Liu and S. Jun and Y.F. Zhang, Reproducing Kernel Particle Methods, *Int. J. Numer. Meth. Fluids* **20** (1995), pp. 1081-1106
13. W.K. Liu and S. Jun and S. Li and J. Adee and T. Belytschko, Reproducing Kernel Particle Methods for Strucural mechanics, *Int. J. Numer. Meth. Engrg.* **38** (1995), pp. 1655-1679.
14. R. E. Rudd and J. Q. Broughton, Coarse-Grained Molecular Dynamics and the Atomic Limit of Finite Elements, *Physical Review B*, **58** (1998), pp. R5893–R5896.
15. D. Shepard, A two-dimensional function for irregularly spaced data, *ACM National Conference*, (1968), pp. 517-524.
16. G. J. Wagner and W. K. Liu, Coupling of Atomistic and Continuum Simulations using a Bridging Scale Decomposition, *J. Comput. Phys.*, 190, 249-274 (2003).
17. S. P. Xiao and T. Belytschko, A Bridging Domain Method for Coupling Continua with Molecular Dynamics, *Comput. Meth. Appl. Mech. Engrg.*, 193, 1645-1669 (2004).

Multiscale Approach for Quantum Systems

Wei Hu and Jiun-Shyan Chen

Department of Civil & Environmental Engineering
University of California, Los Angeles
Los Angeles, CA 90095, USA
weih@seas.ucla.edu, jschen@seas.ucla.edu

Summary. In this work we propose an iterative multiscale scheme for eigen solutions of the Schrödinger equation with application to the quantum dot array (QDA) systems. The asymptotic expansion predictor and inverse iteration and Rayleigh quotient corrector are introduced. The predictor multiscale formulation is constructed by introducing an asymptotic approach of the original problem and an auxiliary problem. The predictor multiscale solution is corrected by the inverse iteration and the Rayleigh quotient iteration. The numerical results show that the multiscale formulation offers comparable accuracy compared to the solution of the single fine scale model with substantial CPU-time reduction for the tested QDA systems. With the additional corrections by the inverse iteration and Rayleigh quotient, the solution accuracy can be further enhanced.

Key words: Asymptotic expansion, Multiscale, Quantum dot array

1 Introduction

Understanding the electron density distribution is crucial to the design of a semiconductor device composed of a large number of quantum dots such as quantum dot laser generator. Traditional single-scale ab initio calculation is very time consuming especially in the cases where multiscale quantum features are involved, such as rapidly oscillating confinement potential fields.

The Schrödinger equation for quantum dot array represents an eigenvalue problem with oscillating coefficients. Bensoussan [1] first introduced the mathematical foundation of the asymptotic analysis for periodic structures and the application for eigenvalue problems involving various types of differential operators associated with different physical problems. Kesavan [6] [7] used a two-scale asymptotic expansion in homogenization of elliptic eigenvalue problems. Paine [8] computed a sequence of eigenvalues of a continuous eigenvalue problem with general boundary conditions based on an asymptotic expansion algorithm, which is extended to discrete eigenvalue problems by Hoog and Anderssen [4].

This work intends to introduce the asymptotic expansion based iterative multiscale method for solving the Schrödinger equations of quantum dot array systems. It is shown that the asymptotic method achieves comparable solution accuracy with significant CPU time saving compared to the singlescale method. We consider using the solution of the asymptotic expansion as the predictor, and enhancing the high scale solution by an inverse iteration and Rayleigh quotient iteration.

This paper is organized as follows. In Section 2, an asymptotic expansion method for the multiscale Schrödinger equation is introduced. The iterative scheme for enhancing fine scale eigen solutions is presented in Section 3. Numerical examples are presented in Section 4 to demonstrate the effectiveness of the proposed method. Concluding remarks are given in Section 5.

2 Multiscale Decomposition for the Schrödinger Equation

Consider the following eigenvalue problem of the Schrödinger equation for quantum dot array (QDA):

$$H\Theta = \varepsilon\Theta \quad \text{in} \quad \Omega, \tag{2.1}$$

$$\Theta = 0 \quad \text{on} \quad \Gamma, \tag{2.2}$$

$$\frac{1}{m^+}\frac{\partial\Theta}{\partial n^+} + \frac{1}{m^-}\frac{\partial\Theta}{\partial n^-} = 0 \quad \text{on} \quad \Gamma_I, \tag{2.3}$$

where Θ is the wave function, ε is the energy function, m is the effective mass, superscripts " $+$ " and " $-$ " denote quantities on the opposite sides of the interface Γ_I in the QDA, V is the confinement potential, and the normalized Hamiltonian differential operator is

$$H = -\frac{\partial}{\partial x_i}\left(\frac{1}{m}\frac{\partial}{\partial x_i}\right) + V. \tag{2.4}$$

In the Hamiltonian operator of Eq. (4), the spatial variation of the effective mass m and confinement potential V exists the fine scale features in QDA as shown in Figure 2.1[1]. We consider a two-scale description of the problem, and introduce $\mathbf{x}$ and $\mathbf{y}$ as the coarse scale and fine scale coordinate systems, respectively. The two coordinate systems are related by a small asymptotic parameter λ as:

$$y_i = \frac{x_i}{\lambda} \quad i = 1, 2, 3. \tag{2.5}$$

A point $\mathbf{x}$ in the coarse scale domain Ω is associated with a unit cell of domain $\Omega_{\mathbf{y}}$. We denote the wave function by Θ and the energy function by ε, which

[1] http://www.physik.tu-berlin.de/IFP/richter/new/research/lowdim.shtml

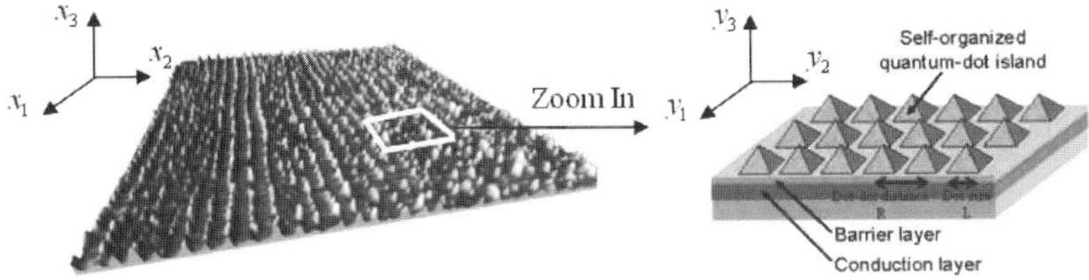

Figure 2.1. A two-scale model of QDA with features in different length scales. (*See also* Color Plate on page 368)

possess coarse and fine scale properties with superscript λ. Therefore, we write the eigenvalue problem as

$$H_\lambda \Theta^\lambda = \varepsilon^\lambda \Theta^\lambda \quad \text{in} \quad \Omega, \tag{2.6}$$

$$\Theta^\lambda = 0 \quad \text{on} \quad \Gamma, \tag{2.7}$$

$$\frac{1}{m^+}\frac{\partial \Theta^\lambda}{\partial n^+} + \frac{1}{m^-}\frac{\partial \Theta^\lambda}{\partial n^-} = 0 \;\; \text{on} \;\; \Gamma_I, \tag{2.8}$$

where

$$H_\lambda = -\frac{\partial}{\partial x_i}\left(\frac{1}{m\left(\mathbf{x},\mathbf{y}\right)}\frac{\partial}{\partial x_i}\right) + \lambda^{-1}V\left(\mathbf{x},\mathbf{y}\right). \tag{2.9}$$

The weak form of the above problem is to find the eigen pair of energy and wave function $\{\varepsilon^\lambda, \Theta^\lambda\} \in \mathrm{R} \times H_0^1(\Omega)$ such that $\forall v \in H_0^1(\Omega)$, we have

$$a_\lambda\left(v,\Theta^\lambda\right) = \varepsilon^\lambda\left(v,\Theta^\lambda\right), \tag{2.10}$$

where

$$a_\lambda\left(v,\Theta^\lambda\right) = \int_\Omega \frac{1}{m}\frac{\partial v}{\partial x_i}\frac{\partial \Theta^\lambda}{\partial x_i} d\Omega + \int_\Omega v\lambda^{-1}V\Theta^\lambda d\Omega, \tag{2.11}$$

$$\left(v,\Theta^\lambda\right) = \int_\Omega v\Theta^\lambda d\Omega. \tag{2.12}$$

Note that a boundary integral term associated with the interface condition has been dropped due to the interface condition (8) in the strong form. The total solution is expressed by an asymptotic expansion as:

$$\Theta^\lambda\left(\mathbf{x},\mathbf{y}\right) = \Theta^{[0]}\left(\mathbf{x}\right) + \lambda\Theta^{[1]}\left(\mathbf{x},\mathbf{y}\right) + O\left(\lambda^2\right), \tag{2.13}$$

$$\varepsilon^\lambda = \varepsilon^{[0]} + \lambda\varepsilon^{[1]} + O\left(\lambda^2\right). \tag{2.14}$$

If a coarse scale problem exists, $\Theta^{[0]}$ and $\varepsilon^{[0]}$ are assumed to satisfy the following coarse scale problem:

$$H_0\Theta^{[0]} = \varepsilon^{[0]}\Theta^{[0]} \quad \text{in} \quad \Omega, \tag{2.15}$$

$$\Theta^{[0]} = 0 \quad \text{on} \quad \Gamma, \tag{2.16}$$

$$\frac{1}{m_{ij}^{[0]+}}\frac{\partial \Theta^{[0]}}{\partial x_i}n_j^+ + \frac{1}{m_{ij}^{[0]-}}\frac{\partial \Theta^{[0]}}{\partial x_i}n_j^- = 0 \;\; \text{on} \;\; \Gamma_I^{[0]}, \tag{2.17}$$

where

$$H_0 = -\frac{\partial}{\partial x_i}\left(\frac{1}{m_{ij}^{[0]}}\frac{\partial}{\partial x_j}\right) + V^{[0]}. \tag{2.18}$$

Here, $m_{ij}^{[0]}$ and $V^{[0]}$ are the homogenized counterparts of m and V, respectively, that are yet to be obtained. A tensor form $m_{ij}^{[0]}$ is used to reflect the fact that the homogenized effective mass of m could have directional properties due to the microstructural effects, and $m_{ij}^{[0]} = m_{ji}^{[0]}$. $\Gamma_I^{[0]}$ is the material interface associated with heterojunctions at the coarse scale level resulting from the heterogeneous QDA microstructures. The weak form associated with Eqns. (15)-(17) is to find a coarse-scale eigen pair of energy and wave function $\{\varepsilon^{[0]}, \Theta^{[0]}\} \in \mathrm{R} \times H_0^1(\Omega)$ such that $\forall v \in H_0^1(\Omega)$, we have

$$a_0\left(v, \Theta^{[0]}\right) = \varepsilon^{[0]}\left(v, \Theta^{[0]}\right), \tag{2.19}$$

where

$$a_0\left(v, \Theta^{[0]}\right) = \int_\Omega \frac{1}{m_{ij}^{[0]}}\frac{\partial v}{\partial x_i}\frac{\partial \Theta^{[0]}}{\partial x_j} d\Omega + \int_\Omega v V^{[0]} \Theta^{[0]} d\Omega, \tag{2.20}$$

$$\left(v, \Theta^{[0]}\right) = \int_\Omega v \Theta^{[0]} d\Omega. \tag{2.21}$$

To construct the coarse scale eigenvalue problem from the fine scale eigenvalue problem, we define an auxiliary problem as follows

$$H_\lambda w^\lambda = \varepsilon^{[0]}\Theta^{[0]} \quad \text{in} \quad \Omega, \tag{2.22}$$

$$w^\lambda = 0 \quad \text{on} \quad \Gamma, \tag{2.23}$$

$$\frac{1}{m^+}\frac{\partial w^\lambda}{\partial n^+} + \frac{1}{m^-}\frac{\partial w^\lambda}{\partial n^-} = 0 \quad \text{on} \quad \Gamma_I, \tag{2.24}$$

where w^λ is an auxiliary function expressed as

$$w^\lambda(\mathbf{x}, \mathbf{y}) = \Theta^{[0]}(\mathbf{x}) + \lambda w^{[1]}(\mathbf{x}, \mathbf{y}) + O\left(\lambda^2\right). \tag{2.25}$$

As λ approaches zero, the auxiliary problem approaches the coarse scale problem, and $w^{[0]}$ approaches $\Theta^{[0]}$. The weak form of the auxiliary problem is to find an auxiliary function $w^\lambda \in H_0^1(\Omega)$ such that $\forall v \in H_0^1(\Omega)$, we have

$$a_\lambda\left(v, w^\lambda\right) = \varepsilon^{[0]}\left(v, \Theta^{[0]}\right). \tag{2.26}$$

Since the weight function v in both Eqs. (2.10) and (2.26) can be arbitrary, one can introduce the auxiliary function w^λ and the wave function Θ^λ as the weight functions in Eqs. (2.10) and (2.26), respectively, and obtain the following condition using the symmetric property of $a_\lambda(.,.)$:

$$\varepsilon^{\lambda}\left(w^{\lambda},\Theta^{\lambda}\right)=\varepsilon^{[0]}\left(\Theta^{\lambda},\Theta^{[0]}\right). \tag{2.27}$$

Equation (2.27) plays an important role in this multiscale solution process. We further consider normalization of eigen function as

$$\left(\Theta^{\lambda},\Theta^{\lambda}\right)=1. \tag{2.28}$$

Introducing the asymptotic expansion (2.13) into (2.28) leads to

$$\left(\Theta^{[0]},\Theta^{[0]}\right)-1, \tag{2.29}$$

$$\left(\Theta^{[0]},\Theta^{[1]}\right)=0. \tag{2.30}$$

Subsitituting Eqs. (2.13), (2.14), (2.25), (2.29) and (30) into Eq. (2.27), the leading order equation associated with $O\left(\lambda\right)$ is

$$\varepsilon^{[1]}=-\varepsilon^{[0]}\left(\Theta^{[0]},w^{[1]}\right). \tag{2.31}$$

Equation (2.31) shows that the fine scale energy $\varepsilon^{[1]}$ can be calculated from the coarse scale eigen pair $\{\varepsilon^{[0]},\Theta^{[0]}\}$ and the fine scale auxiliary function $w^{[1]}$. Therefore, two tasks are to be performed:

1. Obtain the effective mass $m^{[0]}$ and the confinement potential $V^{[0]}$ for solving Eqns. (15)-(17) for the coarse scale eigen pair $\{\varepsilon^{[0]},\Theta^{[0]}\}$.
2. Solve for the fine scale auxiliary function $w^{[1]}$.

To start, introduce the following chain rule:

$$\frac{\partial\left(.\right)}{\partial x_i}=\left.\frac{\partial\left(.\right)}{\partial x_i}\right|_{\mathbf{y}}+\left(\left.\frac{\partial\left(.\right)}{\partial y_i}\right|_{\mathbf{x}}\right)\left(\frac{\partial y_i}{\partial x_i}\right)=\left.\frac{\partial\left(.\right)}{\partial x_i}\right|_{\mathbf{y}}+\lambda^{-1}\left.\frac{\partial\left(.\right)}{\partial y_i}\right|_{\mathbf{x}}. \tag{2.32}$$

By using the chain rule in Eq. (2.32), the total Hamiltonian operator H_{λ} is expressed as:

$$H_{\lambda}\left(.\right)=\lambda^{-2}H_{-2}\left(.\right)+\lambda^{-1}H_{-1}\left(.\right)+\lambda^{0}H_{0}\left(.\right)+O\left(\lambda\right), \tag{2.33}$$

where

$$H_{-2}\left(.\right)=-\frac{\partial}{\partial y_i}\left(\frac{1}{m\left(\mathbf{x},\mathbf{y}\right)}\frac{\partial\left(.\right)}{\partial y_i}\right), \tag{2.34}$$

$$H_{-1}\left(.\right)=-\frac{1}{m\left(\mathbf{x},\mathbf{y}\right)}\frac{\partial^2\left(.\right)}{\partial x_i\partial y_i}-\frac{\partial}{\partial y_i}\left(\frac{1}{m\left(\mathbf{x},\mathbf{y}\right)}\frac{\partial\left(.\right)}{\partial x_i}\right)+V\left(\mathbf{x},\mathbf{y}\right), \tag{2.35}$$

$$H_{0}\left(.\right)=-\frac{1}{m\left(\mathbf{x},\mathbf{y}\right)}\frac{\partial^2\left(.\right)}{\partial x_i\partial x_i}. \tag{2.36}$$

We introduce the following scale coupling for the auxiliary function [1]

$$w^{[1]}(\mathbf{x},\mathbf{y}) = \alpha_j(y_j)\frac{\partial \Theta^{[0]}(\mathbf{x})}{\partial x_j} + \beta(\mathbf{y})\Theta^{[0]}(\mathbf{x}), \tag{2.37}$$

where α and β are scale coupling functions to be computed. In order to preserve symmetry of the coarse scale Hamiltonian operator H_0, the separation of variables in Eq. (2.37) is modified and it differs from the traditional asymptotic expansion approach [1] by setting the j-th coupling function α_j to be dependent on y_j instead of $\mathbf{y}$.

Substituting Eqs. (2.33) and (2.37) into the auxiliary problem in (2.22)-(24) results in a set of leading order equations of λ. First consider those associated with $O(\lambda^{-1})$:

$$H_{-2}\alpha_j(y_j) = -H_{-2}\,y_j, \tag{2.38}$$

$$H_{-2}\beta(\mathbf{y}) = -V. \tag{2.39}$$

The auxiliary function w^λ is required to satisfy the interface condition on Γ_I:

$$\frac{1}{m^+}\frac{\partial w^\lambda}{\partial n^+} + \frac{1}{m^-}\frac{\partial w^\lambda}{\partial n^-} = 0. \tag{2.40}$$

By using the asymptotic expansion of w^λ in Eq. (2.25) and the separation of variable of $w^{[1]}$ in Eq. (2.37), equation (2.40) yields

$$\frac{1}{m^+}\frac{\partial \alpha_j}{\partial n^+} + \frac{1}{m^-}\frac{\partial \alpha_j}{\partial n^-} = 0, \tag{2.41}$$

$$\frac{1}{m^+}\frac{\partial \beta}{\partial n^+} + \frac{1}{m^-}\frac{\partial \beta}{\partial n^-} = 0. \tag{2.42}$$

The associated weak form is to find the coupling functions $\alpha_j, \beta \in H^1(\Omega_\mathbf{y})$ such that $\forall v_1, v_2 \in H^1(\Omega_\mathbf{y})$

$$a_\lambda(v_1,\alpha_j) = -a_\lambda(v_1, y_j), \tag{2.43}$$

$$a_\lambda(v_2,\beta) = -(v_2, V). \tag{2.44}$$

Next, a coarse scale equation is to be established for computing $\Theta^{[0]}$ and $\varepsilon^{[0]}$. The weight function v^λ is expressed as

$$v^\lambda(\mathbf{x},\mathbf{y}) = v^{[0]}(\mathbf{x}) + \lambda v^{[1]}(\mathbf{x},\mathbf{y}) + O(\lambda^2), \tag{2.45}$$

and a scale coupling relation is used

$$v^{[1]}(\mathbf{x},\mathbf{y}) = \alpha_j(y_j)\frac{\partial v^{[0]}(\mathbf{x})}{\partial x_j} + \beta(\mathbf{y})v^{[0]}(\mathbf{x}). \tag{2.46}$$

Introducing the asymptotic expansion and scale coupling equation of test and trial functions in Eqns. (2.37) and (2.46), performing an integration over the associated unit cell in the weak form (2.26), and letting $\lambda \to 0$, we obtain

$$\int_{\Omega} \frac{\partial v^{[0]}}{\partial x_i} \left[\int_{\Omega_{\mathbf{y}}} \frac{\partial (\alpha_i + y_i)}{\partial y_k} \frac{1}{m(\mathbf{x}, \mathbf{y})} \frac{\partial (\alpha_j + y_j)}{\partial y_k} d\Omega_{\mathbf{y}} \right] \frac{\partial \Theta^{[0]}}{\partial x_j} d\Omega$$

$$+ \int_{\Omega} \left[\int_{\Omega_{\mathbf{y}}} \beta V(\mathbf{x}, \mathbf{y}) \, d\Omega_{\mathbf{y}} \right] v^{[0]} \Theta^{[0]} d\Omega = A_{\Omega_{\mathbf{y}}} \varepsilon^{[0]} \int_{\Omega} \left(\Theta^{[0]} v^{[0]} \right) d\Omega, \quad (2.47)$$

where $\Omega_{\mathbf{y}}$ denotes the unit cell domain, and $A_{\Omega_{\mathbf{y}}} = \int_{\Omega_{\mathbf{y}}} d\Omega_{\mathbf{y}}$. To arrive the above condensed form, we also have used conditions

$$\int_{\Omega_{\mathbf{y}}} \left[\frac{1}{m(\mathbf{x}, \mathbf{y})} \frac{\partial (\alpha_i + y_i)}{\partial y_j} \frac{\partial \beta}{\partial y_j} \right] d\Omega_{\mathbf{y}} = 0, \quad (2.48)$$

$$\int_{\Omega_{\mathbf{y}}} \left[\frac{1}{m(\mathbf{x}, \mathbf{y})} \frac{\partial \beta}{\partial y_i} \frac{\partial \beta}{\partial y_i} + \beta V(\mathbf{x}, \mathbf{y}) \right] d\Omega_{\mathbf{y}} = 0, \quad (2.49)$$

which are obtained by introducing the weight function $v_1 = v_2 = \beta$ into Eqns. (2.43) and (44). Equation (2.47) is the weak form of the homogenized eigenvalue problem (2.19), where the homogenized effective mass and confinement potential are expressed as follows

$$\frac{1}{m_{ij}^{[0]}(\mathbf{x})} = \int_{\Omega_{\mathbf{y}}} \frac{\partial (\alpha_i + y_i)}{\partial y_k} \frac{1}{m(\mathbf{x}, \mathbf{y})} \frac{\partial (\alpha_j + y_j)}{\partial y_k} d\Omega_{\mathbf{y}}, \quad (2.50)$$

$$V^{[0]}(\mathbf{x}) = \frac{1}{A_{\Omega_{\mathbf{y}}}} \int_{\Omega_{\mathbf{y}}} \beta V(\mathbf{x}, \mathbf{y}) \, d\Omega_{\mathbf{y}}. \quad (2.51)$$

Note that the off diagonal terms in $m_{ij}^{[0]}$ vanish.

3 Iterative Multiscale Eigenvalue Solution

In this section, we use the multiscale solution obtained in Section 2 as the predictor, and employ Rayleigh quotient iteration as a correction of the multiscale solution. The proposed iterative scheme aims to approach the eigen pairs $\{\varepsilon^{\lambda}, \Theta^{\lambda}\}$ of Eq. (2.6). Let $\{\Theta_1^{\lambda}, \Theta_2^{\lambda}, ..., \Theta_n^{\lambda}\}$ denote the set of eigen functions of the Hamiltonian H_{λ}. Let the approximation of Θ^{λ} obtained from the two-scale method be denoted as $\bar{\Theta}^{\lambda}$, and by eigen function expansion it can be expressed as

$$\bar{\Theta}^{\lambda} = \sum_{i=1}^{n} c_i \Theta_i^{\lambda}, \quad (3.52)$$

where c_i are the coefficients. In the standard power method [3], assuming that eigenvalues are ordered as $\varepsilon_n^{\lambda} > \varepsilon_{n-1}^{\lambda} \geq ... \geq \varepsilon_2^{\lambda} > \varepsilon_1^{\lambda}$, one has

$$\frac{\left(H_\lambda - \bar{\varepsilon}_1^\lambda \mathbf{I}\right)^{-m} \bar{\Theta}^\lambda}{\left\| \left(H_\lambda - \bar{\varepsilon}_1^\lambda \mathbf{I}\right)^{-m} \bar{\Theta}^\lambda \right\|}$$

$$= \frac{c_1}{|c_1|} \left[1 + \sum_{i=2}^{n} \left(\frac{c_i}{c_1}\right)^2 \left(\frac{\varepsilon_i^\lambda - \bar{\varepsilon}_1^\lambda}{\varepsilon_1^\lambda - \bar{\varepsilon}_1^\lambda}\right)^{-2m} \right]^{-\frac{1}{2}} \Theta_1^\lambda \tag{3.53}$$

$$+ \sum_{i=2}^{n} \left\{ \frac{c_i}{|c_1|} \left(\frac{\varepsilon_i^\lambda - \bar{\varepsilon}_1^\lambda}{\varepsilon_1^\lambda - \bar{\varepsilon}_1^\lambda}\right)^{-m} \left[1 + \sum_{j=2}^{n} \left(\frac{c_j}{c_1}\right)^2 \left(\frac{\varepsilon_j^\lambda - \bar{\varepsilon}_1^\lambda}{\varepsilon_1^\lambda - \bar{\varepsilon}_1^\lambda}\right)^{-2m} \right]^{-\frac{1}{2}} \Theta_i^\lambda \right\},$$

where $\bar{\varepsilon}_1^\lambda$ is the approximation of ε_1^λ, $\mathbf{I}$ is the identity matrix, and m is an integer. If $\bar{\varepsilon}_1^\lambda$ is very close to ε_1^λ and c_1 is not too small, one can show that

$$\lim_{m \to \infty} \frac{\left(H_\lambda - \bar{\varepsilon}_1^\lambda \mathbf{I}\right)^{-m} \bar{\Theta}^\lambda}{\left\| \left(H_\lambda - \bar{\varepsilon}_1^\lambda \mathbf{I}\right)^{-m} \bar{\Theta}^\lambda \right\|} = \Theta_1^\lambda. \tag{3.54}$$

Approximating solution using finite m of the above method with iteration as shown in Figure 3 is the inverse iteration method. For large-scale quantum system, one can combine the above method with Rayleigh quotient iteration [3] that has cubic rate of convergence. By employing a similar inverse iteration procedures for Θ_2^λ, and Gram-Schmidt orthogonaliztion, the second eigenfunction Θ_2^λ corresponding to ε_2^λ can be obtained. These procedures are repeated until all eigenfunctions of Eq. (2.6) are obtained.

The flowchart of numerical implementation is given in Figure 3.2, and the inverse iteration and the Rayleigh quotient iteration for the k-th eigen pair are illustrated in Figure 3.3 and Figure 3.4, respectively. Note that in each

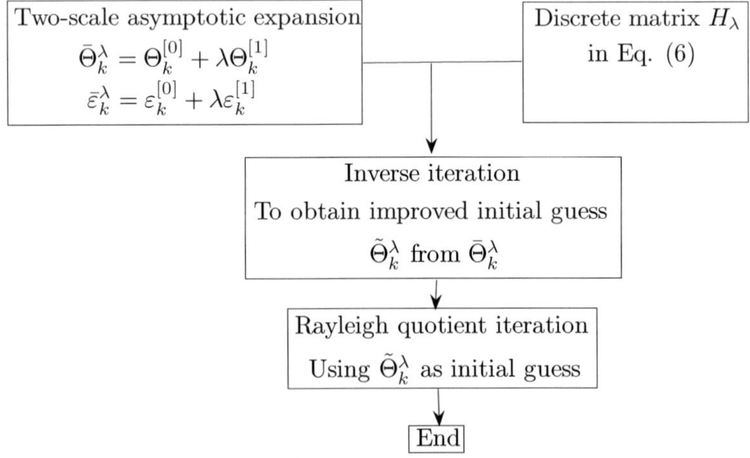

Figure 3.2. Main flowchart of iterative multiscale method.

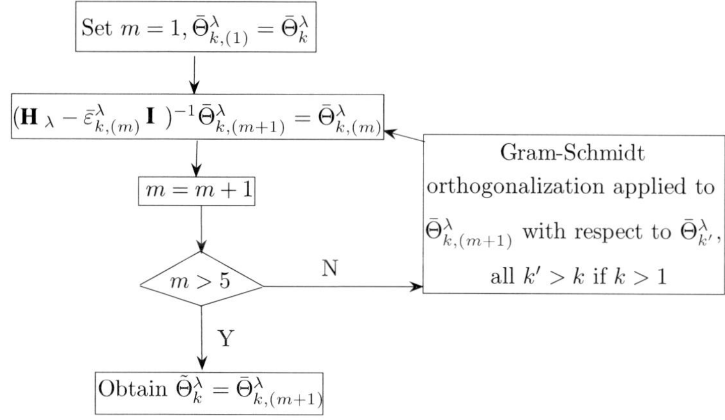

Figure 3.3. Flowchart of inverse iteration.

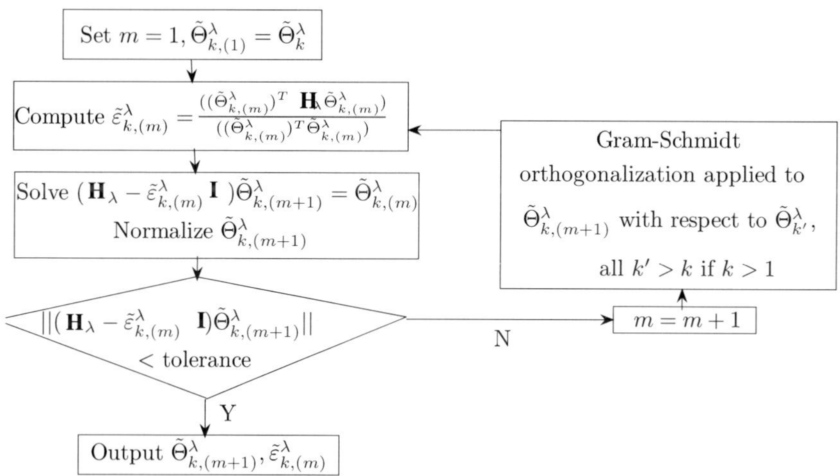

Figure 3.4. Flowchart of Rayleigh quotient iteration.

iteration step of the inverse iteration and Rayleigh quotient iteration, only a linear system of equations with the matrix associated with the discretization of H_λ is solved.

4 Numerical Results

Three QDAs with different complexities in oscillating effective mass and confinement potential are analyzed using single fine scale approach and the multiscale method. For the first two problems where the oscillating effective mass and confinement potential are uniform and slightly nonuniform, inverse iteration and Rayleigh quotient corrections are not used. Inverse iteration and Rayleigh quotient corrections are used only for the case where oscillating effective mass and confinement potential are nonuniform. Reproducing Kernel Particle Method (RKPM) [2] [5] with linear polynomial bases is used for discretization of the weak forms.

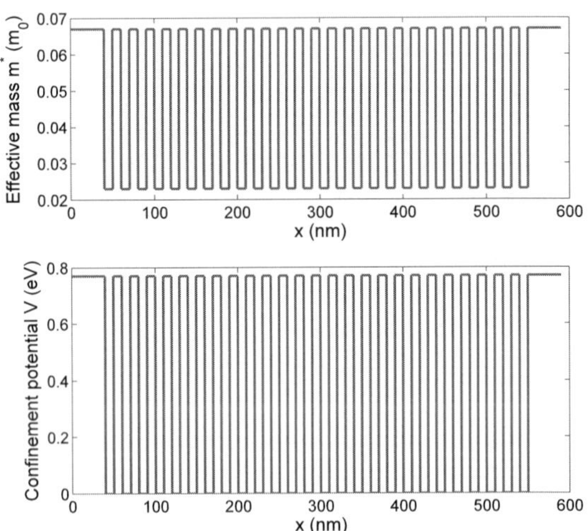

Figure 4.5. 1-D QDA model with uniform oscillating effective mass and confinement potential. (*See also* Color Plate on page 369)

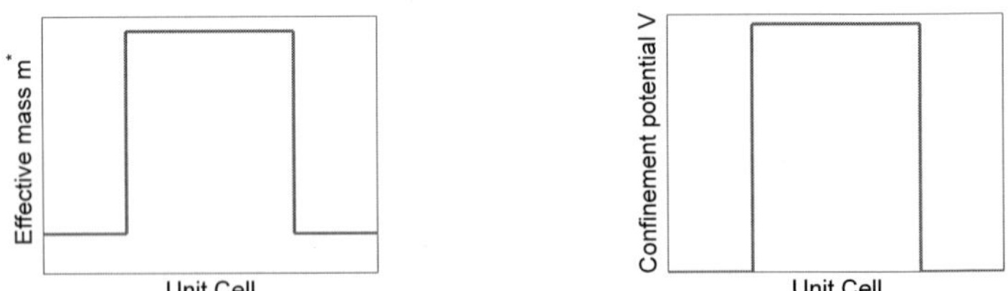

Figure 4.6. Effective mass and confinement potential in the fine scale unit cell. (*See also* Color Plate on page 369)

4.1 QDA model with uniform oscillating effective mass and confinement potential

Consider a one-dimensional QDA model with uniform oscillating effective mass and confinement potential as shown in Figure 4.5. A discretization of the fine scale system with 237 nodes is used for a reference solution of the first 12 pairs of energy levels and the corresponding wave functions. For the multiscale method, 60 nodes are used to solve the coarse scale equation. Due to the uniform oscillating effective mass and confinement potential, every point in the coarse scale domain shares the same homogenized effective mass and confinement potential. That is, only one coupling function is computed in the fine scale unit cell with the effective mass and confinement potential shown in Figure 4.6, which is discretized by 9 nodes. Note that the multiscale approach reduces the cost from solving an eigenvalue problem with dimension 237×237 to solving a eigenvalue problem with dimension 60×60 and a linear system of equations with dimension 9×9. Further, solving eigenvalue problem of coarse scale model is much better conditioned than solving eigenvalue problem of a fine scale model. Table 4.1 compares CPU times and condition numbers of each method.

Table 4.1. Comparison of CPU times and condition numbers in solving QDA model with uniform oscillating effective mass and confinement potential.

	Single-scale Model	**Multiscale Model**	
	Eigenvalue problem	Eigenvalue problem	Unit cell linear system
CPU time (seconds)	0.14	0.031	≈ 0.00
Condition number	259	33.0	

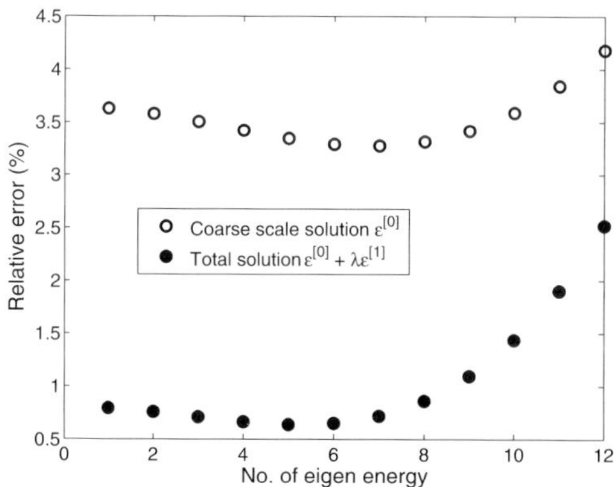

Figure 4.7. Relative error of the first 12 energy levels.

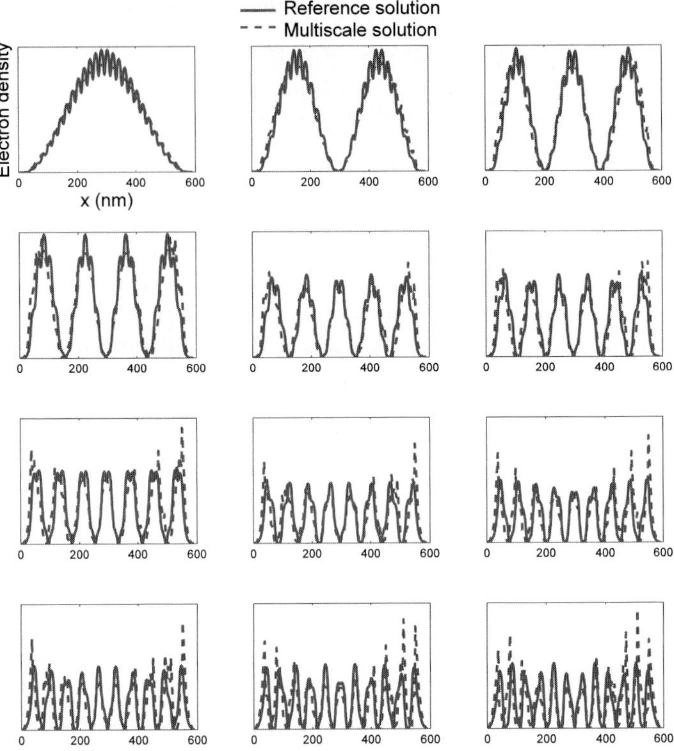

Figure 4.8. Comparison of the electron density distribution associated with the first 12 energy levels between the reference solution and the multiscale solution. (*See also* Color Plate on page 370)

Figure 4.7 shows the relative error of the first 12 eigenvalues, where the coarse scale solution is compared with the multiscale solution. Figure 4.8 shows the comparison of the electron density distributions (eigen function) calculated by the multiscale method and the reference solution. The multiscale method achieves very good accuracy in the first 12 energy levels and the corresponding wave functions.

4.2 QDA with slightly nonuniform oscillating effective mass and confinement potential

Consider a series of 1-D QDA with slightly nonuniform oscillating effective mass and confinement potential, as shown in Figure 4.9. A fine scale model discretized by 921 nodes is used as a reference solution of the first 12 pairs of energy levels and the corresponding wave functions. A total of 101 nodes are used to discretize the coarse scale domain. Since each QDA in this 10-QDA series has different oscillating effective mass and confinement potential

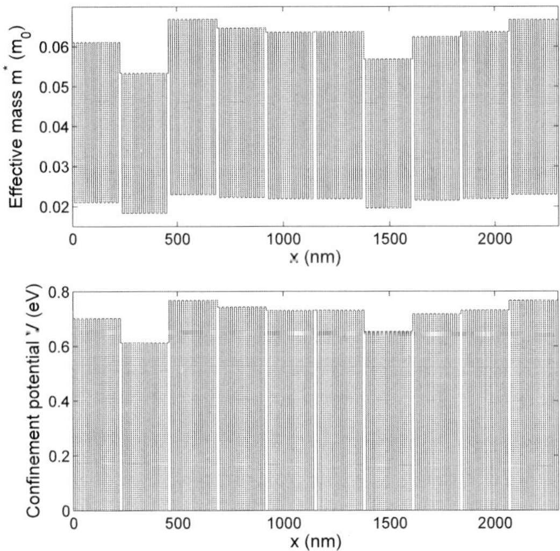

Figure 4.9. 1-D model of a series of QDA with slightly nonuniform oscillating effective mass and confinement potential. (*See also* Color Plate on page 371)

Table 4.2. Comparison of CPU times and condition numbers in solving a series of QDA with slightly nonuniform oscillating effective mass and confinement potential.

	Single-scale Model	**Multiscale Model**	
	Eigenvalue problem	Eigenvalue problem	Unit cell linear system
CPU time (seconds)	54.0	0.0781	≈ 0.00
Condition number	419	21.7	

patterns, 10 fine scale unit cells are modeled. Each unit cell with pattern similar to Figure 4.6 is discretized by 9 nodes. Hence, the multiscale approach reduces the cost from solving a fine scale eigenvalue problem with dimension 921×921 to solving a coarse scale eigenvalue problem with dimension 101×101 and ten linear systems of equations, each with dimension 9×9. Table 4.2 compares CPU times and condition numbers of each method. Figure 4.10 shows the relative errors of the coarse scale solution and the multiscale solution of the first 12 eigenvalues. The electron density distributions calculated by the multiscale approach are shown in Figure 4.11.

4.3 QDA with randomly oscillating effective mass and confinement potential

Consider a series of QDA with randomly oscillating effective mass and confinement potential in one dimension as shown in Figure 4.12, where the ran-

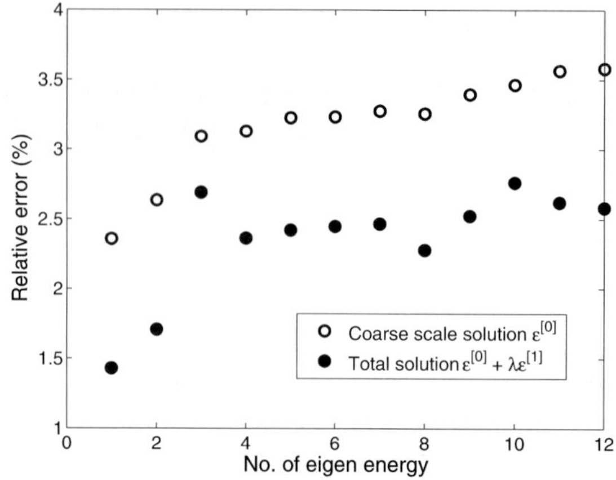

Figure 4.10. Relative error of the first 12 energy levels.

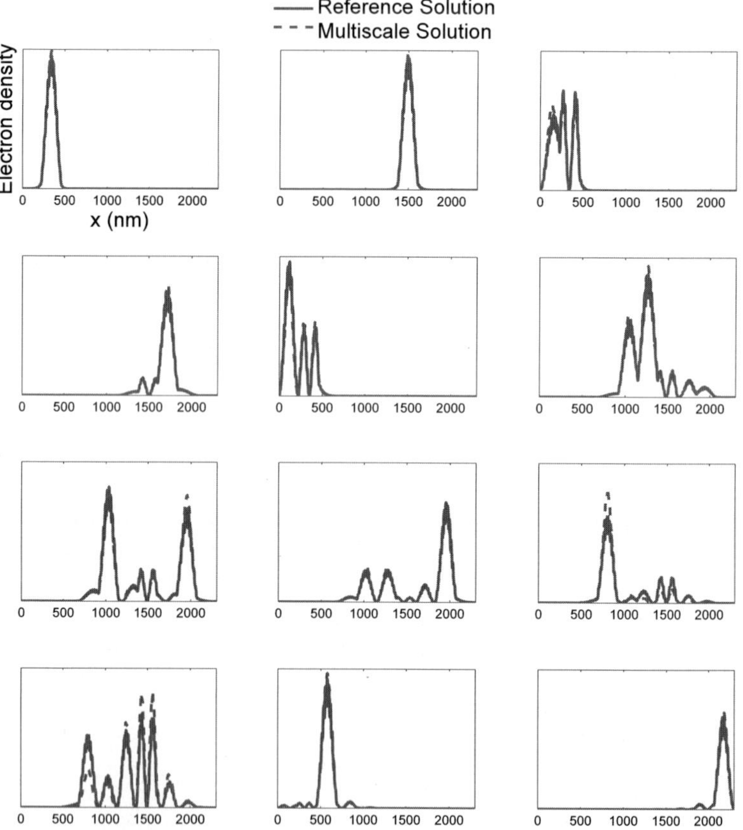

Figure 4.11. Comparison of the electron density distribution associated with the first 12 energy levels between the reference solution and the multiscale solution. (*See also* Color Plate on page 372)

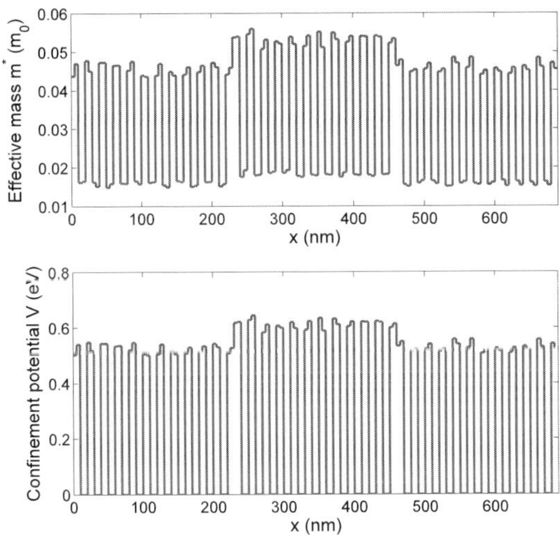

Figure 4.12. 1-D model of a series of QDA with randomly oscillating effective mass and confinement potential. (*See also* Color Plate on page 373)

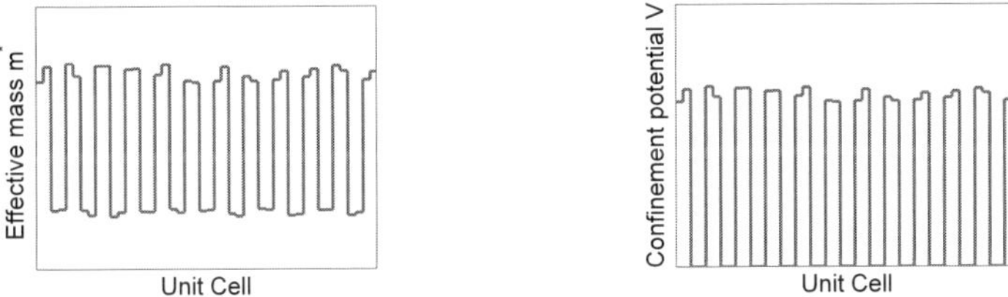

Figure 4.13. Effective mass and confinement potential in the fine scale unit cell. (*See also* Color Plate on page 373)

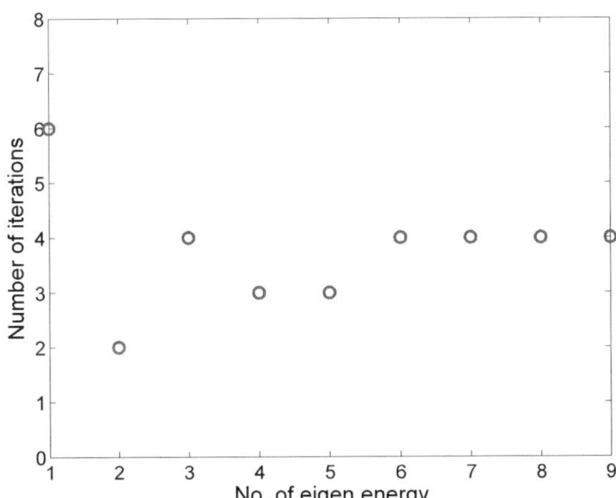

Figure 4.14. Number of iteration steps for the first 9 eigen pairs by using the iterative multiscale method. (*See also* Color Plate on page 374)

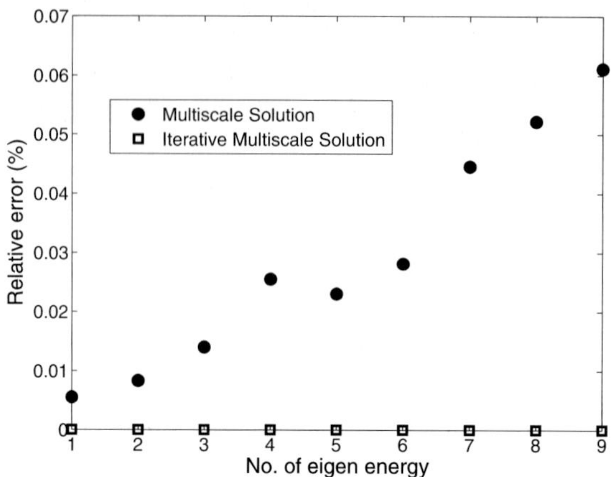

Figure 4.15. Relative error of the first 9 energy levels.

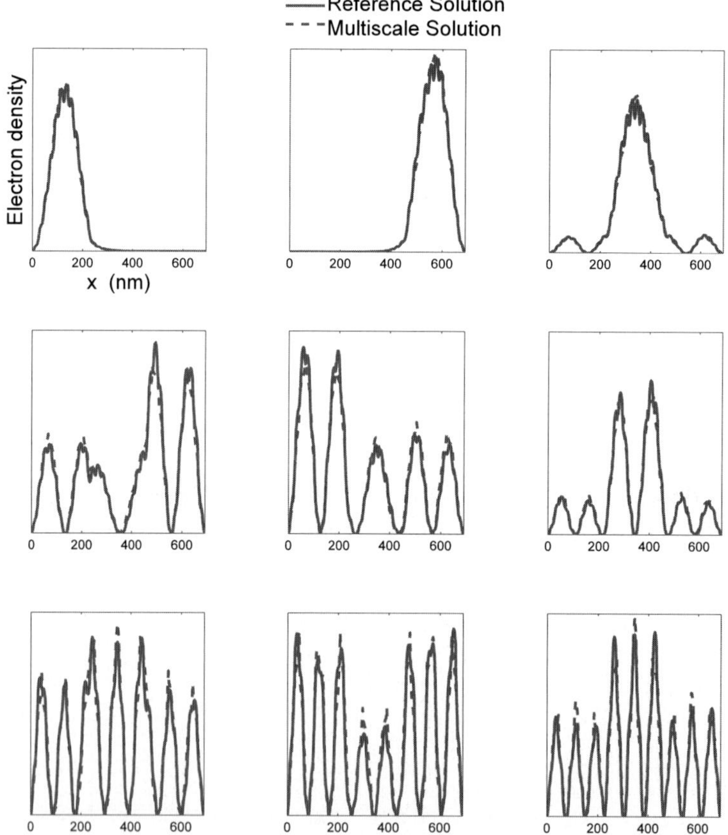

Figure 4.16. Comparison of the electron density distribution associated with the first 9 energy levels between the reference solution and the multiscale solution. (*See also* Color Plate on page 375)

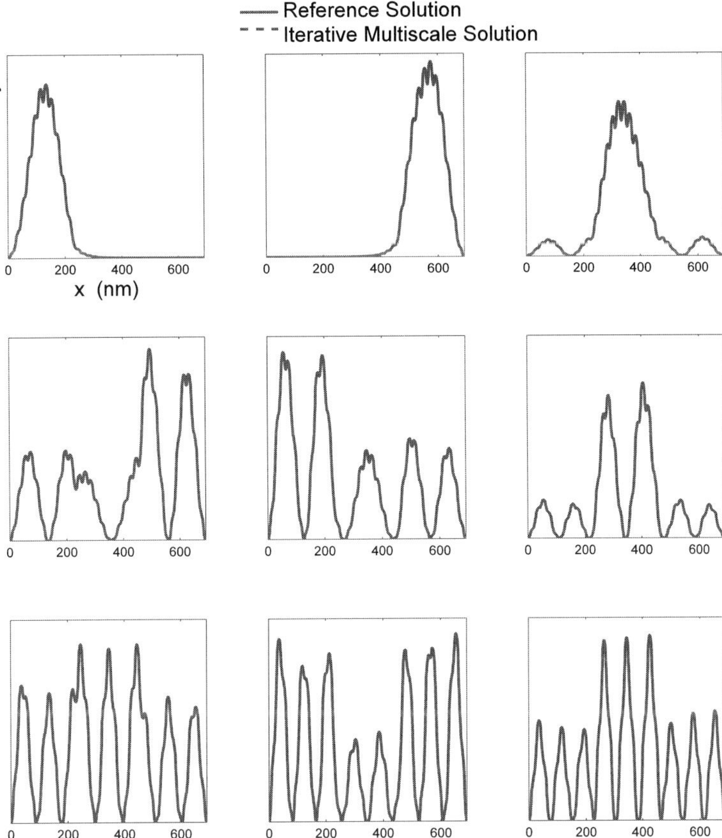

Figure 4.17. Comparison of the electron density distribution associated with the first 9 energy levels between the reference solution and the iterative multiscale solution. (*See also* Color Plate on page 376)

domly oscillating effective mass and confinement potential mimic the intermixed junction between different semiconductor materials. A fine scale model discretized by 553 RKPM nodes with quadratic bases is used for obtaining the reference solution. In the coarse scale domain of the multiscale method, 31 RKPM nodes with linear basis is used to solve the coarse scale equation. Since this series of QDA can be approximately divided into three QDA domains based on the homogenized effective mass and confinement potential, three unit cells are solved separately. Considering that the intermixed junction is randomly distributed, every unit cell with oscillating m and V shown in Figure 4.13 is discretized by 93 nodes. Thus, the multiscale approach reduces the cost from solving a fine scale eigenvalue problem with dimension 553×553 to solving a coarse scale eigenvalue problem with dimension 31×31 and solution of three linear systems of equations with dimension 93×93. Figure

Table 4.3. Comparison of CPU times and condition numbers in solving a series of QDA with randomly oscillating effective mass and confinement potential.

	Singlescale Model	**Iterative Multiscale Model**		
	Eigenvalue problem	Eigenvalue problem	Unit cell linear system	Iteration
CPU time (seconds)	6.3	0.031	≈ 0.00	1.6
Condition number	569	24.0		

4.14 shows that the iterative multiscale method converges very fast with less than 7 iterations for each of the first 9 eigen pairs. The proposed multiscale method without inverse iteration and Rayleigh quotient corrections generates reasonable solution accuracy in eigen pairs as shown in Figure 4.15 and Figure 4.16. As a comparison, the solution of the first 9 eigen pairs obtained with additional inverse iteration and Rayleigh quotient corrections matches the reference solution almost perfectly as shown in Figure 4.17. Table 4.3 compares CPU times and condition numbers of each method.

5 Conclusion

This work presents an asymptotic expansion based multiscale formulation as a predictor in conjunction with the inverse iteration and Rayleigh quotient corrections for analysis of large-scale quantum-dot array. Following [1], an auxiliary problem has been introduced for evaluating the fine scale component of the solution based on the coarse scale solution and the auxiliary function. Proper coarse-fine scale coupling functions for electron energy and wave function have been proposed, and they were solved for obtaining the fine scale information of effective mass and confinement potential. Consequently, the homogenized effective mass and confinement potential have been obtained using the scale coupling functions. Numerical examples demonstrated that a substantial CPU time saving can be achieved with sufficient accuracy with the proposed multiscale method for QDA systems. Additional accuracy can be gained by the inverse iteration and Rayleigh quotient corrections. The numerical examples also showed a fast convergence using the proposed iterative multiscale method.

References

1. A. Bensoussan, J.L. Lions, and G. Papanicolaou, *Asymptotic analysis for periodic structures*, New York, Oxford, 1978.
2. J.S. Chen, C. Pan, C.T. Wu, and W.K. Liu, *Reproducing Kernel Particle Methods for Large Deformation Analysis of Nonlinear Structures*, Computer Methods in Applied Mechanics and Engineering, **139** (1996), 195–229.

3. G.H. Golub, and C.F. Van Loan, *Matrix computations*, The Johns Hopkins University Press, Baltimore, 1996.
4. D.R. Hoog, and R.S. Anderssen, *Asymptotic formulas for discrete eigenvalue problems in Liouville normal form*, Mathematical Models and Methods in Applied Sciences, **11(1)** (2001), 43–56.
5. W.K. Liu, S. Jun, and Y.F. Zhang, *Reproducing Kernel Particle Method*, Int. J. Numer. Methods Fluids, **20** (1995), 1081–1106.
6. S. Kesavan, *Homogenization of elliptic eigenvalue problems, part 1*, Appl. Math. Optim., **5** (1979), 153–167.
7. S. Kesavan, *Homogenization of elliptic eigenvalue problems, part 2*, Appl. Math. Optim., **5** (1979), 197–216.
8. J.W. Paine, *Numerical approximation of Sturm-Liouville eigenvalues*, PhD thesis, Australian National University, Canberra, Australia, 1979.

A Meshless Technique Based on Integrated Radial Basis Function Networks for Elliptic Partial Differential Equations

N. Mai-Duy[1] and T. Tran-Cong[2]

[1] University of Southern Queensland, Toowoomba, QLD 4350, Australia
maiduy@usq.edu.au
[2] University of Southern Queensland, Toowoomba, QLD 4350, Australia
trancong@usq.edu.au

Summary. This paper presents a meshless technique based on radial basis function networks (RBFNs) for solving Dirichlet boundary value problems governed by the Poisson and biharmonic equations. The technique employs integrated RBFNs (IRBFNs) to approximate the field variable and point collocation to discretize the PDE. The boundary conditions are incorporated into IRBFNs via integration constants, which occurs prior to the transformation of the network-weight spaces into the physical space. Several linear and nonlinear test problems are considered to demonstrate the attractiveness of the present meshless technique.

Key words: Radial basis functions, Meshless discretization, Integral collocation formulation

1 Introduction

Engineering and science problems have usually been modelled by partial differential equations (PDEs). Analytical solutions to PDEs can only be obtained for some specific cases. Numerical techniques have thus been developed to convert PDEs into sets of algebraic equations where a solution becomes obtainable. Principal discretization techniques include finite-difference (FDM), finite-element (FEM), finite-volume (FVM) and boundary-element (BEM) methods. These discrete approximation methods require a mesh/grid to support the approximation of the field variables and the integration of the PDEs. For problems involving complex geometries, generating a mesh is known to be the most costly and time-consuming part of the solution process. As a result, the development of the so-called meshless methods has received a great deal of attention from scientific and engineering research communities in recent decades. The reader is referred to, for example, [1,2,6], for a detailed discussion of meshless methods.

RBFNs are considered as a powerful numerical tool for the approximation of a function and its derivatives [8]. One distinguishing feature of RBFNs is that they rely on a set of discrete points, which can be randomly distributed, to represent the approximate function. According to Micchelli's theorem, there is a large class of RBFs, e.g., multiquadrics, inverse multiquadrics and Gaussian functions, whose interpolation matrices are invertible, provided that the data points are distinct. It has been proven that RBFNs are capable of representing any continuous function to a prescribed degree of accuracy in the L_p norm, $p \in [1, \infty]$. Furthermore, according to Cover's theorem, the higher the number of neurons (RBFs) is used, the more accurate the approximation will be, indicating the property of "mesh convergence" of RBFNs. Over the past fifteen years, RBFNs have emerged as an attractive solver for PDEs [6]. The field variable is approximated using RBFNs, while the PDE is discretized by means of point collocation. These RBFN methods can thus be considered as a truly meshless solver.

Integrated RBFNs (IRBFNs) proposed by Mai-Duy and Tran-Cong [14] have several advantages over differentiated RBFNs: (i) they avoid the reduction of convergence rate caused by differentiation, and (ii) they provide an effective way to incorporate "extra" input information (e.g. both function and derivative values given at a point). IRBFNs have been developed for the solution of PDEs, e.g. [11–13, 15–17]. This article presents a meshless IRBFN method for solving the Poisson and biharmonic equations with Dirichlet boundary conditions in rectangular and non-rectangular domains. The emphasis is placed on the technique of imposing the boundary conditions that takes place before the conversion of the network-weight spaces into the physical space. Several linear and non-linear test problems are considered, including a well-known lid-driven cavity viscous flow. An outline of the paper is as follows. A brief review of RBFNs is given in Section 2. The application of IRBFNs for the solution of PDEs is described in Section 3. Numerical results are presented in Section 4. Section 5 concludes the paper.

2 RBFNs

A function y, to be approximated, can be represented by an RBFN as

$$y(\mathbf{x}) \approx f(\mathbf{x}) = \sum_{k=1}^{m} w^{(k)} g^{(k)}(\mathbf{x}), \tag{2.1}$$

where $\mathbf{x}$ the input vector, m the number of RBFs, $\{w^{(k)}\}_{k=1}^{m}$ the set of network weights to be found, and $\{g^{(k)}(\mathbf{x})\}_{k=1}^{m}$ the set of RBFs.

Among RBFs, multiquadrics (MQ), which have spectral approximation power [9, 10, 19], tend to result in the most accurate approximations. We will thus employ the MQ function whose form is

$$g^{(k)}(\mathbf{x}) = \sqrt{(\mathbf{x} - \mathbf{c}^{(k)})^T(\mathbf{x} - \mathbf{c}^{(k)}) + a^{(k)2}}, \tag{2.2}$$

where $\mathbf{c}^{(k)}$ and $a^{(k)}$ are the centre and width of the kth MQ-RBF, respectively.

Differentiated RBFNs: RBFNs (2.1) are used to represent the original function y; derivatives of y are then obtained through differentiation

$$y(\mathbf{x}) \approx f(\mathbf{x}) = \sum_{k=1}^{m} w^{(k)} g^{(k)}(\mathbf{x}), \tag{2.3}$$

$$\frac{\partial^p y(\mathbf{x})}{\partial x_i^p} \approx \frac{\partial^p f(\mathbf{x})}{\partial x_i^p} = \frac{\partial^p \left(\sum_{k=1}^{m} w^{(k)} g^{(k)}(\mathbf{x})\right)}{\partial x_i^p} = \sum_{k=1}^{m} w^{(k)} h_{[x_i]}^{(k)[p]}(\mathbf{x}), \tag{2.4}$$

where the subscript $[x_i]$ is used to denote the process of differentiation with respect to x_i, and $h_{[x_i]}^{(k)[p]}(\mathbf{x}) = \partial^p g^{(k)}(\mathbf{x})/\partial x_i^p$ are basis functions for the approximation of pth-order derivatives of y.

Integrated IRBFNs: RBFNs (2.1) are used to represent pth-order derivatives of y; lower-order derivatives and the function itself are then obtained through integration

$$\frac{\partial^p y(\mathbf{x})}{\partial x_i^p} \approx \frac{\partial^p f(\mathbf{x})}{\partial x_i^p} = \sum_{k=1}^{m} w_{[x_i]}^{(k)} g^{(k)}(\mathbf{x}) = \sum_{k=1}^{m} w_{[x_i]}^{(k)} H_{[x_i]}^{(k)[p]}(\mathbf{x}), \tag{2.5}$$

$$\frac{\partial^{p-1} y(\mathbf{x})}{\partial x_i^{p-1}} \approx \frac{\partial^{p-1} f(\mathbf{x})}{\partial x_i^{p-1}} = \sum_{k=1}^{m+q_1} w_{[x_i]}^{(k)} H_{[x_i]}^{(k)[p-1]}(\mathbf{x}), \tag{2.6}$$

$$\cdots \quad \cdots \quad \cdots \quad \cdots \quad \cdots \quad \cdots \quad \cdots$$

$$y(\mathbf{x}) \approx f(\mathbf{x}) = \sum_{k=1}^{m+q_p} w_{[x_i]}^{(k)} H_{[x_i]}^{(k)[0]}(\mathbf{x}), \tag{2.7}$$

where the subscript $[x_i]$ is used to denote the process of integration with respect to x_i; q_1 the number of new centres in a subnetwork that is employed to approximate a set of nodal integration constants, $q_2 = 2q_1, \cdots, q_p = pq_1$; and $g^{(k)}(\mathbf{x}) = H_{[x_i]}^{(k)[p]}(\mathbf{x})$, $H_{[x_i]}^{(k)[p-1]} = \int H_{[x_i]}^{(k)[p]} dx_i, \cdots, H_{[x_i]}^{(k)[0]} = \int H_{[x_i]}^{(k)[1]} dx_i$. It is noted that the new centres and their associated known basis functions in subnetworks are also denoted by the notations $w^{(k)}$ and $H^{(k)}(\mathbf{x})$, respectively, but with $k > m$. IRBFNs are said to be of order p if their starting points are pth-order derivatives.

For two and higher dimensional problems, it would be convenient to work in the physical space. The evaluation of (2.5)–(2.7) at a set of collocation points $\{\mathbf{x}^{(l)}\}_{l=1}^{m}$, which is selected to coincide with the set of centres $\{\mathbf{c}^{(l)}\}_{l=1}^{m}$, yields

$$\frac{\partial^{\mathbf{p}}\mathbf{f}}{\partial \mathbf{x_i^p}} = \mathbf{H}^{[p]}_{[x_i]}\mathbf{w}_{[x_i]}, \tag{2.8}$$

$$\frac{\partial^{\mathbf{p-1}}\mathbf{f}}{\partial \mathbf{x_i^{p-1}}} = \mathbf{H}^{[p-1]}_{[x_i]}\mathbf{w}_{[x_i]}, \tag{2.9}$$

$$\cdots \quad \cdots$$

$$\mathbf{f} = \mathbf{H}^{[0]}_{[x_i]}\mathbf{w}_{[x_i]}, \tag{2.10}$$

where $\mathbf{w}_{[x_i]} = \left[w^{(1)}_{[x_i]}, w^{(2)}_{[x_i]}, \cdots, w^{(m+q_p)}_{[x_i]}\right]^T$, $\mathbf{f} = [f(\mathbf{x}^{(1)}), f(\mathbf{x}^{(2)}), \cdots, f(\mathbf{x}^{(m)})]^T$, $\cdots$, $\frac{\partial^{\mathbf{p}}\mathbf{f}}{\partial \mathbf{x_i^p}} = \left[\frac{\partial^p f(\mathbf{x}^{(1)})}{\partial x_i^p}, \frac{\partial^p f(\mathbf{x}^{(2)})}{\partial x_i^p}, \cdots, \frac{\partial^p f(\mathbf{x}^{(m)})}{\partial x_i^p}\right]^T$ and the matrices $\mathbf{H}^{[.]}_{[x_i]}$ have entries $\left(\mathbf{H}^{[.]}_{[x_i]}\right)_{lk} = H^{(k)[.]}_{[x_i]}(\mathbf{x}^{(l)})$. In (2.8)-(2.10), the matrices $\mathbf{H}^{[p]}_{[x_i]}$, $\mathbf{H}^{[p-1]}_{[x_i]}, \cdots, \mathbf{H}^{[1]}_{[x_i]}$ are augmented using zero-submatrices so that they have the same size as the matrix $\mathbf{H}^{[0]}_{[x_i]}$. Through (2.10), one can express the network weights in terms of the nodal function values as

$$\mathbf{w}_{[x_i]} = \mathbf{H}^{[0]-1}_{[x_i]}\mathbf{f}, \tag{2.11}$$

where $\mathbf{H}^{[0]-1}_{[x_i]}$ is the Moore-Penrose pseudo-inverse.

Substitution of (2.11) into (2.5)–(2.7) yields

$$f(\mathbf{x}) = \frac{1}{N}\sum_{i=1}^{N}\left(\left[H^{(1)[0]}_{[x_i]}(\mathbf{x}), H^{(2)[0]}_{[x_i]}(\mathbf{x}), \cdots\right]\mathbf{H}^{[0]-1}_{[x_i]}\mathbf{f}\right), \tag{2.12}$$

$$\frac{\partial f(\mathbf{x})}{\partial x_i} = \left[H^{(1)[1]}_{[x_i]}(\mathbf{x}), H^{(2)[1]}_{[x_i]}(\mathbf{x}), \cdots\right]\mathbf{H}^{[0]-1}_{[x_i]}\mathbf{f}, \tag{2.13}$$

$$\cdots \quad \cdots \quad \cdots \quad \cdots \quad \cdots \quad \cdots \quad \cdots$$

$$\frac{\partial^p f(\mathbf{x})}{\partial x_i^p} = \left[H^{(1)[p]}_{[x_i]}(\mathbf{x}), H^{(2)[p]}_{[x_i]}(\mathbf{x}), \cdots\right]\mathbf{H}^{[0]-1}_{[x_i]}\mathbf{f}, \tag{2.14}$$

where N is the dimension of the problem and the approximate function $f(\mathbf{x})$ is taken to be the average value of the $f_{[x_i]}(\mathbf{x})$ due to numerical errors.

The calculation of cross derivatives of f is based on the following relation

$$\frac{\partial^p f}{\partial x_i^r \partial x_j^s} = \frac{1}{2}\left(\frac{\partial^r}{\partial x_i^r}\left(\frac{\partial^s f}{\partial x_j^s}\right) + \frac{\partial^s}{\partial x_j^s}\left(\frac{\partial^r f}{\partial x_i^r}\right)\right), \quad p = r + s,\ i \neq j, \tag{2.15}$$

which reduces the computation of mixed derivatives to that of lower-order pure derivatives for which IRBFNs involve integration with respect to x_i or x_j only. In terms of nodal function values, (2.15) can be expressed as

$$\frac{\partial^p f(\mathbf{x})}{\partial x_i^r \partial x_j^s} = \frac{1}{2}\Big(\left[H^{(1)[r]}_{[x_i]}(\mathbf{x}), H^{(2)[r]}_{[x_i]}(\mathbf{x}), \cdots\right]\mathbf{H}^{[0]-1}_{[x_i]}\left(\mathbf{H}^{[s]}_{[x_j]}\mathbf{H}^{[0]-1}_{[x_j]}\mathbf{f}\right) + \left[H^{(1)[s]}_{[x_j]}(\mathbf{x}), H^{(2)[s]}_{[x_j]}(\mathbf{x}), \cdots\right]\mathbf{H}^{[0]-1}_{[x_j]}\left(\mathbf{H}^{[r]}_{[x_i]}\mathbf{H}^{[0]-1}_{[x_i]}\mathbf{f}\right)\Big). \tag{2.16}$$

3 Solution of PDEs

The solution procedure of the present IRBFN method for solving an elliptic PDE consists of two main steps: (i) the incorporation of the boundary conditions into IRBFNs via integration constants, and (ii) the enforcement of IRBFNs to satisfy the PDE. In what follows, Poisson and biharmonic equations are considered.

3.1 Poisson equation

Consider the Poisson equation

$$\frac{\partial^2 u}{\partial x_1^2} + \frac{\partial^2 u}{\partial x_2^2} = b(x_1, x_2), \tag{3.17}$$

in a two-dimensional bounded domain Ω with Dirichlet boundary conditions. The present domain Ω is replaced by a set of discrete points (Figure 3.1a). The IRBFN expressions for derivatives of u with respect to x_i can be constructed as follows. For any interior point $\boldsymbol{\gamma}$, through the vertical or horizontal line, there are at least two boundary points $\boldsymbol{\alpha}$ and $\boldsymbol{\beta}$ that have $\alpha_j = \beta_j = \gamma_j$ (Figure 3.1b). Using second-order IRBFNs, the values of u at $\boldsymbol{\gamma}$, $\boldsymbol{\alpha}$ and $\boldsymbol{\beta}$ will be computed by

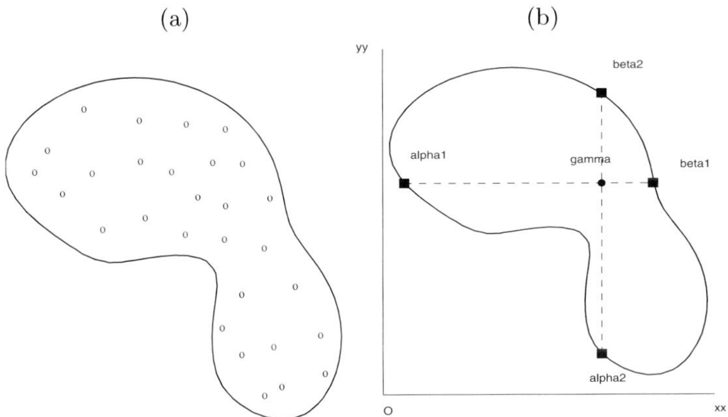

Figure 3.1. Geometry and discretization.

$$u(\gamma) = \sum_{k=1}^{m} w_{[x_i]}^{(k)} H_{[x_i]}^{(k)[0]}(\gamma) + \gamma_i C1_{[x_i]}(\gamma_j) + C2_{[x_i]}(\gamma_j), \tag{3.18}$$

$$u(\boldsymbol{\alpha}) = \sum_{k=1}^{m} w_{[x_i]}^{(k)} H_{[x_i]}^{(k)[0]}(\boldsymbol{\alpha}) + \alpha_i C1_{[x_i]}(\gamma_j) + C2_{[x_i]}(\gamma_j), \tag{3.19}$$

$$u(\boldsymbol{\beta}) = \sum_{k=1}^{m} w_{[x_i]}^{(k)} H_{[x_i]}^{(k)[0]}(\boldsymbol{\beta}) + \beta_i C1_{[x_i]}(\gamma_j) + C2_{[x_i]}(\gamma_j), \tag{3.20}$$

where $C1_{[x_i]}$ and $C2_{[x_i]}$ are the constants of integration that are univariate functions of x_j.

A set of two equations (3.19) and (3.20) is linear and one can solve it for two unknowns $C1_{[x_i]}$ and $C2_{[x_i]}$

$$\begin{aligned} C1_{[x_i]} =& -\frac{1}{\alpha_i - \beta_i} \sum_{k=1}^{m} w_{[x_i]}^{(k)} \left(H_{[x_i]}^{(k)[0]}(\boldsymbol{\alpha}) - H_{[x_i]}^{(k)[0]}(\boldsymbol{\beta}) \right) \\ &+ \frac{1}{\alpha_i - \beta_i} u(\boldsymbol{\alpha}) - \frac{1}{\alpha_i - \beta_i} u(\boldsymbol{\beta}), \end{aligned} \tag{3.21}$$

$$\begin{aligned} C2_{[x_i]} =& \frac{1}{\alpha_i - \beta_i} \sum_{k=1}^{m} w_{[x_i]}^{(k)} \left(\beta_i H_{[x_i]}^{(k)[0]}(\boldsymbol{\alpha}) - \alpha_i H_{[x_i]}^{(k)[0]}(\boldsymbol{\beta}) \right) \\ &- \frac{\beta_i}{\alpha_i - \beta_i} u(\boldsymbol{\alpha}) + \frac{\alpha_i}{\alpha_i - \beta_i} u(\boldsymbol{\beta}). \end{aligned} \tag{3.22}$$

Substitution of (3.21) and (3.22) into (3.18) yields

$$\begin{aligned} u(\gamma) =& \sum_{k=1}^{m} w_{[x_i]}^{(k)} \left(H_{[x_i]}^{(k)[0]}(\gamma) + \frac{\beta_i - \gamma_i}{\alpha_i - \beta_i} H_{[x_i]}^{(k)[0]}(\boldsymbol{\alpha}) - \frac{\alpha_i - \gamma_i}{\alpha_i - \beta_i} H_{[x_i]}^{(k)[0]}(\boldsymbol{\beta}) \right) \\ &- \frac{\beta_i - \gamma_i}{\alpha_i - \beta_i} u(\boldsymbol{\alpha}) + \frac{\alpha_i - \gamma_i}{\alpha_i - \beta_i} u(\boldsymbol{\beta}). \end{aligned} \tag{3.23}$$

Defining

$$D_{[x_i]}^{(k)[0]}(\gamma) = H_{[x_i]}^{(k)[0]}(\gamma) + \frac{\beta_i - \gamma_i}{\alpha_i - \beta_i} H_{[x_i]}^{(k)[0]}(\boldsymbol{\alpha}) - \frac{\alpha_i - \gamma_i}{\alpha_i - \beta_i} H_{[x_i]}^{(k)[0]}(\boldsymbol{\beta}),$$

and

$$M_{[x_i]}^{[0]} = -\frac{\beta_i - \gamma_i}{\alpha_i - \beta_i}, \quad N_{[x_i]}^{[0]} = \frac{\alpha_i - \gamma_i}{\alpha_i - \beta_i},$$

expression (3.23) becomes

$$u(\gamma) = \sum_{k=1}^{m} w_{[x_i]}^{(k)} D_{[x_i]}^{(k)[0]}(\gamma) + M_{[x_i]}^{[0]}(\gamma) u(\boldsymbol{\alpha}) + N_{[x_i]}^{[0]}(\gamma) u(\boldsymbol{\beta}). \tag{3.24}$$

Collocating (3.24) at the interior points $\{\mathbf{x}^{(l)}\}_{l=1}^{n_{ip}}$ (n_{ip}–the number of interior points) results in the following system

$$\mathbf{u}_{ip} = \mathbf{D}_{[x_i]}^{[0]} \mathbf{w}_{[x_i]} + \mathbf{M}_{[x_i]}^{[0]} \mathbf{u}_\alpha + \mathbf{N}_{[x_i]}^{[0]} \mathbf{u}_\beta, \tag{3.25}$$

where

$$\mathbf{D}_{[x_i]}^{[0]} = \begin{bmatrix} D_{[x_i]}^{(1)[0]}(\mathbf{x}^{(1)}) & D_{[x_i]}^{(2)[0]}(\mathbf{x}^{(1)}) & \cdots & D_{[x_i]}^{(m)[0]}(\mathbf{x}^{(1)}) \\ D_{[x_i]}^{(1)[0]}(\mathbf{x}^{(2)}) & D_{[x_i]}^{(2)[0]}(\mathbf{x}^{(2)}) & \cdots & D_{[x_i]}^{(m)[0]}(\mathbf{x}^{(2)}) \\ \vdots & \vdots & \ddots & \vdots \\ D_{[x_i]}^{(1)[0]}(\mathbf{x}^{(n_{ip})}) & D_{[x_i]}^{(2)[0]}(\mathbf{x}^{(n_{ip})}) & \cdots & D_{[x_i]}^{(m)[0]}(\mathbf{x}^{(n_{ip})}) \end{bmatrix},$$

$$\mathbf{M}_{[x_i]}^{[0]} = \begin{bmatrix} M_{[x_i]}^{(1)[0]} & 0 & \cdots & 0 \\ 0 & M_{[x_i]}^{(2)[0]} & \cdots & 0 \\ \vdots & \vdots & \ddots & \vdots \\ 0 & 0 & \cdots & M_{[x_i]}^{(n_{ip})[0]} \end{bmatrix}, \quad \mathbf{N}_{[x_i]}^{[0]} = \begin{bmatrix} N_{[x_i]}^{(1)[0]} & 0 & \cdots & 0 \\ 0 & N_{[x_i]}^{(2)[0]} & \cdots & 0 \\ \vdots & \vdots & \ddots & \vdots \\ 0 & 0 & \cdots & N_{[x_i]}^{(n_{ip})[0]} \end{bmatrix},$$

$$\mathbf{u}_{ip} = \begin{pmatrix} u(\mathbf{x}^{(1)}) \\ u(\mathbf{x}^{(2)}) \\ \vdots \\ u(\mathbf{x}^{(n_{ip})}) \end{pmatrix}, \quad \mathbf{u}_\alpha = \begin{pmatrix} u(\boldsymbol{\alpha}^{(1)}) \\ u(\boldsymbol{\alpha}^{(2)}) \\ \vdots \\ u(\boldsymbol{\alpha}^{(n_{ip})}) \end{pmatrix}, \quad \text{and} \quad \mathbf{u}_\beta = \begin{pmatrix} u(\boldsymbol{\beta}^{(1)}) \\ u(\boldsymbol{\beta}^{(2)}) \\ \vdots \\ u(\boldsymbol{\beta}^{(n_{ip})}) \end{pmatrix}.$$

From (3.25), the multispaces of network weights can be transformed into the single space of nodal variable values

$$\mathbf{w}_{[x_i]} = \mathbf{D}_{[x_i]}^{[0]-1} \left(\mathbf{u}_{ip} - \mathbf{M}_{[x_i]}^{[0]} \mathbf{u}_\alpha - \mathbf{N}_{[x_i]}^{[0]} \mathbf{u}_\beta \right), \tag{3.26}$$

where $\mathbf{D}_{[x_i]}^{[0]-1}$ is the Moore-Penrose pseudoinverse of $\mathbf{D}_{[x_i]}^{[0]}$.

Making use of (3.26), second-order derivatives of u at an arbitrary point $\mathbf{x}$ will be computed by

$$\frac{\partial^2 u(\mathbf{x})}{\partial x_i^2} = \left[g^{(1)}(\mathbf{x}), g^{(2)}(\mathbf{x}), \cdots \right] \mathbf{D}_{[x_i]}^{[0]-1} \left[\mathbf{I}, -\mathbf{M}_{[x_i]}^{[0]}, -\mathbf{N}_{[x_i]}^{[0]} \right] \begin{pmatrix} \mathbf{u}_{ip} \\ \mathbf{u}_\alpha \\ \mathbf{u}_\beta \end{pmatrix}, \tag{3.27}$$

where $\mathbf{I}$ is the $n_{ip} \times n_{ip}$ identity matrix.

Expression (3.27) can be further reduced to

$$\frac{\partial^2 u(\mathbf{x})}{\partial x_i^2} = \left[E_{[x_i]}^{(1)[2]}, E_{[x_i]}^{(2)[2]}, \cdots, E_{[x_i]}^{(n_{ip})[2]} \right] \mathbf{u}_{ip} + k_{[x_i]}^{[2]}, \tag{3.28}$$

where $E_{[x_i]}^{(.)[2]}$ and $k_{[x_i]}^{[2]}$ are known functions of $\mathbf{x}$, $\boldsymbol{\alpha}$, $\boldsymbol{\beta}$, $\mathbf{u}_\alpha$ and $\mathbf{u}_\beta$.

The evaluation of (3.28) at the interior points $\{\mathbf{x}^{(l)}\}_{l=1}^{n_{ip}}$ yields

$$\frac{\partial^2 \mathbf{u}}{\partial \mathbf{x}_i^2} = \mathbf{E}_{[x_i]}^{[2]} \mathbf{u}_{ip} + \mathbf{k}_{[x_i]}^{[2]}, \tag{3.29}$$

where $\frac{\partial^2 \mathbf{u}}{\partial \mathbf{x}_i^2} = \left[\frac{\partial^2 u(\mathbf{x}^{(1)})}{\partial x_i^2}, \frac{\partial^2 u(\mathbf{x}^{(2)})}{\partial x_i^2}, \cdots, \frac{\partial^2 u(\mathbf{x}^{(n_{ip})})}{\partial x_i^2}\right]^T$, the matrix $\mathbf{E}^{[2]}_{[x_i]}$ has entries $\left(\mathbf{E}^{[2]}_{[x_i]}\right)_{lk} = E^{(k)[2]}_{[x_i]}(\mathbf{x}^{(l)})$ and $\mathbf{k}^{[2]}_{[x_i]} = \left[k^{[2]}_{[x_i]}(\mathbf{x}^{(1)}), k^{[2]}_{[x_i]}(\mathbf{x}^{(2)}), \cdots, k^{[2]}_{[x_i]}(\mathbf{x}^{(n_{ip})})\right]^T$. It is noted that the IRBFN approximations (3.29) already include information about the boundary conditions.

Using (3.29), (3.17) can be transformed into the following determinate system of algebraic equations

$$\left(\mathbf{E}^{[2]}_{[x_1]} + \mathbf{E}^{[2]}_{[x_2]}\right)\mathbf{u}_{ip} = \mathbf{b} - \mathbf{k}^{[2]}_{[x_1]} - \mathbf{k}^{[2]}_{[x_2]}, \tag{3.30}$$

where $\mathbf{b} = \left\{b(\mathbf{x}^{(1)}), b(\mathbf{x}^{(2)}), \cdots, b(\mathbf{x}^{(n_{ip})})\right\}^T$.

3.2 Biharmonic equation

Consider the biharmonic equation

$$\frac{\partial^4 v}{\partial x_1^4} + 2\frac{\partial^4 v}{\partial x_1^2 \partial x_2^2} + \frac{\partial^4 v}{\partial x_2^4} = b(x_1, x_2), \tag{3.31}$$

in a two-dimensional bounded domain with Dirichlet boundary conditions (v and $\partial v/\partial n$). Fourth-order IRBFNs are employed to approximate the field variable v and its derivatives. The solution process for solving the biharmonic equation is similar to that for solving the Poisson equation. Only critical steps are presented here. Consider the process of integration with respect to x_i. For each interior point, e.g. $\boldsymbol{\gamma}$, one can form the following system of four algebraic equations

$$v(\boldsymbol{\alpha}) = \sum_{k=1}^{m} w^{(k)}_{[x_i]} H^{(k)[0]}_{[x_i]}(\boldsymbol{\alpha}) + \frac{\alpha_i^3}{6} C1_{[x_i]}(\gamma_j) + \frac{\alpha_i^2}{2} C2_{[x_i]}(\gamma_j) + \alpha_i C3_{[x_i]}(\gamma_j) + C4_{[x_i]}(\gamma_j), \tag{3.32}$$

$$\frac{\partial v(\boldsymbol{\alpha})}{\partial x_i} = \sum_{k=1}^{m} w^{(k)}_{[x_i]} H^{(k)[1]}_{[x_i]}(\boldsymbol{\alpha}) + \frac{\alpha_i^2}{2} C1_{[x_i]}(\gamma_j) + \alpha_i C2_{[x_i]}(\gamma_j) + C3_{[x_i]}(\gamma_j), \tag{3.33}$$

$$v(\boldsymbol{\beta}) = \sum_{k=1}^{m} w^{(k)}_{[x_i]} H^{(k)[0]}_{[x_i]}(\boldsymbol{\beta}) + \frac{\beta_i^3}{6} C1_{[x_i]}(\gamma_j) + \frac{\beta_i^2}{2} C2_{[x_i]}(\gamma_j) + \beta_i C3_{[x_i]}(\gamma_j) + C4_{[x_i]}(\gamma_j), \tag{3.34}$$

$$\frac{\partial v(\boldsymbol{\beta})}{\partial x_i} = \sum_{k=1}^{m} w^{(k)}_{[x_i]} H^{(k)[1]}_{[x_i]}(\boldsymbol{\beta}) + \frac{\beta_i^2}{2} C1_{[x_i]}(\gamma_j) + \beta_i C2_{[x_i]}(\gamma_j) + C3_{[x_i]}(\gamma_j), \tag{3.35}$$

which can be solved analytically for four integration constants, namely $C1_{[x_i]}$, $C2_{[x_i]}$, $C3_{[x_i]}$ and $C4_{[x_i]}$. The field variable v and its derivatives will thus be expressed in terms of network weights and boundary conditions as

$$v(\gamma) = \sum_{k=1}^{m} w_{[x_i]}^{(k)} D_{[x_i]}^{(k)[0]} + M_{[x_i]}^{[0]} v(\boldsymbol{\alpha}) + N_{[x_i]}^{[0]} v(\boldsymbol{\beta}) + P_{[x_i]}^{[0]} \frac{\partial v(\boldsymbol{\alpha})}{\partial x_i} + Q_{[x_i]}^{[0]} \frac{\partial v(\boldsymbol{\beta})}{\partial x_i}, \tag{3.36}$$

$$\frac{\partial v(\gamma)}{\partial x_i} = \sum_{k=1}^{m} w_{[x_i]}^{(k)} D_{[x_i]}^{(k)[1]} + M_{[x_i]}^{[1]} v(\boldsymbol{\alpha}) + N_{[x_i]}^{[1]} v(\boldsymbol{\beta}) + P_{[x_i]}^{[1]} \frac{\partial v(\boldsymbol{\alpha})}{\partial x_i} + Q_{[x_i]}^{[1]} \frac{\partial v(\boldsymbol{\beta})}{\partial x_i}, \tag{3.37}$$

$$\frac{\partial^2 v(\gamma)}{\partial x_i^2} = \sum_{k=1}^{m} w_{[x_i]}^{(k)} D_{[x_i]}^{(k)[2]} + M_{[x_i]}^{[2]} v(\boldsymbol{\alpha}) + N_{[x_i]}^{[2]} v(\boldsymbol{\beta}) + P_{[x_i]}^{[2]} \frac{\partial v(\boldsymbol{\alpha})}{\partial x_i} + Q_{[x_i]}^{[2]} \frac{\partial v(\boldsymbol{\beta})}{\partial x_i}, \tag{3.38}$$

$$\frac{\partial^3 v(\gamma)}{\partial x_i^3} = \sum_{k=1}^{m} w_{[x_i]}^{(k)} D_{[x_i]}^{[3](k)} + M_{[x_i]}^{[3]} v(\boldsymbol{\alpha}) + N_{[x_i]}^{[3]} v(\boldsymbol{\beta}) + P_{[x_i]}^{[3]} \frac{\partial v(\boldsymbol{\alpha})}{\partial x_i} + Q_{[x_i]}^{[3]} \frac{\partial v(\boldsymbol{\beta})}{\partial x_i}, \tag{3.39}$$

where $D_{[x_i]}^{[.](k)}$, $M_{[x_i]}^{[.]}$, $N_{[x_i]}^{[.]}$, $P_{[x_i]}^{[.]}$ and $Q_{[x_i]}^{[.]}$ are known functions of $\boldsymbol{\gamma}, \boldsymbol{\alpha}$ and $\boldsymbol{\beta}$.

Through (3.36), one can express the network weights in terms of the nodal variable values and the boundary conditions as

$$\mathbf{w}_{[x_i]} = \mathbf{D}_{[x_i]}^{[0]-1} \left(\mathbf{v}_{ip} - \mathbf{M}_{[x_i]}^{[0]} \mathbf{v}_{\alpha} - \mathbf{N}_{[x_i]}^{[0]} \mathbf{v}_{\beta} - \mathbf{P}_{[x_i]}^{[0]} \frac{\partial \mathbf{v}_{\alpha}}{\partial \mathbf{x_i}} - \mathbf{Q}_{[x_i]}^{[0]} \frac{\partial \mathbf{v}_{\beta}}{\partial \mathbf{x_i}} \right), \tag{3.40}$$

where $\mathbf{M}_{[x_i]}^{[0]}$, $\mathbf{N}_{[x_i]}^{[0]}$, $\mathbf{P}_{[x_i]}^{[0]}$ and $\mathbf{Q}_{[x_i]}^{[0]}$ are $n_{ip} \times n_{ip}$ diagonal matrices. In the physical space, the values of derivatives of v will be computed by

$$\frac{\partial^4 \mathbf{v}}{\partial \mathbf{x_i^4}} = \mathbf{E}_{[x_i]}^{[4]} \mathbf{v}_{ip} + \mathbf{k}_{[x_i]}^{[4]}, \tag{3.41}$$

$$\frac{\partial^3 \mathbf{v}}{\partial \mathbf{x_i^3}} = \mathbf{E}_{[x_i]}^{[3]} \mathbf{v}_{ip} + \mathbf{k}_{[x_i]}^{[3]}, \tag{3.42}$$

$$\frac{\partial^2 \mathbf{v}}{\partial \mathbf{x_i^2}} = \mathbf{E}_{[x_i]}^{[2]} \mathbf{v}_{ip} + \mathbf{k}_{[x_i]}^{[2]}, \tag{3.43}$$

$$\frac{\partial \mathbf{v}}{\partial \mathbf{x_i}} = \mathbf{E}_{[x_i]}^{[1]} \mathbf{v}_{ip} + \mathbf{k}_{[x_i]}^{[1]}, \tag{3.44}$$

where $\mathbf{E}_{[x_i]}^{[.]}$ and $\mathbf{k}_{[x_i]}^{[.]}$ are respectively known matrices and vectors, whose entries are functions of $\mathbf{x}$, $\boldsymbol{\alpha}$, $\boldsymbol{\beta}$ and the prescribed boundary conditions.

For the calculation of mixed partial derivatives of v, $\partial v^4(\mathbf{x})/\partial x_i^2 \partial x_j^2$, the following relation is employed

$$\frac{\partial^4 v}{\partial x_i^2 \partial x_j^2} = \frac{1}{2} \left[\frac{\partial^2}{\partial x_i^2} \left(\frac{\partial^2 v}{\partial x_j^2} \right) + \frac{\partial^2}{\partial x_j^2} \left(\frac{\partial^2 v}{\partial x_i^2} \right) \right]. \tag{3.45}$$

In (3.45), differential operators $\partial^2()/\partial x_i^2$ and $\partial^2()/\partial x_j^2$ are approximated by means of second-order IRBFNs, while $\partial^2 v/\partial x_j^2$ and $\partial^2 v/\partial x_i^2$ are simply taken from (3.43).

Making use of (3.41) and (3.45), the discrete form of the governing equation (3.31) will be obtained. It can be seen that the IRBFN formulation leads to a square system of linear equations in n_{ip} unknowns $\{v(\mathbf{x}^{(l)})\}_{l=1}^{n_{ip}}$.

4 Numerical results

Several linear and nonlinear examples are considered here. For scattered data, collocation points are generated using the PDETool of MATLAB with the "mesh size" h estimated as $h = \sqrt{A/\triangle/2}$, where A is the area of the domain and $\triangle$ is the number of triangular elements. The MQ width $a^{(k)}$ is simply chosen to be the distance between the centre $\mathbf{c}^{(k)}$ and its closest neighbour. In the case that the exact solution is available, the accuracy of a numerical scheme is measured through the discrete relative L_2 error of the solution defined as

$$N_e = \sqrt{\frac{\sum_{l=1}^{n_{ip}} \left[f_e(\mathbf{x}^{(l)}) - f(\mathbf{x}^{(l)})\right]^2}{\sum_{l=1}^{n_{ip}} f_e(\mathbf{x}^{(l)})^2}}, \tag{4.46}$$

where f_e and f are the exact and calculated solutions, respectively. Another important measure is the convergence rate of the solution with respect to the refinement of spatial discretization, defined by

$$N_e(h) \approx \theta h^{\nu} = O(h^{\nu}), \tag{4.47}$$

where θ and ν are the exponential model's parameters and h is the "mesh size". Given a set of observations $(N_e - h)$, these parameters can be found by the general linear least squares technique.

4.1 Example 1

This example is concerned with the Poisson equation in a unit circular domain with its centre located at the origin. The driving function and exact solution are

$$b(x_1, x_2) = -2\pi^2 \sin(\pi x_1)\sin(\pi x_2), \tag{4.48}$$

$$u_e(x_1, x_2) = \sin(\pi x_1)\sin(\pi x_2). \tag{4.49}$$

Both gridded and scattered data are employed (Figure 4.2a). For the former, the interior points are obtained from uniform grids with densities of $5\times5, 8\times8, \cdots, 50\times50$. The boundary points are formed by $2(n_{x_1}+n_{x_2})$ nodes uniformly distributed along the boundary (n_{x_i}–the number of points along the x_i direction). The "mesh size" h here is chosen to be the grid spacing. Results concerning N_e are presented in Figure 4.2b. For both discretizations, the present technique yields a fast rate of convergence, i.e. $O(h^{4.05})$ for scattered data and $O(h^{4.23})$ for gridded data.

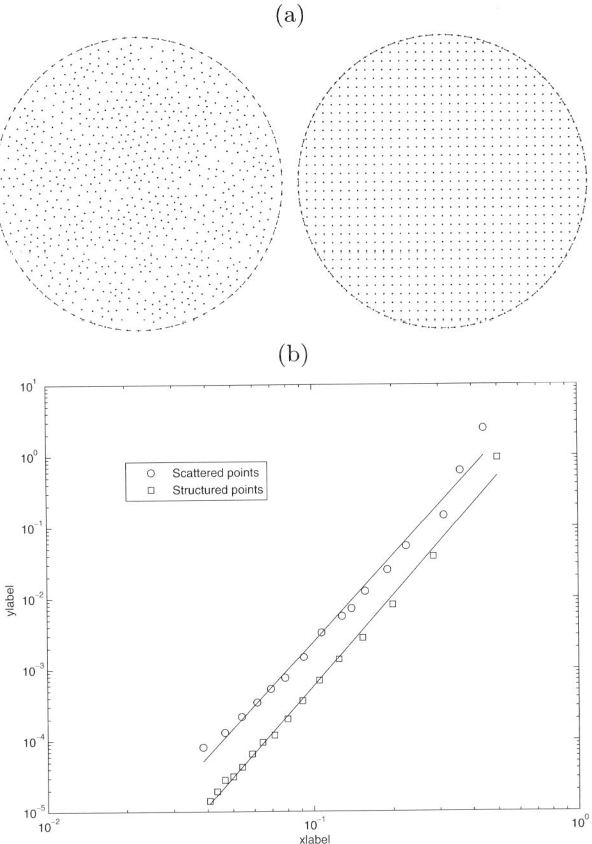

Figure 4.2. Poisson equation, circular domain: (a) Scattered and gridded data, and (b) Error N_e versus the "mesh size" h. Results with gridded data are more accurate than those with scattered data.

4.2 Example 2

This example is governed by the biharmonic equation in a square domain $[-1,-1] \times [1,1]$. The exact solution here is the same as that of the previous problem. The problem domain is discretized using scattered and gridded points (Figure 4.3a). For gridded data, there are 12 uniform grids employed, namely $5 \times 5, 8 \times 8, \cdots, 38 \times 38$. Figure 4.3b shows the convergence behaviour of the present method. The solution converges as $O(h^{4.65})$ for scattered data and $O(h^{4.84})$ for gridded data.

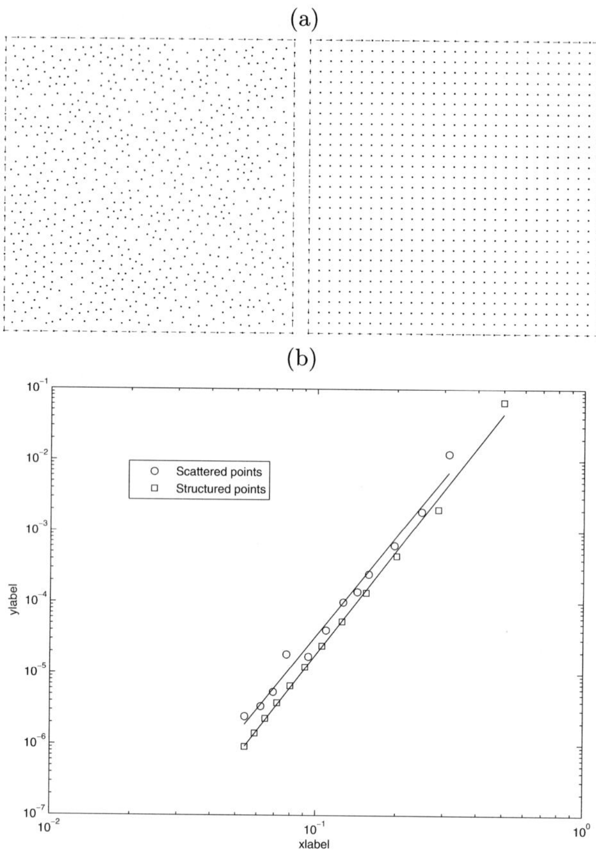

Figure 4.3. Biharmonic equation, square domain: (a) Scattered and gridded data, and (b) Error N_e versus the "mesh size" h. Results with gridded data are more accurate than those with scattered data.

4.3 Example 3

The present IRBFN method is applied to the simulation of a well-known lid-driven cavity viscous flow. This problem has usually been used as a model for the testing of new numerical schemes in CFD. The lid-driven cavity flow possesses physically unrealistic characteristics (discontinuous velocity) at the edges of the lid. This leads to a rapid change in stress near those points, thereby making the numerical simulation difficult.

In the context of Newtonian-fluid flow, Ghia, Ghia and Shin [7] have reported accurate solutions for a wide range of the Reynolds number using a multigrid FD scheme with very dense grids. These results have often been cited in the literature for comparison purposes. Recently, by using the Chebyshev collocation technique, which exhibits exponential convergence/spectral

Table 4.1. Lid-driven cavity flow, $Re = 1000$: Extrema of the vertical and horizontal velocity profiles through the centre of the cavity. It is noted that cpi. stands for consistent physical interpolation; stagg. staggered; $\psi - \omega$ the stream function and vorticity formulation; and $\mathbf{u} - p$ the velocity-pressure formulation.

Method	Density	v_{1min}	(error %)	x_2	v_{2max}	(error %)	x_1	v_{2min}	(error %)	x_1
Present	21 × 21	-0.39819	(2.480)	0.149	0.37826	(0.351)	0.159	-0.54847	(4.061)	0.934
	27 × 27	-0.39433	(1.485)	0.164	0.38027	(0.883)	0.157	-0.54446	(3.300)	0.916
	31 × 31	-0.39312	(1.175)	0.167	0.38008	(0.833)	0.156	-0.53993	(2.440)	0.913
	37 × 37	-0.39156	(0.774)	0.169	0.37901	(0.549)	0.157	-0.53450	(1.409)	0.910
	41 × 41	-0.39067	(0.544)	0.170	0.37833	(0.369)	0.157	-0.53207	(0.948)	0.910
	47 × 47	-0.38967	(0.287)	0.170	0.37752	(0.154)	0.157	-0.52972	(0.503)	0.909
FVM,stagg.	128 × 128	-0.38050	(2.077)	—	0.36884	(2.149)	—	-0.51727	(1.861)	—
FVM,cpi. [5]	128 × 128	-0.38511	(0.890)	—	0.37369	(0.862)	—	-0.52280	(0.812)	—
FDM($\psi - \omega$) [7]	129 × 129	-0.38289	(1.462)	0.171	0.37095	(1.589)	0.156	-0.51550	(2.197)	0.906
FDM($\mathbf{u} - p$) [4]	256 × 256	-0.3764	(3.132)	0.160	0.3665	(2.770)	0.152	-0.5208	(1.192)	0.910
Benchmark [3]		-0.38856		0.171	0.37694		0.157	-0.52707		0.909

accuracy, for the calculation of a regular part of the solution, and by using analytical formulae to obtain the singular part, Botella and Peyret [3] have provided benchmark spectral results on the flow at $Re = 1000$. It will be shown that the IRBFN results are in better agreement with the spectral solutions than the FD ones.

Unlike our previous works that study this flow, the present work takes the stream function formulation as the governing equation

$$-\frac{\partial \psi}{\partial x_2}\left(\frac{\partial^3 \psi}{\partial x_1^3} + \frac{\partial^3 \psi}{\partial x_1 \partial x_2^2}\right) + \frac{\partial \psi}{\partial x_1}\left(\frac{\partial^3 \psi}{\partial x_2 \partial x_1^2} + \frac{\partial^3 \psi}{\partial x_2^3}\right) =$$
$$\frac{1}{Re}\left(\frac{\partial^4 \psi}{\partial x_1^4} + 2\frac{\partial^4 \psi}{\partial x_1^2 \partial x_2^2} + \frac{\partial^4 \psi}{\partial x_2^4}\right), \tag{4.50}$$

where ψ is the stream function and Re is the Reynolds number. The vorticity and stream function are defined by

$$\omega = \frac{\partial v_2}{\partial x_1} - \frac{\partial v_1}{\partial x_2}, \tag{4.51}$$

$$\frac{\partial \psi}{\partial x_1} = -v_2, \quad \frac{\partial \psi}{\partial x_2} = v_1, \tag{4.52}$$

where v_1 and v_2 are two components of the velocity vector in the x_1- and x_2-directions, respectively.

The lid slides toward the right at unit velocity, while the other walls remain stationary:

$$\psi = 0, \quad \frac{\partial \psi}{\partial x_1} = 0, \quad \text{on} \quad x_1 = 0 \text{ and } x_1 = 1, \tag{4.53}$$

$$\psi = 0, \quad \frac{\partial \psi}{\partial x_2} = 0, \quad \text{on} \quad x_2 = 0, \tag{4.54}$$

$$\psi = 0, \quad \frac{\partial \psi}{\partial x_2} = 1, \quad \text{on} \quad x_2 = 1. \tag{4.55}$$

The obtained nonlinear system of equations is solved using a trust region technique [18]. Six uniform grids, namely $21 \times 21, 27 \times 27, \cdots, 47 \times 47$, are employed to study the convergence behaviour of the method. Results concerning extrema of the velocity profiles along the vertical and horizontal centrelines for the flow at $Re = 1000$ are summarized in Table 4.1. The corresponding results obtained by the pseudospectral method [3], FDM [4, 7] and FVM [5] are included for comparison. The IRBFN results are in better agreement with the spectral solutions than those predicted by FDMs and FVMs.

Unlike the pseudospectral technique [3], the presented global method does not require any special treatments for the singularity in stress at the two top corners. This attractive feature is probably owed to (i) the optimality of RBFNs [6] and (ii) the use of integrated basis functions.

5 Concluding remarks

This paper presents a meshless IRBFN collocation method for the solution of Dirichlet boundary value problems governed by second- and fourth-order elliptic PDEs. Integration is applied here to construct the RBFN approximations for the field variable, which helps to stabilize the numerical solution. All relevant integrals are obtained in analytic forms, and hence there is no need for the division of the problem domain into a set of elements. Moreover, prior to the transformation process, the constants of integration are utilized to incorporate the boundary conditions into IRBFNs, providing an effective way to implement the boundary conditions. Numerical results obtained show that the present technique achieves a high level of accuracy with relatively low numbers of points.

Acknowledgement

This research is supported by the Australian Research Council.

References

1. S.N. Atluri and S. Shen, *The Meshless Local Petrov-Galerkin Method*, Tech Science Press, Encino, 2002.

2. T. Belytschko, Y. Krongauz, D. Organ, M. Fleming and P. Krysl, *Meshless methods: an overview and recent developments*, Computer Methods in Applied Mechanics and Engineering, **139** (1996) 3–47.
3. O. Botella and R. Peyret, *Benchmark spectral results on the lid-driven cavity flow*, Computers & Fluids, **27**(4) (1998) 421–433.
4. C.-H. Bruneau and C. Jouron, *An efficient scheme for solving steady incompressible Navier-Stokes equations*, Journal of Computational Physics, **89**(2) (1990) 389–413.
5. G.B. Deng, J. Piquet, P. Queutey and M. Visonneau, *Incompressible flow calculations with a consistent physical interpolation finite volume approach*, Computers & Fluids, **23**(8) (1994) 1029–1047.
6. G.E. Fasshauer, *Meshfree Approximation Methods With Matlab*, Interdisciplinary Mathematical Sciences - Vol. 6, World Scientific Publishers, Singapore, 2007.
7. U. Ghia, K.N. Ghia and C.T. Shin, *High-Re solutions for incompressible flow using the Navier-Stokes equations and a multigrid method*, Journal of Computational Physics, **48** (1982) 387–411.
8. S. Haykin, *Neural Networks: A Comprehensive Foundation*, Prentice-Hall, New Jersey, 1999.
9. W.R. Madych and S.A. Nelson, *Multivariate interpolation and conditionally positive definite functions*, Approximation Theory and its Applications, **4** (1988) 77–89.
10. W.R. Madych and S.A. Nelson, *Multivariate interpolation and conditionally positive definite functions, II*, Mathematics of Computation, **54**(189) (1990) 211–230.
11. N. Mai-Duy and R.I. Tanner, *A collocation method based on one-dimensional RBF interpolation scheme for solving PDEs*, International Journal of Numerical Methods for Heat & Fluid Flow, **17**(2) (2007) 165–186.
12. N. Mai-Duy and T. Tran-Cong, *Numerical solution of differential equations using multiquadric radial basis function networks*, Neural Networks, **14**(2) (2001) 185–199.
13. N. Mai-Duy and T. Tran-Cong, *Numerical solution of Navier-Stokes equations using multiquadric radial basis function networks*, International Journal for Numerical Methods in Fluids, **37** (2001) 65–86.
14. N. Mai-Duy and T. Tran-Cong, *Approximation of function and its derivatives using radial basis function networks*, Applied Mathematical Modelling, **27** (2003) 197–220.
15. N. Mai-Duy and T. Tran-Cong, *An efficient indirect RBFN-based method for numerical solution of PDEs*, Numerical Methods for Partial Differential Equations, **21** (2005) 770–790.
16. N. Mai-Duy and T. Tran-Cong, *Solving biharmonic problems with scattered-point discretisation using indirect radial-basis-function networks*, Engineering Analysis with Boundary Elements, **30**(2) (2006) 77–87.
17. N. Mai-Duy and T. Tran-Cong, *A Cartesian-grid collocation method based on radial-basis-function networks for solving PDEs in irregular domains*, Numerical Methods for Partial Differential Equations, **23**(5) (2007) 1192–1210.
18. J.J. More and D.C. Sorensen, *Computing a trust region step*, SIAM Journal on Scientific and Statistical Computing, **3** (1983) 553–572.
19. H. Wendland, *Scaterred Data Approximation*, Cambridge University Press, Cambridge, 2005.

A Higher-Order Finite Volume Method Using Multiresolution Reproducing Kernels

Xesús Nogueira[1], Luis Cueto-Felgueroso[1,2], Ignasi Colominas[1], Fermín Navarrina[1], and Manuel Casteleiro[1]

[1] Group of Numerical Methods in Engineering, GMNI
Dept. of Applied Mathematics
Civil Engineering School, University of A Coruña
Campus de Elviña, 15071, A Coruña, Spain
xnogueira@udc.es, icolominas@udc.es, fnavarrina@udc.es, casteleiro@udc.es

[2] Department of Aeronautics and Astronautics
Massachusetts Institute of Technology
77 Massachusetts Ave 37-451, Cambridge, MA 02139, USA
lcueto@mit.edu

Summary. In this contribution we describe a numerical method based on the application of a meshfree interpolation technique (Moving Least Squares (MLS)), for the development of a higher-order finite volume discretization useful on structured and unstructured grids. With this procedure it is possible to build a higher-order scheme in which the computation of the derivatives is performed in a truly three-dimensional way. We use a MLS approach to compute the successive derivatives needed for the approximation of variables at element interfaces using Taylor series. Due to the use of cubic (or higher) reconstructions with the MLS technique, viscous fluxes are also approximated with higher-order accuracy and can be directly computed at edges.

The higher-order accuracy achieved by this method makes it suitable for aeroacoustics problems and DNS and LES of turbulent flows. One of the advantages of the use of the meshfree method on a finite volume framework is that it is possible to exploit all the shock capturing techniques developed for finite volume methods, allowing the computation of compressible flows. Moreover, we also present the application of this interpolation technique to shock detection. We make use of its connection with wavelets, and develop a technique we believe superior to traditional shock capturing methods. This application could be included in a very natural way in the computations with the method presented herein.

Key words: Moving Least Squares, higher-order methods, shock capturing, compressible flow.

1 Introduction

Higher-order accuracy and absence of unphysical oscillations in the vicinity of large gradients: this two contradictory requirements have attracted the attention of researchers since the early ages of numerical methods. The research aimed to address this problem has made possible the development of many "*higher-order*" accurate methods in the last decade. Nevertheless, the case of non smooth solutions for the compressible Navier-Stokes equations on unstructured grids, is still a challenging area of research.

In the field of Computational Fluid Dynamics (CFD), the increase of the order of accuracy in finite volume methods on general unstructured grids, has been limited due to the difficulty in the evaluation of field derivatives. These derivatives are required for the reconstruction of field variables by Taylor expansions [1, 2]. Moreover, simple extension of the 1D algorithm for multi-dimensional problems introduces numerical viscosity in case of waves moving in an oblique direction to the grid.

This difficulty may be overcome by meshfree methods. In this context, the Smooth Particle Hydrodynamics (SPH) method was first introduced in astrophysical applications [3]. The Moving Least Squares (MLS) method was proposed by Lancaster and Salkauskas [4] for smoothing and interpolating data. The good behavior of this method for the interpolation of scattered data has focused the attention of the meshfree community, and it has been widely analyzed [5].

In [6–8] a higher-order finite volume method using the MLS approximation has been proposed. In that method (FV-MLS in the sequel), the spatial finite volume discretization uses the MLS approximation to generate an underlying approximation space similar to that generated by shape functions in the finite element method. MLS approximation is used to compute the successive derivatives of the field variables. The meshfree nature of the MLS technique allows the development of a truly multidimensional higher-order approach with a finite volume scheme. The continuous spatial representation given by MLS is "*broken*" inside each cell into piecewise polynomial expansions, to make use of the finite volume technology for hyperbolic problems.

When a higher-order scheme is used and the solution is not smooth, there is a need of some kind of dissipation to maintain stability. The most usual technique in finite volume computations are flux limiters. For second order finite volume schemes well behaved limiters have been developed, but for higher-order reconstructions, the problem is still open. If the limiter is active over the whole domain, the higher-order accuracy of the reconstruction in smooth regions is lost. In this framework, selective limiting is critical. It is possible to use the multiresolution properties of MLS approximants to design a shock detector [5, 7]. It can be easily implemented in a code using the FV-MLS method. Preliminary tests are very promising.

In this contribution, we revisit and describe the FV-MLS method and some of its features, including several numerical examples. The outline of the paper is as follows. Section 2 presents the governing equations and section 3 is devoted to numerical formulations. Shock detection is introduced in section 4, and several numerical examples are presented in section 5.

2 Governing equations

The Navier-Stokes equations can be written in conservative form as:

$$\frac{\partial \boldsymbol{U}}{\partial t} + \frac{\partial(\boldsymbol{F}_x - \boldsymbol{F}_x^V)}{\partial x} + \frac{\partial(\boldsymbol{F}_y - \boldsymbol{F}_y^V)}{\partial y} = 0 \tag{2.1}$$

being

$$\boldsymbol{U} = \begin{pmatrix} \rho \\ \rho v_x \\ \rho v_y \\ \rho E \end{pmatrix} \qquad \boldsymbol{F}_x = \begin{pmatrix} \rho v_x \\ \rho v_x^2 + p \\ \rho v_x v_y \\ \rho v_x H \end{pmatrix} \qquad \boldsymbol{F}_y = \begin{pmatrix} \rho v_y \\ \rho v_x v_y \\ \rho v_y^2 + p \\ \rho v_y H \end{pmatrix} \tag{2.2}$$

$$\boldsymbol{F}_x^V = \begin{pmatrix} 0 \\ \tau_{xx} \\ \tau_{xy} \\ v_x\tau_{xx} + v_y\tau_{xy} - q_x \end{pmatrix} \qquad \boldsymbol{F}_y^V = \begin{pmatrix} 0 \\ \tau_{xy} \\ \tau_{yy} \\ v_x\tau_{xy} + v_y\tau_{yy} - q_y \end{pmatrix} \tag{2.3}$$

and

$$\rho E = \rho e + \frac{1}{2}\rho\,(\boldsymbol{v}\cdot\boldsymbol{v}) \qquad H = E + \frac{p}{\rho} \tag{2.4}$$

The stress tensor is given by

$$\tau_{ij} = \mu\left(\frac{\partial v_i}{\partial x_j} + \frac{\partial v_j}{\partial x_i} - \frac{2}{3}\boldsymbol{\nabla}\cdot\boldsymbol{v}\delta_{ij}\right) \tag{2.5}$$

where $\boldsymbol{\mathcal{F}} = (\boldsymbol{F}_x, \boldsymbol{F}_y)$ is the inviscid flux term, and $\boldsymbol{\mathcal{F}}^V = (\boldsymbol{F}_x^V, \boldsymbol{F}_y^V)$ the viscous flux. On the other hand, $\boldsymbol{U}$ is the vector of conservative variables, $\boldsymbol{v} = (v_x, v_y)$ is the velocity, μ is the effective viscosity of the fluid, H is the enthalpy, E is the total energy, e is the total internal energy and ρ is the density.

The thermal flux $\boldsymbol{q} = (q_x, q_y)$ is calculated by using Fourier's law although alternative formulations are possible. The authors are working on developing formulations for the calculation of this term based on Cattaneo's law [9]. It is assumed that the viscosity depends on the temperature according to Sutherland's law:

$$\mu = \mu_{ref}\frac{T_{ref} + S_0}{T + S_0}\left(\frac{T}{T_{ref}}\right)^{1.5} \tag{2.6}$$

where T is the temperature, the subindex ref refers to a reference value and $S_0 = 110.4K$ is an empirical constant (Sutherland's temperature).

3 Numerical schemes

3.1 Finite Volume Method

In order to solve equation (2.1), a finite volume scheme with the numerical flux proposed by Roe [10] has been used. The equation to be solved is the one resulting by the application of the finite volume discretization:

$$A_I \frac{\partial \boldsymbol{U}_I}{\partial t} = \sum_{iedge=1}^{nedge_I} \sum_{iq=1}^{nq_I} \left[\left(\boldsymbol{\mathcal{F}}^V - \boldsymbol{\mathcal{F}} \right) \cdot \boldsymbol{n} \right]_{iq} \mathcal{W}_{iq} \tag{3.7}$$

A_I is the area of cell I, $nedge_I$ the number of cell edges, $\boldsymbol{U}_I$ the average value of $\boldsymbol{U}$ over the cell I (associated to the cell centroid), iq is the index for integration points and $\mathcal{W}$ are the integration weights.

The MLS approximation has been used to compute the derivatives and the viscous fluxes as it is explained in the following sections. For convective terms, higher-order accuracy is achieved by using a Taylor expansion in the reconstruction of variables at the edges, whereas viscous fluxes are directly computed at quadrature points.

3.2 Moving Least Squares (MLS)

The aim of meshless methods is to solve the continuum equations in a particle framework, by using the information stored at certain nodes without reference to any underlying mesh. If this particle approach is used in combination with classical discretization procedures (e.g. the weighted residuals method), then a spatial approximation is required (similar to "shape functions", as in the finite element method). Such an interpolation scheme should accurately *reproduce* or *reconstruct* a certain function and its successive derivatives, using the *nodal* (particle) *values* and "low-level" geometrical information about the grid, such as the distance between particles. Furthermore, in order to achieve computationally efficient algorithms, the interpolation should have a *local* character, i.e. the reconstruction process should involve only a few "neighbor" nodes. Let us consider a function $u(\boldsymbol{x})$ defined in a domain Ω. The basic idea of the MLS approach is to approximate $u(\boldsymbol{x})$, at a given point $\boldsymbol{x}$, through a weighted least-squares fitting of $u(\boldsymbol{x})$ in a neighborhood of $\boldsymbol{x}$ as

$$u(\boldsymbol{x}) \approx \widehat{u}(\boldsymbol{x}) = \sum_{i=1}^{m} p_i(\boldsymbol{x})\, \boldsymbol{\alpha}_i(\boldsymbol{z}) \Big|_{\boldsymbol{z}=\boldsymbol{x}} = \boldsymbol{p}^T(\boldsymbol{x})\, \boldsymbol{\alpha}(\boldsymbol{z}) \Big|_{\boldsymbol{z}=\boldsymbol{x}} \tag{3.8}$$

$\boldsymbol{p}^T(\boldsymbol{x})$ is an m-dimensional polynomial basis and $\boldsymbol{\alpha}(\boldsymbol{z})\Big|_{\boldsymbol{z}=\boldsymbol{x}}$ is a set of parameters to be determined, such that they minimize the following error functional

$$J\left(\boldsymbol{\alpha}(\boldsymbol{z})\Big|_{\boldsymbol{z}=\boldsymbol{x}} \right) = \int_{\boldsymbol{y}\in\Omega_{\boldsymbol{x}}} W(\boldsymbol{z}-\boldsymbol{y},h)\Big|_{\boldsymbol{z}=\boldsymbol{x}} \left[u(\boldsymbol{y}) - \boldsymbol{p}^T(\boldsymbol{y})\boldsymbol{\alpha}(\boldsymbol{z})\Big|_{\boldsymbol{z}=\boldsymbol{x}} \right]^2 d\Omega_{\boldsymbol{x}} \tag{3.9}$$

being $W(\boldsymbol{z}-\boldsymbol{y},h)\Big|_{\boldsymbol{z}=\boldsymbol{x}}$ a *kernel* with compact support (denoted by $\Omega_{\boldsymbol{x}}$) centered at $\boldsymbol{z}=\boldsymbol{x}$, frequently chosen among the kernels used in standard SPH. The parameter h is the smoothing length, which is a measure of the size of the support $\Omega_{\boldsymbol{x}}$. Following [7], the interpolation structure can be identified as

$$\widehat{u}(\boldsymbol{x})=\boldsymbol{p}^T(\boldsymbol{x})\boldsymbol{M}^{-1}(\boldsymbol{x})\boldsymbol{P}_{\Omega_{\boldsymbol{x}}}\boldsymbol{W}(\boldsymbol{x})\boldsymbol{u}_{\Omega_{\boldsymbol{x}}}=\boldsymbol{N}^T(\boldsymbol{x})\boldsymbol{u}_{\Omega_{\boldsymbol{x}}}=\sum_{j=1}^{n_{\boldsymbol{x}}}N_j(\boldsymbol{x})u_j \quad (3.10)$$

where the vector $\boldsymbol{u}_{\Omega_{\boldsymbol{x}}}$ contains the pointwise values of the function u to be reproduced at the $n_{\boldsymbol{x}}$ particles (nodes) inside $\Omega_{\boldsymbol{x}}$. We define the $(m\times n_{\boldsymbol{x}})$ matrix $\boldsymbol{P}_{\Omega_{\boldsymbol{x}}}=(\boldsymbol{p}(\boldsymbol{x}_1)\,\boldsymbol{p}(\boldsymbol{x}_2)\cdots\boldsymbol{p}(\boldsymbol{x}_{n_{\boldsymbol{x}}}))$ and the $(n_{\boldsymbol{x}}\times n_{\boldsymbol{x}})$ diagonal matrix $\boldsymbol{W}=diag\left[W_i\left(\boldsymbol{x}-\boldsymbol{x_i}\right)\right]$ with $i=1,\ldots,n_{\boldsymbol{x}}$. Moreover, $\boldsymbol{M}=\boldsymbol{P}_{\Omega_{\boldsymbol{x}}}\boldsymbol{W}\boldsymbol{P}_{\Omega_{\boldsymbol{x}}}^T$ is the moment matrix (see [7]).

The approximation is written in terms of the MLS "shape functions" $\boldsymbol{N}^T(\boldsymbol{x})=\boldsymbol{p}^T(\boldsymbol{x})\boldsymbol{C}(\boldsymbol{x})$, where $\boldsymbol{C}(\boldsymbol{x})$ is defined as $\boldsymbol{C}(\boldsymbol{x})=\boldsymbol{M}^{-1}(\boldsymbol{x})\boldsymbol{P}_{\Omega_{\boldsymbol{x}}}\boldsymbol{W}(\boldsymbol{x})$. In this work the following polynomial cubic basis is used:

$$\boldsymbol{p}(\boldsymbol{x})=\begin{pmatrix}1 & x & y & xy & x^2 & y^2 & x^2y & xy^2 & x^3 & y^3\end{pmatrix}^T \quad (3.11)$$

which provides cubic completeness. In the above expression, (x,y) denotes the cartesian coordinates of $\boldsymbol{x}$. It is frequent to use scaled and locally defined monomials in the basis [7]. The particles needed for the application of the method are identified with the centroids of every cell of the grid. For boundary cells, we add nodes (ghost nodes) placed in the middle of the edge wall.

A very important point of this method is the construction of the stencils that determine the clouds of points to be used for the approximation. These stencils are built at the beginning of the computation process. If the mesh is not moving, stencils will not change in time and will be only computed once. In section 3.4 we present some comments about stencils. Complete details about this method can be found in [6–8]. We remark that the calculation of the parameters $\boldsymbol{\alpha}(\boldsymbol{z})\Big|_{\boldsymbol{z}=\boldsymbol{x}}$ leads to "shape functions" that allow us to write an approximate solution:

$$\widehat{u}(\boldsymbol{x})=\sum_{j=1}^{n_{\boldsymbol{x}_I}}u_jN_j(\boldsymbol{x}) \quad (3.12)$$

Then, the gradient of $\widehat{u}(\boldsymbol{x})$ is evaluated as

$$\boldsymbol{\nabla}\widehat{u}(\boldsymbol{x})=\sum_{j=1}^{n_{\boldsymbol{x}_I}}u_j\boldsymbol{\nabla}N_j(\boldsymbol{x}) \quad (3.13)$$

For the case of unlimited reconstructions, where only smooth solutions are calculated, the derivatives of the field variables are computed directly at centroids using moving least-squares. Thus, the approximate gradients of a vector-valued function evaluated at the centroid of a cell I ($\boldsymbol{U}_I$) read:

$$\nabla \boldsymbol{U}_I = \sum_{j=1}^{n_{\boldsymbol{x}_I}} \boldsymbol{U}_j \boldsymbol{\nabla} N_j(\boldsymbol{x}_I) \tag{3.14}$$

whereas the second derivatives can be written as follows:

$$\begin{gathered}\frac{\partial^2 \boldsymbol{U}_I}{\partial x} = \sum_{j=1}^{n_{\boldsymbol{x}_I}} \boldsymbol{U}_j \frac{\partial^2 N_j(\boldsymbol{x}_I)}{\partial x^2}, \frac{\partial^2 \boldsymbol{U}_I}{\partial y^2} = \sum_{j=1}^{n_{\boldsymbol{x}_I}} \boldsymbol{U}_j \frac{\partial^2 N_j(\boldsymbol{x}_I)}{\partial y^2} \\ \frac{\partial^2 \boldsymbol{U}_I}{\partial x \partial y} = \sum_{j=1}^{n_{\boldsymbol{x}_I}} \boldsymbol{U}_j \frac{\partial^2 N_j(\boldsymbol{x}_I)}{\partial x \partial y}\end{gathered} \tag{3.15}$$

Higher-order derivatives are obtained in a straightforward way. In this work, first-order derivatives are computed as full MLS derivatives, whereas second and third-order derivatives are approximated by the diffuse ones [6–8]. The diffuse derivatives result of neglect the successive derivatives of matrix $\boldsymbol{C}(\boldsymbol{x})$, when computing the derivatives of matrix $\boldsymbol{N} = \boldsymbol{p}^T \boldsymbol{C}$. It is possible to show that diffuse derivatives converge optimally to the complete ones [11]. Although this approach greatly simplifies the computing of the MLS approximants, in the case of problems involving rough grids may require the use of full derivatives.

3.3 FV-MLS method.

Standard higher-order schemes are constructed through the substitution of a piecewise constant representation for a *piecewise continuous*(usually polynomial) reconstruction of the flow variables inside each cell. Due to the fact that the reconstructed fields are still discontinuous across interfaces, the discretization of viscous terms is done by different procedures. This "bottom-up" methodology is quite different to the way in that the FV-MLS method works. Here, we work with *pointwise* values of the conserved variables, associated to the cell centroids. The spatial representation provided by the MLS approximants is *continuous* and higher-order accurate. Within this framework, the discretization of elliptic terms is straightforward. In order to deal with convection-dominated problems and to take the most of the finite volume technology for hyperbolic terms, we *break* the continuous representation inside each cell by means of Taylor expansions ("top-down" procedure). The resulting scheme is like a Godunov method [12] in the convective terms, but with a much clearer and more accurate discretization of elliptic terms. In the following sections, it is shown how the reconstruction process is made.

Reconstruction of Inviscid Fluxes.

A reconstruction scheme to evaluate the value of the variables at the edges of the element and compute the numerical flux is needed to apply a higher-order finite volume method. Using a Taylor expansion, the linear component-wise reconstruction of the field variables inside each cell I reads:

$$\boldsymbol{U}(\boldsymbol{x}) = \boldsymbol{U}_I + \boldsymbol{\nabla U}_I \cdot (\boldsymbol{x} - \boldsymbol{x}_I) \tag{3.16}$$

$\boldsymbol{U}_I$ is the average value of $\boldsymbol{U}$ over I (associated to the centroid), $\boldsymbol{x}_I$ denotes the cartesian coordinates of the centroid and $\boldsymbol{\nabla U}_I$ is the gradient of the variable at the centroid. The aforementioned gradient is assumed to be constant inside each cell and, therefore, the reconstructed variable is still discontinuous across interfaces. Note that we have broken the continuity of the spatial representation of the variable. This allows to connect the method with classical higher-order finite volume schemes. The first, second and third-order derivatives of the field variables will be computed using MLS approximations (equations 3.14 and 3.15).

Analogously, the quadratic reconstruction reads

$$\boldsymbol{U}(\boldsymbol{x}) = \boldsymbol{U}_I + \boldsymbol{\nabla U}_I \cdot (\boldsymbol{x} - \boldsymbol{x}_I) + \frac{1}{2}(\boldsymbol{x} - \boldsymbol{x}_I)^T \boldsymbol{H}_I (\boldsymbol{x} - \boldsymbol{x}_I) \tag{3.17}$$

where $\boldsymbol{H}_I$ is the centroid Hessian matrix.

In case of cubic reconstruction:

$$\boldsymbol{U}(\boldsymbol{x}) = \boldsymbol{U}_I + \boldsymbol{\nabla U}_I \cdot (\boldsymbol{x} - \boldsymbol{x}_I) + \frac{1}{2}(\boldsymbol{x} - \boldsymbol{x}_I)^T \boldsymbol{H}_I \boldsymbol{x} - \boldsymbol{x}_I + \frac{1}{6}\boldsymbol{\Delta}^2 \boldsymbol{x}_I^T \boldsymbol{T}_I\,(\boldsymbol{x} - \boldsymbol{x}_I) \tag{3.18}$$

where

$$\boldsymbol{\Delta}^2 \boldsymbol{x}_I^T = \left((x - x_I)^2 \, (y - y_I)^2 \right) \tag{3.19}$$

and

$$\boldsymbol{T}_I = \begin{pmatrix} \dfrac{\partial^3 \boldsymbol{U}_I}{\partial x^3} & 3\dfrac{\partial^3 \boldsymbol{U}_I}{\partial x^2 \partial y} \\ \\ 3\dfrac{\partial^3 \boldsymbol{U}_I}{\partial x \partial y^2} & \dfrac{\partial^3 \boldsymbol{U}_I}{\partial y^3} \end{pmatrix} \tag{3.20}$$

For unsteady problems, additional terms must be introduced in (3.17) and (3.18) to enforce conservation of the mean, i.e.

$$\frac{1}{A_I} \int_{\boldsymbol{x} \in \Omega_I} \boldsymbol{U}(\boldsymbol{x}) d\Omega = \boldsymbol{U}_I \tag{3.21}$$

Thus, the quadratic reconstruction for unsteady problems reads:

$$\begin{aligned} \boldsymbol{U}(\boldsymbol{x}) = \boldsymbol{U}_I + \boldsymbol{\nabla U}_I \cdot (\boldsymbol{x} - \boldsymbol{x}_I) + \frac{1}{2}(\boldsymbol{x} - \boldsymbol{x}_I)^T \boldsymbol{H}_I (\boldsymbol{x} - \boldsymbol{x}_I) - \\ \frac{1}{2A_I}\left[I_{xx} \frac{\partial^2 \boldsymbol{U}}{\partial x^2} + 2 I_{xy} \frac{\partial^2 \boldsymbol{U}}{\partial x \partial y} + I_{yy} \frac{\partial^2 \boldsymbol{U}}{\partial y^2} \right] \end{aligned} \tag{3.22}$$

with:

$$I_{xx} = \int_{\Omega} (x - x_I)^2 d\Omega, \quad I_{xy} = \int_{\Omega} (x - x_I)(y - y_I) d\Omega, \quad I_{yy} = \int_{\Omega} (y - y_I)^2 d\Omega$$

In case of non smooth solutions many shock limiter techniques may be used, including slope limiters techniques (e.g. [13]). Those limit the value of the derivatives in order to impose monotonicity on the solution. Here, monotonicity means that the value of the reconstruction of a variable in one cell can not exceed the maximum and minimum values of the variable in a set of neighboring cells. This set is composed of the first neighbors, but other options are possible [7].

Reconstruction of Viscous Fluxes.

Viscous terms discretization is a great problem for methods that use piecewise polynomial approximations. For example, second order schemes often use the average of the derivatives of the flow variables on each side of the interface to compute the viscous fluxes. This is not acceptable for higher-order approximations. The FV-MLS method performs a centered reconstruction of the viscous fluxes at the quadrature points on the edges using information from neighboring cells. The evaluation of the viscous stresses and heat fluxes requires an approximation of the velocity vector, the temperature, and their corresponding gradients, $\nabla \boldsymbol{v}$ and ∇T, at each quadrature point $\boldsymbol{x}_{iq}$. Using the MLS approximation, these quantities are readily computed as:

$$\boldsymbol{v}_{iq} = \sum_{j=1}^{n_I} \boldsymbol{v}_j N_j(\boldsymbol{x}_{iq}), \quad T_{iq} = \sum_{j=1}^{n_I} T_j N_j(\boldsymbol{x}_{iq}) \tag{3.23}$$

and

$$\boldsymbol{\nabla v}_{iq} = \sum_{j=1}^{n_I} \boldsymbol{v}_j \otimes \boldsymbol{\nabla} N_j(\boldsymbol{x}_{iq}), \quad \boldsymbol{\nabla} T_{iq} = \sum_{j=1}^{n_I} T_j \boldsymbol{\nabla} N_j(\boldsymbol{x}_{iq}) \tag{3.24}$$

where n_I is the number of neighbor centroids given by the stencil. Then, the diffusive fluxes can be computed according to equation (2.3). We note that complete derivatives have been used for the computation of viscous fluxes.

In the application of a higher-order finite volume method, the accuracy of the numerical integration of fluxes at edges is crucial. Thus, it is required to use more than one quadrature points to compute the flux integrals. In this work, 2D calculations have been performed by using three quadrature points per edge for the cubic reconstruction, whereas 3D calculations have been computing with four quadrature points per face. The order of convergence achieved with this method for the reconstruction of viscous terms is the same than the one for the reconstruction of inviscid fluxes. If m is the order of the polynomial basis, then the order of convergence of the global scheme is $m + 1$ [7,14].

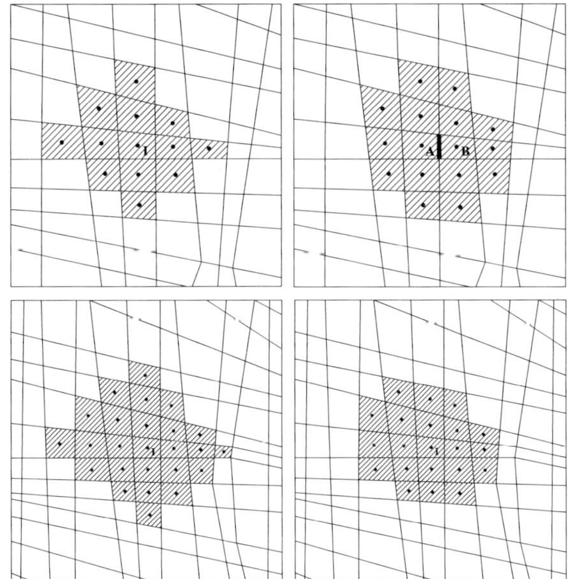

Figure 3.1. Upper figure: Stencil for interior points (left) and edges (right) for MLS interpolation. Lower figure: Full 4th order FV-MLS stencil: Euler and Navier-Stokes (left) and elliptic problems (right).

3.4 The full stencil of FV-MLS.

Figure 3.1 shows the stencils to compute the MLS shape functions for the fourth order FV-MLS method. A complete description of the stencils can be found in [7, 8]. For inviscid problems, the stencil of a cell I is obtained as the union of its MLS stencil and the MLS stencils of its first neighbors. As it can be seen in figure 3.1 the full stencil comprises 25 cells. Analogously, the stencil of the "viscous" discretization is obtained as the union of the MLS stencils associated to all the edges of cell I, and it comprises 21 cells. This is in fact a quite compact stencil, despite the preconceived ideas about the non-compactness of this approach. In [8] it is shown that the average number of entries in each row of the coefficient matrix (what can be considered as a measure of the compactness of a scheme) is fewer for the FV-MLS method than for several DG methods for the same order of accuracy.

4 Shock detection

An interesting feature of MLS approximations is its connection with wavelets [5], where it is used for error estimation and adaptivity. Let consider a function $u(\boldsymbol{x})$. Let us define two sets of MLS shape functions $\boldsymbol{N}^h(\boldsymbol{x})$ and $\boldsymbol{N}^{2h}(\boldsymbol{x})$,

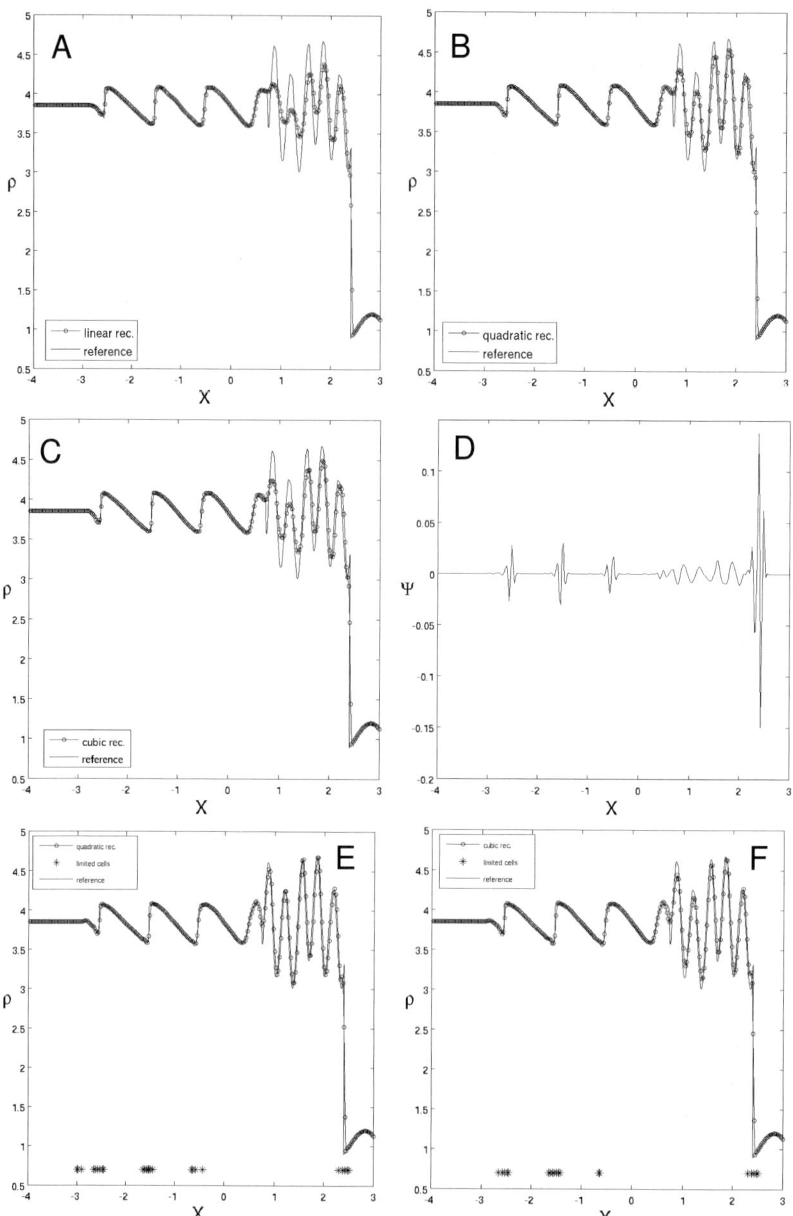

Figure 4.2. Shu-Osher problem, 400 cells, limiters active everywhere for the linear reconstruction (A), quadratic (B) and cubic (C) reconstructions. Selective limiting (at asterisk points) for the quadratic (E) and cubic (F) MLS interpolation. High-scale components of the density are shown in D.

computed with smoothing lengths h and $2h$. We obtain h-scale and $2h$-scale approximations:

$$u^h(\boldsymbol{x}) = \sum_{j=1}^{n_I} u_j N_j^h(\boldsymbol{x}) \qquad u^{2h}(\boldsymbol{x}) = \sum_{j=1}^{n_I} u_j N_j^{2h}(\boldsymbol{x}) \tag{4.25}$$

Thus, it is possible to obtain a set of wavelets functions with:

$$\boldsymbol{\Phi}^{2h}(\boldsymbol{x}) = \boldsymbol{N}^h(\boldsymbol{x}) - \boldsymbol{N}^{2h}(\boldsymbol{x}) \tag{4.26}$$

Then, we can write the h-scale solution as the sum of its low-scale and high-scale complementary parts:

$$u^h(\boldsymbol{x}) = u^{2h}(\boldsymbol{x}) + \boldsymbol{\Psi}^{2h}(\boldsymbol{x}) \tag{4.27}$$

with

$$\boldsymbol{\Psi}^{2h}(\boldsymbol{x}) = \sum_{j=1}^{n_I} u_j \boldsymbol{\Phi}_j^{2h}(\boldsymbol{x}) = \sum_{j=1}^{n_I} u_j (\boldsymbol{N}^h(\boldsymbol{x}) - \boldsymbol{N}^{2h}(\boldsymbol{x})) \tag{4.28}$$

The low-scale $u^{2h}(\boldsymbol{x})$ can be further decomposed using the same rationale. Function $\boldsymbol{\Psi}^{2h}(\boldsymbol{x})$ acts as a smoothness indicator for $u(\boldsymbol{x})$, and so it is possible to use it as shock detector or error sensor for adaptative and multiresolution algorithms. A complete development about this topic can be found in [7]. As an example of the performance of this shock detector,we present the results for the 1D Shu and Osher problem [15] in figure 4.2. In this figure the reference solution has been computed on a grid of 3200 cells with the quadratic reconstruction and the limiter active everywhere. The 1D Euler equations are solved in $[-5, 5]$, with 400 cells and initial conditions $(\rho, v, p) = (3.857, 2.629, 10.333)$ if $x < -4$ and $(\rho, v, p) = (1 + 0.2\sin(5x), 0, 1)$ if $x \geq -4$.

5 Numerical examples

In this section we present two examples: a shock wave impingement on a spatially evolving mixing layer, and a non viscous supersonic flow. For the following calculations we use the cubic kernel:

$$W(s) = \begin{cases} 1 - \frac{3}{2}s^2 + \frac{3}{4}s^3 & s \leq 1 \\ \frac{1}{4}(2-s)^3 & 1 < s \leq 2 \\ 0 & s > 2 \end{cases} \tag{5.29}$$

The smoothing length is the same in each direction. Further examples and discussion can be seen in [7, 8, 14].

5.1 Shock wave impingement on a spatially evolving mixing layer.

We reproduce the example proposed in [16] by Yee et al. An oblique shock impacts on a spatially evolving mixing layer. The problem domain is the rectangle $[0, 200]\times[-20, 20]$. The inflow is set with a hyperbolic tangent profile:

$$v_x = 2.5 + 0.5\tanh(2y) \tag{5.30}$$

Hence, the velocity of the upper stream is $v_{x1} = 3$, and the velocity of the lower stream is $v_{x2} = 2$. The convective Mach number defined by $\frac{v_{x1}-v_{x2}}{c_1+c_2}$, where c_1 and c_2 are the free stream sound speeds, is equal to 0.6. The shear layer is excited by adding a periodic fluctuation to the vertical component of the velocity inflow, as:

$$v_y' = \sum_{k=1}^{2} a_k \cos\left(\frac{2\pi kt}{T} + \phi_k\right) e^{\left(\frac{-y^2}{b}\right)} \tag{5.31}$$

where $T = \frac{\lambda}{u_c}$, $u_c = 2.68$ is the convective velocity defined by $u_c = \frac{v_{x1}c_2+v_{x2}c_1}{c1+c2}$, wavelength $\lambda = 30$, $b = 10$. For $k = 1$ we take the values of $a_1 = 0.05$ and $\phi_1 = 0$. For $k = 2$, the values are $a_2 = 0.05$ and $\phi_2 = \frac{\pi}{2}$. The reference density is taken as the average of the two free streams and a reference pressure is given by:

$$p_R = \frac{(\rho_1 + \rho_2)\,(v_{x1} - v_{x2})^2}{2} \tag{5.32}$$

Under the assumption that both streams have equal stagnation enthalpies, the local speed of sound reads:

$$c^2 = {c_1}^2 + \frac{(\gamma-1)}{2}\left({v_{x1}}^2 - {v_{x2}}^2\right). \tag{5.33}$$

Equal pressure through the mixing layer is assumed. In the simulation, we have taken the following values for the freestream:

$$p_0 = 0.3327, \quad H_0 = 5.211, \quad \mu_0 = 5\times10^{-4} \tag{5.34}$$

On the upper boundary we have set the values according to the properties behind an oblique shock with angle $\beta = 12°$. Thus, we set the following values:

$$v_x = 2.9709, \quad v_y = -0.1367, \quad \rho = 2.1101, \quad p = 0.4754 \tag{5.35}$$

On the lower boundary, we impose standard slip-wall BC. With this problem setup, an oblique shock originates in the top left corner, impacting the shear layer. The shock wave reflects at the lower wall and passes back through the deflected shear layer. The problem has been executed by using a fourth order FV-MLS scheme on both, a 400×100 and a 600×300 cartesian grids to compare with a very fine mesh. Figures 5.3, 5.4 and 5.5 show the contours of density, pressure and temperature. The scheme is capable of capturing the fine scale features of the flow, such as the formation of shocklets or the splitting in two of the vortex core located at $x = 148$, caused by its interaction with the reflected shock wave.

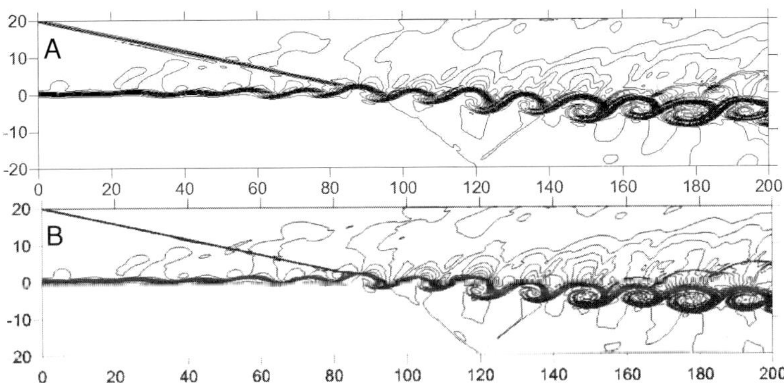

Figure 5.3. Solution for the shock wave impingement on a mixing layer at $t = 120$. Results for the density on the 400×100 grid (A) and on the 600×300 grid (B).

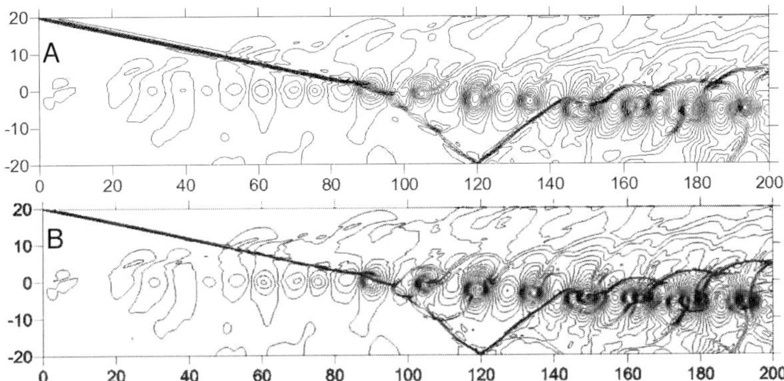

Figure 5.4. Solution for the shock wave impingement on a mixing layer at $t = 120$. Results for the pressure on the 400×100 grid (A) and on the 600×300 grid (B).

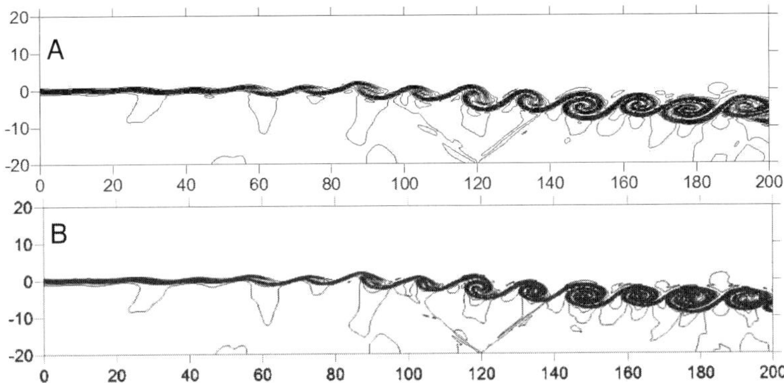

Figure 5.5. Solution for the shock wave impingement on a mixing layer at $t = 120$. Results for the temperature on the 400×100 grid (A) and on the 600×300 grid (B).

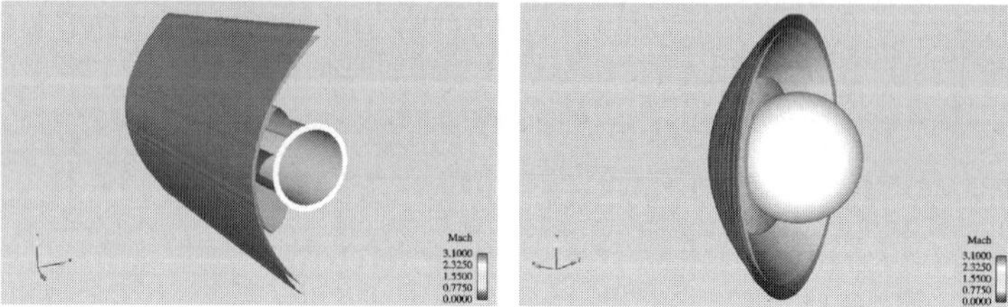

Figure 5.6. Mach isosurfaces for flow past a cylinder, third order reconstruction (left) and past a sphere, second order reconstruction (right). (*See also* Color Plate on page 376)

5.2 Supersonic flow. 3D Euler equations.

In this section we present results for the 3D Euler equations. We have computed the supersonic flow past a cylinder and past a sphere. Mach number is 3 for both examples. Calculations have been performed on a 3D grid (although it is clear that there exist some kind of symmetry). The limiter developed by Barth and Jespersen [13] is used. A cubic polynomial basis has been used for the reconstruction. Figure 5.6 shows Mach isosurfaces for both cases.

6 Conclusions

A higher-order finite volume based method (FV-MLS) coupled with a meshfree technique (Moving Least Squares) has been presented. The good behavior of the MLS for interpolation for scattered point valued data is applied to the computation of successive derivatives for the reconstruction of the convective term at interfaces between elements. MLS is also used to compute the viscous fluxes directly at integration points. This is a very important feature of the method being possible the computation of these terms with high accuracy. Moreover, the multiscale properties of MLS allow the construction of a shock detector easily introduced in a code using FV-MLS. We have presented a 1D test of this detector for the Shu-Osher problem. The results are very good, only comparable to those given by ENO/WENO methods. Numerical examples have been presented, and its results show that it is possible to use this method for real 3D problems.

Acknowledgements

This work has been partially supported by the *Ministerio de Educación y Ciencia* of the Spanish Government (#DPI2006-15275 and #DPI2007-61214) cofinanced with FEDER funds, the *Secretaría Xeral de I+D* of the *Xunta de Galicia* (Grants #PGIDIT05PXIC118002PN and # PGDIT06TAM11801PR) and the *University of A Coruña*. X. Nogueira was granted by the *Fundación*

Caixa Galicia. Dr. Cueto-Felgueroso gratefully acknowledge the support provided by *Ministerio de Educación y Ciencia* through the postdoctoral fellowships program.

References

1. A. Harten, P. Lax and B. Van Leer. *On upstream differencing and Godunov-type schemes for hyperbolic conservation laws*, SIAM Rev. **25**:35-61, (1983).
2. T. J. Barth. *Aspects of unstructured grids and finite-volume solvers for the Euler and Navier-Stokes equations*, VKI Lecture Series 1994-05, (1995).
3. R. A. Gingold and J. J. Monaghan. *Smoothcd Particlc Hydrodynamics: theory and application to non-spherical stars*, Mon. Not. R. Astron. Soc., **181**: 378, (1977).
4. P. Lancaster and K. Salkauskas. *Surfaces generated by moving least squares methods*, Mathematics of Computation, **87**: 141-158, (1981).
5. W. K. Liu, W. Hao, Y. Chen, S. Jun, J. Gosz. *Multiresolution Reproducing Kernel Particle Methods*, Computational Mechanics, **20**: 295-309, (1997).
6. L. Cueto-Felgueroso, I. Colominas, J. Fe, F. Navarrina, M. Casteleiro. *High order finite volume schemes on unstructured grids using Moving Least Squares reconstruction. Application to shallow waters dynamics*, Int.J.Num.Meth.Eng, **65**: 295-331, (2006).
7. L. Cueto-Felgueroso, I. Colominas, X. Nogueira, F. Navarrina, M. Casteleiro. *Finite volume solvers and Moving Least-Squares approximations for the compressible Navier-Stokes equations on unstructured grids*, Comp. Meth. Appl. Mech. Engrg., **196**: 4712-4736, (2007).
8. L. Cueto-Felgueroso and I. Colominas. *High-order finite volume methods and multiresolution reproducing kernels*, Archives of Computational Methods in Engineering. DOI 10.1007/s11831-008-9017-y, June 2008
9. H. Gómez, I. Colominas, F. Navarrina, M. Casteleiro. *A finite element formulation for a convection-diffusion equation based on Cattaneo's law*, Comp. Meth. Appl. Mech. Engrg., doi:10.1016/j.cma.2006.09.016, **155**(9-12): 1757-1766, 2007.
10. P. L. Roe. *Approximate Riemann solvers, parameter vectors and difference schemes*, J. Comput. Physics, **43**: 357-372, (1981).
11. A. Huerta, Y. Vidal, P. Villon. *Pseudo-divergence-free Element Free Galerkin method for incompressible fluid flow*, Journal of Computational Physics, **155**: 54-74, (1999).
12. S. K. Godunov. *A Difference Method for the Numerical Calculation of Discontinuous Solutions of Hydrodynamic Equations*, Mat. Sbornik. **47**(3): 271-306, (1959).
13. T. J. Barth, D. C. Jespersen. *The design and application of upwind schemes on unstructured meshes*, AIAA-89-0366, (1989).
14. X. Nogueira, L. Cueto-Felgueroso, I. Colominas, F. Navarrina, M. Casteleiro. *On the accuracy of Finite Volume and Discontinuous Galerkin discretizations for compressible flow on unstructured grids*, submitted to Int.J.Num.Meth.Eng.
15. C. W. Shu and S. Osher. *Efficient Implementation of Essentially Non-Oscillatory Shock-Capturing Schemes, II*, J. Comput. Physics, **83**: 32-78, (1989).
16. H. C. Yee, N. D. Sandham, M. J. Djomehri. *Low-Dissipative High-Order Shock-Capturing Methods using Characteristic-Based Filters*, J. Comput. Physics, **150**: 199-238, (1999).

Interface Tracking in Meshfree Methods and its Applications

Hirohisa Noguchi[1] and Zhiqian Zhang[2]

[1] Department of System Design and Engineering, Keio University, 3-14-1 Hiyoshi, Kohoku-ku, Yokohama 223-8522, Japan, noguchi@sd.keio.ac.jp
[2] Faculty of Science and Technology, Keio University, 3-14-1 Hiyoshi, Kohoku-ku, Yokohama 223-8522, Japan, zhangzhiqian@noguchi.sd.keio.ac.jp

Summary. An enhanced meshfree method, moving least squares approximation with discontinuous derivative basis functions (MLSA-DBF), has been proposed in order to accurately track the derivative discontinuities of continuum or structures. Firstly, quadratic basis functions for MLSA-DBF in three dimensions are presented and the meshfree formulations in Cartesian coordinates are introduced for the analysis of shell structures with slope discontinuities. Numerical examples demonstrate the validity, accuracy, and convergence properties of the proposed method. Secondly, topology optimization with nonlinear materials under large deformation is established based on MLSA-DBF and the level set method. MLSA-DBF can achieve accurate stress and strain fields and obtain accurate sensitivity analysis in the topology optimization problems with fixed/moving material interfaces. The numerical results give faster convergence rate than the method without treatments for material interfaces, and show superior advantages for large deformation problems. It is shown that MLSA-DBF, which is a simple, universal and accurate method without extra parameters, can accurately track not only the material interfaces but also the slope discontinuities and even moving interfaces.

Key words: Meshfree Methods, Discontinuous Derivative Basis Function, Shell, Slope Discontinuity, Topology Optimization

1 Introduction

Various meshfree methods have been proposed, enhanced and are now widely used in lots of engineering and scientific fields due to the common features involving simple preprocessing, versatile in solving large deformation problems, smooth approximation etc.. However, the smooth meshfree approximation leads to inaccurate strain/stress solutions of the problems with derivative discontinuities. Many efforts have been made to propose the enrichments of meshfree methods in order to obtain accurate discontinuous strain/stress solutions. By Patch Method [1,2], the strain/stress solutions show oscillations in vicinity of the interfaces. The wedge-type enrichment of Element-free Galerkin Method (EFG) was proposed in [3] with additional degrees of freedom. Researches in [4] proposed a jump function enrichment on RKPM without additional degrees of freedom. The method combining MLPG5 and MLPG2 was proposed in [5], which is deficient in consistency due to the inconsistent basis functions. Moving Least-Squares Approximation with Discontinuous Derivative Basis Function (MLSA-DBF) was presented in [6] which adopts different sets of consistent basis functions in the subdomains split by the interfaces. In this work, MLSA-DBF is extended to analyze shells with slope discontinuities, and applied to the level-set based topology optimization.

So far, to the best of the authors' knowledge, current research literatures only document shell analyzes with academic smooth surfaces [7–12]. The parametric coordinates are employed to construct approximation functions except researches in [10] which indicated that neither Moving Least-Squares nor Reproducing Kernel approximations can be constructed when the shell geometry is described by multiple parametric domains. In this research, MLSA-DBF formulations in Cartesian coordinates [10] are proposed for shear deformable shell structures with slope discontinuities.

Another topic of this work is the application of MLSA-DBF to the topology optimization. As an attractive methodology, the level set method [13] was widely used for modeling and tracking the moving boundaries with topology changes [14], and was developed to solve various topology optimization problems [15–17] in recent years. In most of the past researches on the level-set based topology optimization, the structural responses are solved by conventional FEM with structured mesh. The structural responses and sensitivity analysis will lose accuracy if no special treatment is adopted to handle the material interfaces which may cross the element during the topology evolution. Researches in [16] proposed an topology optimization method combined with extended FEM (XFEM) method to deal with the moving material interfaces. In this study, MLSA-DBF is employed to track the fixed/moving material interfaces in the level-set based topology optimization and to solve the problems with large deformation.

In the following discussion, 3-D MLSA-DBF with quadratic basis functions are given in section 2. In section 3, meshfree formulations in Cartesian coordinate for shear deformable shell structures with slope discontinuities are proposed. Level set based topology optimization and the techniques to track the moving interfaces are proposed in section 4. Following some numerical examples in section 5, the conclusions are drawn in section 6.

2 MLSA-DBF with Quadratic Basis Functions

In this paper the following abbreviations are used: **MLSA-CBF**: MLSA with Conventional Basis Functions; **MLSA-DBF**: MLSA with Discontinuous Derivative Basis Functions; **MLSA-QDBF-3D**: MLSA-DBF with Quadratic

Basis Functions for 3-D problem; **MLSA-LDBF-3D**: MLSA-DBF with Linear Basis Functions for 3-D problem.

Following the researches on 2-D MLSA-DBF in [6], the following two sets of complete quadratic polynomial functions are adopted for MLSA-QDBF-3D to make the local approximation $u(\boldsymbol{x})$ in a small neighborhood of $\bar{\boldsymbol{x}}$

$$\begin{aligned} u_1^h(\boldsymbol{x};\bar{\boldsymbol{x}}) &= \sum_{\alpha,\beta,\gamma=0}^{2} a_{\alpha\beta\gamma}(x-\bar{x}_d)^{\alpha}(y-\bar{y}_d)^{\beta}(z-\bar{z}_d)^{\gamma} \\ u_2^h(\boldsymbol{x};\bar{\boldsymbol{x}}) &= \sum_{\alpha,\beta,\gamma=0}^{2} b_{\alpha\beta\gamma}(x-\bar{x}_d)^{\alpha}(y-\bar{y}_d)^{\beta}(z-\bar{z}_d)^{\gamma} \end{aligned} \tag{2.1}$$

where $\bar{\boldsymbol{x}}_d = (\bar{x}_d, \bar{y}_d, \bar{z}_d)$ is the point on the interface surface with the shortest distance to $\bar{\boldsymbol{x}} = (\bar{x}, \bar{y}, \bar{z})$. The unknown coefficients, $a_{\alpha\beta\gamma}$ and $b_{\alpha\beta\gamma}$ are associated with $\bar{\boldsymbol{x}}$. The interface Γ_{12} is expressed by a function $h(\boldsymbol{x};\bar{\boldsymbol{x}}_d) = 0$, and the sub-domains beside Γ_{12} are denoted as Ω_1 and Ω_2, where $h(\boldsymbol{x};\bar{\boldsymbol{x}}_d) > 0\ for\ \boldsymbol{x} \in \Omega_1$ and $h(\boldsymbol{x};\bar{\boldsymbol{x}}_d) < 0\ for\ \boldsymbol{x} \in \Omega_2$. The interface in 3-D is taken to be a plane. If the interface is a curved surface, the tangent plane at $\bar{\boldsymbol{x}}_d$ is constructed as the pseudo interface. Fig. 2.1 plots the discontinuity interfaces Γ_{12}, the sub-domains Ω_1 and Ω_2, and the points $\bar{\boldsymbol{x}}$ and $\bar{\boldsymbol{x}}_d$.

Quadratic basis functions for MLSA-QDBF-3D are derived by considering the following four interface orientations.

(*S1*) $h(\boldsymbol{x};\bar{\boldsymbol{x}}_d) = 0\ :(z-\bar{z}_d) = k_A(x-\bar{x}_d) + k_B(y-\bar{y}_d), (k_A)^2 + (k_B)^2 \neq 0$
(*S2*) $h(\boldsymbol{x};\bar{\boldsymbol{x}}_d) = 0\ :\ x-\bar{x}_d = 0$
(*S3*) $h(\boldsymbol{x};\bar{\boldsymbol{x}}_d) = 0\ :\ y-y_d = 0$
(*S4*) $h(\boldsymbol{x};\bar{\boldsymbol{x}}_d) = 0\ :\ z-\bar{z}_d = 0$

Here the procedure for the situation (*S1*) is shown to derive the corresponding basis functions of MLSA-QDBF-3D. The following five points on the interfaces are used to enforce the displacement continuity condition, i.e., $u_1^h(\boldsymbol{x};\bar{\boldsymbol{x}}) = u_2^h(\boldsymbol{x};\bar{\boldsymbol{x}})$:

$\boldsymbol{x}_{d1} : (\bar{x}_d, \bar{y}_d, \bar{z}_d);$
$\boldsymbol{x}_{d2} : (\bar{x}_d + \Delta x, \bar{y}_d, \bar{z}_d + \Delta z), \Delta x\Delta z \neq 0, \Delta z = k_A\Delta x;$
$\boldsymbol{x}_{d3} : (\bar{x}_d + \Delta\tilde{x}, y_d, \bar{z}_d + \Delta\tilde{z}), \Delta\tilde{x}\Delta\tilde{z} \neq 0, \Delta\tilde{x} \neq \Delta x;$
$\boldsymbol{x}_{d4} : (\bar{x}_d, \bar{y}_d + \Delta y, \bar{z}_d + \Delta z), \Delta y\Delta z \neq 0, \Delta z = k_B\Delta y;$

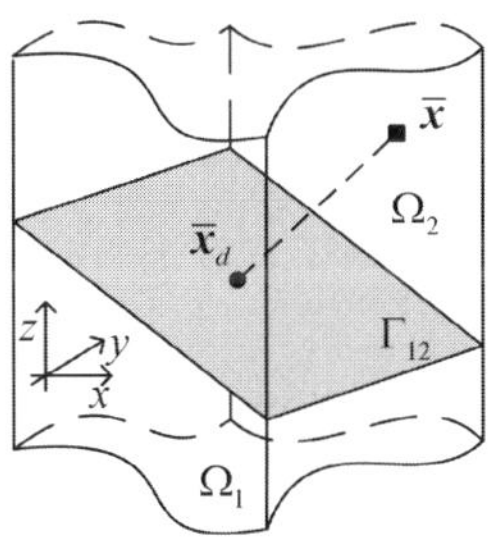

Figure 2.1. Conceptual interpretation of discontinuity interface of 3D MLSA-DBF

$\boldsymbol{x}_{d5} : (\bar{x}_d, \bar{y}_d + \Delta\tilde{y}, \bar{z}_d + \Delta\tilde{z}), \Delta\tilde{y}\Delta\tilde{z} \neq 0, \Delta\tilde{y} \neq \Delta y.$

To satisfy $u_1^h(\boldsymbol{x}_{d1}; \bar{\boldsymbol{x}}) = u_2^h(\boldsymbol{x}_{d1}; \bar{\boldsymbol{x}})$ yields

$$a_{000} = b_{000} \tag{2.2}$$

To satisfy $u_1^h(\boldsymbol{x}_{d2} : \bar{\boldsymbol{x}}) = u_2^h(\boldsymbol{x}_{d2} : \bar{\boldsymbol{x}})$ and $u_1^h(\boldsymbol{x}_{d3} : \bar{\boldsymbol{x}}) = u_2^h(\boldsymbol{x}_{d3} : \bar{\boldsymbol{x}})$ yields

$$\begin{aligned} b_{100} &= a_{100} + a_{001}k_A - b_{001}k_A \\ b_{200} &= a_{200} + a_{002}k_A^2 + a_{101}k_A - b_{002}k_A^2 - b_{101}k_A \end{aligned} \tag{2.3}$$

To satisfy $u_1^h(\boldsymbol{x}_{d4} : \bar{\boldsymbol{x}}) = u_2^h(\boldsymbol{x}_{d4} : \bar{\boldsymbol{x}})$ and $u_1^h(\boldsymbol{x}_{d5} : \bar{\boldsymbol{x}}) = u_2^h(\boldsymbol{x}_{d5} : \bar{\boldsymbol{x}})$ yields

$$\begin{aligned} b_{010} &= a_{010} + a_{001}k_B - b_{001}k_B \\ b_{020} &= a_{020} + a_{002}k_B^2 + a_{011}k_B - b_{002}k_B^2 - b_{011}k_B \end{aligned} \tag{2.4}$$

For the situations (*S2*), (*S3*) and (*S4*), the relationships can also be derived by the similar procedure with some proper points on the interface. By substituting the relationships in (2.3)-(2.4) to (2.1), $u_2^h(\boldsymbol{x}; \bar{\boldsymbol{x}})$ can be recast leading to the quadratic basis functions $\tilde{\boldsymbol{p}}_1(\boldsymbol{x} - \bar{\boldsymbol{x}}_d)$ and $\tilde{\boldsymbol{p}}_2(\boldsymbol{x} - \bar{\boldsymbol{x}}_d)$ for MLSA-QDBF-3D in Table 2.1. The linear basis functions of MLSA-LDBF-3D can be obtained by taking the first 5 elements of $\tilde{\boldsymbol{p}}_1(\boldsymbol{x} - \bar{\boldsymbol{x}}_d)$ and $\tilde{\boldsymbol{p}}_2(\boldsymbol{x} - \bar{\boldsymbol{x}}_d)$.

The approximated $u^h(\boldsymbol{x}; \bar{\boldsymbol{x}})$ at the point $\bar{\boldsymbol{x}}$ can be given by

$$u^h(\bar{\boldsymbol{x}}) = \sum_I \Phi_I(\bar{\boldsymbol{x}}) u_I \tag{2.5}$$

where u_I is the nodal parameter and $\Phi_I(\bar{\boldsymbol{x}})$ is the shape function given by

$$\Phi_I(\bar{\boldsymbol{x}}) = \begin{cases} \tilde{\boldsymbol{p}}_1^T(\bar{\boldsymbol{x}} - \bar{\boldsymbol{x}}_d)\left(\boldsymbol{A}^{-1}(\bar{\boldsymbol{x}})\boldsymbol{B}(\bar{\boldsymbol{x}})\right)_I, \ \boldsymbol{x} \in \Omega_1 \bigcup \Gamma_{12} \\ \tilde{\boldsymbol{p}}_2^T(\bar{\boldsymbol{x}} - \bar{\boldsymbol{x}}_d)\left(\boldsymbol{A}^{-1}(\bar{\boldsymbol{x}})\boldsymbol{B}(\bar{\boldsymbol{x}})\right)_I, \ \boldsymbol{x} \in \Omega_2 \end{cases} \tag{2.6}$$

where

Table 2.1. Quadratic basis functions of MLSA-QDBF-3D

$\tilde{\boldsymbol{p}}_1(\boldsymbol{x} - \bar{\boldsymbol{x}}_d)$	$\tilde{\boldsymbol{p}}_2(\boldsymbol{x} - \bar{\boldsymbol{x}}_d)$			
S(1)-S(4)	*S(1)*	*S(2)*	*S(3)*	*S(4)*
1	1	1	1	1
Δx_d	Δx_d	Δx_d	Δx_d	Δx_d
Δy_d	Δy_d	Δy_d	Δy_d	Δy_d
Δz_d	$k_A\Delta x_d + k_B\Delta y_d$	Δz_d	Δz_d	Δz_d
0	$\Delta z_d - k_A\Delta x_d - k_B\Delta y_d$	0	0	0
$(\Delta x_d)^2$	$(\Delta x_d)^2$	$(\Delta x_d)^2$	$(\Delta x_d)^2$	$(\Delta x_d)^2$
$(\Delta y_d)^2$	$(\Delta y_d)^2$	$(\Delta y_d)^2$	$(\Delta y_d)^2$	$(\Delta y_d)^2$
$(\Delta z_d)^2$	$k_A^2(\Delta x_d)^2 + k_B^2(\Delta y_d)^2$	$(\Delta z_d)^2$	$(\Delta z_d)^2$	$(\Delta z_d)^2$
$\Delta x_d\Delta y_d$	0	$\Delta x_d\Delta y_d$	$\Delta x_d\Delta y_d$	$\Delta x_d\Delta y_d$
$\Delta y_d\Delta z_d$	$k_B(\Delta y_d)^2$	$\Delta y_d\Delta z_d$	$\Delta y_d\Delta z_d$	$\Delta y_d\Delta z_d$
$\Delta x_d\Delta z_d$	$k_A(\Delta x_d)^2$	$\Delta x_d\Delta z_d$	$\Delta x_d\Delta z_d$	$\Delta x_d\Delta z_d$
0	$-k_A^2(\Delta x_d)^2 - k_B^2(\Delta y_d)^2 + (\Delta z_d)^2$	0	0	0
0	$\Delta x_d\Delta y_d$	0	0	0
0	$\Delta y_d\Delta z_d - k_B(\Delta y_d)^2$	0	0	0
0	$\Delta x_d\Delta z_d - k_A(\Delta x_d)^2$	0	0	0

$$\begin{aligned} \boldsymbol{A}(\bar{\boldsymbol{x}}) = & \textstyle\sum_{I=1}^{m_1} w(r_I)\, \tilde{\boldsymbol{p}}_1(\boldsymbol{x}_I - \bar{\boldsymbol{x}}_d)\, \tilde{\boldsymbol{p}}_1^T(\boldsymbol{x}_I - \bar{\boldsymbol{x}}_d) \\ & + \textstyle\sum_{I=m_1+1}^{m} w(r_I)\, \tilde{\boldsymbol{p}}_2(\boldsymbol{x}_I - \bar{\boldsymbol{x}}_d)\, \tilde{\boldsymbol{p}}_2^T(\boldsymbol{x}_2 - \bar{\boldsymbol{x}}_d) \end{aligned} \tag{2.7}$$

$$\begin{aligned} \boldsymbol{B}(\bar{\boldsymbol{x}}) = & \left[w(r_1)\, \tilde{\boldsymbol{p}}_1(\boldsymbol{x}_1 - \bar{\boldsymbol{x}}), \ldots, w(r_{m_1})\, \tilde{\boldsymbol{p}}_1(\boldsymbol{x}_{m_1} - \bar{\boldsymbol{x}}), \right. \\ & \left. w(r_{m_1+1})\, \tilde{\boldsymbol{p}}_2(\boldsymbol{x}_{m_1+1} - \bar{\boldsymbol{x}}), \ldots \right] \end{aligned} \tag{2.8}$$

where $\Phi_I(\bar{\boldsymbol{x}})$ is the shape function, and $\boldsymbol{A}(\bar{\boldsymbol{x}})$ is the moment matrix. The nodes $1 \ldots m_1$ are in $\Omega_1 \bigcup \Gamma_{12}$, and the nodes $m_1 + 1 \ldots m$ are in Ω_2.

The moment matrix $\boldsymbol{A}$ could yield a singular matrix due to the following two possible reasons. Type A: Insufficient influence nodes [6, 18]. Type B: Linear dependency between the geometry function of the shell surface and basis functions [10]. Under certain conditions, a singular system can be solved by a Moore-Penrose pseudoinverse [19, 20]. The Moore-Penrose pseudoinverse provides a least squares solution to $\boldsymbol{A}(\boldsymbol{x})\, \boldsymbol{a}(\boldsymbol{x}) = \boldsymbol{b}(\boldsymbol{x})$. Note that if $\boldsymbol{A}$ is nonsingular, Moore-Penrose pseudoinverse is just the normal inverse of $\boldsymbol{A}$.

3 Application to Shells with Slope Discontinuity

The meshfree formulations in Cartesian Coordinates for shear deformable shells proposed in [10] are employed in this work. The Reissner-Mindlin shear deformable shell with kinematics is shown in Fig. 3.2.

The local displacement $\hat{\boldsymbol{u}}(\boldsymbol{x}) = \left[\, \hat{u}_1\ \hat{u}_2\ \hat{u}_3 \,\right]^T$ can be expressed as

$$\{\hat{\boldsymbol{u}}(\boldsymbol{x})\} = \left\{ \begin{array}{c} \hat{u}_1 \\ \hat{u}_2 \\ \hat{u}_3 \end{array} \right\} = \left\{ \begin{array}{c} \hat{u}_1^t \\ \hat{u}_2^t \\ \hat{u}_3^t \end{array} \right\} + \left\{ \begin{array}{c} \hat{x}_3 \beta \\ -\hat{x}_3 \alpha \\ 0 \end{array} \right\} \tag{3.9}$$

where $\hat{u}_i^t$ are the translational degree of freedom, and α and β are the rotational degree of freedom.

The global displacement $\boldsymbol{u}(\boldsymbol{x})$ in Cartesian coordinates is given by

$$\boldsymbol{u}(\boldsymbol{x}) = \sum_{I=1}^{N} \Phi_I(\boldsymbol{x}) \left\{ \begin{array}{c} u_{I1} \\ u_{I2} \\ u_{I3} \end{array} \right\} + \sum_{I=1}^{N} \frac{t\zeta}{2} \Phi_I(\boldsymbol{x}) \left[\, \hat{\mathbf{e}}_1^I\ -\hat{\mathbf{e}}_2^I \,\right] \left\{ \begin{array}{c} \beta_I \\ \alpha_I \end{array} \right\} \tag{3.10}$$

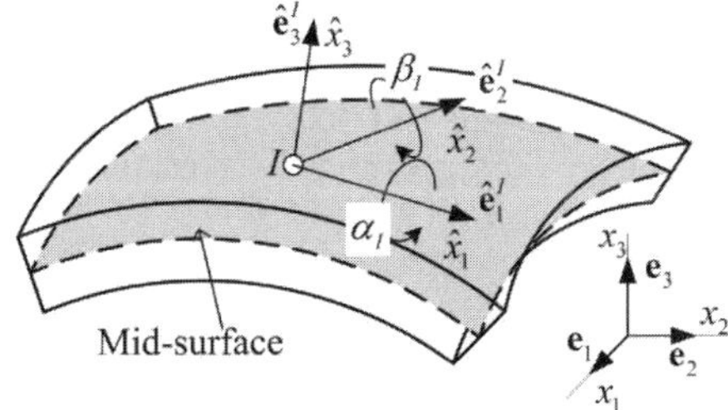

x_i : global Cartesian coordinaes
$\mathbf{e}_i$: orthonormal base vectors in global coordinates
$\hat{x}_i$: local coordinates
$\hat{\mathbf{e}}_i^I$: orthonormal base vectors in local coordinate system at node $\mathbf{x}_I$
α_I, β_I : rotations of director vector about $\hat{\mathbf{e}}_1^I, \hat{\mathbf{e}}_2^I$

Figure 3.2. Global Cartesian coordinates, local coordinates and sign conventions

where t is the shell thickness, and $\hat{x}_3 = t\zeta/2$. The local displacement $\hat{\boldsymbol{u}}(\boldsymbol{x})$ is obtained by coordinate transformation

$$\hat{\boldsymbol{u}}(\boldsymbol{x}) = \sum_{I=1}^{N} \Phi_I(\boldsymbol{x}) \left[\boldsymbol{Q}(\boldsymbol{x}), \hat{x}_3 \bar{\boldsymbol{Q}}(\boldsymbol{x}) \right] \{\boldsymbol{d}_I\} \tag{3.11}$$

where $\boldsymbol{d}_I = \left\{ u_{I1}, u_{I2}, u_{I3}, \beta_I, \alpha_I \right\}^T$ and

$$\boldsymbol{Q}(\boldsymbol{x}) = \begin{bmatrix} \hat{\mathbf{e}}_1^T(\boldsymbol{x}) \\ \hat{\mathbf{e}}_2^T(\boldsymbol{x}) \\ \hat{\mathbf{e}}_3^T(\boldsymbol{x}) \end{bmatrix} = \begin{bmatrix} \hat{e}_{11}\ \hat{e}_{12}\ \hat{e}_{13} \\ \hat{e}_{21}\ \hat{e}_{22}\ \hat{e}_{23} \\ \hat{e}_{31}\ \hat{e}_{32}\ \hat{e}_{33} \end{bmatrix} \quad , \quad \bar{\boldsymbol{Q}}(\boldsymbol{x}) = \boldsymbol{Q}(\boldsymbol{x}) \left[\hat{\mathbf{e}}_1^I\ -\hat{\mathbf{e}}_2^I \right] \tag{3.12}$$

where $\hat{e}_{ij}$ is the direction cosines between $\hat{\mathbf{e}}_i$ and $\boldsymbol{e}_j$.

The local strain tensor $\hat{\varepsilon}$ includes membrane strain $\hat{\varepsilon}^m$, shear strain $\hat{\varepsilon}^s$ and bending strain $\hat{\varepsilon}^b$, as follows

$$\hat{\varepsilon} = \begin{Bmatrix} \hat{\varepsilon}^m + \hat{\varepsilon}^s \\ \hat{\varepsilon}^b \end{Bmatrix} \tag{3.13}$$

$$\hat{\varepsilon}^m = \begin{bmatrix} \frac{\partial \hat{u}_1}{\partial \hat{x}_1} \\ \frac{\partial \hat{u}_2}{\partial \hat{x}_2} \\ \frac{\partial \hat{u}_1}{\partial \hat{x}_2} + \frac{\partial \hat{u}_2}{\partial \hat{x}_1} \end{bmatrix} \quad \hat{\varepsilon}^b = \hat{x}_3 \begin{bmatrix} \frac{\partial \beta}{\partial \hat{x}_1} \\ -\frac{\partial \alpha}{\partial \hat{x}_2} \\ \frac{\partial \beta}{\partial \hat{x}_2} - \frac{\partial \alpha}{\partial \hat{x}_1} \end{bmatrix} \quad \hat{\varepsilon}^s = \begin{bmatrix} \frac{\partial \hat{u}_3}{\partial \hat{x}_1} + \beta \\ \frac{\partial \hat{u}_3}{\partial \hat{x}_2} - \alpha \end{bmatrix} \tag{3.14}$$

where

$$\frac{\partial \hat{u}_i}{\partial \hat{x}_j} = \sum_{I=1}^{N} \frac{\partial \Phi_I}{\partial \hat{x}_j} u_{Ij} = \sum_{I=1}^{N} \frac{\partial \Phi_I}{\partial x_k} \frac{\partial x_k}{\partial \hat{x}_j} \tag{3.15}$$

Then one can reach the following stiffness matrix

$$\boldsymbol{K} = t \int_{\boldsymbol{A}} \left[\boldsymbol{B}^{m-s} \right]^T \left[\hat{\boldsymbol{C}} \right] \left[\boldsymbol{B}^{m-s} \right] dA + \frac{t^3}{12} \int_A \left[\boldsymbol{B}^b \right]^T \left[\hat{\boldsymbol{C}} \right] \left[\boldsymbol{B}^b \right] dA \tag{3.16}$$

See [10] for the detailed expressions of $\hat{\boldsymbol{C}}$, $\boldsymbol{B}_I^{m-s}(\boldsymbol{x})$ and $\boldsymbol{B}_I^b(\boldsymbol{x})$. In this work it is assumed that the thickness t of shells is constant. In order to remove shear and membrane locking in thin shells, stabilized conforming nodal integration [10] are employed to integrate the second term of $\boldsymbol{K}$, whereas a direct nodal integration is used to calculate the first term.

For shell structures with slope discontinuities, local coordinates associated with each tangent plane are defined as shown in Fig. 3.3.

In Fig. 3.3, at a node K located on Γ_{12}, ambiguity exists in the rotational degrees of freedom associated with multiple local coordinates. To overcome this difficulty, a reference to the global Cartesian rotational degrees of freedom is employed [21] as given in (3.17).

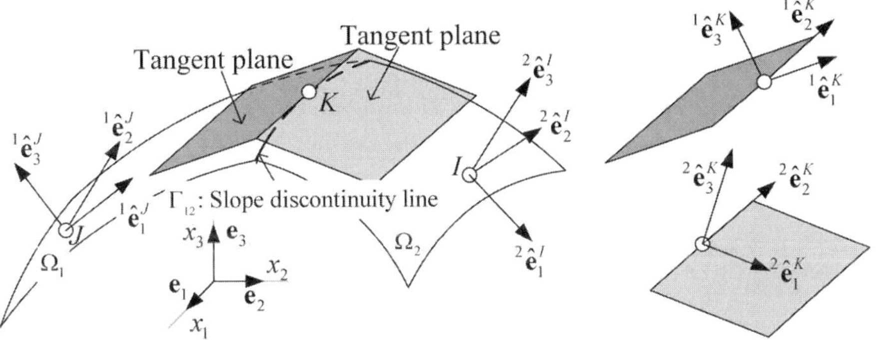

Figure 3.3. Local coordinates along slope discontinuous line

$$\left\{ \begin{matrix} {}^s\alpha_K \\ {}^s\beta_K \end{matrix} \right\} = \begin{bmatrix} \left({}^s\hat{\mathbf{e}}_1\right)^T \\ \left({}^s\hat{\mathbf{e}}_2\right)^T \end{bmatrix} \left\{ \begin{matrix} \theta_1^K \\ \theta_2^K \\ \theta_3^K \end{matrix} \right\} = \begin{bmatrix} {}^s\hat{e}_{11}^K & {}^s\hat{e}_{12}^K & {}^s\hat{e}_{13}^K \\ {}^s\hat{e}_{21}^K & {}^s\hat{e}_{22}^K & {}^s\hat{e}_{23}^K \end{bmatrix} \left\{ \begin{matrix} \theta_1^K \\ \theta_2^K \\ \theta_3^K \end{matrix} \right\}, \quad s = 1, 2 \tag{3.17}$$

where $\{{}^s\alpha_K, {}^s\beta_K\}_{s=1}^n$ are the n sets of local rotational degrees of freedom, and $\{\theta_1^K, \theta_2^K, \theta_3^K\}$ are the global rotational degrees of freedom associated with node K. Note that the above transformation is only performed for the nodes on the slope-discontinuity line, otherwise, the stiffness matrix will be singular.

When MLSA-DBF approximations in Cartesian coordinates is used to analyze shell structures, pseudo interface is required and defined as the angle-bisecting plane of the tangent planes of Ω_1 and Ω_2 at $\bar{x}_d$.

4 Application to Level-set Based Topology Optimization

The topology optimization problems in current researches are given as [15]

$$\begin{aligned} \underset{\varphi}{\text{Minimize}} \quad & J(\boldsymbol{u}, \varphi) = \int_D F(\boldsymbol{u}) H(\varphi)\, d\Omega \\ s.t. \quad & \boldsymbol{Q}(\boldsymbol{u}, \varphi) = \boldsymbol{T} \\ & G(\varphi) = \int_D H(\varphi)\, d\Omega - V^* = 0 \end{aligned} \tag{4.18}$$

where, φ is the level set function taken as signed distance function to the topological boundary, $J(\boldsymbol{u}, \varphi)$ is the objective function, $F(\boldsymbol{u})$ is the objective function density, $H(\varphi)$ is the Heaviside function, $\boldsymbol{u}$ is the displacement, $G(\varphi)$ is the volume constraint. $\boldsymbol{Q}(\boldsymbol{u}, \varphi)$ is the internal force vector, and $\boldsymbol{T}$ is the external force vector. The following regularized Heaviside function $\bar{H}(\varphi)$ is employed [16]

$$\bar{H}(\varphi) = \left\{ \begin{matrix} \Delta & \varphi \leq -l \\ \frac{1}{4}\left(1 + \sin\frac{\pi\varphi}{2l}\right)^2 & -l < \varphi < l \\ 1 & \varphi \geq l \end{matrix} \right. \tag{4.19}$$

where $l = 2h$ and h is the average nodal distance.

Adjoint Method [22] is employed to perform sensitivity analysis of the objective function $J(\boldsymbol{u}, \varphi)$ under the constraint condition $\boldsymbol{Q}(\boldsymbol{u}, \varphi) = \boldsymbol{T}$ in (4.18). Construct a functional π without considering volume constrain condition as follows

$$\pi(\boldsymbol{u}, \boldsymbol{w}, \varphi) = J(\boldsymbol{u}, \varphi) - \left\{\boldsymbol{w}^T \boldsymbol{Q}(\boldsymbol{u}, \varphi) - \boldsymbol{w}^T \boldsymbol{T}\right\} \tag{4.20}$$

where $\boldsymbol{w}$ is the adjoint variable vector. Taking variation of (4.20) and considering the stationary condition lead to

$$\boldsymbol{Q}(\boldsymbol{u}, \varphi) = \boldsymbol{T} \tag{4.21}$$

$$\left(\frac{\partial \boldsymbol{Q}(\boldsymbol{u}, \varphi)}{\partial \boldsymbol{u}}\right)^T \boldsymbol{w} = \frac{\partial J(\boldsymbol{u}, \varphi)}{\partial \boldsymbol{u}} \tag{4.22}$$

$$\frac{\partial J(\boldsymbol{u}, \varphi)}{\partial \varphi} = \frac{\partial \left(\boldsymbol{w}^T Q(\boldsymbol{u}, \varphi)\right)}{\partial \varphi} \tag{4.23}$$

(4.21) is the system equilibrium equation. The discrete form of (4.23) is given by

$$\frac{\partial J(\boldsymbol{u}, \varphi)}{\partial \varphi} = \int_D \Pi(\boldsymbol{u}, w)\, \delta(\varphi)\, d\Omega = \int_D \{\boldsymbol{w}\}^T [\boldsymbol{B}]^T \{\boldsymbol{S}\}\, \delta(\varphi)\, d\Omega \tag{4.24}$$

where $[\boldsymbol{B}]$ is the Green-Lagrange strain-displacement relation matrix, $\{\boldsymbol{S}\}$ is the second Piola-Kichhoff stress vector. The sensitivity analysis of the volume constraint can be performed by

$$\frac{\partial G(\varphi)}{\partial \varphi} = \int_D \frac{\partial H(\varphi)}{\partial \varphi} d\Omega = \int_D \delta(\varphi)\, d\Omega \tag{4.25}$$

In order to satisfy the volume constraint, Lagrange Multiplier is employed in the following Lagrange functional

$$\hat{\pi}(\boldsymbol{u}, \boldsymbol{w}, \varphi) = J(\boldsymbol{u}, \varphi) + \lambda \int_D \left(H(\varphi) - V^*\right) d\Omega \tag{4.26}$$

where λ is Lagrange Multiplier for the volume constraint, and it can be calculated by gradient projection method [23] as follows

$$\lambda = \frac{\int_D \Pi(\boldsymbol{u}, \boldsymbol{w})\, \delta^2(\varphi)\, |\nabla\varphi|\, d\Omega}{\int_D \delta^2(\varphi)\, |\nabla\varphi|\, d\Omega} \tag{4.27}$$

Therefore, the normal velocity of the boundary movement will be given as

$$V_n = \begin{cases} -\left[\Pi(\boldsymbol{u}, \boldsymbol{w}) + \lambda\right] \delta(\varphi) & G(\varphi) > 0 \\ -\Pi(\boldsymbol{u}, \boldsymbol{w})\, \delta(\varphi) & G(\varphi) \le 0 \end{cases} \tag{4.28}$$

where is V_n the normal velocity. The Hamilton-Jacobi equation for updating level-set function is given as follows

$$\frac{\partial \varphi}{\partial t} + V_n \left| \nabla \varphi (x) \right| = 0 \tag{4.29}$$

(4.29) is solved by finite difference method with up-wind scheme. From (4.28), it is clear that the accuracy of stress/strain solutions will influence the accuracy of V_n, and influence updating level set function consequently.

Generally, the level set function is taken as the signed distance function to the boundaries, which can be written as follows for 2-D problem.

$$\varphi (\boldsymbol{x}) = \begin{cases} 0 & \boldsymbol{x} \in \partial \Omega \\ - \left| \boldsymbol{x} - \boldsymbol{x}_d \right| = - \left[(x - x_d)^2 + (y - y_d)^2 \right]^{1/2} & \boldsymbol{x} \in \Omega^- \\ + \left| \boldsymbol{x} - \boldsymbol{x}_d \right| = + \left[(x - x_d)^2 + (y - y_d)^2 \right]^{1/2} & \boldsymbol{x} \in \Omega^+ \end{cases} \tag{4.30}$$

where x_d has the same definition in MLSA-DBF. $\partial \Omega$ is the boundary or interface, Ω^- is the interior region, and Ω^+ is the exterior region. $\boldsymbol{x}_d$ on the moving boundary or interfaces can be tracked by the following equations.

$$x_d = x - \varphi (x, y) \frac{\partial \varphi (x, y)}{\partial x} \quad , \quad y_d = y - \varphi (x, y) \frac{\partial \varphi (x, y)}{\partial y} \tag{4.31}$$

The slope of the interface of MLSA-DBF at $\boldsymbol{x}_d$ can be calculated as follows

$$k = -n_1 (\boldsymbol{x}_d) / n_2 (\boldsymbol{x}_d) \quad , \quad \boldsymbol{n} (\boldsymbol{x}_d) = \nabla \varphi (\boldsymbol{x}_d) \tag{4.32}$$

With $\boldsymbol{x}_d$ and k one can implement 2-D MLSA-DBF [6] at every step of topology evolution. In order to keep φ as the signed distance function, reinitialization [14, 15] is required after solving (4.29).

5 Numerical Examples

Example 1 L-Shape Timoshenko Beam

Timoshenko beam can be regarded as one dimensional shell structure. The problem statement of an L-shape Timoshenko beam structure subjected to a tip load $f = -0.1$ is given in Fig. 5.4. Pseudo discontinuous interface for MLSA-DBF is plotted by the dashed line at the point C. MLSA-CBF and MLSA-DBF with quadratic basis functions are used to solve this problem. Fig. 5.5 shows the axial force and moment solutions, solved by using uniformly distributed 61 nodes and $\rho = 1.5h$, $2.3h$, $3.3h$, where h is the average nodal distance.

It can be observed that the slope discontinuity leads to a discontinuous axial force shown in Fig. 5.5. As displayed in the figures, accurate solutions are

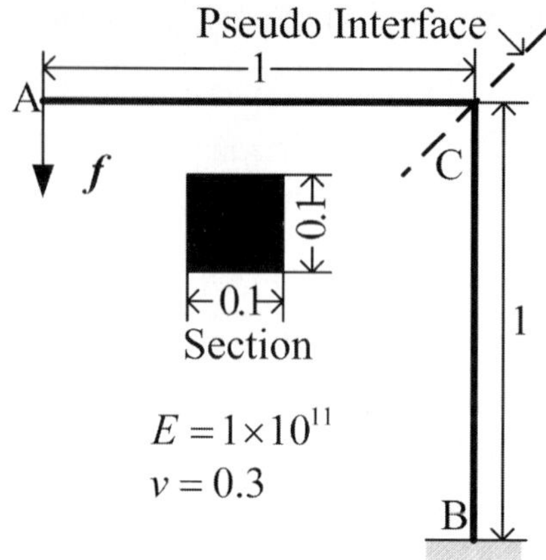

Figure 5.4. Problem Statement of Numerical Example 1

achieved by using MLSA-DBF whereas the solutions solved by MLSA-CBF show oscillations around slope discontinuous point. Results in Fig. 5.5 also illustrate the effect of the influence domain size on the accuracy. The solutions of MLSA-DBF do not seem to be affected by the influence domain size, and the Type A singularity in the moment matrix ($\rho = 1.5h$) is handled properly by the employment of Moore-Penrose Pseudoinverse. The solutions of MLSA-CBF exhibit oscillations in all cases. Fig. 5.6 presents the L2 error norms in the force-moment solutions, which shows a better convergence property in MLSA-DBF than in MLSA-CBF. The relative error is defined as

$$e = \|\{\boldsymbol{F}\}^n - \{\boldsymbol{F}\}^e\|_2 / \|\{\boldsymbol{F}\}^e\|_2 \tag{5.33}$$

where $\{\boldsymbol{F}\}_I = \{\, F_I \; M_I \,\}^T$, $\{\boldsymbol{F}\}^n$ is the numerical solution, and $\{\boldsymbol{F}\}^e$ is the exact solution.

Example 2 T-Shape Shell Structure

The problem statement is given in Fig. 5.7 with the reference lines where the strain solutions at the top shell surfaces are compared. The thickness of the shell is 0.1. The edge GH is fixed, and the prescribed displacements are imposed on the edge CD as $\bar{u}_x|_{GH} = 0.01m$, $\bar{u}_y|_{GH} = \bar{u}_z|_{GH} = 0$, $\bar{\alpha}|_{GH} = \bar{\beta}|_{GH} = 0$. All the other edges are free. Material Properties are: Young's modulus $E = 1 \times 10^{10}$ and Poisson ratio $v = 0.3$. The pseudo discontinuous interface for MLSA-QDBF-3D is the gray plane shown in Fig. 5.7.

FEM with uniform 120,000 4-node elements using selective reduced integration is employed for comparison. The left figure in Fig. 5.8 compares nodal local strain solution ε_{xx} solved by MLSA-QCBF-3D and MLSA-QDBF-3D along the reference line *a-b-c*. The results show a better agreement between FEM and MLSA-QDBF-3D, and MLSA-QCBF-3D loses accuracy in vicinity

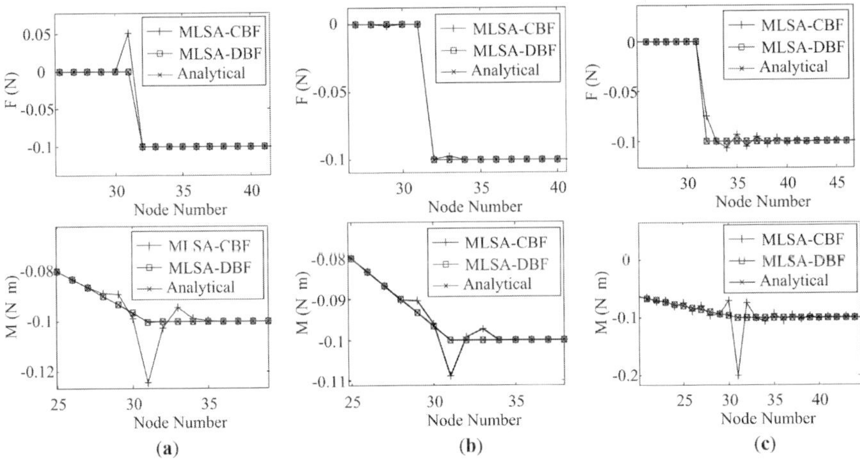

Figure 5.5. Solutions of L-shape frame: (**a**) $\rho = 1.5h$; (**b**) $\rho = 2.3h$; (**c**) $\rho = 3.3h$

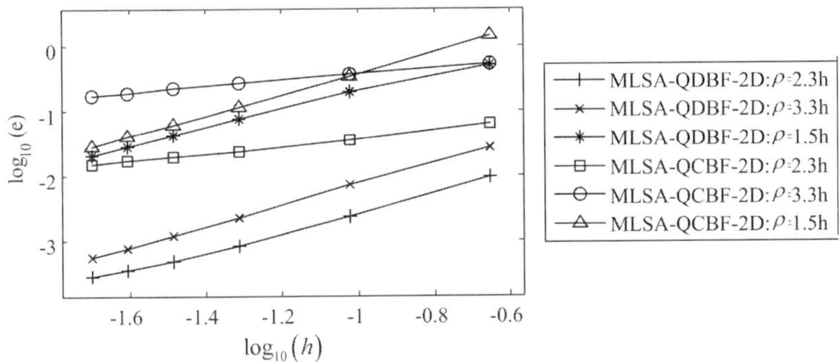

Figure 5.6. L2 norms of errors in force-moment solutions (Example 1)

of the slope discontinuity line. The L2 norms of the following strain error measure are plotted in the right figure in Fig. 5.8.

$$e=\left\|\{\varepsilon\}^n - \{\varepsilon\}^r\right\|_2 / \left\|\{\varepsilon\}^e\right\|_2 \tag{5.34}$$

where $\{\varepsilon\}^n$ is the numerical strain solutions and $\{\varepsilon\}^r$ is FEM strain reference solutions. The superior accuracy and convergence properties n MLSA-QDBF-3D are observed.

Example 3 Topology Optimization of Bi-Material Structure

Fig. 5.9 shows the topology optimization problem of a structure with Mooney-Rivlin materials. The left edge is fixed, and the middle point of the right edge is subjected to a prescribed displacement $\bar{u}_y = 0.3H$.

The following strain energy density of Mooney-Rivlin material is employed:

$$W = A_1\left(\bar{I}_1 - 3\right) + A_2\left(\bar{I}_2 - 3\right) + A\left(J - 1\right)^2/2 \tag{5.35}$$

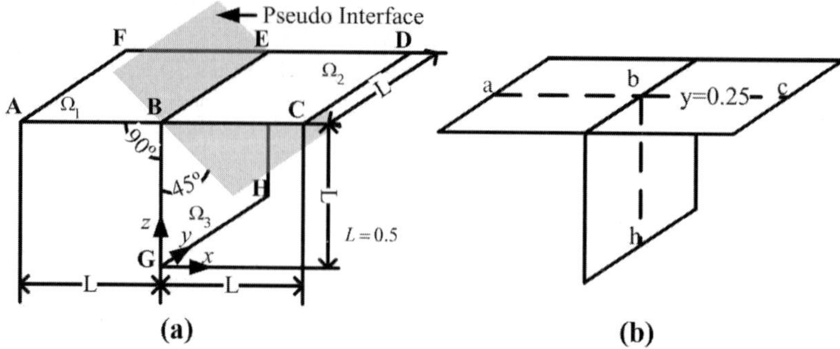

Figure 5.7. Problem statement of Example 2: (**a**) Geometry; (**b**) Reference Lines

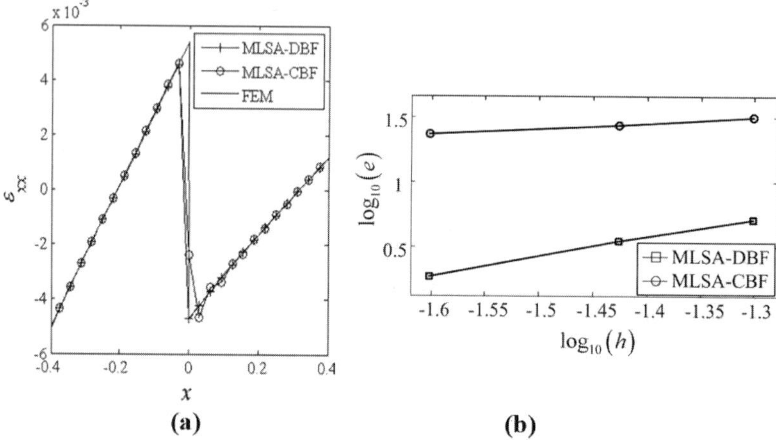

Figure 5.8. Solutions of Example 2: (**a**) Strain solution; (**b**) Convergence properties

The material properties are: Material 1) $A_1 = 10$, $A_2 = 1$ and $A = 1 \times 10^4$, and Material 2) $A_1 = 10 \times 10^6$, $A_2 = 1 \times 10^6$ and $A = 1 \times 10^{10}$. The region occupied by Material 2 is fixed. The structure is discretized by 23×23 uniformly distributed nodes. The objective function is the strain energy of the whole design domain. The volume constraint is the area of the domain occupied by Material 1 is equal to 40% of the whole design domain. The structure responses are solved by MLSA-DBF and MLSA-CBF, respectively. The solutions are given in Fig. 5.10.

In the right figure of Fig. 5.10 the optimal designs obtained by different schemes are compared. In the regions far from the material interface both schemes give almost same optimal design with less influences of material interface. In vicinity of the material interface, the different designs demonstrate that the accurate treatment in the material interface is important to achieve

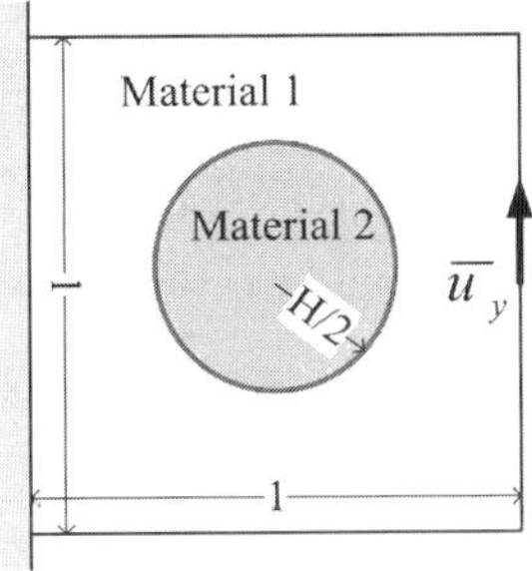

Figure 5.9. Problem statement and initial design (Example 3) (*See also* Color Plate on page 377)

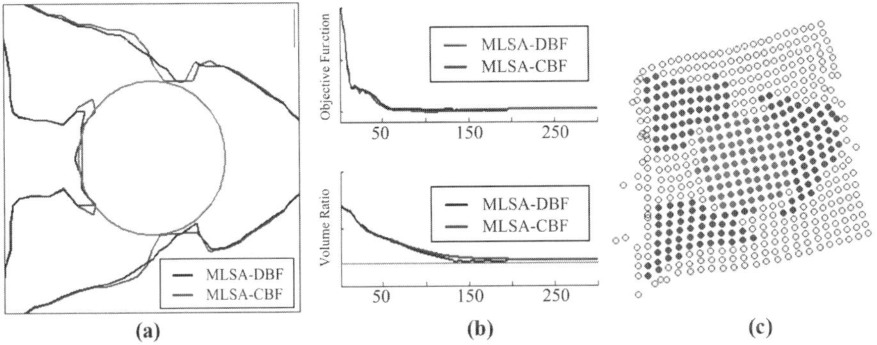

Figure 5.10. Solutions of Example 3: (**a**) Optimal design; (**b**) convergence histories; (**c**) deformation (*See also* Color Plate on page 377)

accurate normal velocity of the boundary movement. The superior convergence rate of the scheme using MLSA-DBF can be observed in the middle figure of Fig. 5.10. The left figure of Fig. 5.10 illustrates the node distributions in the deformed configurations of the structures with the optimal designs. The figure reveals the capability of meshfree method to analyze the structures under large deformations, especially, the extremely large deformations of the nodes without materials (white points). Obviously, it is difficult for the conventional FEM method to analyze such kind of problems due to the severer element distortions.

Example 4 Topology Optimization with Moving Material Interface

The problems statement and the initial design are shown in Fig. 5.11. The structure consists of two kinds of linear elastic materials. The points a and b are fixed, and a vertical force is applied to the point c of the left edge with the amplitude $f = 1$. The objective function is the strain energy of the whole design domain. The volume constraint is the area of material B equals to 40% of the area of the design domain. The moving material interfaces are tracked by the techniques proposed in section 4 and MLSA-DBF and MLSA-CBF are

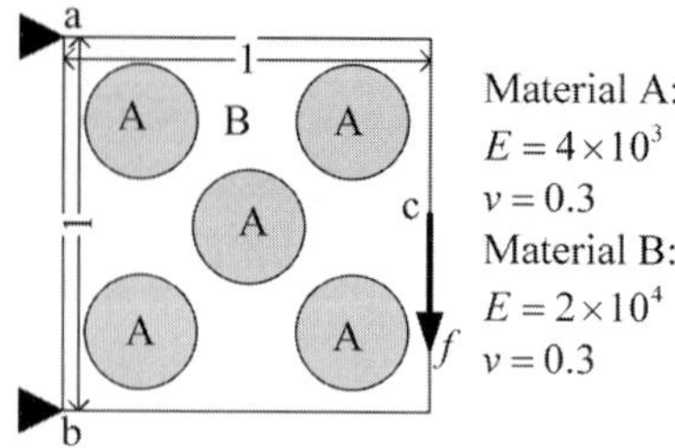

Figure 5.11. Problem statement and the initial design (Example 4)

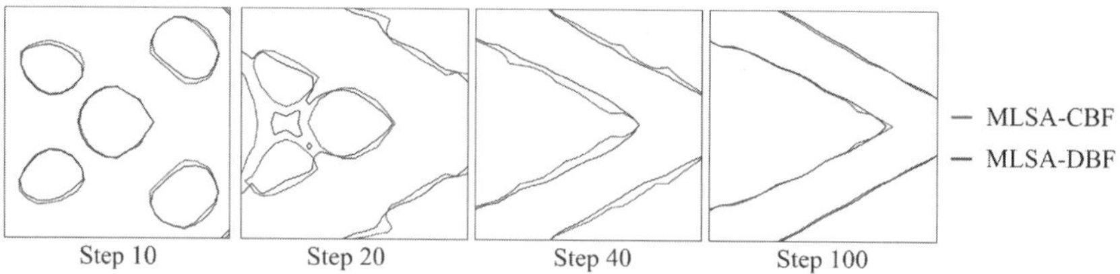

Figure 5.12. Material interfaces at different steps of topology evolution (*See also* Color Plate on page 377)

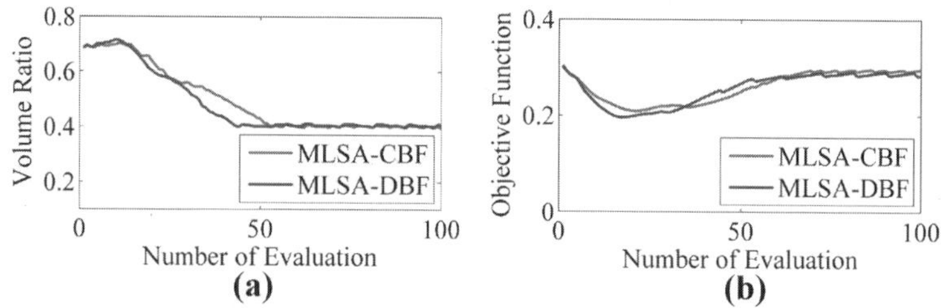

Figure 5.13. Convergence histories: (**a**) objective function; (**b**) volume ratio (*See also* Color Plate on page 378)

employed to solve the structural responses. The design domain is discretized by 41 × 41 uniformly distributed nodes.

Fig. 5.12 shows the comparisons of the material interfaces of the intermediate and the optimal designs. With the treatment of moving material interfaces, the topologies solved by using MLSA-DBF are different from the results solved by using MLSA-CBF. The oscillations of the strain and stress solutions solved by MLSA-CBF can not lead to smooth interfaces as MLSA-DBF. Fig. 5.13 plots the histories of the objective function and the volume ratio. In this figure, the superior convergence rate of the scheme using MLSA-DBF is observed, and the objective function of the optimal design also shows a lower value than MLSA-CBF. Obviously, the higher convergence rate and smooth material interfaces result from the accurate structural response solutions and normal velocity solved by the scheme using MLSA-DBF.

6 Conclusions

In this work, the intrinsically enhanced meshfree method, MLSA-DBF, is extended to three dimensions and the quadratic basis functions are proposed, which can be used to solve the problems with various kinds of discontinuous derivatives accurately.

Formulations of MLSA-DBF in Cartesian coordinates are proposed to solve the shear deformable shell structures with slope discontinuities. The use of Cartesian coordination in the construction of MLS approximation resolves the difficulties associated with junctions in 3D shells. The numerical examples show the less influences of size of the influence domain, and the validity, accuracy and superior convergence properties of the proposed method.

With the technique of tracking moving interfaces based on signed distance function, MLSA-DBF is able to solve the accurate structural response solutions which yields the accurate sensitivity analysis. The numerical examples with fixed and moving material interfaces indicate that accurate structural responses solved by MLSA-DBF lead to a higher convergence rate of the topology evolution and smooth and reasonable boundaries or material interfaces. The application of meshfree method removes the difficulty to cure element distortion in FEM-based method for the problem under large deformation.

References

1. L.W. Cordes, B. Moran, *Treatment of Material Discontinuity in the Element-free Galerkin Method*, Comput. Meth. Appl. Mech. Eng. **139** (1996), 75-89
2. H. Noguchi, T. Kawashima, *Meshfree Analyzes of Cable Reinforced Membrane Structures by ALE-EFG Method*, Eng. Anal. Bound. Elem. **28** (2004), 443-451
3. Y. Krongauz, T. Belytschko, *EFG Approximation with Discontinuous Derivatives*, Int. J. Numer. Meth. Engrg. **41** (1998), 1215-1233
4. D. Wang, J.-S. Chen, L. Sun, *Homogenization of Magnetostrictive Particle-filled Elastomers using an Interface-enriched Reproducing Kernel Particle Method*, Finite Elem. Anal. Des. **39** (2003), 765-782
5. Q. Li, S. Shen, Z.D. Han, S.N. Atluri, *Application of Meshless Local Petrov-Galerkin (MLPG) to Problems with Singularities, and Material Discontinuities, in 3-D Elasticity*, Comput. Model. Eng. Sci. **4** (2003), 571-586
6. S. Masuda, H. Noguchi, *Analysis of Structure with Material Iinterface by Meshfree Method*, Comput. Model. Eng. Sci. **11** (2006), 131-144
7. P. Krysl, T. Belytschko, *Analysis of Thin Shells by the Element-free Galerkin Method*, Int. J. Solids Struct. **33** (1996), 3057-30808.
8. H. Noguchi, T. Kawashima, T. Miyamura T, *Element Free Analyzes of Shell and Spatial Structures*, Int. J. Numer. Meth. Engrg. **47** (2000), 1215-1240
9. T. Jarak, J. Soric J, J. Hoster, *Analysis of Shell Deformation Responses by the Meshless Local Petrov-Galerkin (MLPG) Approach*, Comput. Model. Eng. Sci. **18** (2007), 235-246
10. J.S. Chen, D. Wang, *A Constrained Reproducing Kernel Particle Formulation for Shear Deformable Shell in Cartesian Coordinates*, Int. J. Numer. Meth. Engrg. **68** (2006), 151-172

11. L. Liu, L.P. Chua, D.N. Ghista, *Conforming Radial Point Interpolation Method for Spatial Shell Structures on the Stress-resultant Shell Theory*, Archive of Applied Mechanics. **75** (2006), 248-267
12. L. Liu, L.P. Chua, D.N. Ghista, *Applications of Point Interpolation Method for Spatial General Shells Structures*, Comput. Meth. Appl. Mech. Eng. **196** (2007), 1633-1647
13. S. Osher, J.A. Sethian, *Fronts Propagating with Curvature-dependent Speed: Algorithms based on Hamilton-Jacobi Formulations*, J. Comput. Phys. **79** (1988), 12-49
14. J.A. Sethian. *Level Set Methods and Fast Marching Methods: Evolving Interfaces in Computational Geometry, Fluid Mechanics, Computer Version and Material Science*, Cambridge University Press, 1999.
15. M.Y. Wang, X.M. Wang, D.M. Guo, *A Level Set Method for Structural Topology Optimization*, Comput. Meth. Appl. Mech. Eng. **192** (2003), 227-246
16. T. Belytschko, S.P. Xiao, C. Parimi, *Topology Optimization with Implicit Functions and Regularization*, Int. J. Numer. Meth. Engrg. **57** (2003), 1177–1196
17. J.A. Sethian, A. Wiegmann, *Structural Boundary Design via Level Set and Immersed Interface Methods*, J. Comput. Phys. **163** (2000), 489-528
18. R. Batra, M. Porfiri, D. Spinello, *Free and Forced Vibrations of a Segmented Bar by a Meshless Local Petrov-Galerkin (MLPG) Formulation*, Comput. Mech. **41** (2008), 473-491
19. E.H. Moore, *On the Reciprocal of the General Algebraic Matrix*, Bulletin of the American Mathematical Society. **26** (1920), 394-395
20. R. Penrose, *A Generalized Inverse for Matrices*, Proceedings of the Cambridge Philosophical Society. **51** (1955), 406-413
21. T. Hisada, H. Noguchi, *Basics and Application of Nonlinear Finite Element Method.* Maruzen Press, 1995
22. M. Tanaka, H. Noguchi, *Structural Shape Optimization of Hyperelastic Material by Discrete Force Method*, Theoretical and Applied Mechanics Japan. **53** (2004), 83-91
23. Z. Liu, J.G. Korvink, R. Huang, *Structure Topology Optimization: Fully Coupled Level Set Method via FEMLAB*, Struct. Multidiscip. Optim. **29** (2005), 407-417

A'posteriori Error Estimation Based on Higher Order Approximation in the Meshless Finite Difference Method

Janusz Orkisz[1] and Slawomir Milewski[1]

[1] Division of Computational Methods in Mechanics, Cracow University of Technology, plorkisz@cyf-kr.edu.pl

[2] Division of Computational Methods in Mechanics, Cracow University of Technology, slawek@l5.pk.edu.pl

Summary. The paper presents recent developments in the Higher Order Approximation applied to the Meshless Finite Difference Method MFDM [13]. The concept of the Higher Order Approximation (HOA) [14] is based on considering additional terms in the Taylor expansion of the searched function. Those terms may consist of HO derivatives as well as their jumps and/or singularities. They are used as correction terms to the standard meshless FD operator. Among many applications of the HOA, special emphasis is focused on a'posteriori estimation of the solution and the residual error in both the local and global forms. Thus the HOA approach provides results which may be also used as a high quality reference solution in global or local error estimators. A variety of 1D and 2D tests done indicate clear superiority of such estimation approach over those currently used in the other discrete methods [1].

Key words: Meshless Finite Difference Method, Higher order approximation, correction terms, a'posteriori error estimation

1 Introduction

The Meshless Finite Difference Method [13] belongs to a wide class of the meshless methods. These methods are nowadays one of the more and more popular contemporary tools for analysis of boundary value problems. In the meshless methods approximation of unknown function is described in terms of nodes rather than by means of any imposed structure like elements, regular meshes etc. Therefore, the MFDM, using arbitrarily irregular clouds of nodes and Moving Weighted Least Squares (MWLS) approximation [5,11,13], falls into the category of the Meshless Methods (MM), being in fact the oldest and possibly the most developed one of them.

The objective of the paper is a brief outline of the actual development of the MFDM and presentation of some results of the current research on Higher

Order approximation (HOA) [15-22] in the MFDM, with special emphasis laid on a'posteriori error estimation.

In the standard MFDM differential operators are replaced by the finite difference ones, with a prescribed approximation order. In the HOA approach the rank of the local approximation is raised by considering relevant higher order terms originated from the function expansion intro the Taylor series rather than by introducing new nodes into the MFD operator.

In this paper special attention is laid upon application of the HO MFDM solution to both the local and global a'posteriori solution error estimations [16,18-20,22] needed in adaptive generation of new meshes.

2 Basic MFDM approach

The classical FDM was most commonly used in analysis of boundary value problems posed in the local formulation as a set of differential equations and appropriate boundary conditions. In the considered domain $\Omega \subset \Re^n$ with boundary $\partial\Omega$ a function $u(P)$ is sought at each point P, satisfying equations

$$\mathcal{L}u(P) = f(P) \quad \text{for} \quad P \in \Omega \tag{2.1}$$

$$\mathcal{L}_b u(P) = g(P) \quad \text{for} \quad P \in \partial\Omega \tag{2.2}$$

Global, more complex, formulations may be also analysed by the MFDM nowadays. These may be posed either in the form of optimisation of the functional

$$I(u) = \frac{1}{2}\mathcal{B}(u,u) - \mathcal{L}(u) \tag{2.3}$$

satisfying boundary conditions (2) or as variational principles (e.g. the principle of virtual work)

$$\mathcal{B}(u,v) = \mathcal{L}(v) \quad \text{for} \quad v \in V_{adm} \tag{2.4}$$

In terms of mechanics the first term $\mathcal{B}$ in the energy functional (3) represents internal energy of the system, while the second one, $\mathcal{L}$, is the work done by external forces. In the case of variational principle (4), $v = v(P)$ is a test function from the admissible space V_{adm}. The global approach involves integration over the domain Ω. Mixed, global / local formulations may be also considered. In fact, any formulation of the boundary value problem, which involves the unknown function and its derivatives, can be used here.

When use of arbitrarily irregular cloud of nodes is admitted in the MFDM, a variety of problems arise, that have to be solved [11,13]. Some of them, constituting so called basic MFDM solution approach, will be briefly discussed below.

Nodes generation and mesh topology determination

The MFDM solution approach needs generation of clouds of nodes (arbitrarily irregular mesh). Any mesh generator built for the FEM analysis could be used. However, nodes generator taking advantage of the features specific to the MFDM analysis may better serve this purpose. Therefore, here nodes $x_i = (x_i, y_i), \quad i = 1, 2, ..., N$ are generated using the Liszka type mesh generator [11,12,13], based on the mesh density control. Nodes are not bounded by any type of imposed structure, like element or regular mesh. However, it is convenient to determine afterwards mesh topology information, based on the already generated cloud of nodes. In 2D domain such information includes domain partition into the Voronoi polygons and Delaunay triangles, their interrelations, as well as their neighbourhood determination [12,13]. Without mesh structure imposed, any node can be easily removed, shifted or a new node inserted with only small local changes of the mesh topology. Voronoi tessellation and Delaunay triangulation of the cloud of generated nodes, followed by their topology determination, is needed for analysis of the b.v. problems (e.g. to MFD star selection, numerical integration) posed either in the local (1)-(2) or the global formulation (3)-(4).

MFD star selection and classification

A group of nodes used together as a base for a local FD approximation is called the MFD star. FD stars play similar role in the MFDM as the elements in the FEM, i.e. they are used for spanning a local approximation of the searched function. When dealing with irregular cloud of nodes, both MFD stars and formulas usually differ from node to node. The most important feature of any selection criteria then is to avoid ill conditioned and singular MFD stars. Therefore not only the distance from the central node counts, but also the nodes distribution. MFD star selection at any arbitrary node, and stars classification in a considered domain are based on topology information. The general principles and two criteria of star selection, "cross" and "Voronoi neighbours", considered as the best one, are discussed in [13]. Classification of MFD stars is introduced, based on the notion of "equivalence class" of star configuration. For each class the FDM formulas are generated only once.

MWLS approximation and MFD schemes generation

The Moving Weighted Least Squares approximation [5,11,13], spanned over approximated local MFD stars, is widely used in the MFDM in order to generate MFD formulae as well as in the postprocessing. Consider any of the formulations of a given boundary value problem outlined before (1)-(4). Let us consider a *n-th* order differential operator $\mathcal{L}$. For each MFD star consisting of arbitrarily distributed nodes, the complete set of derivatives up to the assumed *p-th* ($p \geq n$) order is sought. When the MFD formulae are generated, point x

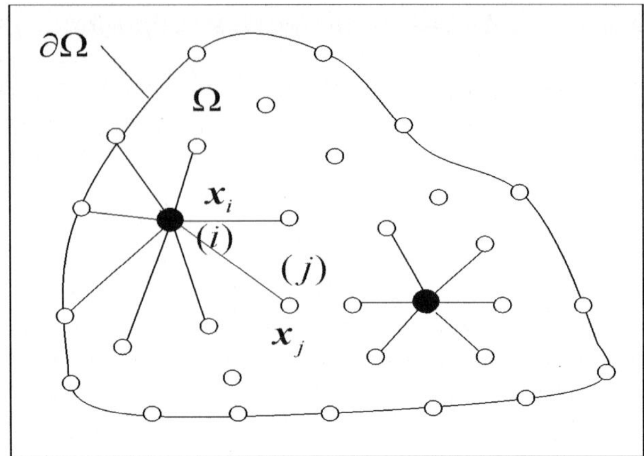

Figure 2.1. Arbitrarily distributed nodes, FD star

is represented either by a mesh node $x_i = (x_i, y_i)$, $i = 1, 2, ..., N$ (for the local formulation (1)) or by an integration point, when using a global formulation (3)-(4). The MFD star at point x_i consists of $r$star nodes $j = 1, 2, ...r$ (Fig. 1).

Local approximation $\hat{u}$ of the sought function $u(x)$ may be written in two equivalent notations. The MFDM approximation [13] is based mainly on the Taylor series expansion of the unknown function at the central point (i) of a MFD star (in 2D)

$$u(x,y) = \hat{u}(x,y) + e = p^T \cdot Du^{(L)} + e \tag{2.5}$$

where

$$p^T \cdot Du^{(L)} = \sum_{j=0}^{p} \frac{1}{j!} \left(h\frac{\partial}{\partial x} + k\frac{\partial}{\partial y} \right)^j u(x,y)\big|_{(x_i,y_i)}, \quad \begin{cases} h = x - x_i \\ k = y - y_i \end{cases} \tag{2.6}$$

whereas the other meshless methods [2,5] use the equivalent polynomial approximation (written here in the incremental form)

$$u(x,y) \approx \hat{u}(x,y) = b_0 + b_1(x - x_i) + b_2(y - y_i) + ... + b_m(y - y_i)^p = p^T \cdot b \tag{2.7}$$

However, the MFDM notation (5)-(6) seems to be more practical, because it offers also information about approximation error e, caused by a truncated part of the Taylor series, as well as providing a simple interpretation of the approximation coefficients considered as function derivatives (local type). Depending on the space dimension we have

$$\underset{(1\times m)}{p^T} = \begin{cases} 1, h, h^2, ..., h^p \\ 1, h, k, ..., \frac{1}{p!}k^p \\ 1, h, k, l, , ..., \frac{1}{p!}l^p \end{cases}, \quad \underset{(m\times 1)}{Du^{(L)}} = \begin{cases} u, u', u'', ...u^{(p)} & \text{in 1D} \\ u, \frac{\partial u}{\partial x}, \frac{\partial u}{\partial y}, ..., u^{(p)}_{yy...y} & \text{in 2D} \\ u, \frac{\partial u}{\partial x}, \frac{\partial u}{\partial y}, \frac{\partial u}{\partial z}, ..., u^{(p)}_{zz...z} & \text{in 3D} \end{cases} \tag{2.8}$$

where m denotes the number of unknown approximation coefficients (e.g. $m = (p+1)(p+2)/2$ for 2D case), p – the local approximation order, p – vector of the local interpolants (8), and $Du^{(L)}$ – vector of all derivatives up to the *p-th* (low) order. Index $^{(L)}$ is assigned to each quantity corresponding to the standard solution i.e. using the low approximation order p. Interpolation conditions imposed at all nodes of the MFD star lead to the overdetermined set of algebraic equations

$$\hat{u}(x_i, y_i) = u_i, \quad \text{for } i = 1, 2, ..., r \quad \rightarrow \quad PDu^{(L)} = q \tag{2.9}$$

For 2D domain we have

$$\underset{(r\times m)}{P} = \begin{bmatrix} 1 & h_1 & k_1 & \frac{1}{2}h_1^2 & h_1k_1 & \frac{1}{2}k_1^2 & ... & \frac{1}{p!}k_1^p \\ 1 & h_2 & k_2 & \frac{1}{2}h_2^2 & h_2k_2 & \frac{1}{2}k_2^2 & ... & \frac{1}{p!}k_2^p \\ ... & ... & ... & ... & ... & ... & ... & ... \\ 1 & h_r & k_r & \frac{1}{2}h_r^2 & h_rk_r & \frac{1}{2}k_r^2 & ... & \frac{1}{p!}k_r^p \end{bmatrix}, \quad \underset{(r\times 1)}{q} = \begin{bmatrix} u_1 \\ u_2 \\ ... \\ u_r \end{bmatrix} \tag{2.10}$$

Here $h_i = x - x_i, \quad k_i = y - y_i, \quad \underset{(r\times m)}{P}$ denotes the matrix of local interpolants ($m \le r$), and $\underset{(r\times 1)}{q}$ - vector of nodal values of a sought function $u(x, y)$. Minimization of the weighted error functional

$$J = (PD\mathbf{u}^{(L)} - q)^T \mathbf{W}^2 (\mathbf{P}D\mathbf{u}^{(L)} - q) \tag{2.11}$$

yields

$$\frac{\partial J}{\partial D\mathbf{u}^{(L)}} = 0 \quad \rightarrow \quad D\mathbf{u}^{(L)} = \mathbf{M} \cdot q, \quad \underset{(m\times r)}{\mathbf{M}} = (\mathbf{P}^T\mathbf{W}^2\mathbf{P})^{-1}\mathbf{P}^T\mathbf{W}^2 \tag{2.12}$$

$$\hat{u} = p^T M q \tag{2.13}$$

namely the complete set of the derivatives $D\mathbf{u}^{(L)}$ up to the *p-th* order, expressed in terms of the MFD formulae matrix $\boldsymbol{M}$ providing the required MWLS approximation $\hat{u}$. Similar results may be obtain when using notation (7).

$$J = (Pb - q)^T \mathbf{W}^2 (\mathbf{P}b - q) \tag{2.14}$$

$$\frac{\partial J}{\partial b} = 0 \rightarrow b = A^{-1}Bq, \quad A = P^TWP, \quad B = P^TW, \quad \hat{u} = p^TA^{-1}Bq \tag{2.15}$$

However, only more convenient notation (12) is used in the following sections.In the above formulas $\underset{(r\times r)}{\mathbf{W}} = diag\,(w_1, w_2, ..., w_r)$ is a diagonal weight matrix. For the weight functions

$$w_j = \frac{1}{\rho_i^{p+1}}, \quad \rho_j = \sqrt{k_j^2 + h_j^2}, \quad j = 1, 2, ..., r \tag{2.16}$$

the matrix $\boldsymbol{W}$ may be singular. It assures, in this way, the delta Kronecker property $w_i(x_j) = \delta_{ij}$, and consequently enforces interpolation $\hat{u}(x_i) = u_i$ at the central node of each MFD star.

MWLS extensions

One may consider various extensions of the MWLS approximation [13] including

- generalised degrees of freedom (e.g. derivatives, operator values, ...),
- singularities and discontinuities of the function and/or its derivatives,
- equality and inequality constraints (global-local approximation [7]),
- Higher Order approximation by means of the correction terms; such approach will be described in the following chapters.

Generation of MFD equations

The following strategy of generation of the MFD operators is adopted. As opposed to the classic FDM approach where the FD operators are developed directly in the final form required, in the MFDM the MFD operators are generated first for the complete set of derivatives $D\mathbf{u}^{(L)}$ (zero-th, first, second,... up to p-th order) needed. Each point, chosen for generation of derivatives $D\mathbf{u}$, may represent either an arbitrary point or a node in the considered domain. The local MWLS approximation, based on development of searched function into the Taylor series is spanned over an appropriate MFD star with a sufficient number of m nodes. Evaluation of the derivatives $D\mathbf{u}$ is based on the formulas (9)-(13), (16), and is described in details in [13].

Having found the MFD operators for the derivatives one may compose any MFD operator required either for a MFD equation, and boundary conditions or for an integrand (global MFD formulation).

Consider e.g. a class of linear differential operators of the second order

$$Lu = c_0 u + c_1 \frac{\partial u}{\partial x} + c_2 \frac{\partial u}{\partial y} + c_3 \frac{\partial^2 u}{\partial x^2} + c_4 \frac{\partial^2 u}{\partial x \partial y} + c_5 \frac{\partial^2 u}{\partial y^2} \approx c^T Du \qquad (2.17)$$

where $c = \{c_0, ..., c_5\}$ are coefficients. A required MFD operator is formed here by a linear combination of derivatives $D\mathbf{u}$.

MFD discretization of boundary conditions

MFDM discretization of boundary conditions may be done in a similar way as described before, or using some other ways presented in [13]. Quality of the MFDM solutions usually depends on the quality of discretization of the boundary conditions.

Solution of simultaneous FD equations (linear or non-linear)

The MFDM approach yields simultaneous algebraic MFD equations (SAE). These SAE may be non-symmetric equations (for local b.v. formulation) or symmetric ones (for global formulations). In the last case they might be solved by means of similar procedures like those for the FEM discretization. Non-symmetric equations may use solvers developed e.g. for the CFD. However, the best approach seams to be development of solvers specific for the MFDM taking advantage of this method's nature. Especially, the adaptive solution approach seems to be effective [12,13].

Postprocessing

The MWLS approximation is a powerful tool in postprocessing because it may provide us with values of a considered function, and its derivatives at any point, based on discrete data (function values or other, generalised d.o.f.). These results may be directly obtained using at each point of interest the approximation approach defined in formulas (5)-(16). It uses the same basis as generation of MFD operators discussed above. Though it may be precise, the MWLS approach is time consuming as solution of the local SLAE equations (12) is needed at each point where approximation is required. Its precision depends on many parameters involved. There are several techniques (extensions) already mentioned before that may essentially raise the quality of the MWLS approximation [23].

General remarks

The basic solution MFDM approach [13], outlined above, has been extended in many ways so far, and is still under development. Among many extensions of the basic MFD solution approach one may mention here

- Various Petrov-Galerkin formulations,
- Higher Order approximation [15-22],
- A'posteriori error analysis [16,18-20,22],
- MFDM oriented node generator,
- Mesh refinement and adaptive (multigrid) solution approach [9,18-20],
- MFDM on the differential manifold,
- MFDM/FEM combinations and unification [8],
- Experimental and numerical data smoothing [7],
- Hybrid experimental / theoretical / numerical approach,
- Software development,
- Engineering applications.

The second and third problems from the above list will be considered here.

One of important problems in the contemporary numerical solution approach is error analysis including effective a'posteriori error estimation [16,18-20,22]. In the MFDM mesh refinement (h-type adaptive approach) is based on estimation of a'posteriori residual error and a'posteriori solution error. In the most common cases estimation of solution error needs a reference solution good enough for using it instead of the true analytical solution. Thus high quality numerical solution has to be found first in order to find a sufficiently good error estimate. Such solutions are obtained here mainly by means of the HOA approach raising approximation order from p to $2p$. Various tests done indicate that its quality is clearly superior to the other reference solutions used so far.

3 Higher Order approximation solution approach

There are several possible ways to improve MFD solutions. Increasing the number of nodes is the most obvious one. This may be done by considering either the regular meshes or arbitrarily irregular clouds of nodes. In the last case they may be generated using a'posteriori error estimation (h-adaptive approach). The number of nodes may be rapidly increased then, whereas the order of the approximation remains unchanged. The other way to improve the MFD solution quality is to raise the approximation order. The concept of the HOA MFD operators, used here together with a multigrid iterative procedure, is somehow different than the one presented in [6]. It presents the HO approximation, proposed in [14], and is still under development [15-22]. It uses the same (low order) MFD operator, but with a modified right hand side of the MFD equations.

Consider boundary value problem of the *n-th* order, given in any of formulations (1)-(4). The local MFD *p-th* ($p \geq n$) order approximation Lu_i of the differential operator value $\mathcal{L}u_i$ is assumed in the form

$$\mathcal{L}u_i = Lu_i - \Delta_i - R_i = f_i - \Delta_i - R_i, \quad i \in I_\Omega \tag{3.18}$$

Here L is a MFD operator, corresponding to differential operator $\mathcal{L}$, R_i is the truncated part of the Taylor series. The correction term

$$\Delta_i = \Delta(u_i^{(p+1)}, ..., u_i^{(2p)}; J_i^{(0)}, ..., J_i^{(2p)}; S_i^{(0)}, ..., S_i^{(2p)}) \tag{3.19}$$

includes (higher order) derivatives of the *s-th* order, where $p < s \leq 2p$. It may also contain discontinuities $J^{(k)}$ and singularities $S^{(k)}$ of the function and/or its *k-th* derivatives. These may be either known a'priori or could be treated as additional unknowns. Higher order derivatives may be calculated by the composition of appropriate formulae and the use of low order (without correction) MFD solution inside the domain. However, they may need a special treatment near the domain boundary. Several other techniques for the boundary MFD operators may be also considered:

- using only internal nodes; the approximation is of low quality then,
- using internal nodes with both the boundary conditions and domain equation specified on the boundary,

- using internal nodes and generalized degrees of freedom,
- using specific or general multipoint approach (presented in [5], and further developed in [17,21]),
- using internal and additional external fictitious nodes,
- combinations of the above techniques.

Correction terms may be used for

- improving MFD approximation inside the domain,
- improving MFD approximation on the boudnary,
- generation of a high quality reference solution,
- estimation of the a'posteriori solution and residual errors in both the local and global forms,
- modification of the new nodes generation criteria in the adaptation process,
- improved HO multigrid approach.

Two step solution procedure is applied, when using HO approximation terms in the solution process. In both steps the basic MFD operator does not change. At first the standard procedure is applied yielding solution $u^{(L)}$ of low approximation order. In the second step the correction terms modify only the right hand side of the MFD equations. The final HO MFD solution $u^{(H)}$ does not depend on the quality of the MFD operator. It depends only on the truncation error of the Taylor series used.

Here, and in the following sections, the upper index $^{(L)}$ is referred to a quantity related to the low order approximation, $^{(H)}$ to the higher order one, and $^{(T)}$ - to the true analytical solution. The MFD equations for the formulations (1)-(4) including HO terms are:

- local

$$\begin{cases} Lu_i = f_i \\ L_b u_j = g_j \end{cases} \rightarrow \quad u_i^{(L)} \quad \rightarrow \begin{cases} Lu_i = f_i - \Delta_i \\ L_b u_j = g_j - \Delta_{b,i} \end{cases} \rightarrow \quad u_i^{(H)} \tag{3.20}$$

- global functional I minimisation

$$I(u) \approx J\sum_i^{N_G} \Omega_i \cdot F(u_i,\mathcal{L}_u u_i) \approx J\sum_i^{N_G} \Omega_i \cdot F(u_i,Lu_i + \Delta_i^{(L)}) \tag{3.21}$$

$$\frac{\partial}{\partial u_j} J\sum_i^{N_G} \Omega_i \cdot F(u_i,Lu_i + \left\{\Delta_i^{(L)}\right\}) = 0, \rightarrow u_i^{(L)}, \left\{u_i^{(H)}\right\} \tag{3.22}$$

- global (variational principle), e.g. Galerkin type or equivalent

$$\int_\Omega (\mathcal{L}_u u \cdot \mathcal{L}_v v - f \cdot v)\, d\Omega \approx$$
$$\approx J \cdot \sum_i^{N_G} \omega_i \cdot \left[\left(L_u u_i + \left\{\Delta_i^{(L)}\right\}\right) L_v v_i - f_i \cdot v_i\right] = 0 \quad \rightarrow \quad u_i^{(L)}, \left\{u_i^{(H)}\right\} \tag{3.23}$$

Here $u_i^{(L)}$ and $u_i^{(H)}$ denote MFD solutions based on the lower, p-th (no correction terms) and higher, $2p$-th order (including correction terms up to the order $2p$ for the MFD operators inside the domain $\Delta_i^{(L)}$, and on its boundary $\Delta_i^{(G)}$) approximation respectively. Symbols J, N_g, ω_i, Ω_i denote quantities involved in the Gaussian integration procedure, J is the transformation matrix (Jacobian matrix), N_g - number of Gauss points, and ω_i, Ω_i are integration weights.

The idea of using higher order terms in the MWLS approximation is based on correction of the local approximation by providing higher order derivatives $Du^{(H)}$ up to the order $p+s$. Usually we have $s = p$; derivatives are calculated in the most accurate manner then. We have

$$u(x) = p^T \cdot Du^{(L)} + \left(p^{(H)}\right)^T \cdot Du^{(H)} \tag{3.24}$$

where

$$\underset{[m'\times 1]}{\mathbf{p}^{(H)}} = \left[\frac{1}{(p+1)!}h^{p+1} \; \dots \; \frac{1}{2p!}k^{p+1}\right]^T, \quad \underset{(m'\times 1)}{Du^{(H)}} = \left[u^{(p+1)}_{xx\dots x} \; \dots \; u^{(2p)}_{yy\dots y}\right]^T \tag{3.25}$$

and $m' = \frac{3p(p+1)}{2}$. By assuming conditions (10) one gets system of equations

$$P \cdot Du^{(L)} + P^{(H)} \cdot Du^{(H)} = u \tag{3.26}$$

where

$$\underset{[r\times m']}{\mathbf{P}^{(H)}} = \begin{bmatrix} \frac{1}{(p+1)!}h_1^{p+1} & \dots & \frac{1}{2p!}k_1^{2p} \\ \frac{1}{(p+1)!}h_2^{p+1} & \dots & \frac{1}{2p!}k_2^{2p} \\ \dots & \dots & \dots \\ \frac{1}{(p+1)!}h_r^{p+1} & \dots & \frac{1}{2p!}k_r^{2p} \end{bmatrix} \tag{3.27}$$

Substituting (12) into (26) yields the improved values of the low order derivatives

$$D\mathbf{u}^{(L)} = \mathbf{M} \cdot \mathbf{u} - \mathbf{\Delta} \quad \rightarrow \quad \underset{(m\times r)}{\mathbf{\Delta}} = \mathbf{M} \cdot \mathbf{P}^{(H)} \cdot D\mathbf{u}^{(H)} \tag{3.28}$$

where $\mathbf{\Delta}$ is the vector of correction terms. Derivatives of higher order than p may be calculated inside the domain using formulae composition, e.g.

$$u^{III} = (u')'' \;\text{ or }\; u^{III} = (u'')' \quad , \quad u^{IV} = (u'')'' \tag{3.29}$$

Iteration procedure may be also performed in order to calculate the final corrected value of (12)

$$^{(k)}\mathbf{Du}^{(L)} = \begin{cases} M \cdot {}^{(0)}\mathbf{u}^{(L)}, & \text{for } k = 1 \\ \mathbf{M} \cdot {}^{(k-1)}\mathbf{u}^{(H)} - {}^{(k-1)}\Delta, & \text{for } k > 1 \end{cases} \tag{3.30}$$

where correction terms $^{(k-1)}\Delta$ are built in an iterative way on more and more accurate values of derivatives $^{(k)}\mathbf{Du}^{(L)}$.

The proposed HO MWLS approach may be applied to generation of the MFD formulae, and solution of Eqns (20)-(23) with HO terms included, as well as to postprocessing, mainly in a'posteriori residual error estimation.

4 A'posteriori error estimation

For regular meshes some a'priori estimations can be made using the local approximation order s and mesh modulus h, where $0 < s \leq p$, and k is the highest order of the differential operator

$$\left\| u^{(L)} - u^{(T)} \right\| \leq h^{p+1-k}, \quad \left\| u^{(L)} - u^{(H)} \right\| \leq h^{p+1-k}, \quad \left\| u^{(H)} - u^{(T)} \right\| \leq h^{p+s+1-k} \tag{4.31}$$

Formulas (31) do suggest high quality of the HO MFDM solutions due to approximation order (p+s) higher than the one provided by the other solution techniques. A'posteriori error estimation, based on the MFD solutions, will be discussed for the arbitrarily irregularly distributed cloud of nodes, and for any formulation of boundary value problem.

Considered are both the solution and residual errors. Correction terms may improve the a'posteriori error estimates, especially the residual error estimation, often used for adaptive mesh generation [2,9,12,13,16,18]. Such estimation is usually done locally at the points located somewhere between nodes, especially when the Liszka type node generator ([11,12,13]) is used, where mesh density is controlled. Error is checked at the points which belong only to the mesh one level denser than the current one.

When a numerical solution is found, and the exact analytical solution is known like in benchmark problems, one may solve appropriate MFD equations for any required point P_i in the domain, and examine solution errors there

$$e^{(LT)} = u^{(L)} - u^{(T)} \tag{4.32}$$

$$e^{(HT)} = u^{(H)} - u^{(T)} \tag{4.33}$$

The exact low order solution error (32) can be estimated as follows

$$e^{(LT)} \approx e^{(LH)} = u^{(L)} - u^{(H)} \tag{4.34}$$

Let $\bar{u}$ denotes an approximate smoothed solution based on the nodal function values. The true residual error is defined then as

$$r^{(T)} = L\bar{u} - f \tag{4.35}$$

In the MFDM $L\bar{u}$ is obtained by expansion of an unknown function $\bar{u}$ at any arbitrary point P_i in the domain into the Taylor series, and use of the MWLS approximation. The residual error may be presented in one of the following formulas:

$$r_i^{(L)} = Lu_i^{(L)} - f_i \tag{4.36}$$

$$r_i^{(H)} = Lu_i^{(H)} + \Delta_i^{(H)} - f_i \tag{4.37}$$

$$r_i^{(T)} = Lu_i^{(H)} + \Delta_i^{(H)} + R_i - f_i \tag{4.38}$$

depending on the approximation order required. Here Lu_i denotes a basic low order MFD operator, $\Delta_i^{(H)}$- considered HO correction term, and R_i- neglected truncation error. It is worth stressing that improved HO residuum form (37) involves only the truncation error of the Taylor series, while the low order one (36) is influenced by both the quality of the FD operator itself, and truncation of the Taylor series.

The MFDM global error analysis may be specially designed for the approach developed earlier [1] for the FEM. The FEM most commonly uses global integral estimators η of the solution error of the variational principle (4), where, in the case of Galerkin formulation

$$\begin{aligned} \mathcal{B}(u,v) &= \int_{\Omega} \mathcal{L}_u u \cdot \mathcal{L}_v v \, d\Omega \\ \mathcal{L}(v) &= \int_{\Omega} v \cdot f \, d\Omega + \int_{\partial\Omega} v \cdot g \, d\partial\Omega \end{aligned} \tag{4.39}$$

Solution of (39), due to Cea lemma, is the optimal approximation of $u^T(x)$, when the bilinear form $\mathcal{B}(u,v)$ is continuos and coercive, and the linear form $\mathcal{L}(v)$is continuos. The choice of reference solution $\bar{u} \approx u^T(x)$ determines the type of estimator.

Hierarchic estimators

The global solution hierarchic estimators are based on the local distribution of the solution error

$$e(x) \approx \bar{e}(x) = \bar{u}(x) - u(x) \tag{4.40}$$

where $\bar{u}(x)$ denotes reference solution. Thus the main task is to find good quality $\bar{u}(x)$. The global indicator

$$\eta_h = \|\bar{e}\|_E = \sqrt{B(\bar{e},\bar{e})} \tag{4.41}$$

is calculated either on the level of the whole mesh or on the single element using energy norm defined by a Galerkin type variational form. Reference solution $\bar{u}(x)$ may be calculated in several ways. The most common ones used in the FEM are h- and p- hierarchic. They use solutions $\bar{u}(x)$ either with number of nodes of doubled density $\left(h \rightarrow {}^{h}\!/_{2}\right)$ or with the increased approximation order $(p \rightarrow p+1)$ respectively, where h and p denote mesh modulus and approximation order of the estimated solution $u(x)$. The HO MFDM solution may be successfully applied in the estimators of hierarchic type developed for the FEM. In that case

$$u = u^{(L)}, \quad \bar{u} = u^{(H)} \tag{4.42}$$

which means that approximation order raises from p to $2p$, without the necessity of considering completely different discretization of the boundary value problem.

Smoothing estimators

Local distribution of the solution error (40) may also be presented in other manners. One of them is based on the difference between an improved $\bar{\sigma}$ and the rough (basic) σderivatives of the solution $u(x)$. Such is e.g. the well-known Zienkiewicz-Zhue error estimator based on the difference

$$e(x) \approx \bar{e}^{\sigma}(x) = \bar{\sigma}(x) - \sigma(x) \tag{4.43}$$

where $\bar{\sigma}$ is a smoothed derivative. Higher order terms may be used here to estimate values of the first derivative of u

$$e' = \left(u^{I}\right)^{(H)} - \left(u^{I}\right)^{(L)} \tag{4.44}$$

In the case of smoothing estimators, the Euclidean integral norm L_2 is applied

$$\eta_s = \|\bar{e}\|_{L_2} = \sqrt{\int_{\Omega} (\bar{e})^2 \, d\Omega} \tag{4.45}$$

Residual estimators

The last of the commonly used type of global estimators mentioned here, are the residual ones, of explicit and implicit character. They are based on the true residual error (35) or, in the MFDM, on one of its finite representations (36)-(38). The explicit residual estimator

$$\eta_r = h\sqrt{\int_{\Omega} \bar{r}^2 d\Omega} \tag{4.46}$$

uses residual error (35) as a measure of the true solution error. The implicit type residual error needs solution of the boundary value problem (4) with the modified right hand side

$$\mathcal{B}(\bar{e}, \bar{e}) = \bar{r} \tag{4.47}$$

The error estimator η_r is defined in the same way as in (41).

The quality of the global estimators may be controlled by the effectivity index i defined as

$$i = 1 + \frac{|\|e\| - \eta|}{\|e\|} \quad , \quad \eta = \|\bar{e}\| \tag{4.48}$$

and tested on chosen benchmark problems.

Hierarchic, smoothing, residual and interpolating (designed for regular meshes and not discussed here) estimators belong to the group of global energy estimators. They give information about quality of the solution treated as a whole. The other concept is presented by the goal-oriented estimators which give estimation of selected, discrete or integral, values.

The main task of the estimators, both local and global kind, is to provide information about the quality of the solution and approximation. These estimators may be applied in particular for mesh refinement in the adaptive solution approach.

5 Mesh adaptation

Analysis of the a'posteriori error, especially the residual error estimation (37) is widely used in the h-adaptive mesh refinement technique. It is especially effective when using mesh density based node generator. The MFDM local residuals are checked then at points [11,12,13], which belong to the one level denser mesh (e.g. midpoints between subsequent nodes in 1D) only. The generation criterion, providing the mesh density control, uses residuals

$$\left\| r_i^{(H)} \right\| > t \cdot \left\| r_{\max}^{(H)} \right\| \quad , \quad 0 \leq t < 1 \tag{5.49}$$

evaluated at each of those points. Here t denotes an imposed threshold value representing the admissible level of the residual error. New nodes are inserted at those points only, where the condition (45) is satisfied. In practical calculations also additional criteria may be applied, e.g. the upper limit for percentage number of new nodes among all possible node locations. Other refinement techniques are also possible, e.g. shifting, inserting or eliminating nodes, if necessary. Criterion (49) may be also used in order to remove nodes located in zones, where the residual error is small enough.

In the adaptive solution approach new nodes are generated in each subsequent mesh, until an admissible error level ε_{adm}, appropriate for actual considered mesh, is reached for all residuals

$$\frac{\left\| r_i^{(H)} \right\|}{\left\| f_{\max} \right\|} < \varepsilon_{adm}^{r} \tag{5.50}$$

examined in the whole domain and for all solution errors (32), (33)

$$\frac{\left\| \bar{e}^{(k+1)} - \bar{e}^{(k)} \right\|}{\left\| \bar{e}^{(k+1)} \right\|} < \varepsilon_{adm}^{e} \tag{5.51}$$

checked in every node, common for each of two subsequent meshes, k and k+1. Afterwards mesh smoothness is also examined. Wherever the smooth transition criterion

$$\frac{|\rho_i - \rho_j|}{\|x_i - x_j\|} < \eta_{adm} \tag{5.52}$$

(ρ_i, ρ_j - local mesh densities of the neighbour nodes x_i, x_j, η_{adm} - admissible level of mesh density change) is violated, new nodes are added.

Beyond the FEM based global error criteria discussed above when used in the MFDM based error analysis, several other, possibly more subtle error measures were developed and applied, especially for arbitrarily irregular meshes. These are so called error indicators proposed and examined in [16]. They determine a pair $(\bar{h}, \bar{e})$ of a local mesh modulus $\bar{h}$, and a local solution or residual error $\bar{e}$ representing the cloud of all (N) examined points in the whole mesh. As it was shown in the previous works [16,18-20], the best results are obtained for the following pairs of the discrete indicators

$$\bar{h} = (\frac{1}{N}\sum_i h_i^2)^{\frac{1}{2}} \quad ,\bar{e} = (\frac{1}{N}\sum_i e_i^2)^{\frac{1}{2}} \tag{5.53}$$

$$\bar{h} = \frac{1}{N}\sum_i h_i \quad ,\bar{e} = \frac{1}{N}\sum_i |e_i| \tag{5.54}$$

Thus in adaptation process each mesh has its own representative pair of $(\bar{h}, \bar{e})$. Distribution of $(\bar{h}, \bar{e})$ provides estimation of the convergence rate of the considered quantity, and tests quality of the error indicators as well.

The HO approximation technique, combined with the h-adaptive mesh generation is expected to work effectively with the multigrid solution approach [4,12,13,16], reducing computational time spent on analysis of large boundary value problems. In the multigrid approach, one deals with the set of meshes, varying from coarse to fine. Usually, each finer mesh includes all nodes of the previous one. Multigrid approach, which was applied in the present work, uses original concepts of prolongation and restriction [12,13,16]. Prolongation procedure extends the solution obtained for a coarse mesh to a finer one. Residuum calculated for the finer mesh is reduced then to a coarser mesh by means of the restriction. The whole solution process needs to be used twice. At first for the low order solution, corresponding to the error (36) $r_i^{(L)} = 0$, and later on for the higher order one, relevant to the HO error (37) $r_i^{(H)} = 0$.

6 Numerical examples

A variety of benchmark tests of 1D and 2D boundary value problems were solved so far, using higher order approximation technique. Many aspects of the proposed approach were tested. The most interesting were:

- application of the approach using higher order terms, to the local solution error and residual error estimation,

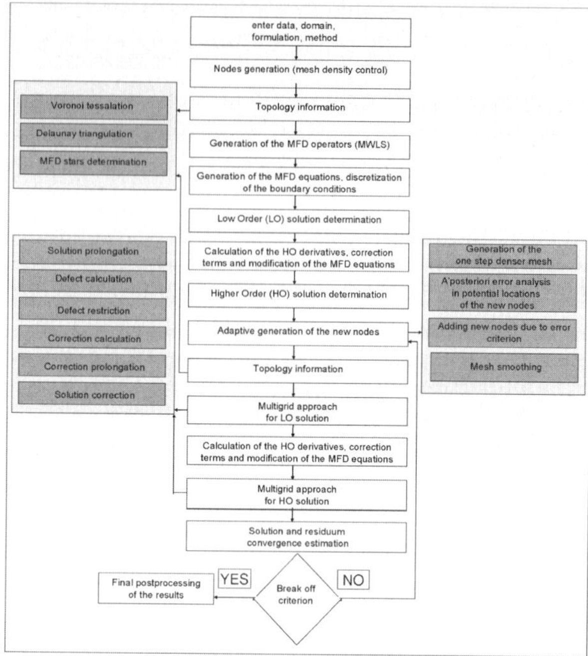

Figure 6.2. HOA multigrid adaptive solution approach - flow chart (*See also* Color Plate on page 379)

- application of the approach using higher order terms to estimation of the global solution and residual errors,
- application of the higher order estimators to appropriate adaptive mesh generation,
- examination of the error indicators, convergence rate, and improvement of solution quality for those meshes.

Results obtained so far are very encouraging and show potential power of the approach. Recent tests concentrate on solving non-linear tasks, as well as on using the approach in the fuzzy sets analysis. The flow chart of the whole solution algorithm is presented below (Fig. 2).

6.1 1D test

Consider the local form of the b.v. problem

$$\begin{aligned} &w''(x) + a \cdot w'(x) = f(x), \quad x \in (0,4) \\ &w(0) = w(4) = 0 \end{aligned} \tag{6.55}$$

Assumed is the right hand side function *f(x)* corresponding to the exact analytical solution (see Fig. 3) given below for $a = 1$

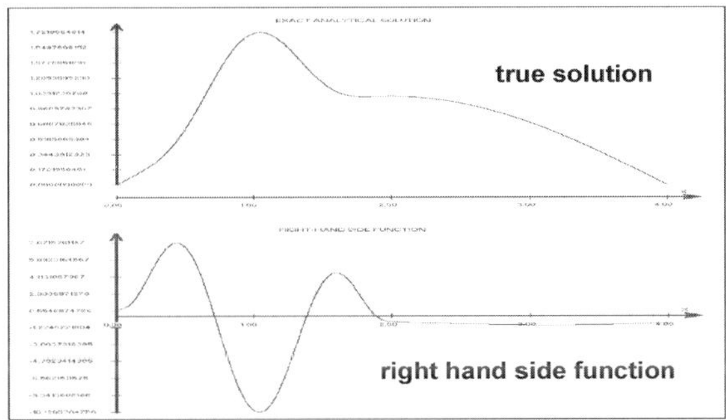

Figure 6.3. Analytical solution and right hand side of 1D test (*See also* Color Plate on page 379)

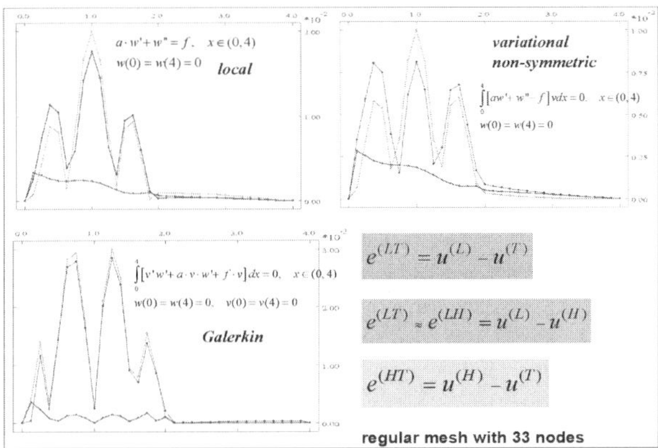

Figure 6.4. Local solution error estimation (*See also* Color Plate on page 380)

$$w(x) = \begin{cases} \sin(\frac{\pi x}{4}) + p_{10}(x), & 0 < x < 2 \\ \sin(\frac{\pi x}{4}), & 2 < x < 4 \end{cases} \tag{6.56}$$

Regular mesh with 33 nodes was used for calculations. Fig. 3 shows the exact low order solution error (32), its higher order estimation (33), and the exact higher order solution error (34), calculated in all nodes for three different formulations (1)-(4) of the b.v. problem (55). In Fig. 4 presented are residual errors evaluated between the nodes, namely the exact residual error (35), and its estimations: the low order (36), as well as the higher order (37) one. A set of adaptive, strongly irregular meshes was generated using criteria (49),(52), starting from a regular mesh. Convergence rates of the MFD solutions (Fig. 6) were estimated using simple error indicator (54). Final observations were done for an irregular mesh with 35 nodes. In Fig. 7 presented are results obtained for the global low order solution error estimators (42) of hierarchic type. The exact low order solution error is presented in the background. In the left column local error distributions are shown, whereas in the right one the global,

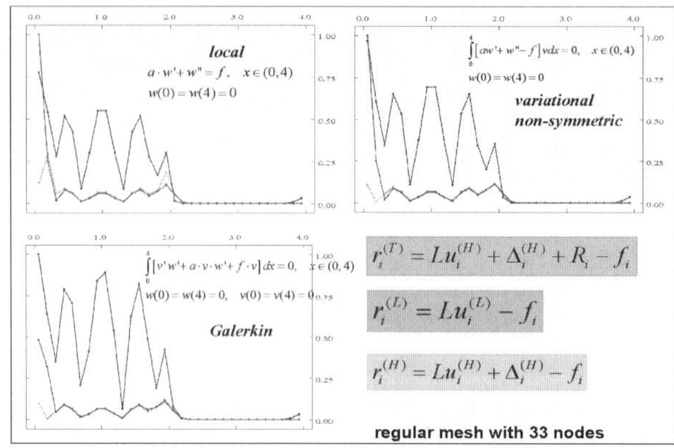

Figure 6.5. Local residual error estimation (*See also* Color Plate on page 380)

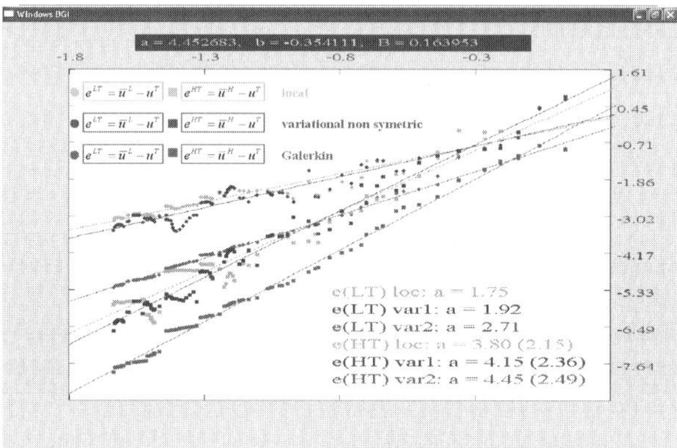

Figure 6.6. Solution convergence rates on the set of adaptive irregular meshes (*See also* Color Plate on page 381)

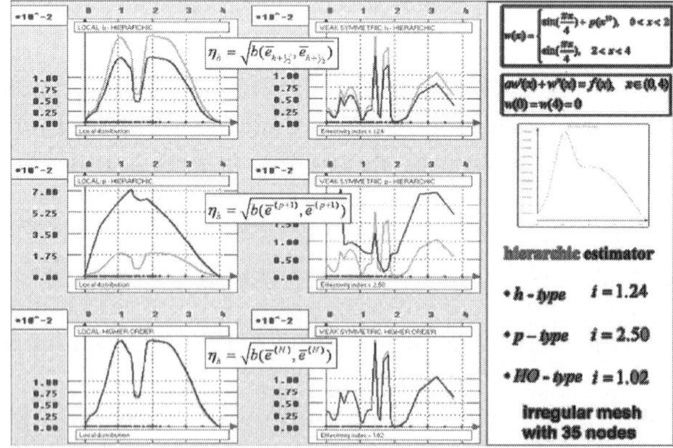

Figure 6.7. Hierarchic estimators on irregular mesh (*See also* Color Plate on page 381)

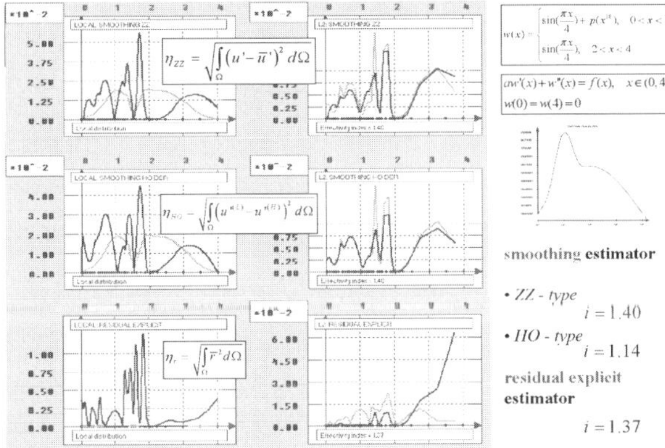

Figure 6.8. Smoothing and residual explicit estimator on irregular mesh (*See also* Color Plate on page 382)

integral forms are given. Figure 8 presents the results obtained for smoothing (45) and residual explicit (46) global estimators.

Quality of the estimator was examined each time using the effectivity index (48). The best results were obtained for the estimators which used higher order solution as a reference one: hierarchic (42) ($i = 1.02$), and smoothing estimator (44) ($i = 1.14$).

6.2 2D test

The Laplace equation with the essential boundary conditions was analysed.

$$\begin{cases} \nabla^2 u = f(x) & \text{in } \Omega \\ u = \bar{u} & \text{on } \partial\Omega \end{cases} \tag{6.57}$$

$$\Omega = \{(x,y), \quad 0 \le x \le 1, \quad 0 \le y \le 1\}$$

The exact solution

$$\begin{array}{l} w(x) = -x^3 - y^3 + e^{-\left(\frac{x-0.5}{0.2}\right)^2 - \left(\frac{y-0.5}{0.2}\right)^2}, \\ 0 \le x \le 1, \quad 0 \le y \le 1 \end{array} \tag{6.58}$$

The right hand side function *f(x)* results from the true analytical solution (58). Fine, regular mesh with 400 nodes and local formulation (57) were used for calculations. In Fig. 9 the low order solution error, and its estimation as well as a higher order solution error are presented. True residual error and its estimations are shown in Fig. 10.

The same mesh was used for comparison of the chosen global estimators of the low order solution error. In Fig. 11 hierarchic estimators and in Fig. 12 smoothing and residual estimators are presented in two manners. The left column is for 3D wire-frame view, and the right one shows a contour map. For

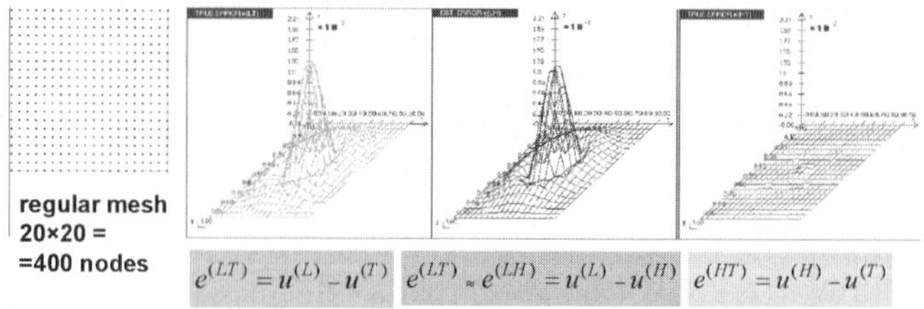

Figure 6.9. Solution error estimation (*See also* Color Plate on page 382)

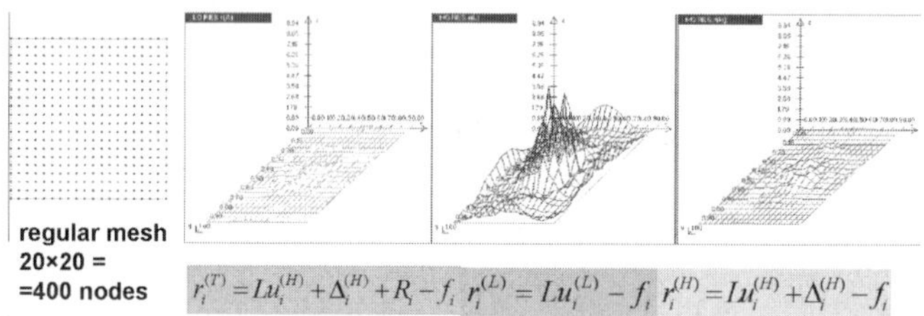

Figure 6.10. Residual error estimation (*See also* Color Plate on page 382)

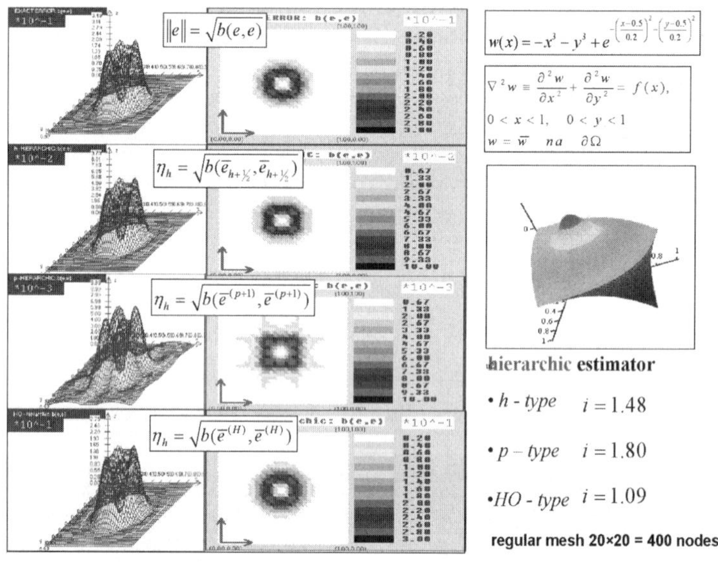

Figure 6.11. Hierarchic estimators (*See also* Color Plate on page 383)

better comparison the global exact low order solution error is presented always in the first row. The lowest values of the effectivity index (48) come from the higher order estimators: hierarchic ($i = 1.09$) and smoothing estimator ($i = 1.34$). In addition, the true analytical solution (58) is presented on the right side of Fig. 11 and Fig. 12.

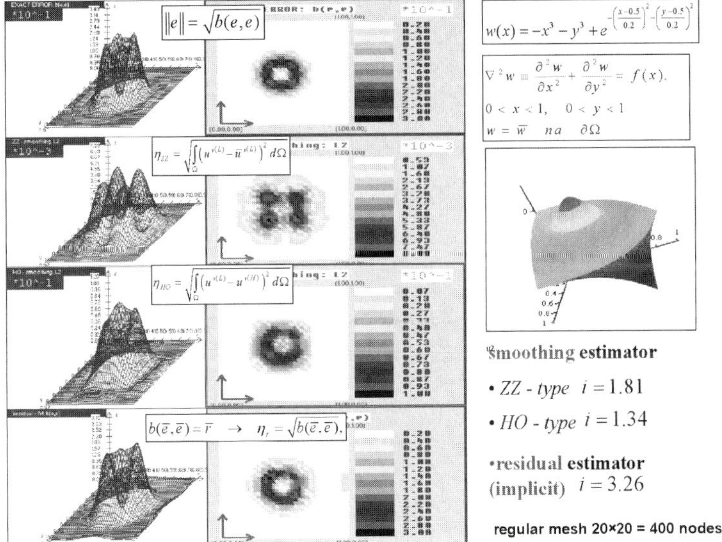

Figure 6.12. Smoothing and residual estimators (*See also* Color Plate on page 383)

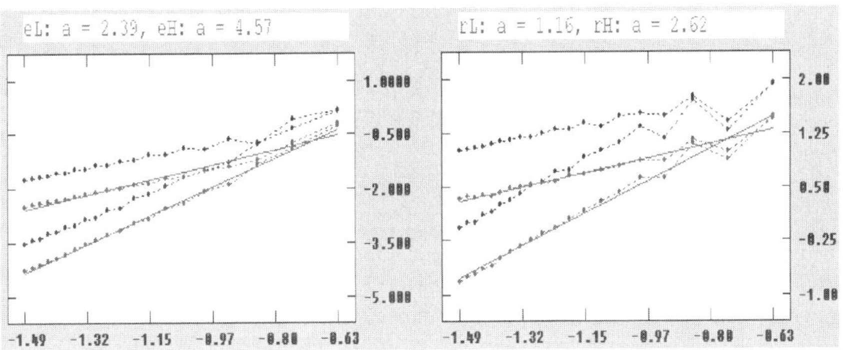

Figure 6.13. Solution and residual convergence on the set of regular meshes (*See also* Color Plate on page 384)

Convergence test was performed on the set of regular meshes at first. The basic, coarsest mesh, consisted of 16 nodes whereas the finest one consisted of 529 nodes. Figure 13 presents the low order and higher order solution convergence in terms of the mesh modulus h (on the left), and the low order as well as higher order residual convergence (on the right), both in the logarithmic scale. The mean values, calculated using (49), and the maximum values are shown. Solution convergence rates are respectively $a^{(L)} = 2.39$ and $a^{(H)} = 4.57$ which gives the following solution improvement rate $a^{(H)}/\ a^{(L)} = 1.91$. Convergence rate for higher order residuum (37) is over 2 times better than for the low order one (36).

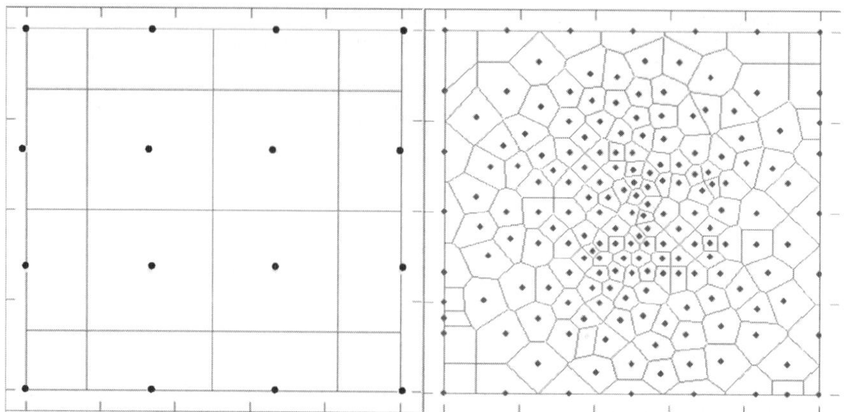

Figure 6.14. Adaptation process, the first regular and last irregular mesh (*See also* Color Plate on page 384)

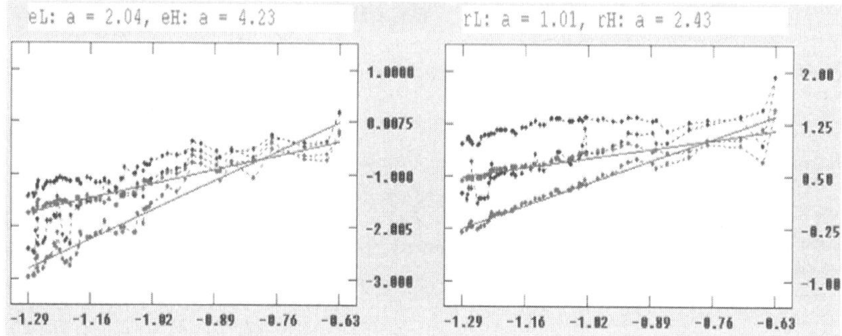

Figure 6.15. Solution and residual convergence on the set of adaptive irregular meshes (*See also* Color Plate on page 384)

Further tests concerned application of the higher order estimators to appropriate adaptive mesh generation. The initial coarse regular mesh with 16 nodes was considered. The subsequent meshes were generated using the criteria (49) and (52) as well as the local estimation of the residual error (37). The adaptation process was stopped after 200 adaptive iterations, when the break off criterion (50) was satisfied.

The final finest, strongly irregular mesh consisted of 179 nodes. During adaptation process, representative pairs $(\bar{h}, \bar{e})$ have been calculated on each mesh separately, using simple (centre of gravity) error indicator (54). Those quantities were used in order to examine the convergence of the solution and residuals on the set of irregular adaptive meshes (Fig. 15). The convergence rates for the low order solution (mean values (54) and maximum values), $a^{(L)} = 2.04$ and for the higher order one $a^{(H)} = 4.23$ give the final solution improvement rate $a^{(H)}/a^{(L)} = 2.07$. In Fig. 15 on the right, residuum based convergence is presented, mean and maximum values of the low order (36), and

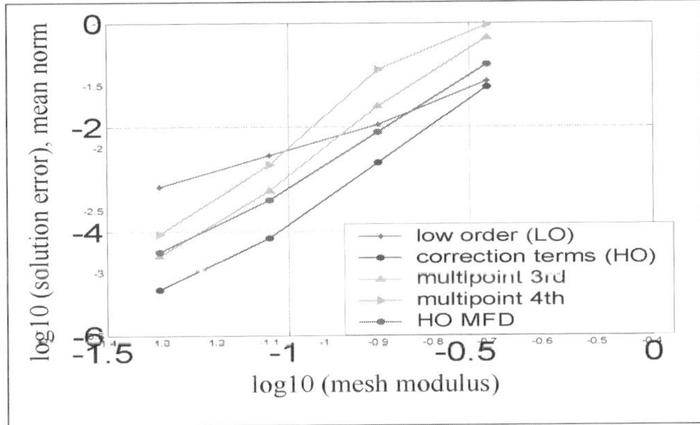

Figure 6.16. Comparison with the other HO techniques (*See also* Color Plate on page 385)

higher order residual errors (37). The last one was used in the error criterion (49) in order to generate new nodes. In order to examine convergence rates, the mean values were additionally subjected to linear approximation, using MWLS technique.

The proposed approach uses the higher order terms in the Taylor series expansion, in order to raise the rank of the local approximation. It was compared with the other higher order techniques, especially with the multipoint approach [17,21]. A comparison was performed first on the set of regular meshes for the considered 2D boundary value problem (57). Examined and compared were solution convergence rates. The results are presented in Fig. 16. The classic MFD low order and higher order solutions, based on the correction terms (20), are compared with two different types of the multipoint solutions, and solution based on the HO MFD operator [6]. HO solution, based on the correction terms, is slightly better in this case than the other HO solutions.

7 Final remarks

An adaptive HO solution approach to analysis of boundary value problems, based on the meshless FDM [13] is presented. Solution process includes original concepts of higher order approximation, a'posteriori error estimation, solution smoothing, nodes generation, and mesh modification, as well as an adaptive multigrid solution procedure. Moving Weighted Least Squares (MWLS) approach [13] is used for local approximation.

The approach is based on the Taylor series expansion and use of relevant correction terms, rather than on adding new nodes into the MFD operators. The approach is very effective – it needs two steps, both using the same basic MFD operator. Quality of the final MFD solution depends only on the truncation error of the Taylor series. It does not depend on the quality of the MFD operator itself. HO approximation terms may be used not only to improve the solution quality, but also to refine, in this way, the estimation of solutions and residuals. Those estimates may be used in the adaptive mesh generation. Specially developed (for irregular cloud of nodes) local error indicators are proposed and used to examine the convergence rate of both the solutions and residuals. Moreover, global criteria developed for error estimation in the FEM analysis may be applied here. When including the local HO MFD error estimates, they provide especially high quality ($2p$-th order) estimation for solution and residual errors, if compared with those obtained by means of the existing smoothing procedures of the $p+1$ order. It is worth stressing here that these error estimates, though developed for the MFDM analysis, may be also used in the other meshless methods or in the FEM.

Many 1D and 2D benchmark tests executed so far indicate the potential power of the approach in fast solving (high convergence rate) of b.v problems, as well as in the error analysis and adaptivity. The total number of nodes in a considered mesh may be reduced, without compromising the quality of the MFD solution improved by raising the rank of the local approximation. However, a lot of work has to be done yet. Besides further testing and solution of b.v. problems and error analysis, future plans include combinations of the approach with the other discrete methods, especially with the Finite Element Method, as well as development of a special MFD node generator, based on mesh density control, higher order approximation technique, and multigrid solution procedure. Considered is also application of the HO MFDM approach to analysis of b.v. problem given in the local Petrov-Galerkin formulation.

References

1. Ainsworth M, Oden JT, A'posteriori error estimation in finite element analysis, Comp. Meth Appl. Mech Engng 142:1-88, 1997.
2. Belytchko T., Meshless methods: An overview and recent developments. Comp Meth Appl Mech Engng 139:3-47, 1996.
3. Benito J.J., Ureña F., Gavete L., Alonso B. Application of the GFDM to improve the approximated solution of 2nd order partial differential equations, ICCES MM 2007.
4. Brandt A., Multi-level Adaptive Solutions To Boundary Value Problems, Math. Comp., 31, 1977, pp 333-390.
5. Collatz L., The Numerical Treatment of Differential Equations, Springer, Berlin, 1966.
6. Hackbush W., Multi-Grid Methods and Applications, Springer – Verlag, Berlin, 1985.
7. Karmowski W., Orkisz J., A physically based method of enhancement of experimental data – concepts, formulation and application to identification of residual stresses, Proc IUTAM Symp on Inverse Problems in Engng Mech, Tokyo 1992; On Inverse Problems in Engineering Mechanics, Springer Verlag, pp 61-70, 1993.

8. Krok J., Orkisz J., A Discrete Analysis of Boundary-Value Problems with Special Emphasis on Symbolic Derivation of Meshless FDM/FEM Models, *Computer Methods in Mechanics CMM*, June 19-22, 2007, Spala, Lódz, Poland.
9. Krok J., A New Formulation of Zienkiewicz-Zhu a Posteriori Error Estimators Without Superconvergence Theory, Proceedings *Third MIT Conference on Computational Fluid and Solid Mechanics*, Cambridge, Ma, USA, June 14-17, 2005.
10. Lancaster P, Salkauskas K, Surfaces generated by moving least-squares method, Math Comp 155(37):141-158, 1981.
11. Liszka T, Orkisz J. The Finite Difference Method at arbitrary irregular grids and its applications in applied mechanics, 1980, Comp Struct 11:83-95.
12. Orkisz J., Lezanski P., Przybylski P., Multigrid approach to adaptive analysis of bv problems by the meshless GFDM. IUTAM/IACM Symposium On Discrete Methods In Structural Mechanics II, Vienna, 1997.
13. Orkisz J., Finite Difference Method (Part III), in Handbook of Computational Solid Mechanics, M.Kleiber (Ed.) Springer-Verlag, Berlin, 1998, 336-431.
14. Orkisz J., Higher Order Meshless Finite Difference Approach, *13th Inter-Institute Seminar for Young Researchers*, Vienna, Austria, October 26-28, 2001.
15. Orkisz J., Milewski S., On higher order approximation in the MDM metod, Elsevier (editor: K.J. Bathe), Proceedings *Third MIT Conference on Computational Fluid and Solid Mechanics*, Cambridge, Ma, USA, June 14-17, 2005.
16. Orkisz J., Milewski S., Higher order approximation approach in meshless finite difference analysis of boundary value problems, *the 16th International Conference on Computer Methods in Mechanics* CMM-2005, June 21–24, 2005, Czestochowa, Poland.
17. Orkisz J., Jaworska I., Milewski S., Meshless finite difference method for higher order approximation, *Third International Workshop Meshfree Methods for Partial Differential Equations*, September 12-15, 2005, Bonn, Germany.
18. Orkisz J.,Milewski S., On a'posteriori error analysis in Higher Order approximation in the Meshless Finite Difference Method, *ICCES Special Symposium On Meshless Methods*, 14-16 June 2006, Dubrovnik, Croatia, - submitted to the Computer Modeling in Engineering and Sciences (CMES, 2006).
19. Orkisz J., Milewski S., Recent advances in the Higher Order Approximation in the Meshless Finite Difference Method, *7th World Congress on Computational Mechanics*, Los Angeles, California, July 16-22, 2006.
20. Orkisz J., Milewski S., Recent advances in a'posteriori error estimation based on the Higher Order correction terms in the Meshless Finite Difference Method, *ICCES Special Symposium On Meshless Methods,* 15-17 June 2007, Patras, Greece.
21. Orkisz J., Jaworska I., Some Concepts of 2D Multipoint HO Operators for the Meshless FDM Analysis, *ICCES Special Symposium On Meshless Methods,* 15-17 June 2007, Patras, Greece.
22. Orkisz J., Milewski S., Higher Order approximation multigrid approach in the Meshless Finite Difference Method, *Computer Methods in Mechanics CMM*, June 19-22, 2007, Spala, Lódz, Poland.
23. Orkisz J., Dobrowolski L., On the Best Approach to Moving Weighted Least Squares Approximation, *Computer Methods in Mechanics CMM*, June 19-22, 2007, Spala, Lódz, Poland.

Exact Bounds for Linear Outputs of the Convection-Diffusion-Reaction Equation Using Flux-Free Error Estimates

Núria Parés, Pedro Díez, and Antonio Huerta

Laboratori de Càlcul Numèric (LaCàN), Universitat Politècnica de Catalunya, Jordi Girona 1-3 E-08034 Barcelona, Spain
{nuria.pares,pedro.diez,antonio.huerta}@upc.edu

Summary. The Flux-free approach is a promising alternative to standard implicit residual error estimators that require the equilibration of hybrid fluxes. The idea is to solve local error problems in patches of elements surrounding one node (*stars*) instead of in single elements [1]. The resulting local problems are flux-free, that is the boundary conditions are natural and hence their implementation is straightforward. This allows precluding the computation and the equilibration of fluxes along the element edges. The domain decomposition is performed using a partition of unity strategy. The resulting estimates are much simpler from the implementation viewpoint, especially in the 3D cases, and provide upper bounds of the energy norm of the error (as well as the standard implicit residual estimators with equilibration of hybrid fluxes).

In the past, the local flux-free problems have been solved using a finite element mesh inside each local subdomain. Consequently, the resulting estimates were asymptotic upper bounds (w.r.t. a reference solution) rather than exact upper bounds (w.r.t. the exact solution). Some effort has been devoted to recover exact upper bounds using the equilibrated hybrid fluxes approach. The idea is to solve the local problem using a dual formulation and to minimize the complementary energy [2]. In this work, the same idea is employed to obtain exact upper bounds using the flux-free approach. The resulting estimates have similar features as their asymptotic version, while providing a guaranteed upper bound. This strategy is applied both to the primal and adjoint problem to recover guaranteed bounds for the quantity of interest.

Key words: Linear-functional outputs; Exact/guaranteed/strict bounds; Error estimation; Goal-oriented adaptivity

1 Problem statement

Consider the steady convection-reaction-diffusion equation posed on a polygonal domain Ω. The weak solution of the problem is $u \in \mathcal{V}$ verifying

$$a(u,v) = \ell(v) \quad \forall v \in \mathcal{V}, \tag{1.1}$$

where

$$a(u,v) := \int_\Omega \Big[\nu \boldsymbol{\nabla} u \cdot \boldsymbol{\nabla} v + (\boldsymbol{\alpha} \cdot \boldsymbol{\nabla} u) v + \sigma u v \Big] \, \mathrm{d}\Omega \ \text{ and } \ \ell(v) := \int_\Omega f v \, \mathrm{d}\Omega.$$

For the sake of simplicity, the presentation concerns only Dirichlet homogeneous boundary conditions, but the methodology is general and it is also applicable to other type of boundary conditions. In this case, the solution and test spaces coincide and are $\mathcal{V} := \{v \in \mathcal{H}^1(\Omega),\ v|_{\Gamma_\mathrm{D}} = 0\}$ where $\mathcal{H}^1$ is the standard Sobolev space of square integrable functions whose first derivative is also square integrable. The data are assumed to be smooth, that is, $f \in \mathcal{H}^{-1}(\Omega)$, $g \in \mathcal{H}^{-\frac{1}{2}}(\Gamma_\mathrm{N})$, $u_\mathrm{D} \in \mathcal{H}^{\frac{1}{2}}(\Gamma_\mathrm{D})$, $\nu \in \mathcal{L}^\infty(\Omega)$, $\sigma \in \mathcal{L}^\infty(\Omega)$ is a non-negative real coefficient and the prescribed vector field $\boldsymbol{\alpha} \in \mathcal{H}(\mathrm{div}; \Omega)$ which is assumed for simplicity to be incompressible, $\boldsymbol{\nabla} \cdot \boldsymbol{\alpha} = 0$.

The goal of the vast majority of finite element simulations is to determine specific quantities (outputs) which depend on the solution of the partial differential equations governing the problem. In most cases, moreover, it is crucial to be able to certify the precision of the approximations of these quantities given by the numerical simulations. Therefore the final goal of most simulations is to provide upper and lower bounds for the exact value of the quantity of interest, or equivalently, to provide a range where the exact value of the output lies.

Here, the quantities of interest are restricted to depend linearly on u and to be of the form

$$\ell^{\mathcal{O}}(u) := \int_\Omega f^{\mathcal{O}} u \, \mathrm{d}\Omega, \tag{1.2}$$

where $f^{\mathcal{O}} \in \mathcal{H}^{-1}(\Omega)$. The goal is then to provide upper and lower bounds for $s = \ell^{\mathcal{O}}(u)$, namely,

$$s^{lb} \leq s \leq s^{ub},$$

and at the same time derive an adaptive refinement procedure to be able to improve the precision of the desired output (i.e. narrow the gap between the bounds $s^{ub} - s^{lb}$).

2 Energy reformulation: representation of the output bounds

Most existing techniques to obtain upper and lower bounds for a quantity of interest $\ell^{\mathcal{O}}(u)$ are based on the fact that bounds for the output may be obtained using available techniques for estimating the error measured in the energy norm [3,4,10,11]. The key point of these strategies is to recover an alternative representation for the output $\ell^{\mathcal{O}}(u)$ (or in the case of non-selfadjoint problems for its bounds) in terms of the energy norms of some continuous functions. This alternative representation does not directly yield a computable expression for the bounds of the output since the energy norms appearing in the expression are non-computable. However, bounds may be easily recovered using existing strategies to bound the error measured in the energy norm.

This section is devoted to detail the representation of bounds for $\ell^{\mathcal{O}}(u)$ in terms of energy norms, for the case of the convection-reaction-diffusion equation which is a non-selfadjoint problem, see [3]. That is, the key ingredients of the methods to obtain bounds for $s = \ell^{\mathcal{O}}(u)$ from upper bounds for the energy norm are briefly summarized.

The non-computable expressions for the upper and lower bounds for $\ell^{\mathcal{O}}(u)$ are recovered from the following steps. First, the finite element approximation of u, u_h, is computed. This approximation is associated to a finite element mesh of the domain Ω and to a finite element interpolation space $\mathcal{V}^h \subset \mathcal{V}$, that is $u_h \in \mathcal{V}^h$. Second, an adjoint problem associated to the selected output is introduced, along with its finite element approximation, $\psi_h \in \mathcal{V}^h \subset \mathcal{V}$. Finally, the error equations for u_h and ψ_h are modified (symmetrizing the l.h.s.) such that bounds of $s = \ell^{\mathcal{O}}(u)$ are obtained from linear combinations of the resulting modified errors.

Indeed, let $\mathcal{V}^h \subset \mathcal{V}$ be the finite element interpolation space associated with a finite element mesh of characteristic size h and degree p for the complete polynomial base. Then, the bounding procedure may be sketched as:

1. Compute the finite element approximation of the primal problem: find $u_h \in \mathcal{V}^h$ such that
$$a(u_h, v) = \ell(v) \quad \forall v \in \mathcal{V}^h.$$
2. Introduce the adjoint problem associated to the selected output: find $\psi \in \mathcal{V}$ such that
$$a(v, \psi) = \ell^{\mathcal{O}}(v) \quad \forall v \in \mathcal{V}.$$
3. Compute the finite element approximation of the adjoint problem: find $\psi_h \in \mathcal{V}^h$ such that
$$a(v, \psi_h) = \ell^{\mathcal{O}}(v) \quad \forall v \in \mathcal{V}^h.$$
4. Consider the errors in the approximations u_h and ψ_h, $e := u - u_h \in \mathcal{V}$ and $\varepsilon := \psi - \psi_h \in \mathcal{V}$, satisfying the residual equations
$$a(e, v) = \ell(v) - a(u_h, v) =: R^{\mathrm{P}}(v) \quad \forall v \in \mathcal{V}, \tag{2.3}$$
and
$$a(v, \varepsilon) = \ell^{\mathcal{O}}(v) - a(v, \psi_h) =: R^{\mathrm{D}}(v) \quad \forall v \in \mathcal{V}, \tag{2.4}$$
where $R^{\mathrm{P}}(\cdot)$ and $R^{\mathrm{D}}(\cdot)$ stand for the weak primal and adjoint residuals associated to the approximations u_h and ψ_h respectively.
5. Introduce the symmetric versions of the residual problems: find e^{s} and $\varepsilon^{\mathrm{s}} \in \mathcal{V}$ such that
$$a^{\mathrm{s}}(e^{\mathrm{s}}, v) = R^{\mathrm{P}}(v) \quad \forall v \in \mathcal{V}, \tag{2.5}$$

and

$$a^{\mathrm{s}}(\varepsilon^{\mathrm{s}}, v) = R^{\mathrm{D}}(v) \quad \forall v \in \mathcal{V}, \tag{2.6}$$

where $a^{\mathrm{s}}(\cdot,\cdot)$ is the symmetric counterpart of $a(\cdot,\cdot)$

$$a^{\mathrm{s}}(w, v) = \int_{\Omega} \Big[\nu \boldsymbol{\nabla} w \cdot \boldsymbol{\nabla} v + \sigma w v \Big] \, \mathrm{d}\Omega. \tag{2.7}$$

6. Compute the upper and lower bounds for s as

$$\ell^{\mathcal{O}}(u_h) - \frac{1}{4}\|\kappa e^{\mathrm{s}} - \frac{1}{\kappa}\varepsilon^{\mathrm{s}}\|_{ub}^2 \leq \ell^{\mathcal{O}}(u) \leq \ell^{\mathcal{O}}(u_h) + \frac{1}{4}\|\kappa e^{\mathrm{s}} + \frac{1}{\kappa}\varepsilon^{\mathrm{s}}\|_{ub}^2,$$

where $\|\cdot\|$ is the energy norm induced by the bilinear form $a^{\mathrm{s}}(\cdot,\cdot)$, $\|v\|^2 = a^{\mathrm{s}}(v,v) = a(v,v)$, $\|\cdot\|_{ub}$ represents an upper bound for the value of the norm and $\kappa \in \mathbb{R}$ is an arbitrary scalar non-zero parameter.

The following theorem shows that indeed bounds for s may be obtained determining upper bounds for the energy norm of the linear combinations of the errors e^{s} and ε^{s}.

Theorem 1. *Let e^{s} and $\varepsilon^{\mathrm{s}} \in \mathcal{V}$ be such that for any $v \in \mathcal{V}$*

$$a^{\mathrm{s}}(e^{\mathrm{s}}, v) = R^{P}(v) \quad \text{and} \quad a^{\mathrm{s}}(\varepsilon^{\mathrm{s}}, v) = R^{D}(v).$$

Then,

$$\ell^{\mathcal{O}}(u_h) - \frac{1}{4}\|\kappa e^{\mathrm{s}} - \frac{1}{\kappa}\varepsilon^{\mathrm{s}}\|^2 \leq \ell^{\mathcal{O}}(u) \leq \ell^{\mathcal{O}}(u_h) + \frac{1}{4}\|\kappa e^{\mathrm{s}} + \frac{1}{\kappa}\varepsilon^{\mathrm{s}}\|^2,$$

and therefore

$$\ell^{\mathcal{O}}(u_h) - \frac{1}{4}\|\kappa e^{\mathrm{s}} - \frac{1}{\kappa}\varepsilon^{\mathrm{s}}\|_{ub}^2 \leq \ell^{\mathcal{O}}(u) \leq \ell^{\mathcal{O}}(u_h) + \frac{1}{4}\|\kappa e^{\mathrm{s}} + \frac{1}{\kappa}\varepsilon^{\mathrm{s}}\|_{ub}^2. \tag{2.8}$$

Proof. Combining the definition of the adjoint residual, equation (2.4) with $v = e$, the Galerkin orthogonality property, the primal error e is orthogonal to the finite element space $\mathcal{V}^h$, and the linearity of the functional $\ell^{\mathcal{O}}(\cdot)$ induces the following relation between $R^{\mathrm{D}}(e)$ and the output

$$R^{\mathrm{D}}(e) = \ell^{\mathcal{O}}(e) - a(e, \psi_h) = \ell^{\mathcal{O}}(e) = \ell^{\mathcal{O}}(u - u_h) = \ell^{\mathcal{O}}(u) - \ell^{\mathcal{O}}(u_h). \tag{2.9}$$

Also, taking $v = e \in \mathcal{V}$ in equation (2.3) yields

$$R^{\mathrm{P}}(e) = a(e, e) = \|e\|^2. \tag{2.10}$$

The proof now follows from a simple algebraic manipulation. Indeed, let κ be a nonzero scalar parameter and consider the obvious inequality

$$\|\frac{1}{2}(\kappa e^{\mathrm{s}} \pm \frac{1}{\kappa}\varepsilon^{\mathrm{s}}) - \kappa e\|^2 \geq 0. \tag{2.11}$$

Expanding the norm, using $v = e$ in the definition of the primal and adjoint errors e^{s} and ε^{s}, i.e. equations (2.5) and (2.6), and equations (2.9) and (2.10) yields

$$\begin{aligned}\|\frac{1}{2}(\kappa e^{\mathrm{s}} \pm \frac{1}{\kappa}\varepsilon^{\mathrm{s}}) - \kappa e\|^2 &= \frac{1}{4}\|\kappa e^{\mathrm{s}} \pm \frac{1}{\kappa}\varepsilon^{\mathrm{s}}\|^2 + \kappa^2\|e\|^2 - \kappa a^{\mathrm{s}}(\kappa e^{\mathrm{s}} \pm \frac{1}{\kappa}\varepsilon^{\mathrm{s}}, e)\\ &= \frac{1}{4}\|\kappa e^{\mathrm{s}} \pm \frac{1}{\kappa}\varepsilon^{\mathrm{s}}\|^2 + \kappa^2\|e\|^2 - \kappa^2 a^{\mathrm{s}}(e^{\mathrm{s}}, e) \mp a^{\mathrm{s}}(\varepsilon^{\mathrm{s}}, e)\\ &= \frac{1}{4}\|\kappa e^{\mathrm{s}} \pm \frac{1}{\kappa}\varepsilon^{\mathrm{s}}\|^2 + \kappa^2\|e\|^2 - \kappa^2 R^{\mathrm{P}}(e) \mp R^{\mathrm{D}}(e)\\ &= \frac{1}{4}\|\kappa e^{\mathrm{s}} \pm \frac{1}{\kappa}\varepsilon^{\mathrm{s}}\|^2 + \kappa^2\|e\|^2 - \kappa^2\|e\|^2\\ &\quad \mp(\ell^{\mathcal{O}}(u) - \ell^{\mathcal{O}}(u_h))\\ &= \pm\ell^{\mathcal{O}}(u_h) + \frac{1}{4}\|\kappa e^{\mathrm{s}} \pm \frac{1}{\kappa}\varepsilon^{\mathrm{s}}\|^2 \mp \ell^{\mathcal{O}}(u).\end{aligned}$$

Finally, joining equation (2.11) along with the previous manipulation yields

$$0 \leq \|\frac{1}{2}(\kappa e^{\mathrm{s}} \pm \frac{1}{\kappa}\varepsilon^{\mathrm{s}}) - \kappa e\|^2 = \pm\ell^{\mathcal{O}}(u_h) + \frac{1}{4}\|\kappa e^{\mathrm{s}} \pm \frac{1}{\kappa}\varepsilon^{\mathrm{s}}\|^2 \mp \ell^{\mathcal{O}}(u),$$

that is

$$\pm\ell^{\mathcal{O}}(u) \leq \pm\ell^{\mathcal{O}}(u_h) + \frac{1}{4}\|\kappa e^{\mathrm{s}} \pm \frac{1}{\kappa}\varepsilon^{\mathrm{s}}\|^2.$$

The proof is concluded by noting that the + sign in the previous equality yields the expression for the upper bound of $\ell^{\mathcal{O}}(u)$, whereas the − sign yields the expression for the lower bound of $\ell^{\mathcal{O}}(u)$.

Next section introduces a methodology to obtain strict upper bounds for the energy norm. This approach is then used to compute $\|\kappa e^{\mathrm{s}} \pm 1/\kappa\,\varepsilon^{\mathrm{s}}\|^2_{ub}$.

3 Upper bounds for the energy norm: complementary energy relaxation

Consider the auxiliary function $z \in \mathcal{V}$ solution of

$$a^{\mathrm{s}}(z, v) = R^*(v) \quad \forall v \in \mathcal{V}, \tag{3.12}$$

where $R^*(v) = \alpha R^{\mathrm{P}}(v) + \beta R^{\mathrm{D}}(v)$ for $\alpha, \beta \in \mathbb{R}$. Note that for $\alpha = 1$ and $\beta = 0$, $R^*(v) = R^{\mathrm{P}}(v)$ and problem (3.12) is the residual problem for e^{s}. Therefore in this case $z = e^{\mathrm{s}}$. Analogously, the choice of $\alpha = 0$ and $\beta = 1$, produces $R^*(v) = R^{\mathrm{D}}(v)$ and the residual problem for ε^{s} is recovered yielding $z = \varepsilon^{\mathrm{s}}$. In particular, $\alpha = \kappa$ and $\beta = \pm 1/\kappa$ will be used later to obtain the required upper bounds for $\|\kappa e^{\mathrm{s}} \pm 1/\kappa\,\varepsilon^{\mathrm{s}}\|^2$.

The purpose of this section is to establish a procedure to compute upper bounds of $\|z\|^2$. Recently, a lot of effort has been focused on the obtainment of exact bounds, that is, bounds guaranteed with respect to the exact solution independently of any underlying mesh (see the series of references [2,5–9]). All these strategies derive strict bounds for $\|z\|$ using the standard complementary energy approach. The key idea is to relax the continuous problem of finding $z \in \mathcal{V}$ fulfilling equation (3.12) into a discrete problem. The relaxed problem consists in obtaining a pair of dual estimates $\hat{\boldsymbol{p}} \in [\mathcal{L}^2(\Omega)]^2$ and $\hat{r} \in \mathcal{L}^2(\Omega)$ such that

$$\int_\Omega \Big[\nu\hat{\boldsymbol{p}}\cdot\boldsymbol{\nabla} v + \sigma\hat{r}v\Big]\, \mathrm{d}\Omega = a^{\mathrm{s}}(z,v) = R^*(v) \quad \forall v \in \mathcal{V}. \tag{3.13}$$

The dual estimates $\hat{\mathbf{p}}$ and $\hat{r}$ are then combined to build an upper bound for $\|z\|$. This is stated in the following theorem.

Theorem 1. *Let $\hat{\mathbf{p}} \in [\mathcal{L}^2(\Omega)]^2$ and $\hat{r} \in \mathcal{L}^2(\Omega)$ be two dual estimates fulfilling equation* (3.13)*. Then, an upper bound for the energy norm of the solution z of* (3.12) *is computed as*

$$\|z\|^2 \leq \int_\Omega \Big[\nu\hat{\mathbf{p}}\cdot\hat{\mathbf{p}} + \sigma\hat{r}^2\Big]\, \mathrm{d}\Omega. \tag{3.14}$$

Proof. The result follows after the following algebraic manipulation

$$\begin{aligned}
0 \leq & \int_\Omega \Big[\nu(\hat{\mathbf{p}} - \boldsymbol{\nabla} z)\cdot(\hat{\mathbf{p}} - \boldsymbol{\nabla} z) + \sigma(\hat{r} - z)^2\Big]\, \mathrm{d}\Omega \\
& = \int_\Omega \Big[\nu\hat{\mathbf{p}}\cdot\hat{\mathbf{p}} + \sigma\hat{r}^2\Big]\, \mathrm{d}\Omega + \int_\Omega \Big[\nu\boldsymbol{\nabla} z\cdot\boldsymbol{\nabla} z + \sigma z^2\Big]\, \mathrm{d}\Omega \\
& -2\int_\Omega \Big[\nu\hat{\mathbf{p}}\cdot\boldsymbol{\nabla} z + \sigma\hat{r}z\Big]\, \mathrm{d}\Omega \\
& = \int_\Omega \Big[\nu\hat{\mathbf{p}}\cdot\hat{\mathbf{p}} + \sigma\hat{r}^2\Big]\, \mathrm{d}\Omega + a^{\mathrm{s}}(z,z) - 2a^{\mathrm{s}}(z,z) \\
& = \int_\Omega \Big[\nu\hat{\mathbf{p}}\cdot\hat{\mathbf{p}} + \sigma\hat{r}^2\Big]\, \mathrm{d}\Omega - \|z\|^2,
\end{aligned}$$

where both equation (3.13) with $v = z$ and the definition of the bilinear form $a^{\mathrm{s}}(\cdot,\cdot)$, equation (2.7) with $w = v = z$, are used.

Theorem 1 allows to compute strict upper bounds for $\|z\|$ recovering two globally equilibrated dual estimates $\hat{\mathbf{p}}$ and $\hat{r}$, i.e. verifying equation (3.13). However, the essential feature of the method is that if the fields f, g, $f^{\mathcal{O}}$ and $g^{\mathcal{O}}$ are piecewise polynomial fields, it is possible to determine — amongst all the dual estimates $\hat{\boldsymbol{p}} \in [\mathcal{L}^2(\Omega)]^2$ and $\hat{r} \in \mathcal{L}^2(\Omega)$ verifying equation (3.13) — two *piecewise polynomial* fields verifying equation (3.13). That is, for a given suitable interpolation degree q, it is possible to find $\hat{\boldsymbol{p}} \in [\widehat{\mathbb{P}^q}(\Omega)]^2$ and $\hat{r} \in \widehat{\mathbb{P}^q}(\Omega)$ verifying equation (3.13) where

$$\widehat{\mathbb{P}^q}(\Omega) := \{v \in \mathcal{L}^2(\Omega),\, v|_{\Omega_k} \in \mathbb{P}^q(\Omega_k)\}.$$

Here a triangulation of the computational domain Ω into $n_{\rm el}$ triangles is considered where Ω_k denote a general triangle, $k = 1, \ldots, n_{\rm el}$.

Therefore, the computation of strict upper bounds for $\|z\|$ is reduced to a discrete problem: determine $\hat{\boldsymbol{p}} \in [\widehat{\mathbb{P}}^q(\Omega)]^2$ and $\hat{r} \in \widehat{\mathbb{P}}^q(\Omega)$ verifying equation (3.13). This problem is a discrete problem posed over the whole domain Ω. Luckily, proper domain decomposition techniques allow to decompose the global discrete problem into local problems. That is, the piecewise polynomial fields $\hat{\boldsymbol{p}}$ and $\hat{r}$ may be computed solving local discrete problems.

In [2, 5–9], the dual estimates $\hat{\boldsymbol{p}}$ and $\hat{r}$ are computed solving local independent problems in each element of the mesh. However, they require the use of flux-equilibration techniques (*hybrid-flux* techniques) to properly set the boundary conditions for the local elementary problems. First, the equilibrated residual method is used to compute the equilibrated fluxes at the interelementary edges of the mesh. These fluxes are then used as local boundary conditions to compute the dual estimates $\hat{\boldsymbol{p}}$ and $\hat{r}$ in each triangle of the mesh.

The next section presents a new approach to compute the dual estimates $\hat{\boldsymbol{p}}$ and $\hat{r}$. The idea is to avoid the use of flux-equilibration techniques and, instead, use the flux-free estimator proposed in [1]. This domain decomposition technique exploits the partition-of-unity property to reduce the problems from Ω to subdomains different than elements. The local problems for the dual estimates $\hat{\boldsymbol{p}}$ and $\hat{r}$ are posed over patches of elements. The advantage of this approach is that the local problems are naturally equilibrated and do not require enforcing equilibrium.

4 Local computation of the dual estimates $\hat{\boldsymbol{p}}$ and $\hat{r}$ using a flux-free approach

This section is devoted to detail the computation of the piecewise polynomial dual estimates $\hat{\boldsymbol{p}}$ and $\hat{r}$ using the flux-free approach proposed in [1]. After introducing some notation, the domain decomposition strategy used to localize the computation of $\hat{\boldsymbol{p}}$ and $\hat{r}$ is presented. The solvability of the local problems is discussed along with the verification that the dual estimates $\hat{\boldsymbol{p}}$ and $\hat{r}$ verify equation (3.13). Finally some computational aspects are discussed.

4.1 Domain decomposition

Let $\boldsymbol{x}^i$ $i = 1, \ldots, n_{\rm np}$ denote the vertices of the elements (triangles) in the computational mesh (thus linked to $\mathcal{V}^h$) and ϕ^i denote the corresponding linear shape functions, which are such that $\phi^i(\boldsymbol{x}^j) = \delta_{ij}$. The support of ϕ^i is denoted by ω^i and is called the star centered in/associated with vertex $\boldsymbol{x}^i$. It is important to recall that the linear shape functions based on the vertices are a *partition of unity*, namely

$$\sum_{i=1}^{n_{\mathrm{np}}} \phi^i = 1. \tag{4.15}$$

Let also $\mathcal{V}(\omega^i)$ and $\widehat{\mathbb{P}}^q(\omega^i)$ denote the local restrictions of the spaces $\mathcal{V}$ and $\widehat{\mathbb{P}}^q(\Omega)$ to the star ω^i. Formally, any function $v \in \mathcal{V}(\omega^i)$ or $v \in \widehat{\mathbb{P}}^q(\omega^i)$ is not defined on the whole domain Ω but only on the star ω^i. However, here any $v \in \mathcal{V}(\omega^i)$ or $v \in \widehat{\mathbb{P}}^q(\omega^i)$ is naturally extended to Ω by setting the values outside ω^i to zero. Thus, functions in $\mathcal{V}(\omega^i)$ are continuous in ω^i but generally discontinuous across the boundary of the star ω^i, whereas functions in $\widehat{\mathbb{P}}^q(\omega^i)$ are piecewise polynomial functions in the triangles contained in ω^i vanishing on the elements outside ω^i.

The dual estimates $\hat{\boldsymbol{p}}$ and $\hat{r}$ are computed as

$$\hat{\boldsymbol{p}} = \sum_{i=1}^{n_{\mathrm{np}}} \hat{\boldsymbol{p}}^i \quad \text{and} \quad \hat{r} = \sum_{i=1}^{n_{\mathrm{np}}} \hat{r}^i \tag{4.16}$$

where the local estimates $\hat{\boldsymbol{p}}^i \in [\widehat{\mathbb{P}}^q(\omega^i)]^2$ and $\hat{r}^i \in \widehat{\mathbb{P}}^q(\omega^i)$, defined inside the star ω^i, verify the local equation

$$\int_{\omega^i} \left[\nu \hat{\boldsymbol{p}}^i \cdot \boldsymbol{\nabla} v + \sigma \hat{r}^i v \right] \mathrm{d}\Omega = R^*(\phi^i v) \quad \forall v \in \mathcal{V}(\omega^i). \tag{4.17}$$

Remark 1 *It is tacitly assumed that the problems given in equation* (4.17) *have at least one solution. A strictly positive reaction term in the l.h.s.,* $\sigma > 0$, *ensures the solvability of local equation* (4.17)*. For* $\sigma|_{\omega^i} = 0$, *the kernel of the bilinear operator appearing in the l.h.s. is the one dimensional space of constants,* $\mathbb{P}^0(\omega^i)$*. Then, equation* (4.17) *is solvable if and only if the compatibility condition holds, namely*

$$R^*(\phi^i c) = c R^*(\phi^i) = 0 \quad \forall c \in \mathbb{P}^0(\omega^i),$$

which follows from the orthogonality of the primal and adjoint residuals to the finite element space $\mathcal{V}^h$ *(i.e.* $R^*(v) = 0 \ \forall v \in \mathcal{V}^h$*), since* $\phi^i \in \mathcal{V}^h$.

Theorem 1. *The dual estimates* $\hat{\boldsymbol{p}} = \sum_{i=1}^{n_{\mathrm{np}}} \hat{\boldsymbol{p}}^i$ *and* $\hat{r} = \sum_{i=1}^{n_{\mathrm{np}}} \hat{r}^i$*, where* $\hat{\boldsymbol{p}}^i$ *and* $\hat{r}^i$ *verify the local problems given in* (4.17)*, verify the hypothesis of theorem 1 and therefore*

$$\|z\|^2 \leq \int_\Omega \left[\nu \hat{\boldsymbol{p}} \cdot \hat{\boldsymbol{p}} + \sigma \hat{r}^2 \right] \mathrm{d}\Omega.$$

Proof. The dual estimates $\hat{\boldsymbol{p}}$ and $\hat{r}$ verify equation (3.13) and therefore theorem 1 is a straightforward particularization of theorem 1. Indeed, let $v \in \mathcal{V}$, then using the definition of the dual estimates, equation (4.16), and the local equations (4.17) — note that if $v \in \mathcal{V}$ then $v|_{\omega^i} \in \mathcal{V}(\omega^i)$ — then,

$$\int_{\Omega} \left[\nu \hat{\boldsymbol{p}} \cdot \boldsymbol{\nabla} v + \sigma \hat{r} v \right] \, \mathrm{d}\Omega = \sum_{i=1}^{n_{\mathrm{np}}} \int_{\omega^i} \left[\nu \hat{\boldsymbol{p}}^i \cdot \boldsymbol{\nabla} v + \sigma \hat{r}^i v \right] \, \mathrm{d}\Omega$$
$$= \sum_{i=1}^{n_{\mathrm{np}}} R^*(\phi^i v) = R^*(\sum_{i=1}^{n_{\mathrm{np}}} \phi^i v) = R^*(v) = a^{\mathrm{s}}(z, v),$$

where in the final equalities the linearity of the residual $R^*(\cdot)$, the partition-of-unity property, equation (4.15), and the definition of z, equation (3.12), have also been used.

5 Bounds for the quantity of interest $s = \ell^{\mathcal{O}}(u)$: an algorithmic summary

According to theorem 1 upper and lower bounds of $s = \ell^{\mathcal{O}}(u)$ are available once upper bounds of the energy norm $\|z\|$ are obtained for the two combinations $(\alpha, \beta) = (\kappa, 1/\kappa)$ and $(\alpha, \beta) = (\kappa, -1/\kappa)$. The general strategy to obtain these upper bounds is described in the previous section. Due to the linearity of the problem, obtaining the estimates for the two values $z = \kappa e^{\mathrm{s}} \pm 1/\kappa \varepsilon^{\mathrm{s}}$ is equivalent to obtain the estimates for $z = e^{\mathrm{s}}$ and $z = \varepsilon^{\mathrm{s}}$, that is for the two combinations $(\alpha, \beta) = (1, 0)$ and $(\alpha, \beta) = (0, 1)$.

The main steps of the procedure to compute bounds for $\ell^{\mathcal{O}}(u)$ are the following:

1. Compute the primal and adjoint solutions u_h and ψ_h respectively.
2. For each star ω^i (associated to the node $\boldsymbol{x}^i$ of the mesh) compute the primal and adjoint dual estimates $\hat{\boldsymbol{p}}_P^i, \hat{\boldsymbol{p}}_D^i \in [\widehat{\mathbb{P}}^q(\omega^i)]^2$ and $\hat{r}_P^i, \hat{r}_D^i \in \widehat{\mathbb{P}}^q(\omega^i)$ such that
$$\int_{\omega^i} \left[\nu \hat{\boldsymbol{p}}_P^i \cdot \boldsymbol{\nabla} v + \sigma \hat{r}_P^i v \right] \, \mathrm{d}\Omega = R^{\mathrm{P}}(\phi^i v) \quad \forall v \in \mathcal{V}(\omega^i),$$
and
$$\int_{\omega^i} \left[\nu \hat{\boldsymbol{p}}_D^i \cdot \boldsymbol{\nabla} v + \sigma \hat{r}_D^i v \right] \, \mathrm{d}\Omega = R^{\mathrm{D}}(\phi^i v) \quad \forall v \in \mathcal{V}(\omega^i).$$
3. Recover the global estimates
$$\hat{\boldsymbol{p}}_P = \sum_{i=1}^{n_{\mathrm{np}}} \hat{\boldsymbol{p}}_P^i, \quad \hat{r}_P = \sum_{i=1}^{n_{\mathrm{np}}} \hat{r}_P^i \quad \text{and} \quad \hat{\boldsymbol{p}}_D = \sum_{i=1}^{n_{\mathrm{np}}} \hat{\boldsymbol{p}}_D^i, \quad \hat{r}_D = \sum_{i=1}^{n_{\mathrm{np}}} \hat{r}_D^i.$$
4. Compute the three scalar quantities
$$(\eta^P)^2 := \sum_{k=1}^{n_{\mathrm{el}}} \eta_k^P = \sum_{k=1}^{n_{\mathrm{el}}} \int_{\Omega_k} \left[\nu \hat{\boldsymbol{p}}_P \cdot \hat{\boldsymbol{p}}_P + \sigma (\hat{r}_P)^2 \right] \, \mathrm{d}\Omega,$$

$$(\eta^D)^2 := \sum_{k=1}^{n_{\text{el}}} \eta_k^D = \sum_{k=1}^{n_{\text{el}}} \int_{\Omega_k} \left[\nu \hat{\boldsymbol{p}}_D \cdot \hat{\boldsymbol{p}}_D + \sigma (\hat{r}_D)^2 \right] \mathrm{d}\Omega,$$

$$\eta^{PD} := \sum_{k=1}^{n_{\text{el}}} \eta_k^{PD} = \sum_{k=1}^{n_{\text{el}}} \int_{\Omega_k} \left[\nu \hat{\boldsymbol{p}}_P \cdot \hat{\boldsymbol{p}}_D + \sigma \hat{r}_P \hat{r}_D \right] \mathrm{d}\Omega,$$

5. Recover the bounds for the output $s^{lb} \leq s \leq s^{ub}$ as

$$s^{lb} := s_h - \frac{1}{2}\eta^P \eta^D + \frac{1}{2}\eta^{PD} \leq s \leq s_h + \frac{1}{2}\eta^P \eta^D + \frac{1}{2}\eta^{PD} =: s^{ub}, \quad (5.18)$$

where $s_h = \ell^{\mathcal{O}}(u_h)$.

6 Numerical examples

In the following, the bound average $s^{ave} := (s^{ub} + s^{lb})/2$ is taken as a new approximation of the quantity of interest and the half bound gap $\Delta = (s^{ub} - s^{lb})/2$ is seen as an error indicator. The relative counterpart of the bound gap $\Delta_{\text{rel}} = \Delta / s^{ave}$ is also used in the presentation.

The meshes are adapted to reduce the half bound gap Δ. In the examples a simple adaptive strategy is used based on the decomposition of Δ into local positive contributions from the elements:

$$\Delta = \sum_{k=1}^{n_{\text{el}}} \Delta_k,$$

where the element contribution to the bound gap Δ_k is

$$\Delta_k := \frac{1}{4}\kappa^2 \eta_k^P + \frac{1}{4\kappa^2}\eta_k^D.$$

Note that this decomposition is valid because

$$\begin{aligned}\Delta &= \frac{s^{ub} - s^{lb}}{2} = \frac{1}{2}\eta^P \eta^D = \frac{1}{4}\kappa^2 (\eta^P)^2 + \frac{1}{4\kappa^2}(\eta^D)^2 \\ &= \sum_{k=1}^{n_{\text{el}}} \left[\frac{1}{4}\kappa^2 \eta_k^P + \frac{1}{4\kappa^2}\eta_k^D \right] = \sum_{k=1}^{n_{\text{el}}} \Delta_k.\end{aligned}$$

The remeshing strategy consists in subdividing the elements with the larger values of Δ_k at each step of the adaptive procedure.

The behavior of the bounds is compared with the three different strategies presented in [1,3,5]. The strategy presented in [1] also solves local problems in subdomains (therefore avoiding the computation of equilibrated fluxes), but the local problems are solved using a local fine submesh. This yields bounds which are only guaranteed in the asymptotic regime. The strategy presented in [3] is the classical equilibrated method where first the (linear) equilibrated fluxes are computed and then the local elementary problems are solved using a local fine submesh. Finally the results are compared to the strategy presented in [5] which also provides strict bounds for the output. This method only differs from [3] in the solution of the local elementary problems. Instead of using a local submesh, dual estimates are computed to recover strict bounds for the output.

The bounds computed using the strategies presented in [1], [3] and [5] will be denoted in the following as asymptotic flux-free bounds, asymptotic equilibrated bounds and strict equilibrated bounds respectively.

6.1 Example 1: uniformly forced square domain

The pure diffusion equation ($\nu = 1$, $\sigma = 0$, $\boldsymbol{\alpha} = \mathbf{0}$ in (1.1)) is solved in the squared domain $\Omega = [0,1] \times [0,1]$. A constant source term $f = \sqrt{10}$ and homogeneous Dirichlet boundary conditions are considered.

The quantity of interest is an average of the solution,

$$\ell^{\mathcal{O}}(u) = \int_{\Omega} \sqrt{10}\; u(x,y)\; \mathrm{d}\Omega,$$

that is $f^{\mathcal{O}} = \sqrt{10}$ in equation (1.2). In this case, the solution ψ of the adjoint problem coincides with the primal solution, $\psi = u$. It is well known that in this case, the finite element approximation of the output is a lower bound for s, $\ell^{\mathcal{O}}(u_h) \leq s$. The present methodology, as well as the strategies presented in [1,3,5], yields $s^{lb} = \ell^{\mathcal{O}}(u_h)$.

Linear triangular elements are used for the computation of the primal and adjoint finite element approximations, and the local dual approximations $\hat{\boldsymbol{p}}^i \in [\widehat{\mathbb{P}}^3(\omega^i)]^2$ and $\hat{r}^i \in \widehat{\mathbb{P}}^3(\omega^i)$ are piecewise third order polynomials, i.e. $q = 3$.

The convergence of the bounds is analyzed for a uniform mesh refinement in a series of structured meshes. The initial mesh is composed of 8 triangular elements (half squares) and in each refinement step every triangle is divided into four similar triangles. The results are displayed in tables 6.1 and in figure 6.1. Since for this particular case all the methodologies yield $s^{lb} = s_h = \ell^{\mathcal{O}}(u_h)$, only the upper bounds are compared. For this problem the exact solution $s = 0.3514425$ is known, and the effectiveness of the bounds are computed as

$$\theta^* = \frac{|s - s^*|}{|s|},$$

where the symbol $*$ stands for ub, lb or ave.

As expected, the upper bounds provided by the asymptotic strategies are lower than the corresponding upper bounds obtained using strict strategies. However, this behavior is non desirable, as can be seen in the asymptotic flux-free results, since for the first three meshes, the upper bound underestimates the exact value s.

Table 6.1. Example 1: series of uniformly h-refined linear triangular meshes.

					asymp. flux-free		strict equil.		asymp. equil.	
n_{el}	s_h	θ^h	s^{ub}	θ^{ub}	s^{ub}	θ^{ub}	s^{ub}	θ^{ub}	s^{ub}	θ^{ub}
8	0.1563	55.54%	0.3582	1.92%	0.3467	1.36%	0.6302	79.33%	0.5774	64.29%
32	0.2881	18.03%	0.3529	0.40%	0.3500	0.42%	0.4457	26.81%	0.4290	22.08%
128	0.3342	4.90%	0.3521	0.18%	0.3514	0.01%	0.3775	7.40%	0.3730	6.13%
512	0.3470	1.26%	0.3516	0.06%	0.3515	0.01%	0.3581	1.90%	0.3570	1.58%
2048	0.3503	0.32%	0.3515	0.02%	0.3513	0.05%	0.3531	0.48%	0.3518	0.10%
8192	0.3512	0.08%	0.3515	0.004%	–	–	0.3519	0.12%	–	–

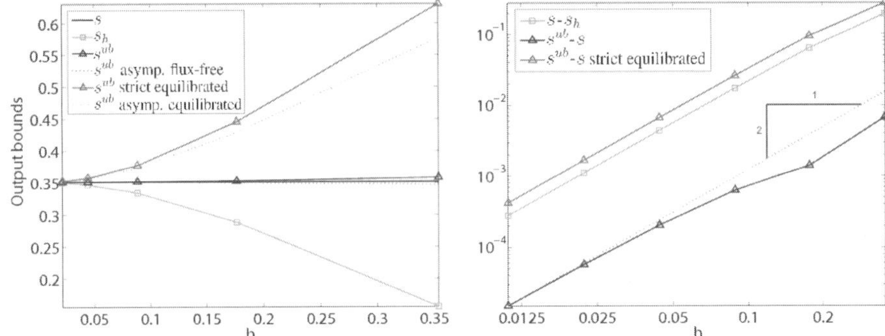

Figure 6.1. Example 1: series of uniformly h-refined linear triangular meshes. Bounds (left) and their convergence (right). (*See also* Color Plate on page 385)

The evolution of the strict bounds (both for the equilibrated and for the flux-free strategies) shows the optimal finite element asymptotic convergence rate of $\mathcal{O}(h^2)$, see figure 6.1. However, the bounds obtained by the flux-free strategy provide much better effectivities.

6.2 Example 2: quasi-2D transport

This example is the quasi-2D transport problem introduced in [5] with known analytical solution. The effect of including the convective term is analyzed in this simple problem for different values of the velocity $\boldsymbol{\alpha}$. Equation (1.1) is solved in the unit square $\Omega = [0,1] \times [0,1]$, for $\nu = 1$ and a uniform horizontal velocity field $\boldsymbol{\alpha} = (\alpha, 0)$. The performance of the introduced estimates is tested for different values of α. The boundary conditions are of Dirichlet type on the lateral sides, homogeneous on the right $u(1,y) = 0$ and set to 1 on the left $u(0,y) = 1$. The boundary condition on both the top and bottom are Neumann homogeneous and the source term is $f = 0$. The degrees of the interpolation spaces are $p = 1$ and $q = 3$.

The quantity of interest is an overall average of the solution, that is

Table 6.2. Example 2: results in a series of uniformly h-refined meshes for $\sigma = 1$.

	$\alpha = 0$		$\alpha = 1$		$\alpha = 5$		$\alpha = 10$	
	$s = 0.462117$		$s = 0.536142$		$s = 0.755101$		$s = 0.862436$	
n_{el}	s^{ave}	Δ	s^{ave}	Δ	s^{ave}	Δ	s^{ave}	Δ
8	0.466353	0.005475	0.532597	0.006295	0.730433	0.107897	0.662229	0.539482
32	0.463163	0.001384	0.535319	0.001635	0.753057	0.028447	0.862177	0.123866
128	0.462380	0.000345	0.535939	0.000411	0.754647	0.007166	0.862363	0.031168
512	0.462183	0.000086	0.536092	0.000103	0.754991	0.001785	0.862418	0.007714
2048	0.462134	0.000022	0.536130	0.000026	0.755074	0.000445	0.862432	0.001911
8192	0.462121	0.000005	0.536139	0.000006	0.755094	0.000111	0.862435	0.000475
8	0.462675	0.009420	0.535946	0.010560	0.741465	0.164952	0.585866	0.782647
32	0.462317	0.002347	0.536002	0.002749	0.754246	0.046232	0.861651	0.193798
128	0.462173	0.000586	0.536099	0.000696	0.754911	0.012078	0.862180	0.051359
512	0.462132	0.000146	0.536131	0.000174	0.755053	0.003060	0.862370	0.013105
2048	0.462121	0.000037	0.536139	0.000044	0.755089	0.000767	0.862419	0.003297
8192	0.462118	0.000009	0.536142	0.000011	0.755098	0.000192	0.862432	0.000826

$$\ell^{\mathcal{O}}(u) = \int_{\Omega} u(x,y)\, \mathrm{d}\Omega,$$

which corresponds to $f^{\mathcal{O}} = 1$.

The error estimation strategies and the computation of bounds are performed for a series of uniformly h-refined meshes for $\sigma = 1$ and different values of α. The results are displayed in table 6.2 and figure 6.2. For all the values of α, the rate of convergence of the bound gap is found to be equal to the expected one for the error, that is $\mathcal{O}(h^2)$. It is worth noting that the bound gap is larger as α increases. For $\alpha = 100$ the bound gap is 4 orders of magnitude larger than for $\alpha = 0$, being the quantity of interest of the same order.

This increment in the bound gap does not correspond to the actual error increment and therefore it has to be concluded that the efficiency of the computed error bounds deteriorate if the convection parameter is large.

Table 6.3 and 6.4 summarize the influence of the Peclet number α on the effectivity of the bounds in the context of the simple adaptive method with a tolerance of $\Delta_{tol} = 0.001s$ and for $\sigma = 10$. Although the method is valid for nonnegative α, the sharpness of the bound degrades significantly with increasing Peclet number, but the bounding property is retained. Since we know the exact output for this example, we can calculate the effectiveness of the bounds as an indicator of the error in the finite element solution using

$$\eta = \frac{s^{ub} - s^{lb}}{2|s - s_h|} = \frac{\Delta}{|s - s_h|} = \frac{\theta^{ub} + \theta^{lb}}{2\theta^h}.$$

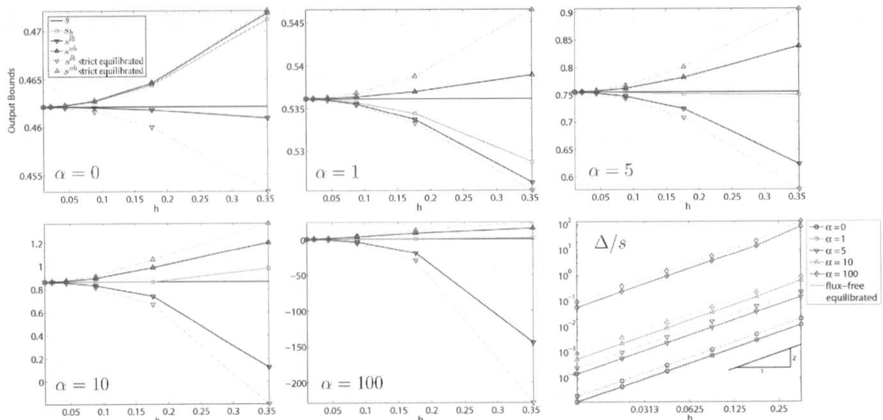

Figure 6.2. Example 2: bounds and convergence of the bound gap for different values of the convection parameter, $\alpha = 0, 1, 5, 10$ and 100 for $\sigma = 1$. (*See also* Color Plate on page 386)

Table 6.3. Example 2: Adaptive mesh refinement results for a final tolerance $\Delta_{tol} = 0.001s$ and various values of α and $\sigma = 10$ using strict flux-free error estimators.

	n_{el}	s	s_h	s^{lb}	s^{ub}	θ^h	θ^{lb}	θ^{ub}	η
$\alpha = 0$	312	0.2905436	0.291503	0.290315	0.291667	0.33%	0.08%	0.39%	70.39%
	734		0.290899	0.290446	0.290969	0.12%	0.03%	0.15%	73.54%
$\alpha = 1$	312	0.3281819	0.329054	0.328050	0.329127	0.27%	0.04%	0.29%	61.77%
	473		0.328640	0.328081	0.328705	0.14%	0.03%	0.16%	68.15%
$\alpha = 5$	312	0.4789324	0.479329	0.478407	0.479826	0.08%	0.11%	0.19%	178.89%
	402		0.479203	0.478642	0.479473	0.06%	0.06%	0.11%	153.56%
$\alpha = 10$	312	0.6182269	0.618311	0.613727	0.622804	0.014%	0.73%	0.74%	5383.16%
	1076		0.618258	0.617690	0.618793	0.005%	0.09%	0.09%	1768.07%

Table 6.4. Example 2: Adaptive mesh refinement results for a final tolerance $\Delta_{tol} = 0.001s$ and various values of α and $\sigma = 10$ using strict equilibrated error estimators.

	n_{el}	s	s_h	s^{lb}	s^{ub}	θ^h	θ^{lb}	θ^{ub}	η
$\alpha = 0$	312	0.2905436	0.291503	0.289512	0.291779	0.33%	0.36%	0.43%	118.08%
	1114		0.290753	0.290298	0.290823	0.07%	0.08%	0.10%	125.26%
$\alpha = 1$	312	0.3281819	0.329054	0.327359	0.329188	0.27%	0.25%	0.31%	104.79%
	853		0.328418	0.327907	0.328493	0.07%	0.08%	0.09%	124.00%
$\alpha = 5$	312	0.4789324	0.479329	0.477642	0.480297	0.08%	0.27%	0.28%	334.65%
	1036		0.479038	0.478474	0.479415	0.02%	0.10%	0.10%	445.34%
$\alpha = 10$	312	0.6182269	0.618311	0.609617	0.626863	0.0136%	1.39%	1.40%	10227.14%
	10619		0.618237	0.617644	0.618812	0.0016%	0.09%	0.09%	5976.13%

7 Concluding remarks

This paper introduces a new technique to compute strict upper and lower bounds for functional outputs. The bounds are guaranteed regardless of the size of the underlying coarse discretization and do not involve uncertain ingredients (in some cases, the bounds may depend on continuity or interpolation constants which are non-computable and have to be approximated, or they may depend on the evaluation of integrals involving analytic functions which have to be numerically integrated ...).

The proposed strategy is an extension of the flux-free technique presented in [1]. In [1], the resulting estimates yield upper bounds of a reference solution associated with a finer reference mesh. This bounds are only strict provided that the error associated to the reference solution is negligible in front of the error associated with the coarse solution. In the present work, this assumption is removed, and the bounds are strict without any further assumption.

As previously proposed flux-free approaches, the implementation of the proposed method is less cumbersome compared to hybrid-flux estimators where flux equilibration algorithms must be implemented. Moreover, the observed accuracy is much better than in the hybrid-flux methods. Also, the new flux-free technique yield much sharper bounds than other previously proposed flux-free techniques.

It is worth noting that although the strategy may be applied to convection-dominated problems, the effectivities of the bounds deteriorate as the convection term becomes dominant. This phenomenon is due to the expression of the bounds of the output given in theorem 1, and is also observed in all the previous works which compute strict or asymptotic bounds for quantities of interest for the convection-reaction-diffusion equation.

Finally, the distribution of the local contributions to the error are well suited to guide goal-oriented adaptive procedures.

References

1. N. Parés, P. Díez, and A. Huerta, *Subdomain-based flux-free a posteriori error estimators*, Comput. Methods Appl. Mech. Engrg. **195** (2006), 297–323.
2. A. M. Sauer-Budge, J. Bonet, A. Huerta, and J. Peraire, *Computing bounds for linear functionals of exact weak solutions to Poisson's equation*, SIAM J. Numer. Anal. **42** (2004), 1610–1630.
3. M. Paraschivoiu, J. Peraire, and A. T. Patera, *A posteriori finite element bounds for linear-functional outputs of elliptic partial differential equations*, Comput. Methods Appl. Mech. Engrg. **150** (1997), 289–312.
4. J. T. Oden, and S. Prudhomme, *Goal-oriented error estimation and adaptivity for the finite element method*, Comput. Math. Appl. **41** (2001), 735–756.
5. A. M. Sauer-Budge, and J. Peraire, *Computing bounds for linear functionals of exact weak solutions to the advection-diffusion-reaction equation*, SIAM J. Sci. Comput. **26** (2004), 636–652.

6. N. Parés, J. Bonet, A. Huerta, and J. Peraire, *The computation of bounds for linear-functional outputs of weak solutions to the two-dimensional elasticity equations*, Comput. Methods Appl. Mech. Engrg. **195** (2006), 406–429.
7. Z. C. Xuan, N. Parés, and J. Peraire, *Computing upper and lower bounds for the J-integral in two-dimensional linear elasticity*, Comput. Methods Appl. Mech. Engrg. **195** (2006), 430–443.
8. N. Parés, P. Díez, and A. Huerta, *Bounds of functional outputs for parabolic problems. Part I: Exact bounds of the Discontinuous Galerkin time discretization*, Comput. Methods Appl. Mech. Engrg. Article in Press. doi:10.1016/j.cma.2007.08.025.
9. N. Parés, P. Díez, and A. Huerta, *Bounds of functional outputs for parabolic problems. Part II: Bounds of the exact solution*, Comput. Methods Appl. Mech. Engrg. Article in Press. doi:10.1016/j.cma.2007.08.024.
10. M. Ainsworth, and J. T. Oden, *A posteriori error estimation in finite element analysis*, John Wiley & Sons, 2000.
11. I. Babuška, and T. Strouboulis, *The finite element method and its reliability*, The Clarendon Press Oxford University Press, 2001.

Preparation of CAD and Molecular Surfaces for Meshfree Solvers

Maharavo Randrianarivony[1] and Guido Brunnett[2]

[1] Institute for Numerical Simulation, Rheinische Friedrich-Wilhelms University of Bonn, 53115 Bonn, Germany. maharavo@informatik.tu-chemnitz.de
[2] Faculty of Computer Science, Technical University of Chemnitz, 09107 Chemnitz, Germany. brunnett@informatik.tu-chemnitz.de

Summary. Raw CAD and molecular data need to be prepared before they can be used in meshfree numerical solvers. After quickly summarizing the different steps required for that processing, we will focus on the practical realization of global continuity. In particular, we will examine error bounds about chord length approximation in which two procedures can yield errors: sample interpolation, length estimation. We will report eventually on some practical results where the initial data are acquired from IGES files for CAD models and from PDB files for molecular ones.

Key words: NURBS, molecule, CAD, IGES, PDB.

1 Introduction and Problem Setting

Many physical phenomena can be formulated in terms of PDE or integral equations [4,6] which can be efficiently solved numerically. Usually, the initial equation [2,11] to be solved is valid on a surface Γ bounding a 3D domain Ω. For CAD models, the surface Γ is generated by a design software or a CAD system whose output is stored in a CAD standard like IGES (Initial Graphics Exchange Specification). As for molecular surfaces [4], the surface Γ represents the solute-solvent boundary. It bounds a cavity Ω containing a distribution of electric charge which interacts with the solvent represented by a continuous dielectric medium. Nowadays, the PDB (Protein Data Bank) format is the standard which is used to digitally store molecular information. Over the last decade, meshfree approaches demonstrated themselves to be very effective. In particular, the capability of the Wavelet Galerkin scheme has been completely proved [2,11] theoretically. Yet, its implementation has not been successfully applied to real geometry data as opposed to mesh-based approaches. In this paper, our purpose is to process raw geometry data into a form acceptable by some meshfree solvers. Note that although the presented preparation of the geometry data is mainly devised for Wavelet Galerkin solvers, we do believe

Figure 1.1. CAD and molecular surfaces: before and after decomposition (*See also* Color Plate on page 386)

that it is general enough to treat some other approaches. Since the whole preparation of the geometry data is very long, we cannot describe it in full detail here. Therefore, we sketch only the parts which have been thoroughly investigated previously [5,9].

Let us suppose that the closed surface Γ which bounds a connected region $\Omega \subset \mathbb{R}^3$ is the union of several parametric surfaces $S_1, \ldots, S_N$. Each surface S_i can be trimmed or untrimmed [1]. For untrimmed surfaces, there are several representations [3] including B-splines, NURBS, surfaces of revolution and tabulated cylinders defined on rectangular domains which will be supposed to be the unit square. For each trimmed surface S_i, we assume that there is a rectangle $\mathcal{R}_i := [a_i, b_i] \times [c_i, d_i]$ containing a multiply connected region $\mathcal{D}_i$. The external and internal (when relevant) boundary curves of the domain $\mathcal{D}_i$ are supposed to be composite curves. That is, there are univariate smooth functions $\boldsymbol{\kappa}_i^j$ defined on $[e_i^j, f_i^j]$ such that

$$\partial\mathcal{D}_i = \bigcup_j \mathrm{Im}(\boldsymbol{\kappa}_i^j). \tag{1.1}$$

We have a parametric function $\boldsymbol{\psi}_i : \mathcal{R}_i \to \mathbb{R}^3$ which is bijective, smooth and regular in the sense that its Jacobian has maximal rank such that $\boldsymbol{\psi}_i(\mathcal{D}_i) = S_i$.

Furthermore, we suppose that the bounding curves $\boldsymbol{\kappa}_i^j$ are sufficiently smooth and that they do not have any cusps [9].
Our objective is to tessellate the surface Γ into a collection of four-sided patches Γ_i, i.e., $\Gamma = \cup_i \Gamma_i$, where the splitting is conforming [5]. We need also some regular functions γ_i such that $\Gamma_i = \gamma_i([0,1]^2)$. Additionally, we require global continuity meaning that for two adjacent patches Γ_i and Γ_j, there is a bijective, affine mapping $\Xi : [0,1]^2 \to [0,1]^2$ such that for all $\mathbf{x} = \gamma_i(s)$ on a common edge of Γ_i and Γ_j it holds that

$$\gamma_i(s) = (\gamma_j \circ \Xi)(s). \tag{1.2}$$

That is, γ_i and γ_j coincide pointwise at common edges up to some reorientation. The whole geometry processing is graphically summarized in Fig. 1.1 where the grids represent the images by γ_i of a uniform grid on the unit square. For real CAD data, we are not able to achieve the exact global continuity (1.2). As a consequence, we will have only matching condition with certain accuracy $\varepsilon > 0$ such that $\mathrm{dist}[\gamma_i(s), \gamma_j(\Xi(s))] < \varepsilon$ for all $s \in \partial[0,1]^2$.

2 Decomposition Procedure

This section will summarize the process of obtaining a conforming decomposition from a set of parametric surfaces. In order to have a unified description, we will define $\mathcal{D}_i := [0,1]^2$ for each untrimmed surface S_i. First, we approximate the whole model by a polyhedral one with nodes $\mathcal{A} = \{\mathbf{x}_k\} \subset \mathbb{R}^3$. This is performed by finding polygonal approximations of the planar domains $\mathcal{D}_i$ as follows. For each surface S_i, we generate a polygon $P^{(i)}$ whose nodes $\mathbf{x}_k^{(i)}$ are taken from the curved boundary of $\mathcal{D}_i$. In order to ensure conformity at an interface curve $\mathcal{C}$ between two adjacent surfaces S_i and S_j, we have to consider that for each vertex $\mathbf{x}_k^{(i)} \in P^{(i)}$ with $\psi_i(\mathbf{x}_k^{(i)}) \in \mathcal{C}$, there must exist a vertex $\mathbf{x}_l^{(j)} \in P^{(j)}$ such that

$$\psi_i(\mathbf{x}_k^{(i)}) = \psi_j(\mathbf{x}_l^{(j)}). \tag{2.3}$$

Using too few vertices will result in polygons with possibly some imperfections such as intersecting edges. But if the polygonal approximation is too fine, this results in too many four-sided patches. Therefore, one has to split the curved edges adaptively while trying to maintain relation (2.3) which requires preimage computations: for a given point $\mathbf{z} \in \mathrm{Im}(\psi_i \circ \boldsymbol{\kappa}_i^j)$, find some $\mathbf{w} = \boldsymbol{\kappa}_i^j(t) \in \partial\mathcal{D}_i$ such that $\psi_i(\mathbf{w}) = \mathbf{z}$. Let us emphasize that only polygons having an even number of boundary vertices can be decomposed into quadrilaterals. In order to convert odd faces into even ones while inserting only few nodes, we assemble the adjacency graph and we use Dijkstra algorithm to search for the shortest path connecting two odd polygons. One can show [9] that the number of odd faces is always even for a closed model and the odd faces can thus be converted to even ones pairwise.

In order to obtain the four-sided patches embedded in space, we split the 2D regions $\mathcal{D}_i$ into four-sided regions $Q_{k,i}$ as $\mathcal{D}_i = \bigcup_k Q_{k,i}$. The final four-sided patches Γ_k are therefore the images by ψ_i of the 2D domains $Q_{k,i}$. To this end, we decompose each polygon $P^{(i)}$ into convex quadrilaterals $q_{k,i}$. We have developed in [9] an approach that performs a quadrilateral decomposition of a polygon with n boundary vertices into $\mathcal{O}(n)$ convex quadrilaterals. Alternatively, one can use advancing front or other techniques based on triangulation conversion [7, 8] such as those in the Q-morph software. During that quadrangulation, we do not use any boundary vertices other than the preimages $\psi_i^{-1}(\mathbf{x}_k)$ of some $\mathbf{x}_k \in \mathcal{A}$. The four-sided domains $Q_{k,i}$ are obtained from $q_{k,i}$ by replacing the straight boundary edges of $q_{k,i}$ by the corresponding curve portion of $\mathcal{D}_i$. Let us note that the process of curve replacement may give rise to three serious problems. First, the boundary curve may intersect an internal edge. Second, sharp corners might be smoothened out by such a replacement. Third, it is possible that the Coons patch is not regular as discussed in the next section. In such cases, we have to make a polygonal refinement. We developed in [9] a method for making only a local rectification while keeping the large part of the quadrangulation in which we guarantee relation (2.3) when inserting new nodes.

3 Transfinite Interpolation and Global Continuity

In this section, we want to use the notion of Coons patches defined on four-sided regions. The mapping γ_i from relation (1.2) will be the composition of the base surface ψ_i from Section 1 and the Coons map of each 2D four sided domain. Let us consider four sufficiently smooth curves $\boldsymbol{\alpha}, \boldsymbol{\beta}, \boldsymbol{\gamma}, \boldsymbol{\delta} : \mathbb{R} \longrightarrow \mathbb{R}^2$. We are interested in their restriction on $[0, 1]$ and we suppose that they fulfill the compatibility conditions at the corners: $\boldsymbol{\alpha}(0) = \boldsymbol{\delta}(0)$, $\boldsymbol{\alpha}(1) = \boldsymbol{\beta}(0)$, $\boldsymbol{\gamma}(0) = \boldsymbol{\delta}(1)$, $\boldsymbol{\gamma}(1) = \boldsymbol{\beta}(1)$. We assume that besides those four corner points, there are no further intersection points. A transfinite interpolation consists in generating a bivariate function $\mathbf{x}$ defined on $[0, 1]^2$ such that

$$\begin{aligned} \mathbf{x}(u,0) &= \boldsymbol{\alpha}(u), \qquad \mathbf{x}(u,1) = \boldsymbol{\gamma}(u) \qquad \forall\, u \in [0,1], \\ \mathbf{x}(0,v) &= \boldsymbol{\delta}(v), \qquad \mathbf{x}(1,v) = \boldsymbol{\beta}(v) \qquad \forall\, v \in [0,1]. \end{aligned}$$

Such an interpolation will be solved by a first order Coons patch which is defined in matrix form as

$$\mathbf{x}(u,v) := - \begin{bmatrix} -1 \\ F_0(u) \\ F_1(u) \end{bmatrix}^T \begin{bmatrix} \mathbf{0} & \mathbf{x}(u,0) & \mathbf{x}(u,1) \\ \mathbf{x}(0,v) & \mathbf{x}(0,0) & \mathbf{x}(0,1) \\ \mathbf{x}(1,v) & \mathbf{x}(1,0) & \mathbf{x}(1,1) \end{bmatrix} \begin{bmatrix} -1 \\ F_0(v) \\ F_1(v) \end{bmatrix},$$

where F_0 and F_1 denote two arbitrary smooth functions satisfying: $F_i(j) = \delta_{ij}$ for $i, j = 0, 1$ and $F_0(t) + F_1(t) = 1$ for all $t \in [0, 1]$.

Since the decomposition process from Section 2 requires frequent regularity checks, we summarize now some efficient method for quickly verifying if a Coons patch is regular. We will examine only different conditions that guarantee the regularity of a planar Coons map with Bézier boundaries while the general case can be treated by using polynomial approximation [5]. That is, we suppose that the boundary curves $\boldsymbol{\alpha}, \boldsymbol{\beta}, \boldsymbol{\gamma}, \boldsymbol{\delta}$ are expressed in terms of their respective control points $\boldsymbol{\alpha}_i, \boldsymbol{\beta}_i, \boldsymbol{\gamma}_i, \boldsymbol{\delta}_i$ $(i = 0, \ldots, n)$ and the Bernstein polynomials $B_i^n(t)$ as follows

$$\boldsymbol{\alpha}(t) = \sum_{i=0}^{n} \boldsymbol{\alpha}_i B_i^n(t), \qquad \boldsymbol{\beta}(t) = \sum_{i=0}^{n} \boldsymbol{\beta}_i B_i^n(t),$$
$$\boldsymbol{\gamma}(t) = \sum_{i=0}^{n} \boldsymbol{\gamma}_i B_i^n(t), \qquad \boldsymbol{\delta}(t) = \sum_{i=0}^{n} \boldsymbol{\delta}_i B_i^n(t).$$

The polynomial blending function F_1 is also expressed in Bézier form $F_1(t) = \sum_{i=0}^{n} \phi_i B_i^n(t) = 1 - F_0(t)$. Furthermore, we suppose that the range of F_0 and F_1 is $[0,1]$ and we define $\mu := \max\{|F_1'(t)| : t \in [0,1]\}$. We let τ denote the minimum of the following expressions over $i, j = 0, \cdots, n$

$$A_{ij} := n^2 \det[\boldsymbol{\alpha}_{i+1} - \boldsymbol{\alpha}_i, \boldsymbol{\delta}_{j+1} - \boldsymbol{\delta}_j], \quad B_{ij} := n^2 \det[\boldsymbol{\alpha}_{i+1} - \boldsymbol{\alpha}_i, \boldsymbol{\beta}_{j+1} - \boldsymbol{\beta}_j],$$
$$C_{ij} := n^2 \det[\boldsymbol{\gamma}_{i+1} - \boldsymbol{\gamma}_i, \boldsymbol{\delta}_{j+1} - \boldsymbol{\delta}_j], \quad D_{ij} := n^2 \det[\boldsymbol{\gamma}_{i+1} - \boldsymbol{\gamma}_i, \boldsymbol{\beta}_{j+1} - \boldsymbol{\beta}_j].$$

We introduce also $G := \max\{G_1, G_2\}$ where

$$G_1 := \max_i \{\mu \|(\boldsymbol{\beta}_i - \boldsymbol{\delta}_i) + \phi_i(\boldsymbol{\gamma}_0 - \boldsymbol{\gamma}_n + \boldsymbol{\alpha}_n - \boldsymbol{\alpha}_0) + (\boldsymbol{\alpha}_0 - \boldsymbol{\alpha}_n)\|\},$$
$$G_2 := \max_i \{\mu \|(\boldsymbol{\gamma}_i - \boldsymbol{\alpha}_i) + \phi_i(\boldsymbol{\gamma}_0 - \boldsymbol{\gamma}_n + \boldsymbol{\alpha}_n - \boldsymbol{\alpha}_0) + (\boldsymbol{\alpha}_0 - \boldsymbol{\gamma}_0)\|\}.$$

Theorem 1. *Let M be a constant such that*

$$\begin{aligned} n\|\phi_j(\boldsymbol{\gamma}_{i+1} - \boldsymbol{\gamma}_i + \boldsymbol{\alpha}_i - \boldsymbol{\alpha}_{i+1}) + (\boldsymbol{\alpha}_{i+1} - \boldsymbol{\alpha}_i)\| \le M, \\ n\|\phi_j(\boldsymbol{\beta}_{i+1} - \boldsymbol{\beta}_i + \boldsymbol{\delta}_i - \boldsymbol{\delta}_{i+1}) + (\boldsymbol{\delta}_{i+1} - \boldsymbol{\delta}_i)\| \le M, \end{aligned} \tag{3.4}$$

for all $i = 0, \ldots, n-1$ and $j = 0, \ldots, n$. If $2MG + G^2 < \tau$ and $\tau > 0$, then $\mathbf{x}$ *is regular.*

Proof
We can express [5, 9] the partial derivatives of $\mathbf{x}$ as $\mathbf{x}_u(u,v) = F_1'(u)\mathbf{S}_1(v) + \mathbf{C}_1(u,v)$ and $\mathbf{x}_v(u,v) = F_1'(v)\mathbf{S}_2(u) + \mathbf{C}_2(u,v)$ where $\mathbf{S}_1(v) := \sum_{i=0}^{n}[(\boldsymbol{\beta}_i - \boldsymbol{\delta}_i) + \phi_i(\boldsymbol{\gamma}_0 - \boldsymbol{\gamma}_n + \boldsymbol{\alpha}_n - \boldsymbol{\alpha}_0) + (\boldsymbol{\alpha}_0 - \boldsymbol{\alpha}_n)]B_i^n(v)$ and $\mathbf{S}_2(u) := \sum_{i=0}^{n}[(\boldsymbol{\gamma}_i - \boldsymbol{\alpha}_i) + \phi_i(\boldsymbol{\gamma}_0 - \boldsymbol{\gamma}_n + \boldsymbol{\alpha}_n - \boldsymbol{\alpha}_0) + (\boldsymbol{\alpha}_0 - \boldsymbol{\gamma}_0)]B_i^n(u)$ while $\mathbf{C}_1$ and $\mathbf{C}_2$ are expressions which verify $\|\mathbf{C}_1(u,v)\| \le M$ and $\|\mathbf{C}_2(u,v)\| \le M$ for each $(u,v) \in [0,1]^2$. We deduce the result from the fact that the Bernstein polynomials form a partition of unity and that $\boldsymbol{\alpha}'(u) = \sum_{i=0}^{n-1} n(\boldsymbol{\alpha}_{i+1} - \boldsymbol{\alpha}_i)B_i^{n-1}(u)$ with similar relations for $\boldsymbol{\beta}$, $\boldsymbol{\gamma}$, $\boldsymbol{\delta}$.

□

In order to employ the technique of adaptive subdivision, let us introduce two notions. First, one can show [9] that a Bézier surface $\sum_{i,j=0}^{n} \mathbf{E}_{ij} B_i^n(u) B_j^n(v)$ has as Jacobian a Bézier function of degree $2n$ with the next control coefficients:

$$J_{pq} := \sum_{\substack{i+k=p \\ j+l=q}} C(i,j,k,l) \frac{\binom{n}{i}\binom{n}{k}}{\binom{2n}{i+k}} \frac{\binom{n}{j}\binom{n}{l}}{\binom{2n}{j+l}}, \qquad p,q = 0,\ldots,2n,$$

where

$$C(i,j,k,l) := \frac{l}{n}\left[\frac{i}{n} D(i-1,j,k,l-1) + \left(1-\frac{i}{n}\right) D(i,j,k,l-1)\right] + \left(1-\frac{l}{n}\right)\left[\frac{i}{n} D(i-1,j,k,l) + \left(1-\frac{i}{n}\right) D(i,j,k,l)\right]$$

and

$$D(i,j,k,l) := n^2 \det[\mathbf{E}_{i+1,j} - \mathbf{E}_{ij}, \mathbf{E}_{k,l+1} - \mathbf{E}_{kl}].$$

On the other hand, a Bézier surface F defined on $[a,b]\times[c,d]$ can be subdivided into four Bézier surfaces F^A, F^B, F^C, F^D which are respectively defined on

$$I^A := [a,(a+b)/2] \times [c,(c+d)/2], \qquad I^B := [a,(a+b)/2] \times [(c+d)/2,d],$$
$$I^C := [(a+b)/2,b] \times [c,(c+d)/2], \qquad I^D := [(a+b)/2,b] \times [(c+d)/2,d],$$

by using the following recursions. Suppose the control points of F are F_{ij}, $i,j = 0,\ldots,n$. We define

$$\left.\begin{array}{lll} F_{ij}^{[0]} := F_{ij} & \text{and} & F_{ij}^{[k]} := 0.5\left(F_{i-1,j}^{[k-1]} + F_{ij}^{[k-1]}\right) \\ P_{ij}^{[0]} := F_{ij}^{[i]} & \text{and} & P_{ij}^{[k]} := 0.5\left(P_{i,j-1}^{[k-1]} + P_{ij}^{[k-1]}\right) \\ Q_{ij}^{[0]} := F_{nj}^{[n-i]} & \text{and} & Q_{ij}^{[k]} := 0.5\left(Q_{i,j-1}^{[k-1]} + Q_{ij}^{[k-1]}\right) \end{array}\right\} \quad i,j = 0,\ldots,n, \quad k \geq 1.$$

The control points of F^A, F^B, F^C and F^D are respectively $A_{ij} := P_{ij}^{[j]}$, $B_{ij} := P_{in}^{[n-j]}$, $C_{ij} := Q_{ij}^{[j]}$, $D_{ij} := Q_{in}^{[n-j]}$. We have in particular

$$F(u,v) = F^r(u,v) \quad \text{for} \quad (u,v) \in I^r, \quad \text{where} \quad r = A,B,C,D.$$

We can apply the same subdivision technique to each of the resulting 4 Bézier surfaces. A recursive application of that subdivision on the unit square generates a uniform grid consisting of σ^2 little squares.

Theorem 2. *Suppose that the Coons patch* $\mathbf{x}$ *defined with* $\boldsymbol{\alpha}, \boldsymbol{\beta}, \boldsymbol{\gamma}, \boldsymbol{\delta}$ *is regular. Suppose that its Jacobian function* J *has been subdivided into* σ^2 *functions* J^{ij} *defined on*

$$I^{ij} := [(i-1)/\sigma, i/\sigma] \times [(j-1)/\sigma, j/\sigma], \qquad i,j = 1,\ldots,\sigma,$$

and with Bézier coefficients J_{pq}^{ij}, $p,q = 0,\ldots,2n$. *Then, for a sufficiently large* σ *all coefficients* J_{pq}^{ij} *have the same sign.*

Proof

We apply [5, 9] the multivariate Taylor expansion to the blossom of J^{ij}, in order to obtain

$$J^{ij}_{pq} = J^{ij}(a_p, c_q) + \mathcal{O}(h^2) \tag{3.5}$$

where $h := 1/(2n\sigma)$, $a_p := a + ph$, $c_q := c + qh$ in which $[a, b] \times [c, d]$ denotes the rectangular domain I^{ij}. As a consequence, we have

$$J^{ij}_{pq} = J^{ij}(a_p, c_q) + J^{ij}_{pq} - J^{ij}(a_p, c_q) \geq J(a_p, c_q) - Ch^2.$$

We deduce the result by letting Ch^2 tends to 0 as σ tends to infinity.

□

We omit the proof of the following results which can be proved in a very similar fashion.

Theorem 3. *Let* $\mathbf{x}$ *be a Coons patch that is not regular. Then, in the situation of Theorem 2 and for sufficiently large* σ*, there must exist* (i_1, j_1) *and* (i_2, j_2) *such that*

$$\left.\begin{array}{l} J^{i_1,j_1}_{pq} > 0 \\ J^{i_2,j_2}_{pq} < 0 \end{array}\right\} \quad \forall\, p, q = 0, \ldots, 2n.$$

In practice, we do not need to subdivide the Jacobian uniformly because we can perform adaptive subdivision. We start from a single Jacobian function written in Bézier form defined on the unit square. Then, we split it recursively by adaptively using the former subdivision techniques. That is, we subdivide only those Bézier functions that have Bézier coefficients J^{ij}_{pq} with different signs. The preceding two theorems serve as abortion conditions for that recursion.

An arbitrary parametrization of the curves $\boldsymbol{\kappa}^j_i$ from (1.1) does not guarantee the global continuity in (1.2). Since we cannot modify the base surfaces $\boldsymbol{\psi}_i$, we want to replace the 2D curves $\boldsymbol{\kappa}^j_i$ by $\widetilde{\boldsymbol{\kappa}}^j_i$ such that they have the same shapes ($\mathrm{im}(\boldsymbol{\kappa}^j_i) = \mathrm{im}(\widetilde{\boldsymbol{\kappa}}^j_i)$) but they have different parametrizations. Let us introduce the length function

$$\chi^j_i(t) := \int_{e^j_i}^{t} \left\| \frac{d\boldsymbol{\rho}^j_i}{dt}(\theta) \right\| d\theta \quad \text{where} \quad \boldsymbol{\rho}^j_i := \boldsymbol{\psi}_i \circ \boldsymbol{\kappa}^j_i.$$

On account of the properties of $\boldsymbol{\kappa}^j_i$ and $\boldsymbol{\psi}_i$, let us observe that

$$\frac{d\chi^j_i}{dt}(t) = \left\| \frac{d\boldsymbol{\rho}^j_i}{dt}(t) \right\| \neq 0 \qquad \forall t \in [e^j_i, f^j_i]. \tag{3.6}$$

Hence, there is an inverse function $\phi^j_i := (\chi^j_i)^{-1}$ and our method consists in replacing the function $\boldsymbol{\kappa}^j_i$ by the chord length parametrization $\widetilde{\boldsymbol{\kappa}}^j_i := \boldsymbol{\kappa}^j_i \circ \phi^j_i$. We have shown [5] that if we use that reparametrization then we have global continuity.

4 Practical Realization

In the following section, we show how to approximate the reparametrization function $\phi = \chi^{-1}$ which we met in (3.6). Additionally, we will examine the accuracy of the approximation. In the description of the method, we suppose that we have a fast method of evaluating the arc length $\mathcal{L}(\mathbf{x}, \mathbf{y})$ between two points $\mathbf{x}$, $\mathbf{y}$ belonging to the curve $\mathcal{C}$.
Let $\{s_i\}_{i=0,...,n} \subset [e, f]$ be some samples such that $s_i < s_{i+1}$ and $s_0 = e$ and $s_n = f$. Let $\{\mathbf{x}_i\}_{i=0,...,n}$ be their images by $\boldsymbol{\psi} \circ \boldsymbol{\kappa}$ and define $L_i := \mathcal{L}(\mathbf{x_i}, \mathbf{x_{i+1}})$ for $i = 0, ..., n-1$. We introduce now $t_i := \sum_{k=0}^{i} L_k \in [0, L]$ where L is the length of the whole curve. Afterwards, we consider the samples $\{(t_i, s_i),\ i = 0, ..., n\}$ and we interpolate them by a composite cubic Bézier curve ϕ_h:

$$\phi_h(t_i) = s_i \qquad \forall\, i = 0, ..., n, \tag{4.7}$$

$$\phi_h(t) = \sum_{j=0}^{3} b_{3i+j} B_j^3(t) \qquad \forall t \in [t_i, t_{i+1}]. \tag{4.8}$$

Lemma 1. *[3] Let f_h be the cubic spline interpolating the samples $[t_i, f(t_i)]$ which are obtained from a function f. By defining $h := \max |t_i - t_{i+1}|$, we have the approximation accuracy $\|f - f_h\|_\infty \leq Ch^4$.*

Since it is practically infeasible to compute the lengths exactly, we want to examine the influence of the length error to the function approximation. More precisely, on account of length evaluation inexactness, let us suppose that we have evaluated $\{\tau_i\}$ instead of $\{t_i\}$. Let $\tilde{\phi}_h$ be the cubic spline which is computed with the help of $\{\tau_i\}$. The following results expressing the accuracy $\|\phi - \tilde{\phi}_h\|_\infty$ is proven from the former lemma.

Theorem 1. *Suppose that the error in length computation is $\max_{i=0,...,n} |t_i - \tau_i| = \mathcal{O}(h^\alpha)$ where $\alpha \geq 1$ is some fixed integer and $h := \max |t_i - t_{i+1}|$. Then, we have the approximation accuracy $\|\phi - \tilde{\phi}_h\|_\infty = \mathcal{O}(h^{\min\{4,\alpha\}})$.*

Proof
Since the proof is easy but lengthy [9], we will sketch only the main points. Consider the piecewise polynomial interpolant σ of degree seven satisfying the following boundary conditions (use higher order Hermite interpolants):

$$\sigma(\tau_i) = t_i, \quad \sigma'(\tau_i) = 1, \quad \sigma^{(k)}(\tau_i) = 0 \qquad k = 2, 3. \tag{4.9}$$

A Taylor development reveals that for $\tau \in [\tau_i, \tau_{i+1}]$, we have:

$$\sigma(\tau) - \tau = \sigma(\tau_i) - \tau_i + \sum_{k=1}^{3} \frac{(\tau - \tau_i)^k}{k!} (\sigma^{(k)}(\tau_i) - \delta_{1,k}) + \mathcal{O}(|\tau_i - \tau_{i+1}|^4). \tag{4.10}$$

As a consequence, we have $\sigma(\tau) - \tau = \mathcal{O}(h^{\min\{4,\alpha\}})$. Since ϕ' is uniformly bounded, we obtain from a second application of a Taylor development that $\|\phi - \tilde{\phi}_h\|_\infty = \mathcal{O}(h^{\min\{4,\alpha\}})$.

□

5 Chord Length Estimation

Since the formerly developed method invokes many evaluations of the length function, our next objective is to design an algorithm for estimating the length of a curve $\mathbf{x}$ inside an interval $[a, b]$:

$$\mathcal{L} := \int_a^b \|\mathbf{x}'(t)\| dt. \tag{5.11}$$

We would like also to investigate the rate of convergence of the approximation. Without loss of generality, we suppose that the curve is defined on $[0, 1]$ and we compute the whole length i.e. $a = 0$, $b = 1$. Without loss of generality, we suppose that the curve is a rational Bézier curve

$$\mathbf{x}(t) := \frac{\sum_{i=0}^{m} \omega_i \mathbf{b}_i B_i^m(t)}{\sum_{i=0}^{m} \omega_i B_i^m(t)}, \tag{5.12}$$

where we have in mind that m is small (say $m \leq 4$). Other curves like NURBS can be converted to piecewise rational Bézier curve. Besides, we assume that the weights ω_i are uniformly bounded:

$$\exists\, R_1 > 0, R_2 > 0 \quad \text{such that} \quad R_1 < \left| \sum_{i=0}^{m} \omega_i B_i^m(t) \right| < R_2 \qquad \forall t \in [0, 1]. \tag{5.13}$$

Let us first introduce some notations related to the successive subdivision of an arbitrary Bézier function $S(t) = \sum_{i=0}^{m} \mathbf{s}_i B_i^m(t)$. Let $\mathbf{s}_i^{(j)}$ be the points which are found by using the de Casteljau algorithm at $t = 0.5$, i.e. $\mathbf{s}_i^{(j+1)} := 0.5(\mathbf{s}_i^{(j)} + \mathbf{s}_{i+1}^{(j)})$ and $\mathbf{s}_i^{(0)} := \mathbf{s}_i$. The function $S^{[0,0]} := S$ can be split into two Bézier functions $S^{[1,1]}$ and $S^{[1,2]}$ which have respectively the control points $\mathbf{s}_i^{[1,1]} := \mathbf{s}_0^{(i)}$ and $\mathbf{s}_i^{[1,2]} := \mathbf{s}_i^{(m-i)}$ and which are defined on $[0, 0.5]$ and $[0.5, 1]$. We can apply that process successively in order to obtain from each Bézier function $S^{[p-1,i]}$ two Bézier functions $S^{[p,2i-1]}$ and $S^{[p,2i]}$. That is, after applying subdivisions n times, we have 2^n Bézier curves $S^{[n,k]}$ whose control points are denoted by $\mathbf{s}_i^{[n,k]}$ for $k = 1, ..., 2^n$ and $i = 0, ..., m$. Each function $S^{[n,k]}$ is defined on the interval $[p_{k-1}, p_k]$ where $p_k := k/2^n$.

Now, we want to apply the above subdivision technique to the numerator $\tilde{\mathbf{x}}(\cdot)$ and the denominator $\omega(\cdot)$ of the function in (5.12). Let us denote the control points of $\tilde{\mathbf{x}}$ by $\tilde{\mathbf{b}}_i := \mathbf{b}_i \omega_i$. The functions $\tilde{\mathbf{x}}(\cdot)$ and $\omega(\cdot)$ will be subdivided into functions $\tilde{\mathbf{x}}^{[n,k]}$ and $\omega^{[n,k]}$ having the control points $\tilde{\mathbf{b}}_i^{[n,k]}$ and $\omega_i^{[n,k]}$. On each subinterval $[p_k, p_{k+1}]$, we introduce the rational Bézier curve $\mathbf{x}^{[n,k]} := \tilde{\mathbf{x}}^{[n,k]}/\omega^{[n,k]}$. Thus, by defining $\mathbf{b}_i^{[n,k]} := \tilde{\mathbf{b}}_i^{[n,k]}/\omega_i^{[n,k]}$, we have for each $\tau \in [p_k, p_{k+1}]$:

$$\mathbf{x}^{[n,k]}(\tau) = \frac{\sum_{i=0}^{m} \omega_i^{[n,k]} \mathbf{b}_i^{[n,k]} B_i^m(s)}{\sum_{i=0}^{m} \omega_i^{[n,k]} B_i^m(s)} \quad \text{where} \quad s = \frac{\tau - p_k}{p_{k+1} - p_k}. \tag{5.14}$$

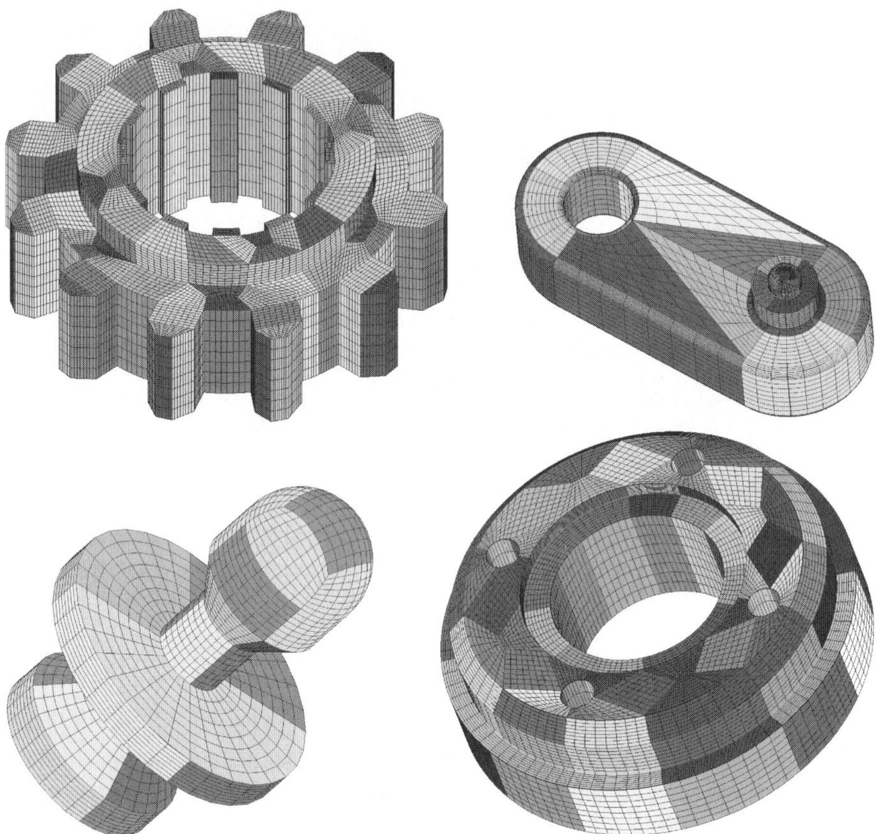

Figure 5.2. Decomposition of simple CAD models (*See also* Color Plate on page 387)

Furthermore, we have the restriction property: $\mathbf{x}^{[n,k]} = \mathbf{x}_{|[p_{k-1},p_k]}$. By considering the interval $[p_{k-1}, p_k]$, we can introduce $\theta_{i,k} := (i/m)p_k + (1-i/m)p_{k-1}$ for $i = 0, ..., m$. We have the following convergence result for the subdivision of a rational Bézier curve.

Theorem 1. *Suppose that the rational Bézier curve in (5.12) has been subdivided n times. Then, we have the following accuracy order for all $k = 1, ..., 2^n$ and $i = 0, ..., m$:*

$$\|\mathbf{x}^{[n,k]}(\theta_{i,k}) - \mathbf{b}_i^{[n,k]}\| = \mathcal{O}(2^{-2n}). \tag{5.15}$$

Proof

Since the function ω is uniformly bounded as specified in (5.13), there exist constants K_1, K_2 such that

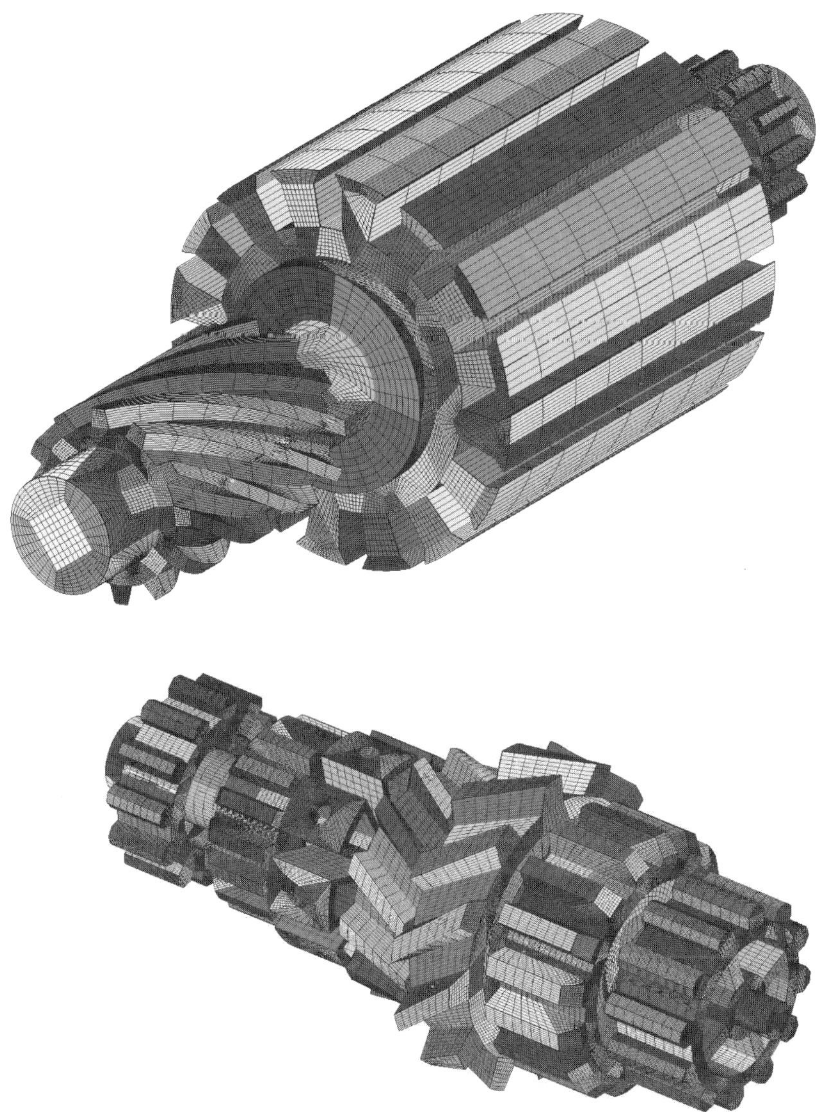

Figure 5.3. Decomposition of realistic CAD models (*See also* Color Plate on page 388)

$$\left\| \mathbf{x}^{[n,k]}(\theta_{i,k}) - \frac{\omega_i^{[n,k]}\mathbf{b}_i^{[n,k]}}{\omega^{[n,k]}(\theta_{i,k})} \right\| \leq K_1 \left\| \tilde{\mathbf{x}}^{[n,k]}(\theta_{i,k}) - \omega_i^{[n,k]}\mathbf{b}_i^{[n,k]} \right\|, \tag{5.16}$$

$$\left\| \mathbf{b}_i^{[n,k]} - \frac{\omega_i^{[n,k]}\mathbf{b}_i^{[n,k]}}{\omega^{[n,k]}(\theta_{i,k})} \right\| \leq K_2 \left| \omega^{[n,k]}(\theta_{i,k}) - \omega_i^{[n,k]} \right|. \tag{5.17}$$

As a consequence, we obtain

$$\begin{aligned} \left\| \mathbf{x}^{[n,k]}(\theta_{i,k}) - \mathbf{b}_i^{[n,k]} \right\| &\leq K_1 \left\| \tilde{\mathbf{x}}^{[n,k]}(\theta_{i,k}) - \omega_i^{[n,k]}\mathbf{b}_i^{[n,k]} \right\| \\ &+K_2 \left| \omega^{[n,k]}(\theta_{i,k}) - \omega_i^{[n,k]} \right|. \end{aligned} \tag{5.18}$$

On the other hand, let us consider the blossom function $\mathcal{P}$ of the polynomial $\tilde{\mathbf{x}}^{[n,k]}$. We have the relation with the control points:

$$\tilde{\mathbf{b}}_i^{[n,k]} = \mathcal{P}(\underbrace{p_{k-1}, ..., p_{k-1}}_{m-i}, \underbrace{p_k, ..., p_k}_{i}). \tag{5.19}$$

Thus, we have the following Taylor development:

$$\begin{aligned} \tilde{\mathbf{b}}_i^{[n,k]} &= \mathcal{P}(\theta_{i,k}, ..., \theta_{i,k}) + \sum_{p=1}^{m-i} (p_{k-1} - \theta_{i,k}) \frac{\partial}{\partial x_p} \mathcal{P}(\theta_{i,k}, ..., \theta_{i,k}) + \\ &\sum_{p=m-i+1}^{m} (p_k - \theta_{i,k}) \frac{\partial}{\partial x_p} \mathcal{P}(\theta_{i,k}, ..., \theta_{i,k}) + \mathcal{O}(|p_k - p_{k-1}|^2). \end{aligned}$$

Since $\mathcal{P}$ is symmetric, all partial derivatives in the above relation are the same. Due to the fact that $(m-i)(p_k - \theta_{i,k}) + i(p_{k-1} - \theta_{i,k}) = 0$, we obtain $\tilde{\mathbf{b}}_i^{[n,k]} = \mathcal{P}(\theta_{i,k}, ..., \theta_{i,k}) + \mathcal{O}(|p_k - p_{k-1}|^2)$. As a consequence, we deduce $\tilde{\mathbf{b}}_i^{[n,k]} = \tilde{\mathbf{x}}^{[n,k]}(\theta_{i,k}) + \mathcal{O}(2^{-2n})$. The same analysis can be repeated to the blossom of the polynomial ω in order to obtain $\omega_i^{[n,k]} = \omega(\theta_{i,k}) + \mathcal{O}(2^{-2n})$. Therefore, we can deduce from (5.18) that $\|\mathbf{x}^{[n,k]}(\theta_{i,k}) - \mathbf{b}_i^{[n,k]}\| = \mathcal{O}(2^{-2n})$.

□

Corollary 1. *Let $\mathcal{L}$ be the length of $\mathbf{x}$. Define*

$$A_k := \sum_{i=0}^{m-1} \|\mathbf{x}^{[n,k]}(\theta_{i,k}) - \mathbf{x}^{[n,k]}(\theta_{i+1,k})\|, \quad \textit{and} \tag{5.20}$$

$$B_k := \sum_{i=0}^{m-1} \|\mathbf{b}_i^{[n,k]} - \mathbf{b}_{i+1}^{[n,k]}\| \qquad \forall k = 0, ..., 2^n - 1. \tag{5.21}$$

We claim that $\mathcal{L}_n := \sum_k (\lambda A_k + (1-\lambda) B_k)$ converges to the true length $\mathcal{L}$ in dyadic order:

$$|\mathcal{L} - \mathcal{L}_n| = \mathcal{O}(2^{-n}). \tag{5.22}$$

Proof

Let l_k be the length of the curve $\mathbf{x}^{[n,k]}(\tau)$. Due to the convex hull property [3], we obtain $A_k \leq l_k \leq B_k$. On the other hand, the difference of the bounds can be estimated as follows

$$|B_k - A_k| \leq \sum_{i=0}^{m-1} \left\| [\mathbf{b}_i^{[n,k]} - \mathbf{x}^{[n,k]}(\theta_{i,k})] \right\| + \left\| [\mathbf{b}_{i+1}^{[n,k]} - \mathbf{x}^{[n,k]}(\theta_{i+1,k})] \right\| . \quad (5.23)$$

By using the previous theorem with the last inequality, we deduce $|B_k - A_k| = \mathcal{O}(2^{-2n})$. As a consequence, we obtain $|B_k - l_k| = \mathcal{O}(2^{-2n})$ and $|A_k - l_k| = \mathcal{O}(2^{-2n})$. Hence, the accuracy of the length estimation is given as $|\mathcal{L} - \mathcal{L}_n| \leq \sum_{k=0}^{2^n} |\lambda(l_k - A_k) + (1 - \lambda)(l_k - B_k)| = \mathcal{O}(2^{-n})$.

□

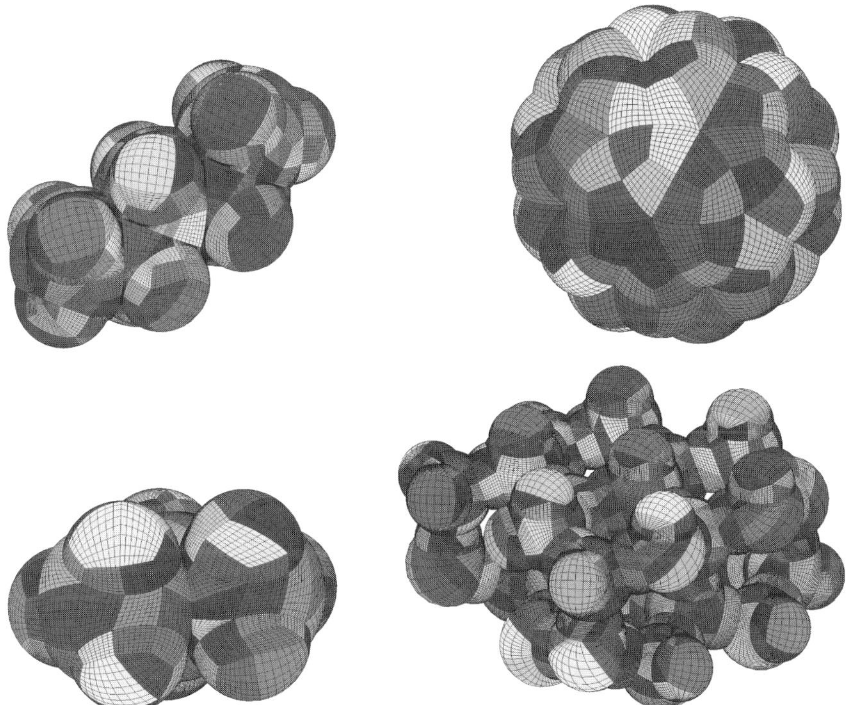

Figure 5.4. Molecules: pentane, fullerene, propane, ice. (*See also* Color Plate on page 389)

6 Similarities and Differences for Molecular Surfaces

Most properties that are valid for CAD surfaces hold true for molecular ones. Still, there are some differences [10] which we want to summarize below. Each constituting atom in a molecule is represented as an imaginary sphere whose radius corresponds to the Van der Walls radius of the atom. The molecular surface which represents a closed surface is the boundary of the union of those spheres. Since a boundary representation structure using parametrizations for the faces is more convenient for the surface decompositions, we have to convert that CSG representation into a boundary representation. That is, we need to find the set of edges, vertices and faces whose parametrization serves as a lifting from planar decompositions. For that, we first perform sphere-sphere intersections and circle-circle intersections. Then, we detect and discard useless geometric entities. Note that all edges can be represented in form of rational Bézier curves because they are circular arcs. As for the parametrization, projective geometry is useful for linearizations. For instance, suppose that a rational Bézier curve $\mathbf{x}$ is given in homogeneous coordinates as $\mathbf{x}(t) = \sum_{i=0}^{n}[\omega_i : \omega_i x_i : \omega_i y_i : \omega_i z_i] B_i^n(t)$. The following curve maps to $\mathbf{x}$ by the stereographic projection $\mathbf{y}(t) := [\sum_{i=0}^{n} \tilde{\omega}_i B_i^n(t) : \sum_{i=0}^{n} \tilde{\omega}_i \tilde{x}_i B_i^n(t) : \sum_{i=0}^{n} \tilde{\omega}_i \tilde{y}_i B_i^n(t) : 0]$ where $\tilde{\omega}_i := \omega_i(1 - z_i)$ and $(\tilde{x}_i, \tilde{y}_i, 0) := (x_i/1 - z_i, y_i/1 - z_i, 0)$. Interestingly, the preceding reparametrization is not required for molecules so as to obtain global continuity. The most complicated process [10] is the creation of the four-sided patches by means of rational Bézier surfaces while achieving an *exact* matching condition.

7 Software Packages and Practical Results

In order to apply the former methods, we have developed a software whose implementation is written in C/C++ and MFC while the visualization is done in OpenGL. We have used two sets of CAD models whose results are displayed in Fig. 5.2 and Fig. 5.3 respectively. The first set consists of simple CAD objects which have relatively few initial surfaces: 136, 40, 18, 28 respectively. The second set is devoted to objects which are already mechanically realistic for complex CAD models without assemblies. The respective numbers of initial surfaces are 243 and 593. Apart from the decomposition, we have also evaluated points on the surfaces where we need $4225 = 65^2$ points per patch. The results are gathered in Table 7.1. As for molecular models, we have implemented a software which accepts PDB files as input. We have applied the geometry preparation method to four molecular surfaces: pentane (17 atoms), fullerene (60 atoms), propane (11 atoms), ice (84 atoms). The final tessellations consist respectively of 388, 360, 231, 1284 four-sided patches as illustrated in Fig. 5.4. Molecular decompositions take longer than CAD ones because of the conversion into boundary representation. The decomposition durations for the four molecules are 11.77 sec, 2.45 sec, 9.65 sec and

Table 7.1. Runtime and point evaluation.

Model	Nb initial surf	Nb patches	Decomposition	Evaluation
Set 1-Object 1	136	352	2.981 sec.	16.456 sec.
Set 1-Object 2	40	96	1.030 sec.	4.456 sec.
Set 1-Object 3	18	46	0.345 sec.	2.898 sec.
Set 1-Object 4	28	212	0.686 sec.	10.124 sec.
Set 2-Object 1	243	727	3.245 sec.	40.987 sec.
Set 2-Object 2	593	2081	16.816 sec.	98.975 sec.

71.97 sec. In the opposite, the point evaluation is faster because we need much fewer control points in molecular surfaces: the point evaluations for the four molecules take 0.361 sec, 0.354 sec, 0.256 sec and 1.345 sec. The results were obtained from a machine with processor Intel Core 2.16GHz having Windows Vista.

References

1. G. Brunnett, *Geometric Design with Trimmed Surfaces*, Computing Supplementum **10** (1995), 101–115.
2. W. Dahmen, A. Kunoth, and K. Urban, *Biorthogonal Spline Wavelets on the Interval: Stability and Moment Conditions*, Appl. Comput. Harmon. Anal. **6**, No. 2 (1999), 132–196.
3. C. de Boor, *A Practical Guide to Splines*, Springer, New York, 1978.
4. M. Griebel, S. Knapek, and G. Zumbusch, *Numerical Simulation in Molecular Dynamics*, Springer, Berlin, 2007.
5. H. Harbrecht and M. Randrianarivony, *From Computer Aided Design till Wavelet BEM*, Berichtsreihe des mathematischen Seminars, Preprint 07-18, University Kiel, October 2007.
6. A. Kunoth, *Adaptive Wavelet Schemes for an Elliptic Control Problem with Dirichlet Boundary Control*, Numer. Algor. **39** (1-3) (2005), 199–220.
7. C. Lee and S. Lo, *A New Scheme for the Generation of a Graded Quadrilateral Mesh*, Comput. Struct. **52** (5) (1994), 847–857.
8. S. Owen, M. Staten, S. Canann, and S. Saigal, *Q-Morph: An Indirect Approach to Advancing Front Quad Meshing*, Int. J. Numer. Methods Eng. **44** (9) (1999), 1317–1340.
9. M. Randrianarivony, *Geometric Processing of CAD Data and Meshes as Input of Integral Equation Solvers*, Ph.D. thesis, Technische Universität Chemnitz, 2006.
10. M. Randrianarivony and G. Brunnett, *Molecular Surface Decomposition Using Graphical Modeling*, Bildverarbeitung für die Medizine (T. Tolxdorff and J. Braun, eds.), Informatik aktuell, Springer, 2008, pp. 197–201.
11. R. Schneider, *Multiskalen- und Wavelet-Matrixkompression: Analysisbasierte Methoden zur Lösung grosser vollbesetzter Gleichungssysteme*, Teubner, Stuttgart, 1998.

3D Meshfree Magnetohydrodynamics

Stephan Rosswog[1] and Daniel Price[2]

[1] Jacobs University Bremen, Campus Ring 1, D-28759 Bremen, Germany
s.rosswog@jacobs-university.de

[2] School of Physics, University of Exeter, Stocker Rd, Exeter EX4 4QL, UK
dprice@astro.ex.ac.uk

Summary. We describe a new method to include magnetic fields into smooth particle hydrodynamics. The derivation of the self-gravitating hydrodynamics equations from a variational principle is discussed in some detail. The non-dissipative magnetic field evolution is instantiated by advecting so-called Euler potentials. This approach enforces the crucial $\nabla \cdot \boldsymbol{B} = 0$-constraint by construction. These recent developments are implemented in our three-dimensional, self-gravitating magnetohydrodynamics code MAGMA. A suite of tests is presented that demonstrates the superiority of this new approach in comparison to previous implementations.

Key words: astrophysics, magnetohydrodynamics, smoothed particle hydrodynamics, magnetic fields, Euler potentials

1 Specific astrophysical requirements

Astrophysical simulations have their specific requirements which differ in many respects from those of other branches of computationally intense research fields. The dynamics of self-gravitating gas masses plays a prominent role throughout astrophysics, but it is usually only one of several ingredients and it is often necessary to account for additional physical processes such as radiative transfer, nuclear burning or the evolution of magnetic fields to address questions of astrophysical interest. These additional processes often involve intrinsic length and time scales that are dramatically different from those of the gas dynamical processes making astrophysical problems prime examples of multi-scale and multi-physics challenges.

Since fixed boundaries are usually absent, flow geometries are determined by the interplay between different physical processes such as gas dynamics and (self-)gravity which often leads to complicated, dynamically changing flow geometries. Therefore, many problems require flexible numerical schemes such as adaptive mesh refinement or completely meshfree, Lagrangian methods.

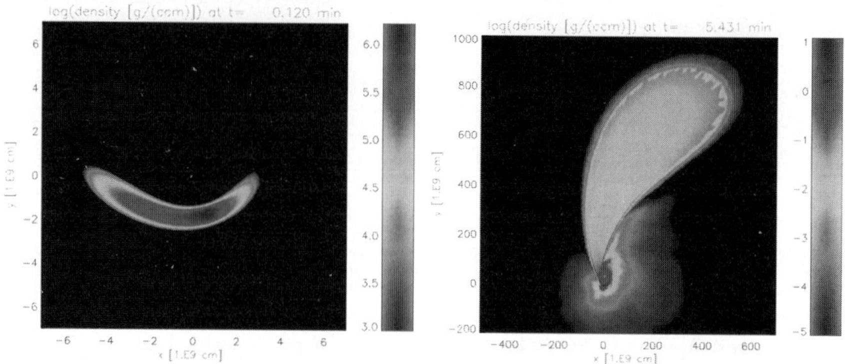

Figure 1.1. Two snapshots from the simulation of the tidal disruption of a white dwarf by a black hole. (*See also* Color Plate on page 389)

Each of these methods has its stengths and weaknesses and the choice of the best-suited method can usually save a tremendous amount of effort.

An astrophysical example of such an intrinsic multi-scale and multi-physics problem is shown in Fig. 1.1[3]. It shows two snapshots from a simulation of the tidal disruption of a white dwarf star by a black hole. The initially spherical star becomes strongly distorted while passing the black hole (left panel), it is heavily compressed compressed and shock-heated which triggers very rapid nuclear reactions whose energy release leads to the thermonuclear explosion of the white dwarf. In order to follow this process for each of the computational fluid particles a nuclear network [24] is evolved on-the-fly together with the hydrodynamics.

In many astrophysical problems the numerical conservation of physically conserved quantities determines the success and the reliability of a numerical simulation. Consider, for example, a molecular gas cloud that collapses under the influence of its own gravity to form stars. If in the simulation angular momentum is artificially dissipated, say due to too coarse a mesh discretization, a collapsing, self-gravitating portion of gas may form just a single stellar object instead of a multiple system of stars and it will thus produce a qualitatively wrong result. The "exact"[4] conservation of mass, energy, linear and angular momentum is –besides its natural adaptivity– one of the main strengths of the smoothed particle hydrodynamics (SPH) method. This exact conservation can be "hardwired" into SPH's evolution equations via symmetries in the fluid particle indices. Originally this was done –successfully, but somewhat arbitrarily– by hand [6, 23, 32], but more recently it was shown [34, 36, 62, 64]

[3] Astrophysical implications of this topic are discussed in [52, 54], details of the numerics can be found in [53].

[4] "Exact" means up to possible effects from the numerical integration of the resulting ODEs or from using approximative forces, say from a tree or some other Poisson-solver.

how the correct symmetries follow elegantly and stringently from a discretized fluid Lagrangian and the Euler-Lagrange variational principle.

In the following, we will review the derivation of the self-gravitating SPH equations from a Lagrangian, see Sect. 2.1. We will also discuss in detail how to implement magnetic fields via so-called Euler potentials, see Sect. 2.2. This approach is similar to evolving a vector potential and enforces the crucial $\nabla \cdot \boldsymbol{B} = 0$-constraint which otherwise poses a severe challenge for particle methods. These new develoments are implemented in our self-gravitating, three-dimensional magnetohydrodynamics code MAGMA, which is described in Sect. 3.

2 Guiding principles

2.1 Ideal smoothed particle hydrodynamics (SPH)

The smoothed particle hydrodynamics method (SPH) had originally been developed in the astrophysical context to simulate the formation of stellar binary systems via fission [29] and the structure of non-sperical stars [19]. While the initial 3D simulations used 80 (Gingold and Monaghan) and 100 SPH particles (Lucy) today's state of the art cosmological SPH simulations have reached particle numbers in excess 10^9, see e.g. [20]. This is only in part due to the increase in hardware performance, also the simulation techniques (in particular the treatment of self-gravity) have become continuously more sophisticated and much effort has been invested to parallelize 3D codes on various computing platforms. Also the formulation of the SPH equations has come a very long way from the initial straight-forward discretisation of the Lagrangian gas dynamics equations to its most recent formulation that follows stringently from a discretized ideal fluid Lagrangian.

Here we will give a brief overview over an older SPH-formulation, but we will mainly focus on an approach that is based on a derivation from a discretised Lagrangian. This latter approach naturally introduces so-called "grad-h" terms that result from changes in the smoothing lengths of the SPH particles.

"Vanilla Ice" SPH

The approximation of function values and derivatives via a kernel summation is at the heart of SPH. If the values of a function f are known at a set of discrete points ("particles") labelled by b, the SPH approximation of the function f at position $\boldsymbol{r}$ is given by [6,32,35][5]

[5] Note that we do not specify at this point which h is used. For this "vanilla ice" SPH the h that enters the kernel should be a symmetric combination of the smoothing lengths of the involved particles. This will be explained in more detail

$$f(\boldsymbol{r}) = \sum_b \frac{m_b}{\rho_b} f_b W(\boldsymbol{r} - \boldsymbol{r}_b, h), \tag{2.1}$$

where m_b is the (usually constant) particle mass, ρ_b is the mass density and W is a kernel function whose width is determined by the smoothing length h. Essentially all astrophysical SPH codes use the cubic spline kernel suggested in [31]. Kernel functions with compact support are preferable since they restrict the SPH-summations to a local set of neighbours. For the conservation properties it is convenient to have "radial" kernels,

$$W(\boldsymbol{r} - \boldsymbol{r}_b, h) = W(||\boldsymbol{r} - \boldsymbol{r}_b||, h), \tag{2.2}$$

so that

$$\nabla_a W_{bk} = \nabla_b W_{kb}(\delta_{ba} - \delta_{ka}), \tag{2.3}$$

and

$$\nabla_a W_{ab} = \frac{\partial W_{ab}}{\partial r_{ab}} \hat{e}_{ab}, \tag{2.4}$$

where $\boldsymbol{r}_{ab} = \boldsymbol{r}_a - \boldsymbol{r}_b$, $r_{ab} = ||\boldsymbol{r}_{ab}||$, $W_{ab} = W(\boldsymbol{r}_{ab}, h)$ and $\hat{e}_{ab} = \boldsymbol{r}_{ab}/r_{ab}$. This immediately leads to

$$\nabla_a W_{ab} = -\nabla_b W_{ab} \tag{2.5}$$

and

$$\frac{dW_{ab}}{dt} = \boldsymbol{v}_{ab} \cdot \nabla_a W_{ab}, \tag{2.6}$$

with $\boldsymbol{v}_{ab} = \boldsymbol{v}_a - \boldsymbol{v}_b$ being the velocity difference between particle a and b.

Eq. (2.1) can be applied in particular to the mass density itself which then reads

$$\rho(\boldsymbol{r}) = \sum_b m_b W(\boldsymbol{r} - \boldsymbol{r}_b, h). \tag{2.7}$$

The gradient of a function is approximated in SPH by taking the exact derivative of the approximant:

$$\nabla f(\boldsymbol{r}) = \sum_b \frac{m_b}{\rho_b} f_b \nabla W(\boldsymbol{r} - \boldsymbol{r}_b, h). \tag{2.8}$$

The most straightforward and historically first taken approch is to apply this set of rules to the Lagrangian form of the ideal hydrodynamics equations:

$$\frac{d\rho}{dt} = -\rho \nabla \cdot \boldsymbol{v}, \tag{2.9}$$

$$\frac{d\boldsymbol{v}}{dt} = -\frac{\nabla P}{\rho} + \boldsymbol{f}, \tag{2.10}$$

$$\frac{du}{dt} = \frac{P}{\rho^2}\frac{d\rho}{dt} = -\frac{P}{\rho}\nabla \cdot \boldsymbol{v}, \tag{2.11}$$

---

below. For simplicity, we are omitting the subscript h in what follows. We also drop the distinction between the function to be interpolated and the interpolant, i.e. we use the same symbol f on both sides of the following equation.

which express the conservation of mass, momentum and energy. Here, P is the thermodynamic pressure, $\boldsymbol{f}$ abbreviates body forces and u is the thermal energy per mass.

To briefly illustrate the dependence of conservation on the symmetry of the particle indices let us apply Eq. (2.8) straightforward to the pressure gradient in Eq. (2.10) (and assume vanishing body forces) to obtain

$$\frac{d\boldsymbol{v}_a}{dt} = -\frac{1}{\rho_a}\sum_b \frac{m_b}{\rho_b} P_b \nabla_a W_{ab} \tag{2.12}$$

for the acceleration of particle a. This form solves the Euler equation to the order of the method, but it does not conserve the total momentum. Consider the force that particle b exerts on particle a

$$\boldsymbol{F}_{ba} = \left(m_a \frac{d\boldsymbol{v}_a}{dt} \right)_b = -\frac{m_a}{\rho_a}\frac{m_b}{\rho_b} P_b \nabla_a W_{ab} \tag{2.13}$$

and similarly, the force from particle a on b

$$\boldsymbol{F}_{ab} = \frac{m_a}{\rho_a}\frac{m_b}{\rho_b} P_a \nabla_a W_{ab}, \tag{2.14}$$

where we have used Eq. (2.5). Since in general $P_a \neq P_b$, the sum over all the momentum derivatives, $\sum_b d(m_b \boldsymbol{v}_b)/dt$, does not vanish and therefore the total momentum is not conserved.

This deficiency can be easily cured by expressing the pressure gradient term as

$$\frac{\nabla P}{\rho} = \frac{P}{\rho^2}\nabla\rho + \nabla\left(\frac{P}{\rho}\right). \tag{2.15}$$

If the gradient formula, Eq. (2.8), is applied to Eq. (2.15), the momentum equation reads

$$\frac{d\boldsymbol{v}_a}{dt} = -\sum_b m_b \left(\frac{P_a}{\rho_a^2} + \frac{P_b}{\rho_b^2} \right) \nabla_a W_{ab}. \tag{2.16}$$

Because the pressure part of the equations is now manifestly symmetric in a and b and $\nabla_a W_{ab} = -\nabla_b W_{ba}$ the forces are now equal and opposite ("actio= reactio") and therefore the total momentum is conserved by construction, i.e. $\sum_a m_a \frac{d\boldsymbol{v}}{dt} = 0$.

Similarly, the total angular momentum is conserved since the sum of all torques vanishes:

$$\begin{aligned}\frac{d\boldsymbol{L}}{dt} &= \sum_{a,b} \boldsymbol{r}_a \times \boldsymbol{F}_{ba} = \frac{1}{2}\left(\sum_{a,b} \boldsymbol{r}_a \times \boldsymbol{F}_{ba} + \sum_{a,b} \boldsymbol{r}_a \times \boldsymbol{F}_{ba} \right) \\ &= \frac{1}{2}\left(\sum_{a,b} (\boldsymbol{r}_a - \boldsymbol{r}_b) \times \boldsymbol{F}_{ba} \right) = 0. \end{aligned} \tag{2.17}$$

Here the summation indices were relabeled and $\boldsymbol{F}_{ab} = -\boldsymbol{F}_{ba}$ was used. The expression finally vanishes, because the forces between particles act along the line joining them, see Eq. (2.4).

A suitable energy equation can be constructed from Eq. (2.11) in a straight forward way. Start from the (adiabatic) first law of thermodynamics

$$\frac{du_a}{dt} = \frac{P_a}{\rho_a^2}\frac{d\rho_a}{dt} \tag{2.18}$$

and insert

$$\frac{d\rho_a}{dt} = \frac{d}{dt}\left(\sum_b m_b W_{ab}\right) = \sum_b m_b \boldsymbol{v}_{ab} \cdot \nabla_a W_{ab}, \tag{2.19}$$

where we have used Eq. (2.6), to find

$$\frac{du_a}{dt} = \frac{P_a}{\rho_a^2} \sum_b m_b \boldsymbol{v}_{ab} \cdot \nabla_a W_{ab}. \tag{2.20}$$

Together with an equation of state the equations (2.7), (2.16) and (2.20) form a complete set of SPH equations.

In the previous derivation it was implicitely assumed that derivatives of the smoothing lengths can be ignored. In a simulation with strongly changing geometry, however, it is advisable to locally adapt the smoothing length. This introduces, in principle, additional terms in the SPH equations. The importance of these extra terms depends very much on the exact application [56,64].

The SPH-equations from a Lagrangian, "grad-h" terms

The Lagrangian and the Euler-Lagrange equations

The SPH equations can be derived by using nothing more than a suitable Lagrangian, the first law of thermodynamics and a prescription on how to obtain the density via summation. The Lagrangian of a perfect fluid [16]

$$L = \int \rho \left(\frac{v^2}{2} - u(\rho, s)\right) dV, \tag{2.21}$$

with s being the specific entropy, can be SPH-discretized in a straightforward way:

$$L_{\mathrm{SPH,h}} = \sum_b m_b \left(\frac{v_b^2}{2} - u(\rho_b, s_b)\right). \tag{2.22}$$

The discretized equations for the fluid are then found by applying the Euler-Lagrange equations

$$\frac{d}{dt}\left(\frac{\partial L}{\partial \boldsymbol{v}_a}\right) - \frac{\partial L}{\partial \boldsymbol{r}_a} = 0. \tag{2.23}$$

The term in brackets yields the canonical particle momentum

$$\frac{\partial L}{\partial \boldsymbol{v}_a} = m_a \boldsymbol{v}_a, \tag{2.24}$$

the potential-type second term in the Lagrangian becomes

$$\frac{\partial L}{\partial \boldsymbol{r}_a} = -\sum_b m_b \frac{\partial u(\rho_b, s_b)}{\partial \boldsymbol{r}_a} = -\sum_b m_b \left.\frac{\partial u_b}{\partial \rho_b}\right|_s \cdot \frac{\partial \rho_b}{\partial \boldsymbol{r}_a}. \tag{2.25}$$

The first derivative can be expressed using the first law of thermodynamics, $du = P/\rho^2 d\rho$, and therefore

$$m_a \frac{d\boldsymbol{v}_a}{dt} = -\sum_b m_b \frac{P_b}{\rho_b^2} \frac{\partial \rho_b}{\partial \boldsymbol{r}_a}. \tag{2.26}$$

The density, its derivatives and the "grad-h"-terms
We will now address the aditional terms resulting from variable smoothing lengths. For a density estimate as "local" as possible we use the smoothing length h_a in

$$\rho_a = \sum_b m_b W(r_{ab}, h_a). \tag{2.27}$$

Adaptivity can be reached by evolving the smoothing length according to

$$h_a = \eta \left(\frac{m_a}{\rho_a} \right)^{1/3}, \tag{2.28}$$

where η is a parameter typically in a range between 1.2 and 1.5 [42]. Since ρ_a and h_a mutually depend on each other, see Eqs. (2.27) and (2.28), an iteration is required for consistency.

If we take the changes of h into account, the Lagrangian time derivative of the density is given by

$$\begin{aligned} \frac{d\rho_a}{dt} &= \sum_b m_b \left\{ \frac{\partial W_{ab}(h_a)}{\partial r_{ab}} \frac{dr_{ab}}{dt} + \frac{\partial W_{ab}(h_a)}{\partial h_a} \frac{dh_a}{dt} \right\} \\ &= \sum_b m_b \boldsymbol{v}_{ab} \cdot \nabla_a W_{ab}(h_a) + \frac{\partial h_a}{\partial \rho_a} \frac{d\rho_a}{dt} \sum_b m_b \frac{\partial}{\partial h_a} W_{ab}(h_a), \end{aligned}$$

where we have used $dr_{ab}/dt = \hat{e}_{ab} \cdot \boldsymbol{v}_{ab}$ and Eq. (2.5). If the $d\rho_a/dt$-terms are collected into the quantity

$$\Omega_a \equiv \left(1 - \frac{\partial h_a}{\partial \rho_a} \cdot \sum_b m_b \frac{\partial}{\partial h_a} W_{ab}(h_a) \right), \tag{2.29}$$

the time derivative of the density reads

$$\frac{d\rho_a}{dt} = \frac{1}{\Omega_a} \sum_b m_b \boldsymbol{v}_{ab} \cdot \nabla_a W_{ab}(h_a). \tag{2.30}$$

This is the generalization of the standard SPH expression, Eq. (2.19). In a similar way the spatial derivatives can be calculated

$$\begin{aligned}\frac{\partial \rho_b}{\partial \boldsymbol{r}_a} &= \sum_k m_k \left\{ \nabla_a W_{bk}(h_b) + \frac{\partial W_{bk}(h_b)}{\partial h_b} \frac{\partial h_b}{\partial \boldsymbol{r}_a} \right\} \\ &= \sum_k m_k \nabla_a W_{bk}(h_b) + \frac{\partial h_b}{\partial \rho_b} \frac{\partial \rho_b}{\partial \boldsymbol{r}_a} \sum_k m_k \frac{\partial W_{bk}(h_b)}{\partial h_b},\end{aligned}$$

or,

$$\frac{\partial \rho_b}{\partial \boldsymbol{r}_a} = \frac{1}{\Omega_b} \sum_k m_k \nabla_a W_{bk}(h_b). \tag{2.31}$$

The SPH equations with "grad-h"-terms
Inserting Eq. (2.30) into Eq. (2.18) yields the "grad-h" energy equation

$$\frac{du_{a,\mathrm{h}}}{dt} = \frac{1}{\Omega_a} \frac{P_a}{\rho_a^2} \sum_b m_b \boldsymbol{v}_{ab} \cdot \nabla_a W_{ab}(h_a). \tag{2.32}$$

With the derivative Eq. (2.31) one can write Eq. (2.26) as

$$m_a \frac{d\boldsymbol{v}_a}{dt} = -\sum_b m_b \frac{P_b}{\rho_b^2} \nabla_a \rho_b = -\sum_b m_b \frac{P_b}{\rho_b^2} \left(\frac{1}{\Omega_b} \sum_k m_k \nabla_a W_{bk}(h_b) \right) \tag{2.33}$$

With Eq. (2.3), the above equation becomes

$$\begin{aligned}m_a \frac{d\boldsymbol{v}_a}{dt} &= -\sum_b m_b \frac{P_b}{\rho_b^2} \frac{1}{\Omega_b} \sum_k m_k \nabla_b W_{kb}(h_b) \, (\delta_{ba} - \delta_{ka}) \\ &= -m_a \sum_b m_b \left(\frac{P_a}{\Omega_a \rho_a^2} \nabla_a W_{ab}(h_a) + \frac{P_b}{\Omega_b \rho_b^2} \nabla_a W_{ab}(h_b) \right),\end{aligned} \tag{2.34}$$

i.e. the final momentum equation reads

$$\frac{d\boldsymbol{v}_{a,\mathrm{h}}}{dt} = -\sum_b m_b \left(\frac{P_a}{\Omega_a \rho_a^2} \nabla_a W_{ab}(h_a) + \frac{P_b}{\Omega_b \rho_b^2} \nabla_a W_{ab}(h_b) \right). \tag{2.35}$$

Together with the density equation, Eq. (2.7), and an equation of state, Eqs. (2.32) and (2.35) form a complete set of "grad-h" SPH-equations.

Self-gravity and gravitational softening
The variational concept can also be applied to derive the gravitational forces

including softening in a self-consistent way [43]. If gravity is taken into account, a gravitational part has to be added to the Lagrangian, $L_{\mathrm{SPH}} = L_{\mathrm{SPH,h}} + L_{\mathrm{SPH,g}}$ with

$$L_{\mathrm{SPH,g}} = -\sum_b m_b \Phi_b, \tag{2.36}$$

where Φ_b is the potential at the particle position b, $\Phi(\boldsymbol{r}_b)$. The potential Φ can be written as a sum over particle contributions

$$\Phi(\boldsymbol{r}) = -G \sum_b m_b \phi(|\boldsymbol{r} - \boldsymbol{r}_b|, h), \tag{2.37}$$

and it is related to the matter density by Poisson's equation

$$\nabla^2 \Phi = 4\pi G \rho. \tag{2.38}$$

If we insert the sum representations of both the potential, Eq. (2.37), and the density, Eq. (2.7), into the Poisson equation, Eq. (2.38), we obtain a relationship between the gravitational softening kernel, ϕ, and the SPH-smoothing kernel W:

$$W(|\boldsymbol{r} - \boldsymbol{r}_b|, h) = -\frac{1}{4\pi} \frac{\partial}{\partial r} \left(r^2 \frac{\partial}{\partial r} \phi(|\boldsymbol{r} - \boldsymbol{r}_b|, h) \right). \tag{2.39}$$

Here we have used that both ϕ and W depend only radially on the position coordinate.

Applying the Euler-Lagrange equations, Eq. (2.23), to $L_{\mathrm{grav,g}}$ yields the particle acceleration due to gravity [43]

$$\begin{aligned} \frac{d\boldsymbol{v}_{a,\mathrm{g}}}{dt} &= -G \sum_b m_b \left[\frac{\phi'_{ab}(h_a) + \phi'_{ab}(h_b)}{2} \right] \hat{e}_{ab} \\ &\quad - \frac{G}{2} \sum_b m_b \left[\frac{\zeta_a}{\Omega_a} \nabla_a W_{ab}(h_a) + \frac{\zeta_b}{\Omega_b} \nabla_a W_{ab}(h_b) \right], \end{aligned} \tag{2.40}$$

where $\phi'_{ab} = \partial \phi / \partial |\boldsymbol{r}_a - \boldsymbol{r}_b|$. The first term in Eq. (2.40) is the gravitational force term usually used in SPH. The second term is due to gradients in the smoothing lengths and contains the quantities

$$\zeta_k \equiv \frac{\partial h_k}{\partial \rho_k} \sum_b m_b \frac{\partial \phi_{kb}(h_k)}{\partial h_k} \tag{2.41}$$

and the Ω_k defined in Eq. (2.29). Formally, it looks very similar to the pressure gradient terms in Eq. (2.35) with $G\zeta_k/2$ corresponding to P_k/ρ_k^2. As ζ_k is a negative definite quantity, these adaptive softening terms act against the gas pressure and therefore tend to increase the gravitational forces. The explicit forms of ϕ, ϕ' and $\partial\phi/\partial h$ for the cubic spline kernel an be found in Appendix A of [43].

2.2 Ideal magnetohydrodynamics

Magnetic fields pervade the Universe in substantial strengths on all scales [57]. They are observed in intra-cluster media in galaxy clusters [13] as well as in individual galaxies [69]. They are thought to be important for the birth of stars [30], they influence the life of stars e.g. via Sun spots or via controlling the angular momentum evolution during a stellar lifetime [21]. Stellar corpses such as neutron stars make themselves known via their magnetic field as pulsars, in a particular breed of neutron stars, so-called "magnetars" [68], the field reaches gigantic field strengths of the order $\sim 10^{15}$ Gauss. On the scale of planets, magnetic fields controle the magnetospheres that can shield the planet from the lethal cosmic rays, a fact that has certainly facilitated the evolution of life on our planet.

Basic equations of ideal MHD

Magnetohydrodynamics is a one-fluid model for a highly conducting plasma. It assumes that electromagnetic fields are highly coupled to the electron-ion component so that if the fields have a typical frequency ω and wave number k, they fulfill $\omega\tau_{\rm h} \sim 1$ and $k\lambda_{\rm h} \sim 1$, where $\tau_{\rm h}$ and $\lambda_{\rm h}$ are the typical hydrodynamic time and length scales. If

$$\frac{1}{\beta_{\rm plas}}\left(\frac{r_{\rm L_i}}{\lambda_{\rm h}}\right)^2 \ll \left(\frac{m_{\rm i}}{m_{\rm e}}\right)^{1/2}\left(\frac{\tau_{\rm i}}{\tau_{\rm h}}\right) \ll 1, \tag{2.42}$$

where $\beta_{\rm plas}$ is the ratio between gas and magnetic pressure, $r_{\rm L_i}$ the Larmor radius of the ions, $m_{\rm i}$ and $m_{\rm e}$ the ion and electron masses and $\tau_{\rm i}$ is the typical ion collision time, is fulfilled, the plasma can be described by the equations of ideal magnetohydrodynamics [8]:

$$\frac{d\rho}{dt} = -\rho\nabla\cdot\boldsymbol{v} \tag{2.43}$$

$$\frac{dv^i}{dt} = \frac{1}{\rho}\frac{\partial S^{ij}}{\partial x^j} \tag{2.44}$$

$$\frac{du}{dt} = -\frac{P}{\rho}\nabla\cdot\boldsymbol{v} \tag{2.45}$$

$$\frac{d\boldsymbol{B}}{dt} = -\boldsymbol{B}(\nabla\cdot\boldsymbol{v}) + (\boldsymbol{B}\cdot\nabla)\boldsymbol{v}, \tag{2.46}$$

where the magnetic stress tensor is given by

$$S^{ij} = -P\delta^{ij} + \frac{1}{\mu_0}\left(B^iB^j - \frac{1}{2}B^2\delta^{ij}\right) \tag{2.47}$$

and the B^k are the components of the magnetic field strength. Note that for ideal magnetohydrodynamics only the momentum equation has to be modified, the energy and the continuity equation are identical to the case of vanishing magnetic field. The form of the momentum equation employed here formally accounts for $\boldsymbol{B}(\nabla \cdot \boldsymbol{B})$ terms which are needed for momentum conservation in shocks but on the other hand can be the cause of numerical instabilities, see [47] for a detailed discussion.

Due to its relative simplicity in comparison to a more sophisticated plasma treatment magnetohydrodynamics and in particular ideal magnetohydrodynamics has been employed throughout a broad range of applications with sometimes not sufficient consideration about its range of applicability. Whether the conditions of applicabilty [8, 27] really hold needs to be checked for each problem individually.

Euler potentials

Being dissipationless the ideal MHD equations are conservative which leads to some important implications, the most powerful of which is probably the *frozen flux theorem* [2] which states that the magnetic field is carried around by the plasma. This kinematic effect is due to the evolution equation of the magnetic field, Eq. (2.46), and represents the conservation of magnetic flux through a fluid element. In reality, i.e. in the presence of dissipative terms, some slippage between the magnetic field and the plasma will occur.

The idea that the magnetic field lines are carried around by the flow is closely related to the concepts of Euler potentials [17] which are sometimes also referred to as Clebsch variables. For a review on Euler potentials we refer to [65, 66]. The basic idea is to present the magnetic field by two scalar variables, α and β such that

$$\boldsymbol{B} = \nabla\alpha \times \nabla\beta. \tag{2.48}$$

From this definition it is obvious that

$$\boldsymbol{B} \cdot \nabla\alpha = 0 = \boldsymbol{B} \cdot \nabla\beta, \tag{2.49}$$

in other words: α and β are constant along each field line and can therefore be used as field line labels. This is graphically represented in Fig. 2.2. The frozen flux property of ideal MHD the simply translates into advecting α and β with a Lagrangian fluid element:

$$\frac{d\alpha_a}{dt} = 0 \quad \text{and} \quad \frac{d\beta_a}{dt} = 0. \tag{2.50}$$

The Euler potentials naturally relate to the magnetic vector potential via

$$\boldsymbol{A} = \alpha\nabla\beta + \nabla\xi \tag{2.51}$$

or

$$\boldsymbol{A} = -\beta\nabla\alpha + \nabla\psi, \tag{2.52}$$

where ξ and ψ are arbitrary smooth functions. It is straightforward to check that both of the above forms of the vector potential yield the magnetic field via

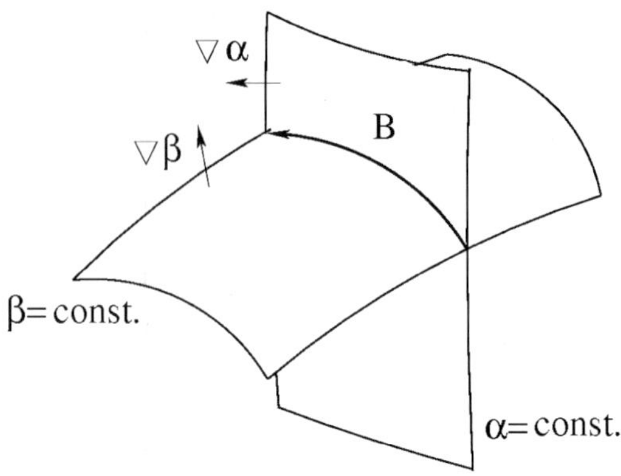

Figure 2.2. The intersection of a plane of constant α with a plane of constant β labels a magnetic field line.

$$\nabla \times \boldsymbol{A} = \nabla\alpha \times \nabla\beta = \boldsymbol{B}. \tag{2.53}$$

Thus, the $\nabla \cdot \boldsymbol{B} = 0$-constraint that is otherwise very hard to fulfill in a particle method [38, 42, 48] can be hard-wired into the numerical scheme by using the advected quantities α and β to construct the magnetic field via Eq. (2.48). This approach has the additional ease that, as long as the magnetic field is not strong enough to substantially influence the dynamics of the plasma, i.e. in the high-$\beta_{\rm plas}$-case, the evolution of different initial field configurations can be explored by just re-processing an existing simulation with different initial values for α and β. To find Euler potential pairs for a given magnetic field configuration is however usually a non-trivial task due to the non-linear nature of Eq. (2.48). The Euler potentials for a dipole field are known, but for more complicated field geometries its usually a challenge to find an analytical expression for the Euler potentials. There are however numerical procedures to find suitable pairs of Euler potentials, see e.g. [25, 40].

In two dimensions, a magnetic field can be represented by

$$\alpha = A_z \quad \beta = z, \tag{2.54}$$

where A_z is the z-component of a vector potential.

Limitations of the Euler potential approach

While the Euler potential approach makes some otherwise rather challenging problems such as magnetic field advection (see below) a trivial task, they have their own difficulties and limitations.

First, the Euler potentials for a given field configuration are not uniquely determined [65]. Assume, for example, that one particular set of Euler potentials, α_1 and β_1, is known. Then for a second set that is a function of the known ones, $\alpha_2 = \alpha_2(\alpha_1, \beta_1)$ and $\beta_2 = \beta_2(\alpha_1, \beta_1)$, one finds

$$\nabla\alpha_2 \times \nabla\beta_2 = \left(\frac{\partial\alpha_2}{\partial\alpha_1}\frac{\partial\beta_2}{\partial\beta_1} - \frac{\partial\beta_2}{\partial\alpha_1}\frac{\partial\alpha_2}{\partial\beta_1}\right)\nabla\alpha_1 \times \nabla\beta_1 \tag{2.55}$$

and therefore α_2 and β_2 will also be a set of Euler potentials for the same field as long as the term in brackets is equal to unity.

Second, by their very nature the Euler potentials are restricted to the purely non-disspative case and thus they are not immediately suited to treat the case of dissipative effects in a plasma.

Third, there are restrictions with respect to the magnetic field geometries that can be represented by Euler potentials. It is, for example, impossible to represent a linked ploidal and toroidal field. Nevertheless, as will be demonstrated in Sec. 3.3, on a large set of standard MHD-test problems the Euler potential approach yields excellent results.

From a numerical point of view they involve higher-order derivatives, see Eqs. (2.44) and (2.48) which is usually numerically challenging. However, as will be shown below, this not necessarily has to degrade the accuracy of the solution.

2.3 Dissipative terms

In both hydrodynamics and magnetohydrodynamics we are interested in principle in the non-dissipative cases, see Sects. 2.1 and 2.2. The corresponding equations, however, allow for discontinuous shock solutions which need to be captured in order to allow for a physically correct and numerically stable solution. This can be done by either making use of the analytical solution by locally solving a Riemann-type problem or by artificially spreading the discontinuities to a numerically resolvable width which means making them continuous. This latter artificial viscosity approach is most often used in the context of smooth particle hydrodynamics, although Riemann-solver-type approaches also do exist [11, 26].

A careful design of artificial dissipation terms is essential to capture physically correct solutions. This was recently demonstrated at the example of Kelvin-Helmholtz instabilities [45]. In the design of artificial dissipation terms we are guided by two principles: a) we want to use a form of the artificial dissipation equations that is oriented at Riemann-solvers [33] and b) we aim at applying dissipative terms only where they are necessary, i.e. near discontinuities, and follow in this respect Morris and Monaghan [37] who suggested to use time dependent dissipation parameters.

Based on the analogy with Riemann solvers Monaghan [33] presented a general formulation of dissipative terms. It was noted that the evolution equations of every conservative quantity should contain dissipative terms to controle discontinuities. This approach has been applied to ultra-relativistic [12] and magnetohydrodynamic shocks [48]. The "discontinuity capturing" term for a variable A is of the form

$$\left(\frac{dA}{dt}\right)_{a,\mathrm{diss}} = \sum_b m_b \frac{\alpha_A v_{\mathrm{sig},A}}{\rho_{ab}} (A_a - A_b) \hat{e}_{ab} \cdot \nabla W_{ab}, \qquad (2.56)$$

where α_A is a number of order unity that specifies the exact amount of dissipation, $v_{\mathrm{sig},A}$ is an appropriate signal velocity and ρ_{ab} the average mass density of particles a and b.

A comparison with the SPH expression for Laplacians [10] shows that the above equation is really an expression for [45]

$$\left(\frac{dA}{dt}\right)_{a,\mathrm{diss}} = \eta \nabla^2 A \qquad (2.57)$$

with $\eta \propto \alpha_A v_{\mathrm{sig}} |r_{ab}|$.

Following [37] the parameter that determines the exact values of the dissipative parameters, α_A, is made time-dependent. This is put into effect by integrating an additional differential equation containing both a source term, S_A, that indicates the necessity of artificial dissipation and a decay term that contains the typical time scale, τ_A, it takes a particle to pass the discontinuity. The evolution equation of the dissipation parameter is given by

$$\frac{d\alpha_{A,a}}{dt} = -\frac{\alpha_{A,a} - \alpha_{\mathrm{min}}}{\tau_{A,a}} + S_{A,a}, \qquad (2.58)$$

where α_{min} is the minimum value to which we allow α_A to decay. The decay time scale is given by

$$\tau_{A,a} = \frac{h_a}{C v_{\mathrm{sig},A}}, \qquad (2.59)$$

where C is a constant of order unity that is chosen after careful numerical experiments at problems with analytically known solutions.

3 The MAGMA code

Collisions between stars are very rare events in the solar neighbourhood. Close to centres of galaxies and globular clusters, however, the number densities of stars are higher by up to a factor of 10^6 [22] and therefore stellar collisions are very common events. In fact, the innermost 0.3 lightyears of our Galaxy can be considered an efficient "stellar collider" [1]. A different type of encounter can occur for stellar binary systems that contain compact stellar objects. If born at close enough separations such systems can be driven towards merger by the emission of gravitional waves. Although rare per space volume these types of encounters release tremendous amounts of gravitational energy and are therefore potentially visibly out to cosmological distances thereby making huge volumes observationally accessible and producing a substantial observational rate.

Some of the most exciting astrophysical objects are thought to form in such encounters and since both neutron stars and white dwarfs are known to be threaded by very large magnetic fields, a careful study of such mergers requires the inclusion of magnetic fields and their evolution.

3.1 Scope and physics modules

The acronym MAGMA stands for *a magnetohydrodynamics code for merger applications* and this code has originally been developed for the study of magnetized neutron stars [44, 49]. A very detailed description of this code can be found in [56].

For astrophysical studies the code contains several physics modules that go beyong the scope of this article and shall only be briefly sketched here. The interested reader is referred to the astrophysical literature.

Equation of state

For the thermodynamic properties of neutron star matter we use a temperature-dependent relativistic mean-field equation of state [59, 60]. It can handle temperatures from 0 to 100 MeV[6], electron fractions from $Y_e = 0$ (pure neutron matter) up to 0.56 and densities from about 10 to more than 10^{15} g cm^{-3}. No attempt is made to include matter constituents that are more exotic than neutrons and protons at high densities. For more details we refer to [50].

Neutrino emission

The code contains a detailed multi-flavor neutrino leakage scheme. An additional mesh is used to calculate the neutrino opacities that are needed for the neutrino emission rates at each particle position. The neutrino emission rates are used to account for the local cooling and the compositional changes due to weak interactions such as electron captures. A detailed description of the neutrino treatment can be found in [55].

Self-gravity

The self-gravity of the fluid is treated in a Newtonian fashion. Both the gravitational forces and the search for the particle neighbors are performed with a binary tree that is based on the one described in [7]. These tasks are the computationally most expensive part of the simulations and in practice they completely dominate the CPU-time usage. Forces emerging from the emission of gravitational waves are treated in a simple approximation. For more details, we refer to the literature [50, 51].

[6] 1 MeV corresponds to $1.16 \cdot 10^{10}$ K.

3.2 The MAGMA equations

Here, we will only briefly summarize the implemented equations, the explicit forms of all the equations can be found in [56].

Instead of explicitely integrating the continuity equation, we calculate the density via summation as in Eq. (2.7). The momentum equation is used in the form

$$\frac{d\boldsymbol{v}_{a,\mathrm{MHD}}}{dt} = \frac{d\boldsymbol{v}_{a,\mathrm{h}}}{dt} + \frac{d\boldsymbol{v}_{a,\mathrm{h,diss}}}{dt} + \frac{d\boldsymbol{v}_{a,\mathrm{g}}}{dt} + \frac{d\boldsymbol{v}_{a,\mathrm{mag}}}{dt} + \frac{d\boldsymbol{v}_{a,\mathrm{mag,diss}}}{dt} \tag{3.60}$$

where $d(\boldsymbol{v}_{a,\mathrm{h}})/dt$ is given in Eq. (2.35), $d(\boldsymbol{v}_{a,\mathrm{g}})/dt$ is given in Eq. (2.40), and the explicit forms of the dissipative terms,$d(\boldsymbol{v}_{a,\mathrm{h,diss}})/dt$ and $d(\boldsymbol{v}_{a,\mathrm{mag,diss}})/dt$, can be found in [56]. The magnetic force term is used in the form

$$\begin{aligned}\frac{d\boldsymbol{v}_{a,\mathrm{mag}}}{dt} = & -\sum_b \frac{m_b}{\mu_0}\left\{\frac{B_a^2/2}{\Omega_a\rho_a^2}\nabla_a W_{ab}(h_a) + \frac{B_b^2/2}{\Omega_b\rho_b^2}\nabla_a W_{ab}(h_b)\right\} \\ & +\sum_b \frac{m_b}{\mu_0}\left\{\frac{\boldsymbol{B}_a(\boldsymbol{B}_a\cdot\overline{\nabla_a W_{ab}}) - \boldsymbol{B}_b(\boldsymbol{B}_b\cdot\overline{\nabla_a W_{ab}})}{\rho_a\rho_b}\right\},\end{aligned} \tag{3.61}$$

where the symmetrized kernel gradient is given by

$$\overline{\nabla_a W_{ab}} = \frac{1}{2}\left[\frac{1}{\Omega_a}\nabla_a W_{ab}(h_a) + \frac{1}{\Omega_b}\nabla_a W_{ab}(h_b)\right]. \tag{3.62}$$

The magnetic field is calculated from the Euler potentials[7]. Note that another form of the magnetic force term is also possible [48, 56]. The gradients of the Euler potentials are calculated in a way that gradients of linear functions are reproduced exactly [42, 56]. To handle magnetic shocks artificial dissipation terms were constructed according to the ideas outlined in Sect. 2.3. They are also applied to the evolution of α_a and β_a. They are not meant to mimic physical dissipation in any way, their exclusive aim is to keep gradients numerically treatable.

The MAGMA energy equation is of the form

$$\frac{du_{a,\mathrm{MHD}}}{dt} = \frac{du_{a,\mathrm{h}}}{dt} + \frac{du_{a,AV}}{dt} + \frac{du_{a,C}}{dt}, \tag{3.63}$$

where $d(u_{a,\mathrm{h}})/dt$ is given in Eq. (2.32), the explicit form of the artificial viscosity term $d(u_{a,AV})/dt$ and the thermal conductivity term $d(u_{a,C})/dt$ can be found in [56].

3.3 Tests and benchmarks

We present here a selection of standard tests used in the hydro- and magnetohydrodynamics community to validate numerical schemes. For a more exhaustive set of benchmarks we refer to [56].

[7] Note that the code also allows to evolve magnetic fields according to a more straightforward SPH discretisation [48].

Hydrodynamics

1D: Sod's shock tube

As a standard test of the shock capturing capability we show the results of Sod's shock tube test [61]. To the left of the origin, the initial state of the fluid is given by $[\rho, P, v_x]_L = [1.0,1.0,0.0]$ whilst to the right of the origin the initial state is $[\rho, P, v_x]_R = [0.125,0.1,0.0]$ with $\gamma = 1.4$. The problem is setup using 900 equal mass particles in one spatial dimension. Rather than adopting the usual practice of smoothing the initial conditions across the discontinuity, we follow [42] in using unsmoothed initial conditions but applying a small amount of artificial thermal conductivity. The results are shown in Figure 3.3, where the points represent the SPH particles. For comparison the exact solution computed using a Riemann solver is given by the solid line.

The shock itself is smoothed by the artificial viscosity term, which in this case can be seen to spread the discontinuity over about 6 particles. The contact discontinuity is smoothed by the application of artificial thermal conductivity which (in particular) eliminates the "wall heating" effect often visible in numerical solutions to this problem. The exact distribution of particle sepa-

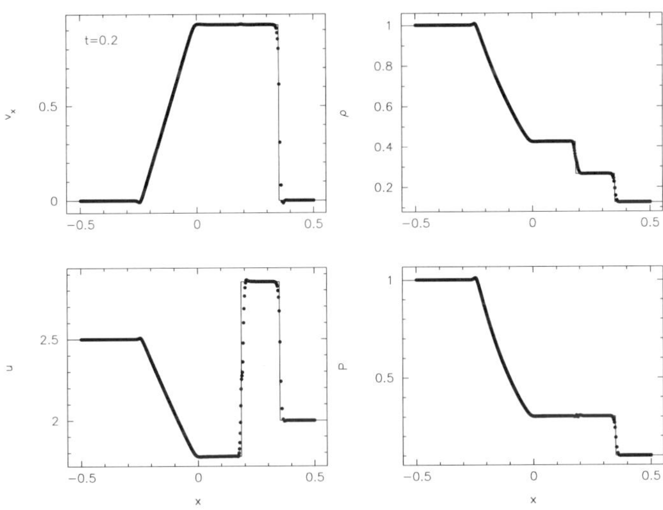

Figure 3.3. Results of the Sod shock tube test in one dimension using 900 SPH particles setup using unsmoothed initial conditions. Artificial viscosity and thermal conductivity are applied to appropriately smooth the shock and contact discontinuity respectively. The exact solution is given by the solid line. The upper row displays the velocity (left) and the density (right), the bottom row shows specific internal energy (left) and the pressure (right).

rations in the contact discontinuity seen in Figure 3.3 is related to the initial particle placement across the discontinuity.

For this test, applying artificial viscosity and thermal conductivity as described, we do not find a large difference between the "grad-h" formulation and other variants of SPH based on averages of the smoothing length. If anything, the "grad-h"-terms tend to increase the order of the method, which, as in any higher order scheme, tends to enhance oscillations which may otherwise be damped, visible in Figure 3.3 as small "bumps" at the head of the rarefaction wave (in the absence of artificial viscosity these bumps appear as small but regular oscillations with a wavelength of a few particle spacings).

3D: Sedov blast wave test

In order to demonstrate that our scheme is capable of handling strong shocks in three dimensions, we have also tested the code on a Sedov blast wave problem both with, see Sect. 3.3, and without magnetic fields. Without magnetic fields the explosion is spherically symmetric, however for a strong magnetic field the blast wave is significantly inhibited perpendicular to the magnetic field lines, resulting in a compression along one spatial dimension. Similar tests for both hydrodynamics and MHD have been used by many authors – for example by [4] in order to benchmark an Adaptive Mesh Refinement

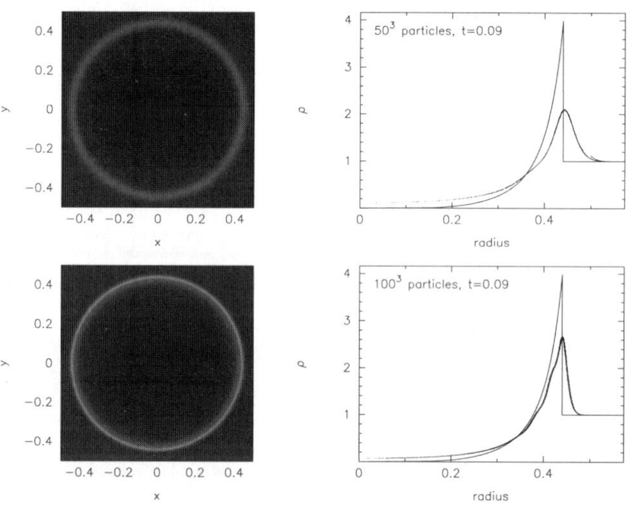

Figure 3.4. Results of the hydrodynamic Sedov blast wave test in 3D at $t = 0.09$ at resolutions of 125,000 (top) and 1 million (bottom) particles respectively. The density and radial position of each SPH particle are shown in each case, which may be compared to the exact solution given by the solid line. (*See also* Color Plate on page 390)

(AMR) code for MHD and by [63] in benchmarking the cosmological SPH code GADGET.

The hydrodynamic version is set up as follows: The particles are placed in a cubic lattice configuration in a three dimensional domain $[-0.5, 0.5] \times [-0.5, 0.5] \times [-0.5, 0.5]$ with uniform density $\rho = 1$ and zero pressure and temperature apart from a small region $r < R$ near the origin, where we initialize the pressure using the total blast wave energy $E = 1$, ie. $P = (\gamma-1)E/(\frac{4}{3}\pi R^3)$. We set the initial blast radius to the size of a single particle's smoothing sphere $R = 2\eta \Delta x$ (where 2 is the kernel radius, $\eta(= 1.5)$ is the smoothing length in units of the average particle spacing as in Eq. (2.28) and Δx is the initial particle spacing on the cubic lattice) such that the explosion is as close to point-like as resolution allows. Boundaries are not important for this problem, however we use periodic boundary conditions to ensure that the particle distribution remains smooth at the edges of the domain.

The results shown in Figure 3.4 at $t = 0.09$ have been obtained with a resolution of 50 and 100 particles^3 (ie. 125,000 and 1 million particles respectively), where we have plotted (left panels) the density in a $z = 0$ cross section slice and (right panels) the density and radial position of each particle (dots) together with the exact self-similar Sedov solution (solid line).

We found that the key to an accurate simulation of this problem in SPH is to incorporate an artificial thermal conductivity term due to the huge initial discontinuity in thermal energy. The importance of such a term for shock problems in SPH has been discussed recently by [42, 45]. In the absence of this term the particle distribution quickly becomes disordered around the shock front and the radial profile appears to be noisy. From Figure 3.4 we see that at a resolution of 1 million particles the highest density in the shock at $t = 0.09$ is $\rho_{\max} = 2.67$ whereas for the lower resolution run $\rho_{\max} = 2.1$, consistent with a factor of 2 change in smoothing length. Using this we can estimate that a resolution of $\sim 345^3 = 41$ million particles is required to fully resolve the density jump in this problem in three dimensions. Note that the minimum density obtained in the post-shock rarefaction also decreases with resolution. Some small-amplitude post-shock oscillations are visible in the solution which we attribute to interaction of the spherical blast wave with particles in the surrounding medium initially placed on a regular (Cartesian) cubic lattice.

Magnetohydrodynamics

1D: Brio-Wu shock tube test

The magnetic shock tube test of [9] has become a standard test case for numerical MHD schemes that has been widely used by many authors to benchmark (mainly grid-based) MHD codes [3, 15, 58, 67]. The Brio-Wu shock test is the MHD analogon to Sod's shock tube problem that was described earlier, but here no analytical solution is known. The MHD Riemann problem allows for much more complex solutions than the hydrodynamic case which can occur because of the three different types of waves (i.e. slow, fast and Alfvén, compared to just the sound waves in hydrodynamics). In the Brio-Wu shock test the solution contains the following components (from left to right in Fig. 3.5): a fast rarefaction fan and a slow compound wave consisting of a slow rarefaction attached to a slow shock (moving to the left) and a contact discontinuity, a slow shock and a fast rarefaction fan (moving to the right). It has been pointed out, however, that the stability of the unusual compound wave may be an artifact of the restriction of the symmetry to one spatial dimension whilst allowing the magnetic field to vary in two dimensions, [5].

The shown results are obtained using Euler potential formulation. Results of this problem using Smoothed Particle Magnetohydrodynamics (SPMHD) have been presented elsewhere [42, 47]. The Euler potentials show a distinct improvement over the standard SPMHD results. The initial conditions on the left side of the discontinuity are $[\rho, P, v_x, v_y, B_y]_{\rm L} = [1, 1, 0, 0, 1]$ and $[\rho, P, v_x, v_y, B_y]_{\rm R} = [0.125, 0.1, 0, 0, -1]$ on the right side. The $x-$component of the magnetic field is $B_x = 0.75$ everywhere and a polytropic exponent of $\gamma = 2.0$ is used. Using the Euler potentials the components are given by $\alpha = -B_y x$ (equivalent to the vector potential A_z) and $\beta = z$ (or more specifically $\nabla\beta = \hat{\mathbf{z}}$) and the B_x component is treated as an external field which requires adding a source term to the evolution equation for α. Particles are restricted to move in one spatial dimension only, whilst the magnetic field is allowed to vary in two dimensions (that is, we compute a v_y but do not use it to move the particles).

We setup the problem using 631 equal mass particles in the domain $x \in [-0.5, 0.5]$ using, as in the hydrodynamic case, purely discontinuous initial conditions. Artificial viscosity, thermal conductivity and resistivity are applied. The results are shown at $t = 0.1$ in Figure 3.5. For comparison the numerical solution from [3] is given by the solid line (no exact solution exists for this problem). The solution is generally well captured by our numerical scheme. Two small defects are worth noting. The first is that a small offset is visible in the thermal energy – this is a result of the small non-conservation introduced by use of the Morris formulation [38] of the magnetic force, Eq. (3.61). Secondly, the rightmost discontinuity is somewhat over-smoothed by the artificial resistivity term. We attribute this to the fact that the dissipative terms involve simply the maximum signal velocity v_{sig} (that is the maximum of all the wave types). Ideally each discontinuity should be smoothed taking account of it's individual characteristic and corresponding v_{sig} (as would occur in a Godunov-MHD scheme). Increasing the total number of particles also decreases the smoothing applied to this wave.

2D: Current loop advection problem

A simple test problem for MHD is to compute the advection of a weak magnetic field loop. This test, introduced by [18] in the development of the *Athena*

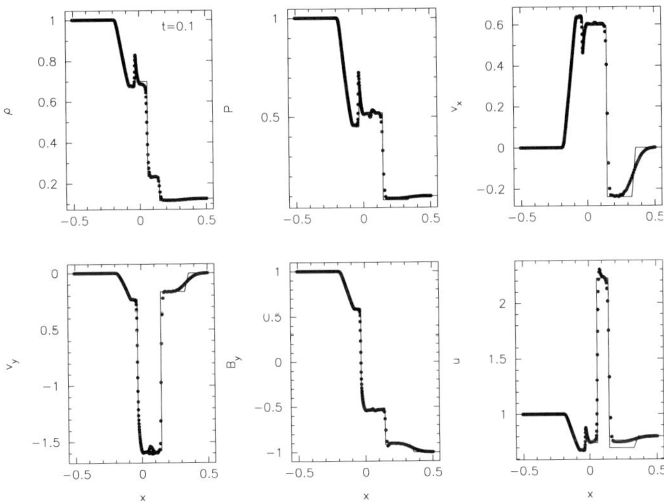

Figure 3.5. Results of the Brio & Wu MHD shock tube test at $t = 0.1$ using 631 particles and the Euler potential formulation. For comparison the numerical solution taken from [3] is given by the solid line. The solution illustrates the complex shock structures which can be formed due to the different wave types in MHD, including in this case a compound wave consisting of a slow shock attached to a rarefaction wave.

MHD code[8], presents a challenging problem for grid-based MHD schemes requiring careful formulation of the advection terms in the MHD equations. For our Lagrangian scheme, this test is straightforward to solve which strongly highlights the advantage of using a particle method for MHD in problems where there is significant motion with respect to a fixed reference frame.

We setup the problem following [18]: the computational domain is two dimensional with $x \in [-1, 1]$, $y \in [-0.5, 0.5]$ using periodic boundary conditions. Density and pressure are uniform with $\rho = 1$ and $P = 1$. The particles are laid down in a cubic lattice configuration with velocity initialized according to $\mathbf{v} = (v_0 \cos\theta, v_0 \sin\theta)$ with $\cos\theta = 2/\sqrt{5}$, $\sin\theta = 1/\sqrt{5}$ and $v_0 = 1$ such that by $t = 1$ the field loop will have been advected around the computational domain once. The magnetic field is two dimensional, initialized using a vector potential given by

$$A_z = \begin{cases} A_0(R - r) & r \leq R, \\ 0 & r > R, \end{cases} \tag{3.64}$$

where $A_0 = 10^{-3}$, $R = 0.3$ and $r = \sqrt{x^2 + y^2}$. The ratio of thermal to magnetic pressure is thus given by $\beta_{\rm plas} = P/(\frac{1}{2}B^2) = 2 \times 10^6$ (for $r < R$) such that the magnetic field is passively advected. [18] show the results of this problem after two crossings of the computational domain, by which time the

[8] http://www.astro.princeton.edu/~jstone/athena.html

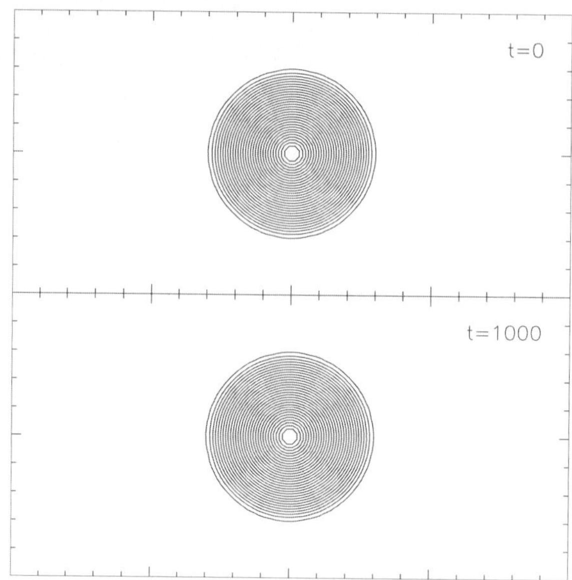

Figure 3.6. Magnetic field lines in the current loop advection test, plotted at $t = 0$ (top) and after 1000 crossings of the computational domain (bottom).

loop has either been significantly diffused or has disintegrated into oscillations depending on details of their particular choice of scheme. The advantages of a Lagrangian scheme are that advection is computed exactly, and using our Euler potential formulation for the magnetic field (which in two dimensions is equivalent to a vector potential formulation with $\alpha = A_z$ and $\beta = z$), this is also true for the evolution of the magnetic field. The result is that the field loop is advected *without change* by our code for as long as one may care to compute it. This is demonstrated in Fig. 3.6 which shows the magnetic field lines at $t = 0$ (top) and after 1000 (!) crossings of the computational domain (bottom), in which the field configuration can be seen to be identical to the top figure. The magnetic energy (not shown) is also maintained exactly, whereas [18] find of order a 10% reduction in magnetic energy after two crossings of the domain.

In a realistic simulation involving MHD shocks there will be some diffusion of the magnetic field introduced by the addition of artificial diffusion terms, which are required to resolve discontinuities in the magnetic field. However the point is that these terms are explicitly added to the SPH calculation and can be turned off where they are not necessary whereas the diffusion present in a grid-based code is intrinsic and always present.

2D: Orszag-Tang test

The evolution of the compressible Orszag-Tang vortex system [39] involves the interaction of several shock waves traveling at different speeds. Originally studied in the context of incompressible MHD turbulence, it has later been extended to the compressible case [14,41]. It is generally considered a good test to validate the robustness of numerical MHD schemes. In the SPH context, this test has been discussed in detail by [42] and [48].

The problem is two dimensional with periodic boundary conditions on the domain $[0,1] \times [0,1]$. The setup consists of an initially uniform state perturbed by periodic vortices in the velocity field, which, combined with a doubly periodic field geometry, results in a complex interaction between the shocks and the magnetic field.

The velocity field is given by $\boldsymbol{v} = v_0[-\sin(2\pi y), \sin(2\pi x)]$ where $v_0 = 1$. The magnetic field is given by $\boldsymbol{B} = B_0[-\sin(2\pi y), \sin(4\pi x)]$ where $B_0 = 1/\sqrt{4\pi}$. Using the Euler potentials this corresponds to $\alpha \equiv A_z = B_0/(2\pi)[\cos(2\pi y) + \frac{1}{2}\cos(4\pi x)]$. The flow has an initial average Mach number of unity, a ratio of magnetic to thermal pressure of 10/3 and we use a polytropic exponent $\gamma = 5/3$. The initial gas state is therefore $P = 5/3B_0^2 = 5/(12\pi)$ and $\rho = \gamma P/v_0 = 25/(36\pi)$. Note that the choice of length and time scales differs slightly between various implementations in the literature. The setup used above follows that of [58] and [28].

We compute the problem using 512×590 particles initially placed on a uniform, close-packed lattice. The density at $t = 0.5$ is shown in Figure 3.7 using both the SPMHD formalism of [48] (left), and the Euler potential approach (right) outlined in Sect. 2.2. The Euler potential formulation is clearly superior to the standard SPMHD method. This is largely a result of the relative

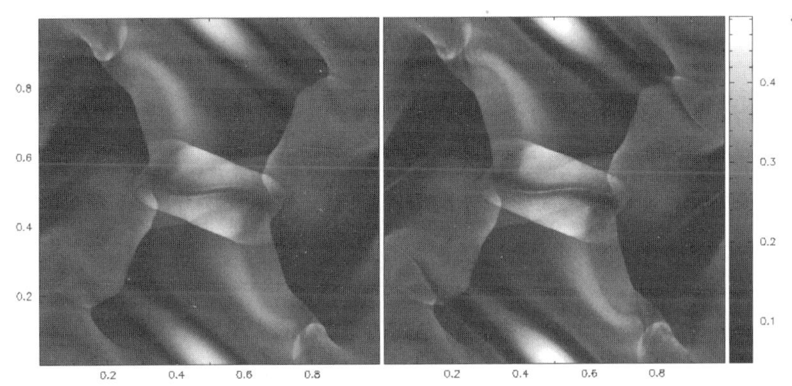

Figure 3.7. Density distribution in the two dimensional Orzsag-Tang vortex problem at $t = 0.5$. The initial vortices in the velocity field combined with a doubly periodic field geometry lead to a complex interaction between propagating shocks and the magnetic field. Results are shown using 512×590 particles using a SPMHD formalism of [48] (left) and using the Euler potentials (right). The reduced artificial resistivity required in the Euler potential formalism leads to a much improved effective resolution. (*See also* Color Plate on page 391)

requirements for artificial resistivity in each case. In the standard SPMHD method the application of artificial resistivity is crucial for this problem (that is, in the absence of artificial resistivity the density and magnetic field distributions are significantly in error). Using the Euler potentials we find that the solution can be computed using zero artificial resistivity, relying only on the "implicit smoothing" present in the computation of the magnetic field using SPH operators. This means that topological features in the magnetic field are much better preserved, which is reflected in the density distribution. For example the filament near the center of the figure is well resolved using the Euler potentials but completely washed out by the artificial resistivity in the standard SPMHD formalism. Also the high density features near the top and bottom of the figure (coincident to a reversal in the magnetic field) are much better resolved using the Euler potentials.

3D: MHD blast wave

The MHD version of the Sedov test is identical to the hydrodynamic test with the addition of a uniform magnetic field in the $x-$direction, that is $\mathbf{B} = [B_0, 0, 0]$ with $B_0 = 3.0$. Initially the surrounding material has zero thermal pressure, meaning that the plasma $\beta_{\rm plas}$ is zero (ie. magnetic pressure infinitely strong compared to thermal pressure). However, this choice of field strength gives a mean plasma $\beta_{\rm plas}$ in the post-shock material of $\beta_{\rm plas} \sim 1.3$, such that the magnetic pressure plays an equal or dominant role in the evolution of the shock. The results of this problem at $t = 0.05$ are shown in Fig. 3.8,

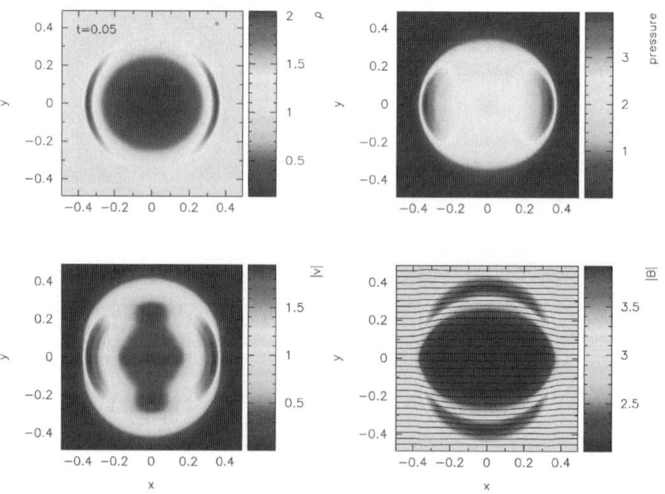

Figure 3.8. Results of the 3D MHD blast wave test at $t = 0.05$ at a resolution of 1 million (100^3) particles. Plots show (left to right, top to bottom) density, pressure, magnitude of velocity and magnetic field strength (with overlaid field lines), plotted in a cross-section slice through the $z = 0$ plane. (*See also* Color Plate on page 391)

where plots show density, pressure, magnitude of velocity and magnetic field strength in a cross section slice taken at $z = 0$. In addition the magnetic field lines are plotted on the magnetic field strength plot.

In this strong-field regime, the magnetic field lines are not significantly bent by the propagating blast wave but rather strongly constrain the blast wave into an oblate spheroidal shape. The density (and likewise pressure) enhancement in the shock is significantly reduced in the $y-$direction (left and top right panels) due to the additional pressure provided by the magnetic field which is compressed in this direction (bottom right panel).

4 Summary and conclusion

We have outlined several recent developments in smooth particle hydrodynamics. The equations of self-gravitating, ideal hydrodynamics were derived explicitely from a Lagrangian thereby yielding the correct particle index symmetries that ensure that the physical conservation laws are hard-wired into the discrete set of SPH equations without any arbitrariness. We have further described the implementation of ideal MHD via so-called Euler potentials. This approach enforces the crucial $\nabla \cdot \boldsymbol{B} = 0$-constraint by construction. All dissipative terms required to capture discontinuities were carefully designed so that they a) have a form suggested in analogy with Riemann-solvers and b) are only active near discontinuities. These principles are implemented in our three-dimensional, Lagrangian magnetohydrodynamics code MAGMA. In a large set of standard test problems that is often used to benchmark numerical (magneto-)hydrodynamics schemes we have demonstrated the excellent performance of the code.

Acknowledgement

The simulations presented here were in part performed on the JUMP computer of the Höchstleistungsrechenzentrum Jülich. DJP is supported by a UK Royal Society University Research Fellowship though much of this work has been funded by a PPARC/STFC postdoctoral fellowship. Some of the results were visualized using SPLASH [46], a publicly available visualisation tool for SPH.

References

1. T. Alexander, *Stellar processes near the massive black hole in the Galactic center*, Phys. Rep., 419 (2005), pp. 65–142.
2. H. Alfven, *Cosmical Electrodynamics*, Oxford University Press, Oxford, 1951.

3. D. S. BALSARA, *Total Variation Diminishing Scheme for Adiabatic and Isothermal Magnetohydrodynamics*, ApJS, 116 (1998), pp. 133–+.
4. D. S. BALSARA, *Divergence-Free Adaptive Mesh Refinement for Magnetohydrodynamics*, J. Comp. Phys., 174 (2001), pp. 614–648.
5. A. A. BARMIN, A. G. KULIKOVSKIY, AND N. V. POGORELOV, *Shock-Capturing Approach and Nonevolutionary Solutions in Magnetohydrodynamics*, J. Comp. Phys., 126 (1996), pp. 77–90.
6. W. BENZ, *Smooth particle hydrodynamics: A review*, in Numerical Modeling of Stellar Pulsations, J. Buchler, ed., Kluwer Academic Publishers, Dordrecht, 1990, p. 269.
7. W. BENZ, R. BOWERS, A. CAMERON, AND W. PRESS, *Dynamic mass exchange in doubly degenerate binaries. i - 0.9 and 1.2 solar mass stars*, ApJ, 348 (1990), p. 647.
8. T. BOYD AND J. SANDERSON, *The Physics of Plasmas*, Cambridge University Press, Cambridge, 2003.
9. M. BRIO AND C. C. WU, *An upwind differencing scheme for the equations of ideal magnetohydrodynamics*, Journal of Computational Physics, 75 (1988), pp. 400–422.
10. L. BROOKSHAW, *A method of calculating radiative heat diffusion in particle simulations*, Proceedings of the Astronomical Society of Australia, 6 (1985), pp. 207–210.
11. S.-H. CHA AND A. P. WHITWORTH, *Implementations and tests of Godunov-type particle hydrodynamics*, MNRAS, 340 (2003), pp. 73–90.
12. J. E. CHOW AND J. MONAGHAN, *Ultrarelativistic sph*, J. Computat. Phys., 134 (1997), p. 296.
13. T. E. CLARKE, P. P. KRONBERG, AND H. BÖHRINGER, *A New Radio-X-Ray Probe of Galaxy Cluster Magnetic Fields*, ApJL, 547 (2001), pp. L111–L114.
14. R. B. DAHLBURG AND J. M. PICONE, *Evolution of the Orszag-Tang vortex system in a compressible medium. I - Initial average subsonic flow*, Physics of Fluids B, 1 (1989), pp. 2153–2171.
15. W. DAI AND P. R. WOODWARD, *Extension of the Piecewise Parabolic Method to Multidimensional Ideal Magnetohydrodynamics*, J. Comp. Phys., 115 (1994), pp. 485–514.
16. C. ECKART, *Variation principles of hydrodynamics*, Physics of Fluids, 3 (1960), p. 421.
17. L. EULER, *De curva hypergeometrica hac aequatione expressa y=...*, Novi Commentarrii Acad. Sci. Petropolitanae, 14 (1769), p. 270.
18. T. A. GARDINER AND J. M. STONE, *An unsplit Godunov method for ideal MHD via constrained transport*, J. Comp. Phys., 205 (2005), pp. 509–539.
19. R. A. GINGOLD AND J. J. MONAGHAN, *Smoothed particle hydrodynamics - Theory and application to non-spherical stars*, MNRAS, 181 (1977), pp. 375–389.
20. S. GOTTLOEBER, G. YEPES, C. WAGNER, AND R. SEVILLA, *The MareNostrum Universe*, ArXiv Astrophysics e-prints, (2006).
21. A. HEGER, S. E. WOOSLEY, AND H. C. SPRUIT, *Presupernova Evolution of Differentially Rotating Massive Stars Including Magnetic Fields*, ApJ, 626 (2005), pp. 350–363.
22. D. HEGGIE AND P. HUT, *The Gravitational Million-Body Problem: A Multidisciplinary Approach to Star Cluster Dynamics*, The Gravitational Million-Body

Problem: A Multidisciplinary Approach to Star Cluster Dynamics, by Douglas Heggie and Piet Hut. Cambridge University Press, 2003, 372 pp., Feb. 2003.

23. L. Hernquist and N. Katz, *Treesph - a unification of sph with the hierarchical tree method*, ApJS, 70 (1989), p. 419.
24. W. R. Hix, A. M. Khokhlov, J. C. Wheeler, and F.-K. Thielemann, *The Quasi-Equilibrium-reduced alpha -Network*, ApJ, 503 (1998), pp. 332–+.
25. C. Ho, T. Huang, and S. Gao, *Contributions to the high-degree multipoles of neptunes magnetic field*, J. Geophys. Res., 102 (1997), p. 393.
26. S.-I. Inutsuka, *Reformulation of Smoothed Particle Hydrodynamics with Riemann Solver*, Journal of Computational Physics, 179 (2002), pp. 238–267.
27. J. Jackson, *Classical Electrodynamics*, Wiley, New York, 3. ed., 1998.
28. P. Londrillo and L. Del Zanna, *High-Order Upwind Schemes for Multidimensional Magnetohydrodynamics*, ApJ, 530 (2000), pp. 508–524.
29. L. Lucy, *A numerical approach to the testing of the fission hypothesis*, The Astronomical Journal, 82 (1977), p. 1013.
30. M.-M. Mac Low and R. S. Klessen, *Control of star formation by supersonic turbulence*, Reviews of Modern Physics, 76 (2004), pp. 125–194.
31. J. Monaghan and J. Lattanzio, *A refined particle method for astrophysical problems*, A&A, 149 (1985), p. 135.
32. J. J. Monaghan, *Smoothed particle hydrodynamics*, Ann. Rev. Astron. Astrophys., 30 (1992), p. 543.
33. J. J. Monaghan, *SPH and Riemann Solvers*, Journal of Computational Physics, 136 (1997), pp. 298–307.
34. J. J. Monaghan, *SPH compressible turbulence*, MNRAS, 335 (2002), pp. 843–852.
35. J. J. Monaghan, *Smoothed particle hydrodynamics*, Reports of Progress in Physics, 68 (2005), pp. 1703–1759.
36. J. J. Monaghan and D. J. Price, *Variational principles for relativistic smoothed particle hydrodynamics*, MNRAS, 328 (2001), pp. 381–392.
37. J. Morris and J. Monaghan, *A switch to reduce sph viscosity*, J. Comp. Phys., 136 (1997), p. 41.
38. J. P. Morris, *Analysis of smoothed particle hydrodynamics with applications*, PhD thesis, Monash University, Melbourne, Australia, 1996.
39. S. Orszag and C. Tang, *Small-scale structure in of two-dimensional magnetohydrodynamic turbulence*, Journ. Fluid Mech., 90 (1979), p. 129.
40. C. Peymirat and D. Fontaine, *A numerical method to compute euler potentials*, Ann. Geophysicae, 17 (1999), p. 328.
41. J. M. Picone and R. B. Dahlburg, *Evolution of the Orszag-Tang vortex system in a compressible medium. II - Supersonic flow*, Physics of Fluids B, 3 (1991), pp. 29–44.
42. D. Price, *Magnetic Fields in Astrophysics*, PhD thesis, University of Cambridge, arXiv:astro-ph/0507472, 2004.
43. D. Price and J. Monaghan, *An energy-conserving formalism for adaptive gravitational force softening in sph and n-body codes*, MNRAS, 374 (2007), p. 1347.
44. D. Price and S. Rosswog, *Producing ultra-strong magnetic fields in neutron star mergers*, Science, 312 (2006), p. 719.

45. D. J. Price, *Modelling discontinuities and Kelvin-Helmholtz instabilities in SPH*, ArXiv e-prints, 709 (2007).
46. D. J. Price, *splash: An Interactive Visualisation Tool for Smoothed Particle Hydrodynamics Simulations*, Publications of the Astronomical Society of Australia, 24 (2007), pp. 159–173.
47. D. J. Price and J. J. Monaghan, *Smoothed Particle Magnetohydrodynamics - I. Algorithm and tests in one dimension*, MNRAS, 348 (2004), pp. 123–138.
48. D. J. Price,, *Smoothed Particle Magnetohydrodynamics - III. Multidimensional tests and the* $\nabla.B = 0$ *constraint*, MNRAS, 364 (2005), pp. 384–406.
49. S. Rosswog, *Last moments in the life of a compact binary*, Rev. Mex. Astron. Astrophys., 27 (2007), pp. 57–79.
50. S. Rosswog and M. B. Davies, *High-resolution calculations of merging neutron stars - I. Model description and hydrodynamic evolution*, MNRAS, 334 (2002), pp. 481–497.
51. S. Rosswog, M. B. Davies, F.-K. Thielemann, and T. Piran, *Merging neutron stars: asymmetric systems*, A&A, 360 (2000), pp. 171–184.
52. S. Rosswog, E.Ramirez-Ruiz, and R. Hix, *Atypical thermonuclear supernovae from tidally crushed white dwarfs*, ApJ, (2008).
53. S. Rosswog, E.Ramirez-Ruiz, and R. Hix, *Simulating black hole white dwarf encounters*, Comp. Phys. Comm, (2008).
54. S. Rosswog, E.Ramirez-Ruiz, and R. Hix, *Tidal disruption and ignition of white dwarfs by intermediate-mass black holes*, in prep., (2008).
55. S. Rosswog and M. Liebendörfer, *High-resolution calculations of merging neutron stars - II. Neutrino emission*, MNRAS, 342 (2003), pp. 673–689.
56. S. Rosswog and D. Price, *Magma: a magnetohydrodynamics code for merger applications*, MNRAS, 379 (2007), pp. 915 – 931.
57. G. Rüdiger and R. Hollerbach, *The magnetic universe : geophysical and astrophysical dynamo theory*, The Magnetic Universe: Geophysical and Astrophysical Dynamo Theory, by Günther Rüdiger, Rainer Hollerbach, pp. 343. ISBN 3-527-40409-0. Wiley-VCH , August 2004., Aug. 2004.
58. D. Ryu and T. W. Jones, *Numerical magetohydrodynamics in astrophysics: Algorithm and tests for one-dimensional flow'*, ApJ, 442 (1995), pp. 228–258.
59. H. Shen, H. Toki, K. Oyamatsu, and K. Sumiyoshi, *Relativistic equation of state of nuclear matter for supernova and neutron star.*, Nuclear Physics, A 637 (1998), p. 435.
60. H. Shen, H. Toki, K. Oyamatsu, and K. Sumiyoshi, *Relativistic equation of state of nuclear matter for supernova explosion*, Progress of Theoretical Physics, 100 (1998), pp. 1013–1031.
61. G. Sod, *A survey of several finite difference methods for systems of nonlinear hyperbolic conservation laws*, J. Comput. Phys., 43 (1978), pp. 1–31.
62. R. Speith, *Untersuchung von Smoothed Particle Hydrodynamics anhand astrophysikalischer Beispiele*, PhD thesis, Eberhard-Karls-Universität Tübingen, 1998.
63. V. Springel, *The cosmological simulation code GADGET-2*, MNRAS, 364 (2005), pp. 1105–1134.
64. V. Springel and L. Hernquist, *Cosmological smoothed particle hydrodynamics simulations: the entropy equation*, MNRAS, 333 (2002), pp. 649–664.
65. D. Stern, *Euler potentials*, American Journal of Physics, 38 (1970), p. 494.

66. D. P. Stern, *The Motion of Magnetic Field Lines*, Space Science Reviews, 6 (1966), p. 147.
67. J. M. Stone, J. F. Hawley, C. R. Evans, and M. L. Norman, *A test suite for magnetohydrodynamical simulations*, ApJ, 388 (1992), pp. 415–437.
68. C. Thompson and R. C. Duncan, *Neutron star dynamos and the origins of pulsar magnetism*, ApJ, 408 (1993), pp. 194–217.
69. L. M. Widrow, *Origin of galactic and extragalactic magnetic fields*, Reviews of Modern Physics, 74 (2002), pp. 775–823.

A Particle-Partition of Unity Method Part VIII: Hierarchical Enrichment

Marc Alexander Schweitzer

Institut für Numerische Simulation, Universität Bonn, Wegelerstr. 6, D-53115 Bonn, Germany
schweitzer@ins.uni-bonn.de

Summary. This paper is concerned with automatic enrichment in the particle-partition of unity method (PPUM). The goal of our automatic enrichment is to recover the optimal convergence rate of the uniform h-version independent of the regularity of the solution. Hence, we employ enrichment not only for modeling purposes but rather to improve the approximation properties of the numerical scheme. To this end we enrich our PPUM function space in an automatically determined enrichment zone hierarchically near the singularities of the solution. To overcome the ill-conditioning of the enriched shape functions we present an appropriate local preconditioner. The results of our numerical experiments clearly show that the hierarchically enriched PPUM recovers the optimal convergence rate globally and even shows a kind of superconvergence within the enrichment zone. The condition number of the stiffness matrix is independent of the employed enrichment and the relative size of the enrichment zone.

Key words: meshfree method, partition of unity method, extrinsic enrichment, preconditioning

1 Introduction

Singular and discontinuous enrichment functions are used for the modeling of e.g. cracks in many meshfree methods [5, 20], the extended finite element method (XFEM) [4, 6, 18], the generalized finite element method (GFEM) [10–12] or the particle-partition of unity method (PPUM) [22]. In most cases, the considered enrichment functions serve the purpose of modeling only. Hence the approximation properties of the resulting numerical scheme are limited by the regularity of the solution and no improvement in the asymptotic convergence rate is obtained. A naive approach to overcome this issue is the use of a predefined constant enrichment zone [21]. This however leads to a fast growth of the condition number of the stiffness matrix and thereby has an adverse effect on the overall accuracy of the approximation. In [7] the use of an additional cut-off function which controls the enrichment zone was suggested. This approach yields some improvement, i.e. a less severe increase of the condition number. However, ultimately an ill-conditioned stiffness matrix arises also in this approach. The observed deterioration of the condition number can be remedied only by an appropriate basis transformation and projection—in essence a special preconditioner.

In this paper we focus on enrichment of the PPUM, however, the presented techniques can be applied also to other PU-based enrichment schemes. In particular we present an hierarchical enrichment procedure which defines an intermediate enrichment zone for the discretization. For each patch within this intermediate enrichment zone we construct a *local* basis transformation and a *local* projection (i.e. a special *local* preconditioner) which eliminates the *global* ill-conditioning due to the enrichment functions completely. The presented scheme attains a stable discretization independent of the employed enrichment functions and an optimal global convergence rate of the uniform h-version; i.e., the uniform h-version converges globally with a rate that is *not* limited by the regularity of the solution. For instance we obtain an $O(h)$ convergence in the energy-norm using linear polynomials globally. Within the enrichment zone the local convergence behavior is even $O(h^{1+\delta})$ with $\delta > 0$ in the energy-norm.

The remainder of this paper is organized as follows. In Section 2 we give a short review of the essential ingredients of the multilevel PPUM. In Section 3 we introduce our hierarchical enrichment scheme and the construction of our local preconditioner which yields a stable basis of the global PPUM space independent of the employed enrichment. The results of our numerical experiments are given in Section 4. These results clearly show that we obtain an optimal convergence behavior of the uniform h-version of the PPUM globally and that the condition number of the stiffness matrix does not suffer from the employed enrichment. Within the enrichment zone we obtain an almost quadratic convergence using linear polynomials only. Finally, we conclude with some remarks in Section 5.

2 Particle–Partition of Unity Method

In this section let us shortly review the core ingredients of the PPUM, see [14, 15, 21] for details. In a first step, we need to construct a PPUM space V^{PU}, i.e., we need to specify the PPUM shape functions $\varphi_i \vartheta_i^n$ where the functions φ_i form a partition of unity (PU) on the domain Ω and the functions ϑ_i^n denote the associated approximation functions considered on the patch $\omega_i := \mathrm{supp}(\varphi_i)$, i.e. polynomials ψ_i^s or enrichment functions η_i^t. With these shape functions, we then set up a sparse linear system of equations $A\tilde{u} = \hat{f}$ via the classical Galerkin method. The linear system is then solved by our multilevel iterative solver [15,17]. However, we need to employ a non-standard variational formulation of the PDE to account for the fact that our PPUM

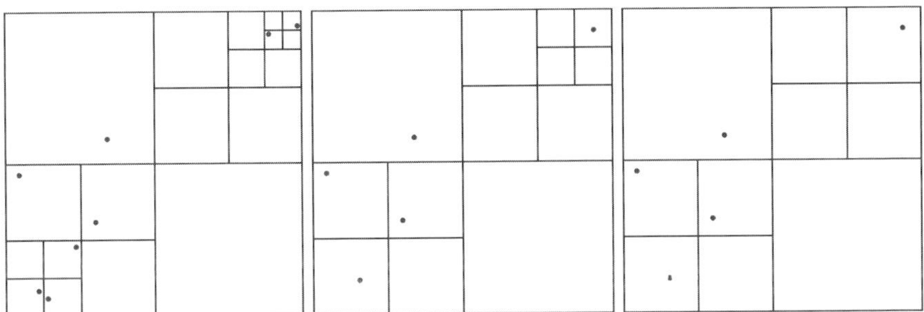

Figure 2.1. Subdivision corresponding to a cover on level $J = 4$ with initial point cloud (left), derived coarser subdivisions on level 3 (center), and level 2 (right) with respective coarser point cloud.

shape functions—like most meshfree shape functions—do not satisfy essential boundary conditions explicitly.

The fundamental construction principle employed in [14] for the construction of the PU $\{\varphi_i\}$ is a d-binary tree. Based on the given point data $P = \{x_i \,|\, i = 1, \ldots, \hat{N}\}$, we sub-divide a bounding-box $\mathcal{C}_\Omega \supset \Omega$ of the domain Ω until each cell

$$\mathcal{C}_i = \prod_{l=1}^{d} (c_i^l - h_i^l, c_i^l + h_i^l)$$

associated with a leaf of the tree contains at most a single point $x_i \in P$, see Figure 2.1. We obtain an overlapping cover $C_\Omega := \{\omega_i\}$ from this tree by defining the cover patches ω_i by

$$\omega_i := \prod_{l=1}^{d} (c_i^l - \alpha h_i^l, c_i^l + \alpha h_i^l), \quad \text{with } \alpha > 1. \tag{2.1}$$

Note that we define a cover patch ω_i for leaf-cells $\mathcal{C}_i$ that contain a point $x_i \in P$ as well as for *empty* cells that do not contain any point from P. The coarser covers C_Ω^k are defined considering coarser versions of the constructed tree, i.e., by removing a complete set of leaves of the tree, see Figure 2.1. For details of this construction see [14, 15, 21].

To obtain a PU on a cover C_Ω^k with $N_k := \text{card}(C_\Omega^k)$ we define a weight function $W_{i,k} : \Omega \to \mathbb{R}$ with $\text{supp}(W_{i,k}) = \omega_{i,k}$ for each cover patch $\omega_{i,k}$ by

$$W_{i,k}(x) = \begin{cases} \mathcal{W} \circ T_{i,k}(x) & x \in \omega_{i,k} \\ 0 & \text{else} \end{cases} \tag{2.2}$$

with the affine transforms $T_{i,k} : \overline{\omega}_{i,k} \to [-1, 1]^d$ and $\mathcal{W} : [-1, 1]^d \to \mathbb{R}$ the reference d-linear B-spline. By simple averaging of these weight functions we obtain the functions

$$\varphi_{i,k}(x) := \frac{W_{i,k}(x)}{S_{i,k}(x)}, \quad \text{with} \quad S_{i,k}(x) := \sum_{l=1}^{N_k} W_{l,k}(x). \tag{2.3}$$

We refer to the collection $\{\varphi_{i,k}\}$ with $i = 1, \ldots, N_k$ as a partition of unity since there hold the relations

$$\begin{aligned} 0 \leq \varphi_{i,k}(x) \leq 1, \qquad & \sum_{i=1}^{N_k} \varphi_{i,k} \equiv 1 \text{ on } \overline{\Omega}, \\ \|\varphi_{i,k}\|_{L^\infty(\mathbb{R}^d)} \leq C_{\infty,k}, \qquad & \|\nabla \varphi_{i,k}\|_{L^\infty(\mathbb{R}^d)} \leq \frac{C_{\nabla,k}}{\operatorname{diam}(\omega_{i,k})} \end{aligned} \tag{2.4}$$

with constants $0 < C_{\infty,k} < 1$ and $C_{\nabla,k} > 0$ so that the assumptions on the PU for the error analysis given in [2] are satisfied by our PPUM construction. Furthermore, the PU (2.3) based on the cover C_Ω^k obtained from the scaling of a tree decomposition with $\alpha > 1$ satisfies

$$\mu(\{x \in \omega_{i,k} \,|\, \varphi_{i,k}(x) = 1\}) \approx \mu(\omega_{i,k}),$$

i.e., the PU has the flat-top property, see [17,22]. This ensures that the product functions $\varphi_{i,k}\vartheta_{i,k}^n$ are linearly independent, provided that the employed local approximation functions $\vartheta_{i,k}^n$ are linearly independent with respect to $\{x \in \omega_{i,k} \,|\, \varphi_{i,k}(x) = 1\}$. Hence, we obtain global stability of the product functions $\varphi_{i,k}\vartheta_{i,k}^n$ from the local stability of the approximation functions $\vartheta_{i,k}^n$.

In general the local approximation space $V_{i,k} := \operatorname{span}\langle\vartheta_{i,k}^n\rangle$ associated with a particular patch $\omega_{i,k}$ of a PPUM space V_k^{PU} consists of two parts: A smooth approximation space, e.g. polynomials $\mathcal{P}^{p_{i,k}}(\omega_{i,k}) := \operatorname{span}\langle\psi_i^s\rangle$, and an enrichment part $\mathcal{E}_{i,k}(\omega_{i,k}) := \operatorname{span}\langle\eta_i^t\rangle$, i.e.

$$V_{i,k}(\omega_{i,k}) = \mathcal{P}^{p_{i,k}}(\omega_{i,k}) + \mathcal{E}_{i,k}(\omega_{i,k}) = \operatorname{span}\langle\psi_i^s, \eta_i^t\rangle.$$

Note that for the smooth space $\mathcal{P}^{p_{i,k}}$ we employ a local basis $\psi_{i,k}^s$ on $\omega_{i,k}$, i.e. $\psi_{i,k}^s = p_s \circ T_{i,k}$ and $\{p_s\}$ denotes a stable basis on $[-1,1]^d$. The enrichment functions $\eta_{i,k}^t$ however are often given as global functions η^t on the computational domain Ω since they are designed to capture special behavior of the solution at a particular location. Therefore, the restrictions $\eta_{i,k}^t := \eta^t|_{\omega_{i,k}}$ of the enrichment functions η^t to a particular patch $\omega_{i,k}$ may be ill-conditioned or even linearly dependent on $\omega_{i,k}$, even if the enrichment functions η^t are well-conditioned on a global scale. Furthermore, the coupling between the spaces $\mathcal{P}^{p_{i,k}}$ and $\mathcal{E}_{i,k}$ on the patch $\omega_{i,k}$ must be considered. The set of functions $\{\psi_{i,k}^s, \eta_{i,k}^t\}$ will also degenerate from a basis of $V_{i,k}$ to a generating system only, if the restricted enrichment functions $\eta_{i,k}^t = \eta^t|_{\omega_{i,k}}$ can be well approximated by polynomials $\psi_{i,k}^s$ on the patch $\omega_{i,k}$.

Remark 1. The elimination of these linear dependencies and the selection of an appropriate basis $\langle\tilde{\vartheta}_{i,k}^m\rangle$ for the space $V_{i,k}(\omega_{i,k})$ is the main challenge in an enriched PPUM computation (and any other numerical method that employs enrichment). To this end we have developed a projection operator or preconditioner

$$\Pi_{i,k}^* : \operatorname{span}\langle\vartheta_{i,k}^n\rangle \to \operatorname{span}\langle\tilde{\vartheta}_{i,k}^m\rangle$$

that maps the ill-conditioned generating system $\langle \psi^s_{i,k}, \eta^t_{i,k}\rangle = \langle \vartheta^n_{i,k}\rangle$ to a stable basis $\langle \tilde{\vartheta}^m_{i,k}\rangle$, see Section 3.

With the help of the shape functions $\varphi_{i,k}\vartheta^n_{i,k}$ we then discretize a PDE in weak form

$$a(u,v) = \langle f, v\rangle$$

via the classical Galerkin method to obtain a discrete linear system of equations $A\tilde{u} = \hat{f}$. Note that the PU functions (2.3) in the PPUM are in general piecewise rational functions only. Therefore, the use of an appropriate numerical integration scheme is indispensable in the PPUM as in most meshfree approaches [1, 3, 8, 9, 15]. Moreover, the functions $\varphi_{i,k}\vartheta^n_{i,k}$ in general do not satisfy the Kronecker property. Thus, the coefficients $\tilde{u}_k := (u^n_{i,k})$ of a discrete function

$$u_k^{\mathrm{PU}} = \sum_{i=1}^{N_k} \varphi_{i,k} \sum_{n=1}^{d_{i,k}} u^n_{i,k}\vartheta^n_{i,k} = \sum_{i=1}^{N_k} \varphi_{i,k}\Big(\sum_{s=1}^{d^{\mathcal{P}}_{i,k}} u^s_{i,k}\psi^s_{i,k} + \sum_{t=1}^{d^{\mathcal{E}}_{i,k}} u^{t+d^{\mathcal{P}}_{i,k}}_{i,k}\eta^t_{i,k}\Big) \qquad (2.5)$$

with $d^{\mathcal{P}}_{i,k} := \dim \mathcal{P}_{i,k}$, $d^{\mathcal{E}}_{i,k} := \dim \mathcal{E}_{i,k}$ and $d_{i,k} := d^{\mathcal{P}}_{i,k} + d^{\mathcal{E}}_{i,k}$ on level k do not directly correspond to function values and a trivial interpolation of essential boundary data is not available.

2.1 Essential Boundary Conditions

The treatment of essential boundary conditions in meshfree methods is not straightforward and a number of different approaches have been suggested. In [16] we have presented how Nitsche's method [19] can be applied successfully in the meshfree context. Here, we give a short summary of this approach. To this end, let us consider the model problem

$$\begin{aligned} -\,\mathbf{div}\,\boldsymbol{\sigma}(u) &= f && \text{in } \Omega \subset \mathbb{R}^d \\ \boldsymbol{\sigma}(u)\cdot n &= g_N && \text{on } \Gamma_N \subset \partial\Omega \\ u\cdot n &= g_{D,n} && \text{on } \Gamma_D = \partial\Omega\setminus\Gamma_N \\ (\boldsymbol{\sigma}(u)\cdot n)\cdot t &= 0 && \text{on } \Gamma_D = \partial\Omega\setminus\Gamma_N \end{aligned} \qquad (2.6)$$

In the following we drop the level subscript $k = 0,\dots,J$ for the ease of notation.

Let us define the cover of the Dirichlet boundary

$$C_{\Gamma_D} := \{\omega_i \in C_\Omega \,|\, \Gamma_{D,i} \neq \emptyset\}$$

where $\Gamma_{D,i} := \omega_i \cap \Gamma_D$ and $\gamma_{D,i} := \operatorname{diam}(\Gamma_{D,i})$. With these conventions we define the cover-dependent functional

$$J_{C_\Omega}(w) := \int_\Omega \boldsymbol{\sigma}(w) : \boldsymbol{\epsilon}(w)\,dx - 2\int_{\Gamma_D} (n\cdot\boldsymbol{\sigma}(w)n)n\cdot w\,ds + \beta\sum_{\omega_i\in C_{\Gamma_D}} \gamma_{D,i}^{-1}\int_{\Gamma_{D,i}} (w\cdot n)^2\,ds$$

with some parameter $\beta > 0$. Minimizing J_{C_Ω} with respect to the error $u - u^{\mathrm{PU}}$ yields the weak formulation

$$a_{C_\Omega}(w, v) = l_{C_\Omega}(v) \quad \text{for all } v \in V^{\mathrm{PU}} \tag{2.7}$$

with the cover-dependent bilinear form

$$\begin{aligned} a_{C_\Omega}(u, v) := & \int_\Omega \boldsymbol{\sigma}(u) : \boldsymbol{\epsilon}(v)\, dx - \int_{\Gamma_D} (n \cdot \boldsymbol{\sigma}(u) n) n \cdot v\, ds \\ & - \int_{\Gamma_D} (n \cdot \boldsymbol{\sigma}(v) n) n \cdot u\, ds + \beta \sum_{\omega_i \in C_{\Gamma_D}} \gamma_{D,i}^{-1} \int_{\Gamma_{D,i}} u \cdot n v \cdot n\, ds \end{aligned}$$

and the corresponding linear form

$$\langle l_{C_\Omega}, v \rangle := \int_\Omega f v + \int_{\Gamma_N} g_N v - \int_{\Gamma_D} g_{D,n} (n \cdot \boldsymbol{\sigma}(v) n) + \beta \sum_{\omega_i \in C_{\Gamma_D}} \gamma_{D,i}^{-1} \int_{\Gamma_{D,i}} g_{D,n} v \cdot n\, ds$$

There is a unique solution u^{PU} of (2.7) if the regularization parameter β is chosen large enough; i.e., the regularization parameter $\beta = \beta_{V^{\mathrm{PU}}}$ is dependent on the discretization space V^{PU}. This solution u^{PU} satisfies optimal error bounds if the space V^{PU} admits the inverse estimate

$$\|(n \cdot \boldsymbol{\sigma}(v) n)\|^2_{-\frac{1}{2}, C_{\Gamma_D}} \le C^2_{V^{\mathrm{PU}}} \|v\|^2_E = C^2_{V^{\mathrm{PU}}} \int_\Omega \boldsymbol{\sigma}(v) : \boldsymbol{\epsilon}(v)\, dx \tag{2.8}$$

for all $v \in V^{\mathrm{PU}}$ with respect to the cover-dependent norm

$$\|w\|^2_{-\frac{1}{2}, C_{\Gamma_D}} := \sum_{\omega_i \in C_{\Gamma_D}} \gamma_{D,i} \|w\|^2_{L^2(\Gamma_{D,i})}$$

with a constant $C_{V^{\mathrm{PU}}}$ depending on the cover C_Ω and the employed local bases $\langle \vartheta_i^n \rangle$ only. If $C_{V^{\mathrm{PU}}}$ is known, the regularization parameter $\beta_{V^{\mathrm{PU}}}$ can be chosen as $\beta_{V^{\mathrm{PU}}} > 2C^2_{V^{\mathrm{PU}}}$ to obtain a symmetric positive definite linear system [19]. Hence, the main task associated with the use of Nitsche's approach in the PPUM context is the efficient and automatic computation of the constant $C_{V^{\mathrm{PU}}}$, see [16, 21]. To this end, we consider the inverse assumption (2.8) as a generalized eigenvalue problem locally on each patch $\omega_i \in C_{\Gamma_D}$ and solve for the largest eigenvalue to obtain an approximation of $C^2_{V^{\mathrm{PU}}}$.

In summary, the PPUM discretization of our model problem (2.6) using the space V^{PU} on the cover C_Ω is carried out in two steps: First, we estimate the regularization parameter $\beta_{V^{\mathrm{PU}}}$ from (2.8). Then, we define the weak form (2.7) and use Galerkin's method to set up the respective symmetric positive definite linear system $A\tilde{u} = \hat{f}$. This linear system is then solved by our multilevel iterative solver [15, 17].

3 Hierarchical Enrichment and Local Preconditioning

The use of smooth polynomial local approximation spaces $V_{i,k} = \mathcal{P}^{p_{i,k}}$ in our PPUM is optimal only for the approximation of a smooth or regular solution u. In the case of a discontinuous and singular solution u there are essentially two approaches we can pursue: First, a classical adaptive refinement process which essentially resolves the singular behavior of the solution by geometric subdivision, see [17, 22]. Second, an algebraic approach that is very natural to the PPUM, the explicit enrichment of the global approximation space by special shape functions η^s. This approach is also pursued in other meshfree methods [5, 20], the XFEM [4, 6, 18] or the GFEM [10–12]. Most enrichment schemes however focus on modeling issues and not on approximation properties or the conditioning of the resulting stiffness matrix.

In this section we introduce an automatic hierarchical enrichment scheme for our PPUM that provides optimal convergence properties and avoids an ill-conditioning of the resulting stiffness matrix due to enrichment. To this end, we consider a reference problem from linear elastic fracture mechanics

$$\begin{aligned} -\,\mathbf{div}\,\boldsymbol{\sigma}(u) &= f \quad \text{in } \Omega = (-1,1)^2, \\ \boldsymbol{\sigma}(u)\cdot n &= g_N \text{ on } \Gamma_N \subset \partial\Omega \cup C, \\ u &= g_D \text{ on } \Gamma_D = \partial\Omega \setminus \Gamma_N. \end{aligned} \tag{3.9}$$

The internal traction-free segment

$$C := \{(x,y) \in \Omega \,|\, x \in (-0.5, 0.5) \text{ and } y = 0\}$$

is referred to as a crack. The crack C induces a discontinuous displacement field u across the crack line C with singularities at the crack tips $c_l := (-0.5, 0)$ and $c_u := (0.5, 0)$. Hence, the local approximation spaces $V_{i,k}$ employed in our PPUM must respect these features to provide good approximation.

The commonly used enrichment strategy employs simple geometric information only. A patch $\omega_{i,k}$ (or an element) is enriched by the discontinuous Haar function if the patch is (completely) cut by the crack C, i.e.

$$\mathcal{E}_{i,k} := H^C_\pm \mathcal{P}^{p_{i,k}} \quad \text{and} \quad V_{i,k} := \mathcal{P}^{p_{i,k}} + H^C_\pm \mathcal{P}^{p_{i,k}}. \tag{3.10}$$

Note that in fact most other enrichment procedures employ $\mathcal{E}_{i,k} = H^C_\pm$ only. If the patch $\omega_{i,k}$ contains a crack tip ξ_{tip}, i.e. $c_l \in \omega_{i,k}$ or $c_u \in \omega_{i,k}$, then the patch is enriched by the respective space of singular tip functions

$$W_{\text{tip}} := \{\sqrt{r}\cos\frac{\theta}{2}, \sqrt{r}\sin\frac{\theta}{2}, \sqrt{r}\sin\theta\sin\frac{\theta}{2}, \sqrt{r}\sin\theta\cos\frac{\theta}{2}\} \tag{3.11}$$

given in local polar coordinates with respect to the tip ξ_{tip}, i.e. $\mathcal{E}_{i,k} = W_{\text{tip}}|_{\omega_{i,k}}$. This yields the local approximation space

$$V_{i,k} := \mathcal{P}^{p_{i,k}} + W_{\text{tip}}$$

for a patch $\omega_{i,k}$ that contains the tip ξ_{tip}. Let us summarize this geometric modeling enrichment scheme in the following classifier function $e^M : C^k_\Omega \to \{\texttt{lower\_tip}, \texttt{upper\_tip}, \texttt{jump}, \texttt{none}\}$

$$e^M(\omega_{i,k}) := \begin{cases} \texttt{lower\_tip} & \text{if } c_l \in \omega_{i,k} \text{ and } c_u \notin \omega_{i,k}, \\ \texttt{upper\_tip} & \text{if } c_l \notin \omega_{i,k} \text{ and } c_u \in \omega_{i,k}, \\ \texttt{jump} & \text{if } \{c_l, c_u\} \cap \omega_{i,k} = \emptyset \text{ and } C \cap \omega_{i,k} \neq \emptyset, \\ \texttt{none} & \text{else.} \end{cases} \tag{3.12}$$

Note that the direct evaluation of e^M for all patches $\omega_{i,k} \in C^k_\Omega$ requires $O(N_k)$ rather expensive geometric operations such as line-line intersections.

Even though this enrichment is sufficient to model a crack and captures the asymptotic behavior of the solution at the tip, this strategy suffers from various drawbacks. With respect to the discontinuous enrichment the main issue is that very small intersections of a patch with a crack cause an ill-conditioned stiffness matrix which can compromise the stability of the discretization; e.g. when the volumes of the sub-patches induced by the cut with the crack differ substantially in size. This is usually circumvented by a predefined geometric tolerance parameter which rejects such small intersections. In the case of a one-dimensional enrichment space $\mathcal{E}_{i,k} = H^C_\pm$ this approach is sufficient—if the tolerance parameter is chosen relative to the diameter of the patch. For a multi-dimensional enrichment space $\mathcal{E}_{i,k} = H^C_\pm \mathcal{P}^{p_{i,k}}$ this approach can be too restrictive to obtain optimal results.

The crack tip enrichment space W_{tip} given in (3.11) models the essential behavior of the solution at the tip. However, the singularity at the tip has a substantially larger zone of influence than just the containing patch. Therefore, the simple geometric modeling enrichment (3.12) is not sufficient to improve the asymptotic convergence behavior of the employed numerical scheme.

These issues can be overcome with the help of our multilevel sequence of covers C^k_Ω and a local preconditioner. Starting on the coarsest level $k = 0$ of our cover sequence we consider the cover $C^0_\Omega = \{\omega_{i,0}\}$ and define the intermediate enrichment classifier $I_0 : C^0_\Omega \to \{\texttt{lower\_tip}, \texttt{upper\_tip}, \texttt{jump}, \texttt{none}\}$ by the geometric/modeling enrichment scheme discussed above

$$I_0(\omega_{i,0}) := e^M(\omega_{i,0}).$$

In the next step we define the associated intermediate enrichment spaces $\mathcal{E}^I_{i,k}$ for $k = 0$

$$\mathcal{E}^I_{i,k} := \begin{cases} W_{c_l} & \text{if } \texttt{lower\_tip} = I_k(\omega_{i,k}), \\ W_{c_u} & \text{if } \texttt{upper\_tip} = I_k(\omega_{i,k}), \\ H^C_\pm \mathcal{P}^{p_{i,k}} & \text{if } \texttt{jump} = I_k(\omega_{i,k}) \text{ and } C \cap \omega_{i,k} \neq \emptyset, \\ 0 & \text{else.} \end{cases} \tag{3.13}$$

with $d^{\mathcal{E}^I}_{i,k} := \text{card}(\{\eta^t_{i,k}\})$ and the respective intermediate approximation spaces

$$V_{i,k}^I := \mathcal{P}^{p_{i,k}} + \mathcal{E}_{i,k}^I = \operatorname{span}\langle \psi_{i,k}^s, \eta_{i,k}^t \rangle = \operatorname{span}\langle \vartheta_{i,k}^n \rangle$$

with $d_{i,k}^{V^I} := d_{i,k}^{\mathcal{P}} + d_{i,k}^{\mathcal{E}^I}$ and $d_{i,k}^{\mathcal{P}} = \dim(\mathcal{P}^{p_{i,k}})$. Using all functions $\vartheta_{i,k}^n$, i.e. $\psi_{i,k}^s$ and $\eta_{i,k}^t$, we setup the local mass matrix $M_{i,k}$ with the entries

$$(M_{i,k})_{m,n} := \int_{\omega_{i,k}\cap\Omega} \vartheta_{i,k}^n \vartheta_{i,k}^m \, dx \quad \text{for all } m,n = 1,\dots,d_{i,k}^{\mathcal{V}^{\mathcal{I}}}. \tag{3.14}$$

From the eigenvalue decomposition

$$O_{i,k}^T M_{i,k} O_{i,k} = D_{i,k} \quad \text{with } O_{i,k}, D_{i,k} \in \mathbb{R}^{d_{i,k}^{V^I} \times d_{i,k}^{V^I}} \tag{3.15}$$

of the matrix $M_{i,k}$ where

$$O_{i,k}^T O_{i,k} - \mathbb{I}_{d_{i,k}^{V^I}}, \qquad (D_{i,k})_{m,n} = 0 \quad \text{for all } m \neq n$$

we can extract a stable basis $\langle \tilde{\vartheta}_{i,k}^m \rangle$ by a simple cut-off of small eigenvalues. To this end let us assume that the eigenvalues $(D_{i,k})_{m,m}$ are given in decreasing order, i.e. $(D_{i,k})_{m,m} \geq (D_{i,k})_{m+1,m+1}$. Then we can easily partition the matrices $O_{i,k}^T$ and $D_{i,k}$ as

$$O_{i,k}^T = \begin{pmatrix} \tilde{O}_{i,k}^T \\ K_{i,k}^T \end{pmatrix}, \quad D_{i,k} = \begin{pmatrix} \tilde{D}_{i,k} & 0 \\ 0 & \kappa_{i,k} \end{pmatrix}$$

where the mth row of the rectangular matrix $\tilde{O}_{i,k}^T$ is an eigenvector of $M_{i,k}$ that is associated with an eigenvalue $(D_{i,k})_{m,m} = (\tilde{D}_{i,k})_{m,m} \geq \epsilon \, (D_{i,k})_{0,0}$ and $K_{i,k}^T$ involves all eigenvectors that are associated with small eigenvalues. Since $(\tilde{D}_{i,k})_{m,m} \geq \epsilon \, (D_{i,k})_{0,0}$ the operator

$$\Pi_{i,k}^* := \tilde{D}_{i,k}^{-1/2} \tilde{O}_{i,k}^T$$

is well-defined and can be evaluated stably. Furthermore, the projection $\Pi_{i,k}^*$ removes the near-null space of $M_{i,k}$ due to the cut-off parameter ϵ and we have

$$\Pi_{i,k}^* M_{i,k} (\Pi_{i,k}^*)^T = \tilde{D}_{i,k}^{-1/2} \tilde{O}_{i,k}^T M_{i,k} O_{i,k} \tilde{D}_{i,k}^{-1/2} = \mathbb{I}_{d_{i,k}^{\Pi}}$$

where $d_{i,k}^{\Pi} := \operatorname{card}\{(D_{i,k})_{m,m} \geq \epsilon \, (D_{i,k})_{0,0}\}$ denotes the row-dimension of $\tilde{O}_{i,k}^T$ and $\Pi_{i,k}^*$. Hence, the operator $\Pi_{i,k}^*$ maps the ill-conditioned generating system $\langle \vartheta_{i,k}^n \rangle = \langle \psi_{i,k}^s, \eta_{i,k}^t \rangle$ to a basis $\langle \tilde{\vartheta}_{i,k}^m \rangle$ that is optimally conditioned — it is an optimal preconditioner.

Assuming that the employed local basis $\langle \psi_{i,k}^s \rangle$ is well-conditioned and that ϵ is small we have $\mathcal{P}^{p_{i,k}} \subset \operatorname{span}\langle \tilde{\vartheta}_{i,k}^m \rangle$ so that if $\dim(\mathcal{P}^{p_{i,k}}) = d_{i,k}^{\Pi}$ we can remove the enrichment functions $\eta_{i,k}^t$ completely from the local approximation space and use $V_{i,k} = \mathcal{P}^{p_{i,k}}$. Therefore, we define our final enrichment indicator $E_k : C_\Omega^k \to \{\texttt{lower\_tip}, \texttt{upper\_tip}, \texttt{jump}, \texttt{none}\}$ on level k as

$$E_k(\omega_{i,k}) := \begin{cases} I_k(\omega_{i,k}) & \text{if } \dim(\mathcal{P}^{p_{i,k}}) \neq d^{\Pi}_{i,k}, \\ \texttt{none} & \text{else.} \end{cases} \tag{3.16}$$

The local approximation space $V_{i,k}$ assigned to an enriched patch $\omega_{i,k}$ is given by

$$V_{i,k} := \Pi^*_{i,k} V^I_{i,k} = \operatorname{span}\langle \tilde{\vartheta}^m_{i,k} \rangle \tag{3.17}$$

On the next finer level $k+1$ we utilize the geometric hierarchy of our cover patches to define our intermediate enrichment indicator I_{k+1}. Recall that for each cover patch $\omega_{i,k+1}$ there exists exactly one cover patch $\omega_{\tilde{i},k}$ such that $\omega_{i,k+1} \subset \omega_{\tilde{i},k}$, compare Figure 2.1. Hence we can define our intermediate enrichment indicator I_{k+1} on level $k+1$ as

$$I_{k+1}(\omega_{i,k+1}) := \begin{cases} E_k(\omega_{\tilde{i},k}) & \text{if } E_k(\omega_{\tilde{i},k}) \neq \texttt{jump}, \\ \texttt{jump} & \text{if } E_k(\omega_{\tilde{i},k}) = \texttt{jump} \text{ and } C \cap \omega_{i,k} \neq \emptyset, \\ \texttt{none} & \text{else} \end{cases}$$

directly from the enrichment indicator E_k on level k and a minimal number of geometric operations. With this intermediate enrichment indicator we apply the above scheme recursively to derive the enrichment indicators E_l for all levels $l = 1, \ldots, J$. Finally, we obtain stable local basis systems $\langle \tilde{\vartheta}^m_{i,l} \rangle$ and the respective approximation spaces $V_{i,l} = \operatorname{span}\langle \tilde{\vartheta}^m_{i,l} \rangle$ for all cover patches $\omega_{i,l} \in C^l_\Omega$ on all levels $l = 0, \ldots, J$. Recalling that our PU functions $\varphi_{i,l}$ satisfy the flat-top condition (see Section 2) this is sufficient to obtain the stability of the global basis $\langle \varphi_{i,l} \tilde{\vartheta}^m_{i,l} \rangle$ for the PPUM space V^{PU}_l on level l.[1]

Remark 1. Note that we do not need to apply the local preconditioner $\Pi^*_{i,k}$ for the evaluation of the basis $\langle \varphi_{i,k} \tilde{\vartheta}^m_{i,k} \rangle$ in each quadrature point during the assembly of the stiffness matrix. It is sufficient to transform the stiffness matrix A^{GS}_k on level k which was assembled using the generating system $\langle \psi^s_{i,k}, \eta^t_{i,k} \rangle$ by the block-diagonal operator Π^*_k with the block-entries

$$(\Pi^*_k)_{g,h} := \begin{cases} \Pi^*_{g,k}, & g = h \\ 0 & \text{else,} \end{cases}$$

for all $g = 1, \ldots, N_k$; i.e., we obtain the stiffness matrix A_k with respect to the basis $\langle \varphi_{i,k} \tilde{\vartheta}^m_{i,k} \rangle$ on level k as the product operator

$$A_k = \Pi^*_k A^{\mathrm{GS}}_k (\Pi^*_k)^T.$$

Remark 2. Note that in the discussion above we have considered the identity operator $\mathbb{I}$ on the local patch $\omega_{i,k}$, i.e. the mass matrix $M_{i,k}$. However, we can construct the respective preconditioner also for different operators e.g. the

---

[1] Actually we need to apply the construction of the preconditioner to the operator $M^{\mathrm{FT}}_{i,k}$ which involves integrals on $\{x \in \omega_{i,k} \mid \varphi_{i,k}(x) = 1\}$ instead of the complete patch $\omega_{i,k}$.

operator $-\Delta + \mathbb{I}$ which corresponds to the H^1-norm. In exact arithmetic and with a cut-off parameter $\epsilon = 0$ changing the operator in the above construction has an impact on the constants only. However, due to our cut-off parameter ϵ we may obtain a different subspace span$\langle \tilde{\vartheta}^m_{i,k} \rangle$ for different operators with the same ϵ.

3.1 Error Bound

Due to our hierarchical enrichment we obtain a sequence of PPUM spaces V_k^{PU} with $k = 0, \ldots, J$ that contain all polynomials up to degree $p_k = \min_i p_{i,k}$ on a particular level k and all enrichment functions η^t (up to the cut-off parameter ϵ) in the enrichment zone E on all levels k. Hence the global convergence rate of our enriched PPUM is not limited by the regularity of the solution u. To confirm this assertion let us consider the splitting

$$u = u_p + \tilde{\chi}_E u_s$$

where u_p denotes the regular part of the solution u, u_s the singular part, and $\tilde{\chi}_E$ is a mollified characteristic function of the enrichment zone E which contains all singular points of u, i.e. of u_s. Multiplication with $1 \equiv \sum_{i=1}^N \varphi_i$ yields

$$u = \sum_{i=1}^N \varphi_i u_p + \sum_{i=1}^N \varphi_i \tilde{\chi}_E u_s.$$

Let us further consider the PPUM function (we drop the level index k for the ease of notation in the following)

$$u^{\mathrm{PU}} := \sum_{E(\omega_i)=\mathtt{none}} \varphi_i \varpi_i + \sum_{E(\omega_i)\neq\mathtt{none}} \varphi_i(\varpi_i + e_i)$$

where $E(\omega_i)$ denotes the enrichment indicator given in (3.16), $\varpi_i \in \mathcal{P}^{p_i}$ and $e_i \in \mathcal{E}_i$. For the ease of notation let us assume that $E(\omega_i) = \mathtt{none}$ holds for all patches ω_i with $i = 1, \ldots, M-1$ and $E(\omega_i) \neq \mathtt{none}$ holds for all patches ω_i with $i = M, \ldots, N$ so that we can write

$$u^{\mathrm{PU}} = \sum_{i=1}^{M-1} \varphi_i \varpi_i + \sum_{i=M}^{N} \varphi_i(\varpi_i + e_i).$$

With the assumption

$$\mathrm{supp}(\tilde{\chi}_E) \cap \bigcup_{i=1}^{M-1} \omega_i = \emptyset, \qquad \text{i.e.} \quad \tilde{\chi}_E \sum_{i=1}^{M-1} \varphi_i \equiv 0,$$

we can write the analytic solution u as

$$u = \sum_{i=1}^{M-1} \varphi_i u_p + \sum_{i=M}^{N} \varphi_i (u_p + \tilde{\chi}_E u_s)$$

and obtain the error with respect to the PPUM function u^{PU} as

$$u^{\mathrm{PU}} - u = \sum_{i=1}^{M-1} \varphi_i (\varpi_i - u_p) + \sum_{i=M}^{N} \varphi_i ((\varpi_i + e_i) - (u_p + \tilde{\chi}_E u_s)). \tag{3.18}$$

By the triangle inequality we have

$$\begin{aligned} \|u - u^{\mathrm{PU}}\| \leq \| &\sum_{i=1}^{M-1} \varphi_i (\varpi_i - u_p)\| \\ &+ \| \sum_{i=M}^{N} \varphi_i ((\varpi_i + e_i) - (u_p + \tilde{\chi}_E u_s))\|. \end{aligned} \tag{3.19}$$

The first term on the right-hand side corresponds to the error of a PPUM approximation of a regular function with polynomial local approximation spaces. For the ease of notation let us assume $h = \mathrm{diam}(\omega_i)$ and $p_i = 1$ for all $i = 1, \dots, N$, then we can bound this error term with the help of the standard PUM error analysis [2] by $O(h)$ in the H^1-norm, i.e.

$$\| \sum_{i=1}^{M-1} \varphi_i (\varpi_i - u_p)\|_{H^1} \leq O(h).$$

To obtain an upper bound for the second term of (3.19)

$$J_E := \| \sum_{i=M}^{N} \varphi_i ((\varpi_i + e_i) - (u_p + \tilde{\chi}_E u_s))\|$$

we consider the equality

$$u_p + \tilde{\chi}_E u_s = u_p + (\tilde{\chi}_E - 1) u_s + u_s$$

and attain an upper bound of J_E again by the triangle inequality

$$\begin{aligned} J_E &= \| \sum_{i=M}^{N} \varphi_i ((\varpi_i + e_i) - (u_p + (\tilde{\chi}_E - 1) u_s + u_s))\| \\ &\leq \| \sum_{i=M}^{N} \varphi_i (\varpi_i - (u_p + (\tilde{\chi}_E - 1) u_s))\| \\ &\quad + \| \sum_{i=M}^{N} \varphi_i (e_i - u_s)\|. \end{aligned}$$

The function $u_p + (\tilde{\chi}_E - 1) u_s$ is regular since $\tilde{\chi}_E = 1$ in the vicinity of the singular points of u_s. Hence, we can bound the first term on the right-hand side

$$\| \sum_{i=M}^{N} \varphi_i(\varpi_i - (u_p + (\tilde{\chi}_E - 1)u_s))\|_{H^1} \leq O(h)$$

again by $O(h)$. Assuming that the enrichment functions resolve the singular part u_s of the solution u we can choose $e_i = u_s$ and so the second term vanishes and we obtain the upper bound

$$\| \sum_{i=M}^{N} \varphi_i((\varpi_i + e_i) - \tilde{\chi}_E(u_p + u_s))\|_{H^1} \leq O(h)$$

for the error in $\mathrm{supp}(\tilde{\chi}_E) \subset E$. This yields the error bound

$$\|u - u^{\mathrm{PU}}\|_{H_1} \leq O(h)$$

for the global error on the domain Ω.

Note however that we can obtain a better estimate for the error in the enrichment zone; i.e., the hierarchically enriched PPUM shows a kind of superconvergence within the enrichment zone. To this end consider the case $u_s = 0$, i.e., the approximation of a regular solution $u = u_p$ by an enriched PPUM. Then, J_E becomes

$$J_E = \| \sum_{i=M}^{N} \varphi_i((\varpi_i + e_i) - u_p))\|$$

and the standard error bound $O(h)$ ignores all degrees of freedom collected in e_i which are associated with the restrictions $\eta^t|_{\omega_{i,k}}$ of the enrichment functions to the local patches. Globally, the functions η^t represent a specific (type of) singularity. The restrictions $\eta^t|_{\omega_{i,k}}$ however are regular functions (if the patch does not contain the singular points of η^t) and can improve the local approximation substantially. Hence, a better bound for J_E for regular solutions u can be attained.

For a singular solution $u_s \neq 0$ we can utilize this observation by considering the splitting of the enrichment part e_i on a particular patch in two local components $e_i = e_i^s + e_i^p$. On each patch ω_i with $i = M, \dots, N$ this splitting can be chosen to balance the two error terms on the right-hand side of the inequality

$$\begin{aligned} J_E &= \| \sum_{i=M}^{N} \varphi_i((\varpi_i + e_i^p + e_i^s) - (u_p + \tilde{\chi}_E u_s))\| \\ &\leq \| \sum_{i=M}^{N} \varphi_i((\varpi_i + e_i^p) - (u_p + (\tilde{\chi}_E - 1)u_s))\| \\ &\quad + \| \sum_{i=M}^{N} \varphi_i(e_i^s - u_s)\|. \end{aligned}$$

This can yield a much smaller error since the regular function $u_p + (\tilde{\chi}_E - 1)u_s$ is now approximated by more degrees of freedom, i.e., by all polynomials and a number of enrichment functions restricted to the local patch ω_i. Hence, the Galerkin solution which minimizes J_E (i.e. minimizes (3.18) with respect to the energy-norm) can show a much better convergence of $O(h^{1+\delta})$ with $\delta > 0$ in the enrichment zone than the global $O(h)$ behavior.

The impact of this observation is that the coefficients of the asymptotic expansion of the solution, e.g. the stress intensity factors, can be extracted from the solution with much higher accuracy and better convergence behavior in the enrichment zone than the global error bound implies.

Remark 3. Note that due to our cut-off parameter $\epsilon > 0$ our local function spaces $V_{i,k}$ may not contain all enrichment functions $\eta^t_{i,k}$. Especially we may encounter the situation that a particular patch $\omega_{i,k+1}$ employs less enrichment functions than its parent patch $\omega_{\tilde{i},k} \supset \omega_{i,k+1}$; i.e., the local approximation spaces $V_{i,k+1}$ and $V_{\tilde{i},k}$ are *nonnested* due to the cut-off. Hence, the parameter δ in the discussion given above may not be constant on all levels and the measured convergence rates can jump due to the cut-off.

4 Numerical Results

In this section we present some results of our numerical experiments using the hierarchically enriched PPUM discussed above. To this end, we introduce some shorthand notation for various norms of the error $u - u^{\mathrm{PU}}$, i.e., we define

$$e_{L^\infty} := \frac{\|u-u^{\mathrm{PU}}\|_{L^\infty}}{\|u\|_{L^\infty}}, e_{L^2} := \frac{\|u-u^{\mathrm{PU}}\|_{L^2}}{\|u\|_{L^2}}, e_{H^1} := \frac{\|u-u^{\mathrm{PU}}\|_{H^1}}{\|u\|_{H^1}}. \quad (4.20)$$

For each of these error norms we compute the respective algebraic convergence rate ρ by considering the error norms of two consecutive levels $l-1$ and l

$$\rho := -\frac{\log\left(\frac{\|u-u_l^{\mathrm{PU}}\|}{\|u-u_{l-1}^{\mathrm{PU}}\|}\right)}{\log\left(\frac{\mathsf{dof}_l}{\mathsf{dof}_{l-1}}\right)}, \quad \text{where } \mathsf{dof}_k := \sum_{i=1}^{N_k} \dim(V_{i,k}). \quad (4.21)$$

Hence the optimal rate ρ_{H^1} of an uniformly h-refined sequence of spaces with $p_{i,k} = p$ for all $i = 1, \ldots, N_k$ and $k = 0, \ldots, J$ for a regular solution u is $\rho_{H^1} = \frac{p}{d}$ where d denotes the spatial dimension of $\Omega \subset \mathbb{R}^d$. This corresponds to the classical $h^{\gamma_{H^1}}$ notation with $\gamma_{H^1} = \rho_{H^1} d = p$.

To assess the quality of our hierarchical enrichment scheme we consider the simple model problem

$$\begin{aligned} -\Delta u &= f \quad \text{in } \Omega = (-1,1)^2 \subset \mathbb{R}^2, \\ u &= g \quad \text{on } \partial\Omega, \end{aligned} \quad (4.22)$$

where we choose f and g such that the analytic solution u is given by

$$u(x,y) = \sqrt{r}(\sin\frac{\theta}{2} + \cos\frac{\theta}{2})(1+\sin\theta) + (x^2-1) + (y^2-1) + 1 \quad (4.23)$$

Table 4.1. Relative errors e (4.20) and convergence rates ρ (4.21) with respect to the complete domain Ω.

J	dof	N	e_{L^∞}	ρ_{L^∞}	e_{L^2}	ρ_{L^2}	e_{H^1}	ρ_{H^1}
1	28	4	7.262_{-2}	–	5.011_{-2}	–	1.663_{-1}	–
2	70	16	4.741_{-2}	0.47	3.096_{-2}	0.53	1.449_{-1}	0.15
3	226	64	1.614_{-2}	0.92	1.098_{-2}	0.88	8.826_{-2}	0.42
4	868	256	5.192_{-3}	0.84	2.974_{-3}	0.97	4.544_{-2}	0.49
5	3400	1024	1.488_{-3}	0.92	7.779_{-4}	0.98	2.296_{-2}	0.50
6	13456	4096	4.491_{-4}	0.87	1.990_{-4}	0.99	1.148_{-2}	0.50
7	53214	16384	1.317_{-4}	0.89	5.034_{-5}	1.00	5.864_{-3}	0.49
8	210036	65536	3.748_{-5}	0.92	1.266_{-5}	1.01	3.010_{-3}	0.49
9	837176	262144	1.042_{-5}	0.93	3.173_{-6}	1.00	1.405_{-3}	0.55
10	3288341	1048576	2.942_{-6}	0.92	8.078_{-7}	1.00	7.367_{-4}	0.47

where $r = r(x,y)$ and $\theta = \theta(x,y)$ denote polar coordinates, see Figure 4.2. This solution is discontinuous along the line

$$C := \{(x,y) \in \Omega \,|\, x \in (-1,0) \text{ and } y = 0\}$$

and weakly singular at the point $(0,0)$. The model problem (4.22) with the considered data f and g is essentially a scalar analogue of a linear elastic fracture mechanics problem such as (3.9). Hence, we employ the enrichment functions (3.10) and (3.11) with respect to the crack C in our computations.

We consider a sequence of uniformly refined covers C_Ω^k with $\alpha = 1.3$ in (2.1) and local polynomial spaces $\mathcal{P}^{p_{i,k}} = \mathcal{P}^1$ on all levels $k = 1,\ldots,J$ for the discretization of (4.22). The number of patches on level k is given by $N_k = 2^{2k}$. On the levels $k \leq 3$ we use the geometric/modeling indicator e^M (3.12) as enrichment indicator, on the finer levels $k > 3$ we use the recursively defined hierarchical enrichment indicator (3.16), see Section 3. Hence, the subdomain

$$E_{\text{tip}} := (-0.25, 0.25)^2 \subset \Omega = (-1,1)^2 \tag{4.24}$$

denotes the initial enrichment zone with respect to the point singularity of (4.23) at $(0,0)$ on the levels $k > 3$, see Figure 4.24. The respective intermediate local enrichment spaces $\mathcal{E}_{i,k}^I$ are defined by (3.13) and the resulting local approximation spaces $V_{i,k}$ by (3.17). The local preconditioner $\Pi_{i,k}^*$ is based on the local mass matrix and employs a cut-off parameter $\epsilon = 10^{-12}$.

Since the solution is singular at $(0,0)$ a classical uniform h-version without enrichment (or just modeling enrichment) yields convergence rates (4.21) of $\rho_{L^2} = \frac{2}{3}$ and $\rho_{H^1} = \frac{1}{3}$ only. Due to our hierarchical enrichment we anticipate to recover the optimal convergence rates $\rho_{L^2} = 1$ and $\rho_{H^1} = \frac{1}{2}$ globally. The convergence behavior inside E_{tip} can be better, see Section 3.1.

We assess the quality of our local preconditioner $\Pi_{i,k}^*$ by studying the convergence behavior of our multilevel solver [13,15,21] applied to the enriched PPUM discretization using the transformed basis $\tilde{\vartheta}_{i,k}^m$ locally. We denote the respective linear system on level k by

$$A_k \tilde{u}_k = \hat{f}_k, \tag{4.25}$$

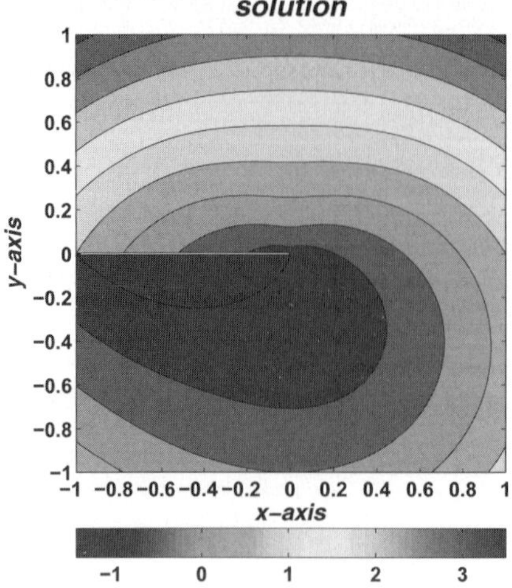

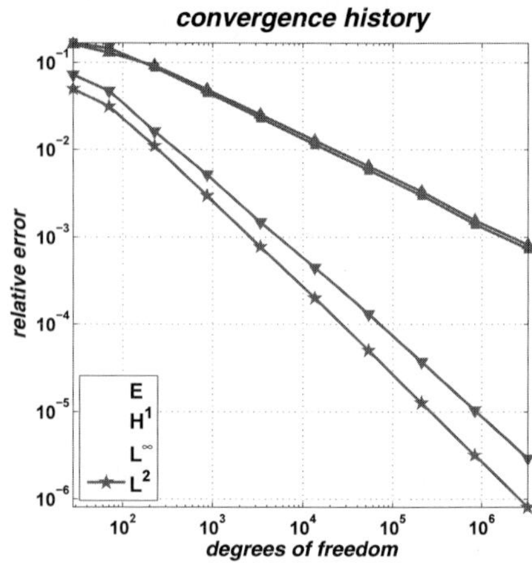

Figure 4.2. Contour plot of the solution (4.23). (*See also* Color Plate on page 392)

Figure 4.3. Convergence history of the measured relative errors e (4.20) with respect to the complete domain Ω in the L^∞-norm, the L^2-norm, the H^1-norm, and the energy-norm on the respective level (denoted by E in the legend). (*See also* Color Plate on page 392)

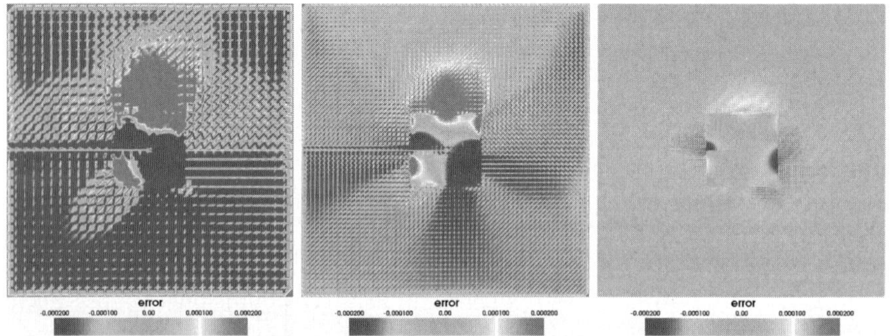

Figure 4.4. Contour plot of the error $u_k^{\mathrm{PU}} - u$ for $k = 5, 6, 7$ (from left to right). (*See also* Color Plate on page 392)

where $\tilde{u}_k := (u_{i,k}^m)$ denotes the coefficient vector associated with the PPUM function

$$u_k^{\mathrm{PU}} = \sum_{i=1}^{N_k} \varphi_{i,k} \sum_{m=1}^{d_{i,k}} u_{i,k}^m \tilde{\vartheta}_{i,k}^m$$

and $\hat{f}$ denotes a moment vector with respect to the PPUM basis functions $\langle \varphi_{i,k}\, \tilde{\vartheta}_{i,k}^m \rangle$ on level k.

We employ a standard $V(1,1)$-cycle with block-Gauß–Seidel smoother [13, 15, 21] for the iterative solution of (4.25). We consider three choices for

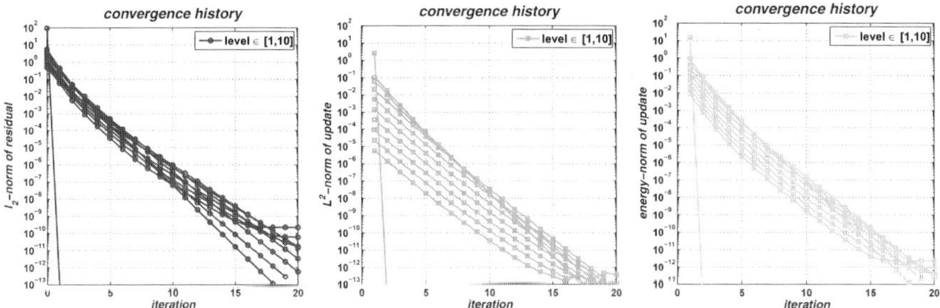

Figure 4.5. Convergence history for a $V(1,1)$-cycle multilevel iteration with block-Gauß–Seidel smoother and nested iteration initial guess (left: convergence of residual vector (4.27) in the l^2-norm, center: convergence of iteration update (4.26) in the L^2-norm, right: convergence of iteration update (4.26) in the energy-norm). (*See also* Color Plate on page 393)

the initial guess: First, a nested iteration approach, where the solution u_k^{PU} obtained on level k is used as initial guess on level $k+1$. Note that this approach avoids unphysical oscillations in the initial guess which can otherwise spoil the convergence of the iterative solution process. Therefore, we also consider the choices of a vanishing initial guess and a random valued initial guess for the iterative solver to enforce these unphysical oscillations in the initial guess. The condition number of the iteration matrix is bounded by a constant if the asymptotic convergence rate of the iterative solver is independent of the number of levels. In our context this also means that we find no deterioration of the condition number due to the enrichment.

In Figure 4.3 we give the plots of the relative errors with respect to the L^∞-norm, the L^2-norm, the H^1-norm, and the energy-norm. From these plots and the respective convergence rates ρ_{L^∞}, ρ_{L^2}, and ρ_{H^1} given in Table 4.1 we can clearly observe the anticipated optimal global convergence of our hierarchically enriched PPUM with $\rho_{L^2} = 1$ and $\rho_{H^1} = \frac{1}{2}$. On level $k = 10$ we obtain 7 digits of relative accuracy in the L^2-norm and 4 digits in the H^1-norm. In Figure 4.4 we have depicted contour plots of the error $u_k^{\mathrm{PU}} - u$ for levels $k = 5, 6, 7$ using the same scaling for all plots. We can clearly see the enrichment zone E_{tip} from these plots and observe the fast reduction of the error due to the refinement.

The convergence behavior of our multilevel solver with a nested iteration initial value is depicted in Figure 4.5. We consider the convergence of the iterative update $c_{k,\mathrm{iter}}^{\mathrm{PU}}$ associated with the coefficient vector

$$\tilde{c}_{k,\mathrm{iter}} := \tilde{u}_{k,\mathrm{iter}} - \tilde{u}_{k,\mathrm{iter}-1} \tag{4.26}$$

with respect to the L^2-norm and the energy-norm as well as the convergence of the residual vector

$$\hat{r}_{\mathrm{iter}} := \hat{f}_k - A_k \tilde{u}_{k,\mathrm{iter}} \tag{4.27}$$

in the l^2-norm. All depicted lines are essentially parallel with a gradient of -0.25 indicating that the convergence rate of our multilevel solver is 0.25 and

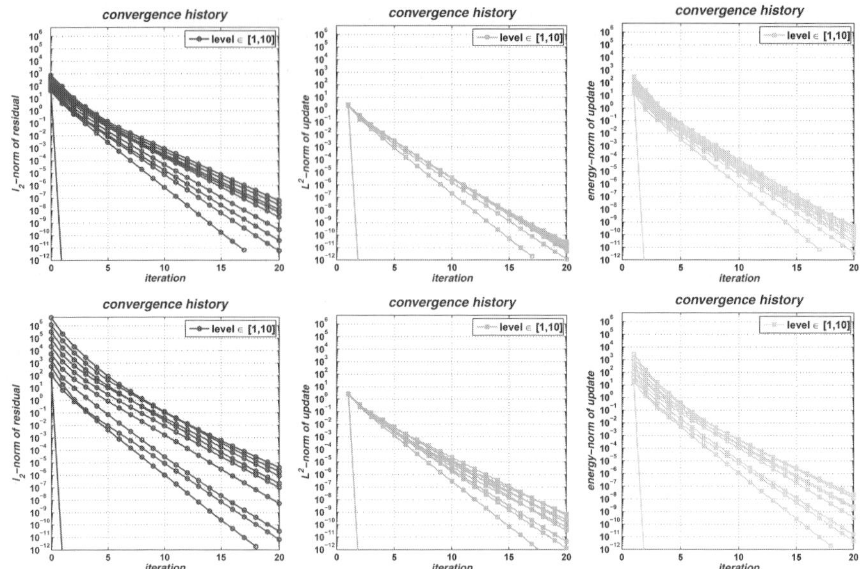

Figure 4.6. Convergence history for a $V(1,1)$-cycle multilevel iteration with block-Gauß–Seidel smoother and zero (upper row) and random (lower row) initial guess (left: convergence of residual vector (4.27) in the l^2-norm, center: convergence of iteration update (4.26) in the L^2-norm, right: convergence of iteration update (4.26) in the energy-norm). (*See also* Color Plate on page 393)

independent of the number of levels and thereby independent of the number of enriched patches and the number of enrichment functions. In Figure 4.6 we give the respective convergence behavior using a vanishing initial guess and a random valued initial guess to enforce the presence of unphysical oscillations in the (early) iterates. Again we observe that all depicted lines have essentially the same gradient -0.25 which is identical to the lines given in Figure 4.5. Hence, there is no amplification of unphysical oscillations and our multilevel solver converges with the same rate of 0.25 independent of the employed initial value. The convergence behavior of our iterative multilevel solver is independent of the number of patches and the number of enrichment degrees of freedom. Therefore, the condition number of the iteration matrix is stable and independent of these parameters. Our hierarchical enrichment scheme with local preconditioning avoids a deterioration of the condition number and thereby a deterioration of the stability due to the enrichment completely.

Finally, let us focus on the local convergence properties of our enriched PPUM discretization within the enrichment zone E_{tip} (4.24). To this end, we define three subdomains

$$E_1 := E_{\text{tip}} = \Big(-\frac{1}{4}, \frac{1}{4}\Big)^2, \quad E_2 := \Big(-\frac{1}{8}, \frac{1}{8}\Big)^2, \quad E_2 := \Big(-\frac{1}{16}, \frac{1}{16}\Big)^2 \tag{4.28}$$

Table 4.2. Relative errors e (4.20) and convergence rates ρ (4.21) with respect to the subsets E_1, E_2, and E_3 from (4.28).

J	dof	N	e_{L^∞}	ρ_{L^∞}	e_{L^2}	ρ_{L^2}	e_{H^1}	ρ_{H^1}
				with respect to E_1				
2	28	4	4.329_{-2}	0.94	3.245_{-2}	1.03	9.040_{-2}	0.72
3	70	16	3.145_{-2}	0.35	2.061_{-2}	0.50	6.313_{-2}	0.39
4	178	36	1.058_{-2}	1.17	5.325_{-3}	1.45	2.559_{-2}	0.97
5	562	100	3.385_{-3}	0.99	1.390_{-3}	1.17	9.671_{-3}	0.85
6	2002	324	1.018_{-3}	0.95	3.534_{-4}	1.08	3.495_{-3}	0.80
7	7248	1156	2.965_{-4}	0.96	8.867_{-5}	1.07	1.272_{-3}	0.79
8	25926	4356	8.477_{-5}	0.98	2.217_{-5}	1.00	1.620_{-4}	0.79
9	100298	16900	2.387_{-5}	0.94	5.539_{-6}	1.03	1.577_{-4}	0.80
10	340007	66564	6.648_{-6}	1.05	1.745_{-6}	0.95	3.933_{-4}	−0.75
				with respect to E_2				
3	28	4	1.659_{-2}	1.23	1.178_{-2}	1.33	3.500_{-2}	1.00
4	112	16	4.200_{-3}	0.99	3.707_{-3}	0.83	1.216_{-2}	0.78
5	252	36	1.209_{-3}	1.54	8.846_{-4}	1.77	4.271_{-3}	1.29
6	700	100	3.040_{-4}	1.35	2.133_{-4}	1.39	1.472_{-3}	1.04
7	2268	324	7.329_{-5}	1.21	5.210_{-5}	1.20	5.182_{-4}	0.89
8	7500	1156	1.819_{-5}	1.17	1.287_{-5}	1.17	1.901_{-4}	0.84
9	26576	4356	4.555_{-6}	1.09	3.202_{-6}	1.10	7.596_{-5}	0.72
10	101194	16900	1.671_{-6}	0.75	1.072_{-6}	0.82	6.574_{-5}	0.11
				with respect to E_3				
4	28	4	3.567_{-3}	1.69	2.447_{-3}	1.80	1.096_{-2}	1.35
5	112	16	9.241_{-4}	0.97	7.710_{-4}	0.83	4.124_{-3}	0.70
6	252	36	2.651_{-4}	1.54	1.803_{-4}	1.79	1.471_{-3}	1.27
7	700	100	6.244_{-5}	1.42	4.338_{-5}	1.39	5.247_{-4}	1.01
8	2268	324	1.495_{-5}	1.22	1.056_{-5}	1.20	1.882_{-4}	0.87
9	7376	1156	3.685_{-6}	1.19	2.612_{-6}	1.18	8.269_{-5}	0.70
10	26470	4356	1.794_{-6}	0.56	9.266_{-7}	0.81	6.490_{-5}	0.19

and measure the relative errors (4.20) and convergence rates (4.21) with respect to E_1, E_2, and E_3 of (4.28). According to Section 3.1 we anticipate to find a faster convergence within the enrichment zone $_{\text{tip}}$ than for the complete domain Ω. The plots given in Figure 4.7 and the measured rates displayed in Table 4.2 clearly show this anticipated behavior. Note that we find about 6 digits of relative accuracy in the L^2-norm and about 4 digits in the H^1-norm on E_1, i.e. in the vicinity of the singularity. Up to level $k = 9$ we find $\rho_{H^1} \approx 0.8$ within the subdomains E_1, E_2, and E_3 whereas globally we have the optimal rate $\rho_{H^1} = 0.5$. Hence, we obtain a convergence behavior better than $O(h^{3/2})$ in the enrichment zone with respect to the energy-norm using enriched linear local approximation spaces only.

On level $k = 10$ however we find a sharp jump in the measured convergence rates. For the H^1-norm we even find an increase in the error on level $k = 10$ compared with level $k = 9$. Recall from Remark 3 that we may not expect the measured convergence rates to be constant due to the employed cut-off in the construction of our projection $\Pi^*_{i,k}$ which can yield nonnested local approximation spaces. The projection operators $\Pi^*_{i,k}$ employed in this computation were based on the identity operator, i.e. on the L^2-norm. Hence, we have eliminated enrichment functions whose contribution to the approximation of the L^2-norm is insubstantial. This however may not be true for the H^1-norm.

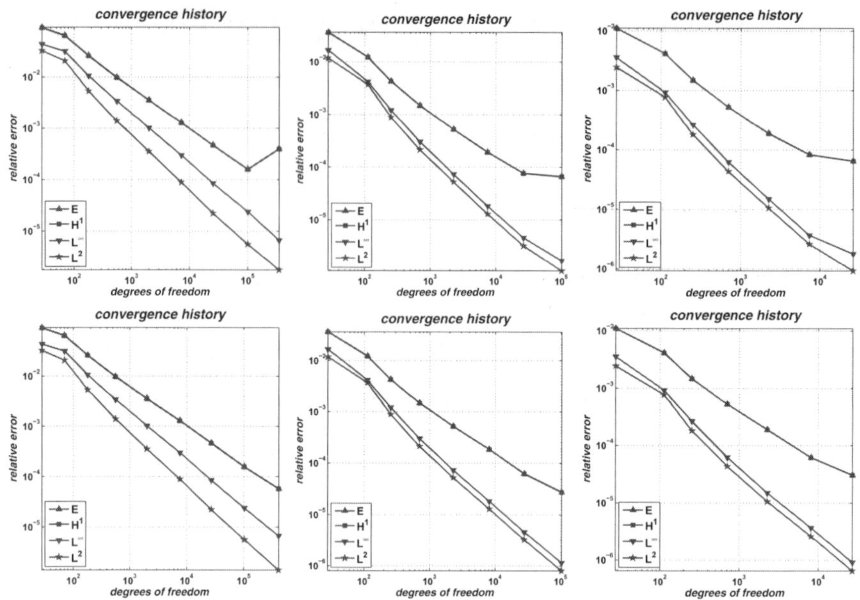

Figure 4.7. Convergence history of the measured relative errors e (4.20) with respect to the subdomains E_1 (left), E_2 (center), and E_3 (right) given in (4.28) with respect to the L^∞-norm, the L^2-norm, the H^1-norm, and the energy-norm on the respective level (denoted by E in the legend). The upper row refers to the enriched PPUM using a preconditioner based on the L^2-norm, i.e., the local mass matrix, the lower row refers to the enriched PPUM using a preconditioner based on the H^1-norm, i.e., the local stiffness matrix. (*See also* Color Plate on page 394)

According to Remark 2 changing the operator in the construction of $\Pi^*_{i,k}$ can impact the cut-off behavior. Constructing the projection $\Pi^*_{i,k}$ based on the operator $-\Delta + \mathbb{I}$ yields an elimination of functions that contribute insubstantially to the H^1-norm. This can eliminate (or will at least reduce) the jumps of the measured convergence rates for the H^1-norm (and weaker norms). From the numbers given in Table 4.3 where we employ a projection $\Pi^*_{i.k}$ based on the H^1-norm we can clearly observe this anticipated improvement, see also Figure 4.7. Now we have $\rho_{H^1} > 0.5$ on all levels and $\rho_{H^1} \approx 0.7$ on level $k = 10$ for E_1. In Figure 4.8 we have depicted the enrichment patterns within E_{tip} for the projections based on the L^2-norm and on the H^1-norm. For each patch $\omega_{i,k} \subset E_{\text{tip}}$ on level $k = 9, 10$ we have plotted the dimension of the local approximation space $V_{i,k}$, i.e., the number of shape functions $\tilde{\vartheta}^m_{i,k}$ after cut-off. From these plots we can clearly observe that more enrichment functions are present in the H^1-based approach on level $k = 10$ than for the L^2-based projection. On level $k = 9$ the enrichment patterns are almost identical and so are the measured errors, compare Table 4.2 and Table 4.3. On level $k = 10$ however we see a substantial reduction in degrees of freedom due to the cut-

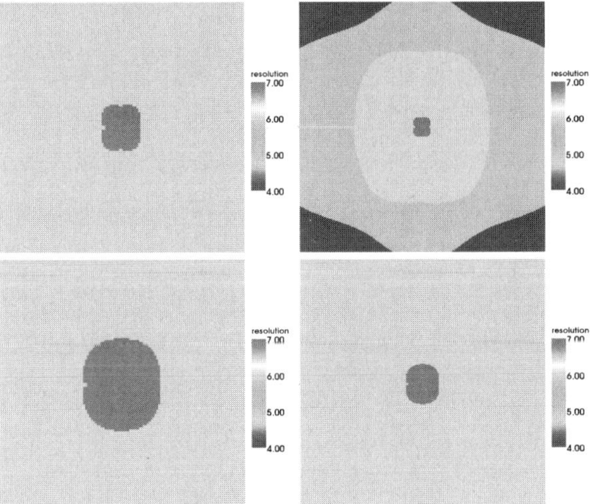

Figure 4.8. Enrichment pattern on levels $k = 9$ (left) and $k = 10$ (right) within the enrichment zone E_{tip}. Color coded is the dimension of the local approximation space $\dim(V_{i,k}) = \text{card}(\{\tilde{\vartheta}^m_{i,k}\})$ (denoted as 'resolution' in the legend). The upper row refers to the enriched PPUM using a preconditioner based on the L^2-norm, i.e., the local mass matrix, the lower row refers to the enriched PPUM using a preconditioner based on the H^1-norm, i.e., the local stiffness matrix. In both cases we used a cut-off parameter of $\epsilon = 10^{-12}$. (*See also* Color Plate on page 395)

off for the L^2-based projection and almost no reduction in degrees of freedom for the H^1-based projection. Hence, the nonnestedness of the respective local approximation spaces is as expected more severe for the L^2-based projection than for the H^1-based projection.

In summary, the presented hierarchical enrichment scheme yields a globally optimal convergence behavior of $O(h)$ in the energy-norm for the uniform h-version of the enriched PPUM without compromising the condition number of the resulting stiffness matrix. This is achieved by a local preconditioner which eliminates the near-null space of an arbitrary local operator. Within the enrichment zone the hierarchically enriched PPUM yields a convergence rate of $O(h^{1+\delta})$ in the energy-norm with $\delta > 0$.

5 Concluding Remarks

We presented an automatic hierarchical enrichment scheme for the PPUM which yields a stable discretization with optimal convergence properties globally and a kind of superconvergence within the employed enrichment zone. The core ingredients of the presented approach are a geometric hierarchy of the cover patches and a special local preconditioner. The construction of the presented preconditioner relies on the flat-top property of the employed PU.

Table 4.3. Relative errors e (4.20) and convergence rates ρ (4.21) with respect to the subsets E_1, E_2, and E_3 from (4.28). The local projections $\Pi^*_{i,k}$ are based on the H^1-norm.

J	dof	N	e_{L^∞}	ρ_{L^∞}	e_{L^2}	ρ_{L^2}	e_{H^1}	ρ_{H^1}
				with respect to E_1				
2	28	4	4.329_{-2}	0.94	3.245_{-2}	1.03	9.040_{-2}	0.72
3	70	16	3.145_{-2}	0.35	2.061_{-2}	0.50	6.313_{-2}	0.39
4	178	36	1.058_{-2}	1.17	5.325_{-3}	1.45	2.559_{-2}	0.97
5	562	100	3.385_{-3}	0.99	1.390_{-3}	1.17	9.671_{-3}	0.85
6	2002	324	1.018_{-3}	0.95	3.534_{-4}	1.08	3.495_{-3}	0.80
7	7570	1156	2.974_{-4}	0.93	8.867_{-5}	1.04	1.270_{-3}	0.76
8	27516	4356	8.477_{-5}	0.97	2.217_{-5}	1.07	4.613_{-4}	0.78
9	101490	16900	2.387_{-5}	0.97	5.539_{-6}	1.06	1.548_{-4}	0.84
10	397538	66564	6.632_{-6}	0.94	1.384_{-6}	1.02	5.701_{-5}	0.73
				with respect to E_2				
3	28	4	1.659_{-2}	1.23	1.178_{-2}	1.33	3.599_{-2}	1.00
4	112	16	4.200_{-3}	0.99	3.707_{-3}	0.83	1.216_{-2}	0.78
5	252	36	1.209_{-3}	1.54	8.846_{-4}	1.77	4.271_{-3}	1.29
6	700	100	3.040_{-4}	1.35	2.133_{-4}	1.39	1.472_{-3}	1.04
7	2268	324	7.327_{-5}	1.21	5.209_{-5}	1.20	5.182_{-4}	0.89
8	8092	1156	1.819_{-5}	1.10	1.287_{-5}	1.10	1.849_{-4}	0.81
9	27768	4356	4.556_{-6}	1.12	3.200_{-6}	1.13	6.193_{-5}	0.89
10	102632	16900	1.142_{-6}	1.06	7.982_{-7}	1.06	2.702_{-5}	0.63
				with respect to E_3				
4	28	4	3.567_{-3}	1.69	2.447_{-3}	1.80	1.096_{-2}	1.35
5	112	16	9.241_{-4}	0.97	7.710_{-4}	0.83	4.124_{-3}	0.70
6	252	36	2.651_{-4}	1.54	1.803_{-4}	1.79	1.471_{-3}	1.27
7	700	100	6.244_{-5}	1.42	4.337_{-5}	1.39	5.247_{-4}	1.01
8	2268	324	1.495_{-5}	1.22	1.056_{-5}	1.20	1.882_{-4}	0.87
9	8092	1156	3.685_{-6}	1.10	2.606_{-6}	1.10	6.133_{-5}	0.88
10	27368	4356	9.081_{-7}	1.15	6.484_{-7}	1.14	3.038_{-5}	0.58

Acknowledgement. This work was supported in part by the Sonderforschungsbereich 611 *Singular phenomena and scaling in mathematical models* funded by the *Deutsche Forschungsgemeinschaft.*

References

1. I. Babuška, U. Banerjee, and J. E. Osborn, *Survey of Meshless and Generalized Finite Element Methods: A Unified Approach*, Acta Numerica, (2003), pp. 1–125.
2. I. Babuška and J. M. Melenk, *The Partition of Unity Method*, Int. J. Numer. Meth. Engrg., 40 (1997), pp. 727–758.
3. S. Beissel and T. Belytschko, *Nodal Integration of the Element-Free Galerkin Method*, Comput. Meth. Appl. Mech. Engrg., 139 (1996), pp. 49–74.
4. T. Belytschko and T. Black, *Elastic crack growth in finite elements with minimal remeshing*, Int. J. Numer. Meth. Engrg., 45 (1999), pp. 601–620.
5. T. Belytschko, Y. Y. Lu, and L. Gu, *Crack propagation by element-free galerkin methods*, Engrg. Frac. Mech., 51 (1995), pp. 295–315.
6. T. Belytschko, N. Moës, S. Usui, and C. Parimi, *Arbitrary discontinuities in finite elements*, Int. J. Numer. Meth. Engrg., 50 (2001), pp. 993–1013.
7. E. Chahine, P. Laborde, J. Pommier, Y. Renard, and M. Salaün, *Study of some optimal xfem type methods*, in Advances in Meshfree Techniques, V. M. A.

Leitao, C. J. S. Alves, and C. A. M. Duarte, eds., vol. 5 of Computational Methods in Applied Sciences, Springer, 2007.

8. J. S. Chen, C. T. Wu, S. Yoon, and Y. You, *A Stabilized Conforming Nodal Integration for Galerkin Mesh-free Methods*, Int. J. Numer. Meth. Engrg., 50 (2001), pp. 435–466.
9. J. Dolbow and T. Belytschko, *Numerical Integration of the Galerkin Weak Form in Meshfree Methods*, Comput. Mech., 23 (1999), pp. 219–230.
10. C. A. Duarte, L. G. Reno, and A. Simone, *A higher order generalized fem for through-the-thickness branched cracks*, Int. J. Numer. Meth. Engrg., 72 (2007), pp. 325–351.
11. C. A. M. Duarte, I. Babuška, and J. T. Oden, *Generalized Finite Element Methods for Three Dimensional Structural Mechanics Problems*, Comput. Struc., 77 (2000), pp. 215–232.
12. C. A. M. Duarte, O. N. H. T. J. Liszka, and W. W. Tworzydlo, *A generalized finite element method for the simulation of three-dimensional dynamic crack propagation*, Int. J. Numer. Meth. Engrg., 190 (2001), pp. 2227–2262.
13. M. Griebel, P. Oswald, and M. A. Schweitzer, *A Particle-Partition of Unity Method—Part VI: A p-robust Multilevel Preconditioner*, in Meshfree Methods for Partial Differential Equations II, M. Griebel and M. A. Schweitzer, eds., vol. 43 of Lecture Notes in Computational Science and Engineering, Springer, 2005, pp. 71–92.
14. M. Griebel and M. A. Schweitzer, *A Particle-Partition of Unity Method—Part II: Efficient Cover Construction and Reliable Integration*, SIAM J. Sci. Comput., 23 (2002), pp. 1655–1682.
15. ———, *A Particle-Partition of Unity Method—Part III: A Multilevel Solver*, SIAM J. Sci. Comput., 24 (2002), pp. 377–409.
16. ———, *A Particle-Partition of Unity Method—Part V: Boundary Conditions*, in Geometric Analysis and Nonlinear Partial Differential Equations, S. Hildebrandt and H. Karcher, eds., Springer, 2002, pp. 517–540.
17. ———, *A Particle-Partition of Unity Method—Part VII: Adaptivity*, in Meshfree Methods for Partial Differential Equations III, M. Griebel and M. A. Schweitzer, eds., vol. 57 of Lecture Notes in Computational Science and Engineering, Springer, 2006, pp. 121–148.
18. N. Moës, J. Dolbow, and T. Belytschko, *A finite element method for crack growth without remeshing*, Int. J. Numer. Meth. Engrg., 46 (1999), pp. 131–150.
19. J. Nitsche, *Über ein Variationsprinzip zur Lösung von Dirichlet-Problemen bei Verwendung von Teilräumen, die keinen Randbedingungen unterworfen sind*, Abh. Math. Sem. Univ. Hamburg, 36 (1970–1971), pp. 9–15.
20. J. T. Oden and C. A. Duarte, *Clouds, Cracks and FEM's*, Recent Developments in Computational and Applied Mechanics, 1997, pp. 302–321.
21. M. A. Schweitzer, *A Parallel Multilevel Partition of Unity Method for Elliptic Partial Differential Equations*, vol. 29 of Lecture Notes in Computational Science and Engineering, Springer, 2003.
22. ———, *An adaptive hp-version of the multilevel particle–partition of unity method*, Comput. Meth. Appl. Mech. Engrg., (2008). accepted.

A Framework For Studying The RKEM Representation of Discrete Point Sets

Daniel C. Simkins, Jr.[1], Nathan Collier[1], and Mario Juha[1], and Lisa B. Whitenack[2]

[1] Department of Civil and Environmental Engineering
University of South Florida
4202 E. Fowler Ave. ENB118
Tampa, FL 33620 USA dsimkins@eng.usf.edu

[2] Department of Biology
University of South Florida
4202 E. Fowler Ave. SCA110
Tampa, FL 33620 USA

Summary. The application of engineering analysis to new areas, such as nanomechanics and the life sciences, often involves geometric problem domains defined by discrete point sets as measured from diagnostic equipment. The development of a suitable mesh for finite element analysis can be a tedious task. One approach to simplifying the geometric description is to use a parametrized set of basis functions, and fit the parameters to the data set. In this paper, we discuss the problem of determining suitable parameters for the Reproducing Kernel Element Method representation of discrete point sets, and in particular the solution of the inverse problem of determining pre-image evaluation points in the parametric space that correspond to a given input point. We justify our solution by posing a theoretical framework and an error indicator.

Key words: RKEM, geometry representation, discrete point sets.

1 Introduction

A topic of recent research has been in improving the representation of the geometric domain for engineering analyses. The motivation for this arises from several sources. First, the polygonal or polyhedral representation of smooth geometries used by finite elements leads to errors inherent in the approximation of the domain. Secondly, the process of converting computer aided design (CAD) geometries based on NURBS into FEM meshes can be time consuming. A unified and improved representation would facilitate the interaction and data exchange between analysis and design. A similar, but substantially different problem arises in other areas of research, e.g. nanomechanics and the life sciences. In these areas, the problem domain often comes in the form of a discrete point set obtained from experimental equipment, e.g. atomic force microscopes (AFM), computed tomography (CT), or magnetic resonance imaging (MRI). The development of suitable FEM meshes is still troublesome, but there is an additional difficulty. This difficulty is that, unlike in CAD, where the geometry is created in terms of a known basis set, the discrete point set has no such existing definition and can not be interrogated for additional information, e.g. derivatives. The absence of an underlying known basis representation introduces a new problem to parametric formulations. In a parametric formulation, the basis functions are defined in terms of coordinates in a parametric space, and when a weighted sum of these functions are evaluated at a point in the parametric space, one yields a point in the problem domain. We call this the forward problem. In the case of an input point set, one must first solve the inverse problem of determining the parametric coordinates that are the pre-image of the input. This paper is organized as follows: in §1.1 we layout the motivating problem; in §2 we review the highlights of the RKEM method, in §3 we summarize geometry representation using RKEM, in §4 we introduce the theoretical framework for studying the method with an error indicator and examples.

1.1 Motivating Problem

The motivation for this research is to reduce the effort and complexity of performing engineering analysis on problems whose geometry is defined by a discrete point set. One of the authors (Whitenack) is studying the mechanical behavior of shark teeth. The geometry is defined by performing CT or micro CT scans of a tooth. These scans consist of a sequence of 2D slices of greyscale pixel images. These slices are processed into a mesh for finite element (FEM) analysis, in this case using the commercial program Strand7. In Table 1.1, we list the various steps performed, the time range rquired and indicate whether it is machine or human time for processing a single tooth. The total time ranges from 8 - 32 hours, and we see that human time accounts for 72% in the worst case. We believe that human time is much more valuable than machine time, hence our goal of reducing the amount of human effort to perform an analysis. Our research focuses on steps 4,5,6 which are geometry representation, and step 8, analysis.The method we pursue falls into the general area now known as isogeometric analysis, [2], which intimately couples the analysis and geometry representation. In this paper, we will only focus on the geometry representation aspects.

2 Overview of RKEM

The Reproducing Kernel Element Method was introduced in a series of four papers, [4], [3], [5] and [9]. Further developments are presented in the PhD

Table 1.1. Shark tooth analysis process

Step	Activity	Time range (min)	Human/Machine
1	CT scan	15-30	machine
2	micro CT scan	up to 180	machine
3	Segmentation	30-60	human
4	Geometry cleanup	120-360	human
5	Meshing	120-960	human
6	Meshing	120-180	machine
7	Loads, boundary cond. etc.	60-120	human
8	Analysis & post processing	15-60	machine

dissertation [7], including an initial exploration of the use of RKEM for geometry representation. Here, we review the basics of RKEM paying particular attention to the properties that prove useful in the present context. For full details, the reader is referred to the above references.

We will denote the RKEM domain by Ω throughout this article.

The RKEM is a hybrid of finite element shape functions and a meshfree kernel in such a way that the following properties hold for the RKEM shape functions:

1. The shape functions are Generalized Hermite interpolants. By this we mean that at each node the primary variable and various of its derivatives are interpolated, in analogy to Hermite polynomial interpolation, [10]. We use the notation for the shape functions at node I: Ψ_I^α, where α is a multi-index indicating the derivative being interpolated. For example, see Eqn. 2.3.
2. The shape functions possess the Higher-order Kronecker-δ property:

$$D^\alpha \Psi_I^{(\beta)} \Big|_{\mathbf{x}=\mathbf{x}_J} = \delta_{IJ}\delta_{\alpha\beta}, \quad \mathbf{x}_I,\ \mathbf{x}_J \in \Omega, \quad |\alpha|, |\beta| \leq m, \tag{2.1}$$

3. The shape functions form a Partition of Unity: $\sum_I \Psi_I^\alpha(x) = 1 \quad \forall x \in \Omega; \quad |\alpha| = 0$
4. The shape functions possess the global reproducing property:

$$\sum_I \left\{ \sum_\alpha \Psi_I^\alpha(\mathbf{x})(D^\alpha \mathbf{x}^\beta)|_{\mathbf{x}=\mathbf{x}_I} \right\} = \mathbf{x}^\beta, \quad \forall\, \mathbf{x} \in \Omega;\ |\beta| \leq P \tag{2.2}$$

   where P is the highest degree complete polynomial in the so-called global partition polynomials used to construct the element. See [9] for complete details.
5. Each shape function has compact support with fixed size, regardless of reproducing order. The support size is dependent solely on the topology of the RKEM mesh.

6. Smoothness: Given a kernel function that is C^n continuous, the resulting RKEM shape functions are also C^n ([7]).

To guarantee the above properties, RKEM meshes must satisfy a *quasi-uniformity* condition as discussed in [3]. In simple terms, this condition places some restriction on the aspect ratio of individual elements and the gradation of element sizes within the mesh.

Throughout this article, we will focus on problems in two dimensions using the the T9P2I1 element. The complete derivation and implementation details of the T9P2I1 triangular element has been published in [7]. The T9P2I1 element globally reproduces quadratic fields and interpolates the first order derivatives. At a node I there are three shape functions,

$$\Psi_I^{00}\,;\Psi_I^{10}\,;\Psi_I^{01}.$$

A function $f(u,v)$ can be interpolated as

$$f(u,v) = \sum_{I=1}^{N} \Psi_I^{00}(u,v)f^I + \Psi_I^{10}(u,v)f^I_{,u} + \Psi_I^{01}(u,v)f^I_{,v}, \tag{2.3}$$

where N is the number of nodes in the mesh and f, $f_{,u}$, $f_{,v}$ are the nodal data. The compact support of the RKEM shape functions limits the number of nodes that will actually contribute at a point, but for notational convenience, we sum over all nodes.

3 Geometry Representation

The shape functions computed via RKEM can be viewed simply as interpolants for use in any problem of interpolation. One such problem is the representation of geometry within a computer. In this section, we discuss our approach to using RKEM shape functions for geometry representation, for complete details, see [8].

3.1 The Abstract Problem

A geometric body can be viewed as a point set $\mathcal{B} \subset \mathbb{R}^n$. To represent it in a computer, we select a finite set of functions $\mathcal{F} : \Omega \to \mathbb{R}$ and a finite set of weights $\mathcal{W}$ such that an approximation to the body $\mathcal{B}$ is obtained by interpolating the points $\mathcal{W}$ with the functions in $\mathcal{F}$. In the present work, the functions in $\mathcal{F}$ consist of RKEM generalized Hermite shape functions. In other representations, the function space $\mathcal{B}$ will consist of other functions. See Eqn. 3.4 below. For convenience, we define the following symbols:

- n is the spatial dimension.
- $\mathcal{B} \subset \mathbb{R}^n$ represents the body

- $\tilde{\mathcal{B}} \subset \mathbb{R}^n$ is the approximate RKEM representation of the body.
- $\partial\mathcal{B} \subset \mathbb{R}^{n-1}$ is the boundary of the body.
- $\partial\tilde{\mathcal{B}} \subset \mathbb{R}^{n-1}$ is the boundary of the approximate body.
- $\mathbf{x}$ is a point in the body, $\mathbf{x} \in \mathcal{B}$.
- $\tilde{\mathbf{x}}$ is a point in the approximate body, $\tilde{\mathbf{x}} \in \tilde{\mathcal{B}}$.
- $\mathcal{M}$ is the RKEM mesh used to approximate the body and consists of both the domain, $\Omega \subset \mathbb{R}^n$ with boundary $\partial\Omega$, and the topology, or connectivity.
- $\mathbf{u}$ is a point in the RKEM domain, $\mathbf{u} \in \Omega$.
- $\mathcal{W}$ is the set of weights associated with the mesh $\mathcal{M}$ for interpolating the body. $\mathcal{W}_i$ denotes the set of weights for interpolating the ith component of a point $\tilde{x}_i$. In particular, the set $\mathcal{W}$ consists of the primary nodal unknowns and their derivatives.
- $\mathcal{I}$ is the RKEM interpolation operator on the RKEM mesh evaluated at a point in Ω for a set of nodal data, $\mathcal{W}$, i.e., $\mathcal{I} : \mathcal{M} \otimes \mathcal{W} \to \tilde{\mathcal{B}}$. While the set $\mathcal{W}$ is subordinate to a particular mesh $\mathcal{M}$, for clarity, we emphasize this by using the notation $\mathcal{I}_{\mathcal{M}}(\mathcal{W}; \mathbf{u})$ for $\mathbf{u} \in \Omega$.

Using these definitions then, we have $\tilde{\mathcal{B}} \approx \mathcal{B}$ and $\mathcal{I}_{\mathcal{M}}(\mathcal{W}; \Omega) = \tilde{\mathcal{B}}$. This idea and the definitions are depicted in Fig. 3.1.

Our approach uses two basic concepts. First, the components of a point in $\tilde{\mathcal{B}}$ are considered to be independent functions of points in Ω, and the RKEM mesh, $\mathcal{M}$, is *conforming*. The first condition is succinctly stated by using standard index notation:

$$\tilde{x}_i(\mathbf{u}) = \mathcal{I}_{\mathcal{M}}(\mathcal{W}_i, \mathbf{u})$$

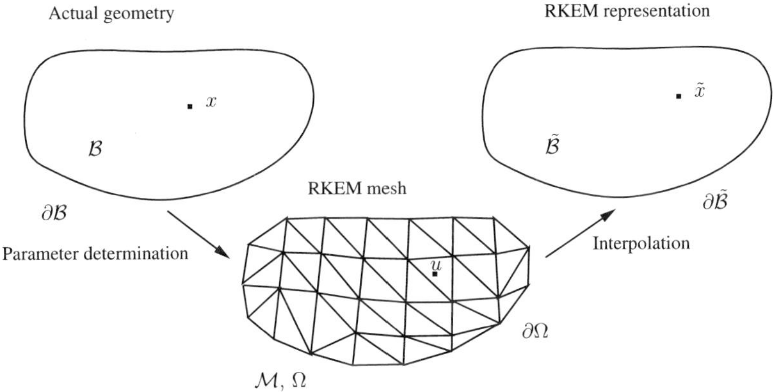

Figure 3.1. Lagrange Analogy for Geometry Representation.

For example, in two dimensions using the T9P2I1 RKEM element to interpolate the point $\mathbf{x} \in \mathcal{B}$ the approximate point, $\tilde{\mathbf{x}}$, is obtained through

$$\tilde{x}_i = \sum_{I=1}^{N} \mathbf{f}_I \mathbf{w}_i^I \quad \mathbf{f}_I \in \mathcal{F}, \quad \mathbf{w}_I \in \mathcal{W} \tag{3.4}$$

where,

$$\mathbf{f}_I = \left[\Psi_I^{00}\ \Psi_I^{10}\ \Psi_I^{01}\right]$$

and

$$\mathbf{w}_i^I = \left[x_i(u,v)|_{\mathbf{u}^I}\ (x_i(u,v)_{,u})|_{\mathbf{u}^I}\ (x_i(u,v)_{,v})|_{\mathbf{u}^I}\right]^T$$

and x_i^I represents the ith component of the body point $\mathbf{x}$ associated with the Ith RKEM node with coordinate $\mathbf{u}^I$. The second concept leads to the following definition:

Definition 1 (Conforming Mesh). *An RKEM mesh, $\mathcal{M}$, used to represent a body $\mathcal{B}$, is said to be* conforming *if every boundary node in the mesh lies exactly on the boundary of the body, i.e., if node I lies on the boundary of $\mathcal{M}$ and has coordinates $\mathbf{x}_I$, then $\mathbf{x}_I \in \partial\mathcal{B}$. See Figs. 3.2 and 3.3.*

These two concepts then lead to the notion that the approximate geometry is the image of the RKEM mesh:

$$\tilde{\mathcal{B}} = \mathcal{I}_{\mathcal{M}}(\mathcal{W}, \Omega) \tag{3.5}$$

in analogy with the Lagrange description of solid bodies where the deformation map is defined by the RKEM interpolation.

The problem, then, is to find an RKEM conforming mesh for the point set and determine the nodal data. One can choose to model an object by its boundary (surface), or as a complete volume. Since our interest is analysis, we

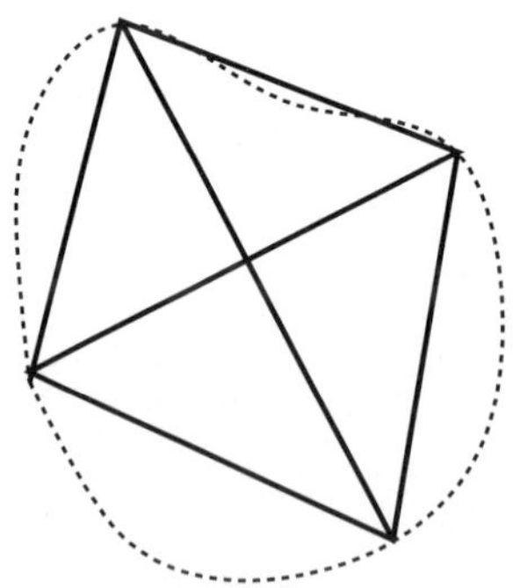

Figure 3.2. Conforming Mesh.

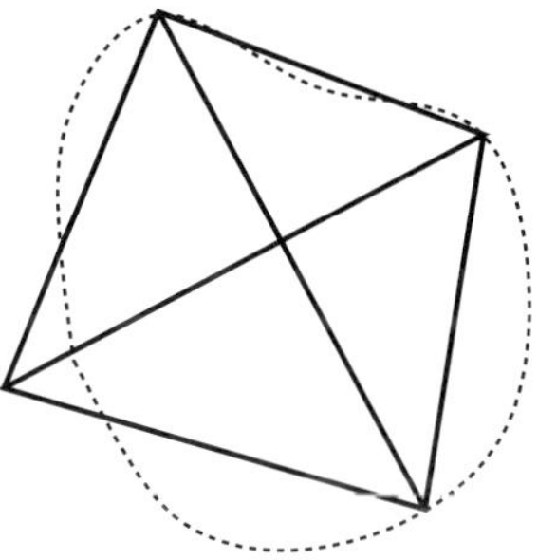

Figure 3.3. Non-conforming Mesh.

focus on volume representations. The nodal data is found by solving a least squares problem, setup as follows:
Given a surface point set in spatial dimension $n-1$, we first construct a conforming mesh in dimension n. We need to construct a set of linear equations of the form in Eqn. 3.6

$$(x_i)_j = (\tilde{x}_i)_j(\mathbf{u}_j) = \mathcal{I}_{\mathcal{M}}(\mathcal{W}_i, \mathbf{u}_j); \quad j = 1, ..., M \ ; \quad i = 1, ..., n \tag{3.6}$$

where M is the number of points to be fit. Two problems arise. The first is finding the pre-images $\mathbf{u}$ of the input boundary points. The assumptions is that

$$\partial\tilde{\mathcal{B}} = \mathcal{I}_{\mathcal{M}}(\mathcal{W}, \partial\Omega)$$

so the input boundary points should map to points on the edges of the RKEM mesh boundary. We have termed the set of body points $\mathbf{x}_i$ and their associated RKEM image points $\mathbf{u}_i$ as the *auxiliary point set.* We investigated several possible solutions to determining the auxiliary points, but found that the simplest methods worked well. The projection in 2D is quite straight forward: simply find the normal projection of the point on the body onto the nearest element edge. Since the mesh is conforming, the projection is well-defined as long as the boundary between the adjacent RKEM nodes does not vary too widely. See Fig. 3.4. One of the main points of this paper is to argue that these projections are satisfactory. The three dimensional case is somewhat more difficult since the direction in which to project the point is not well defined. The input boundary pionts now represent a surface. To solve the inverse problem for the auxiliary points, we assume the surface point set has been triangulated. Each boundary point is the vertex of some number of surface facet triangles and each facet defines an outward normal vector. These normal vectors can be used to determine a direction along which one can project the vertex onto a face of an element in the RKEM mesh. We found that the weighted normal method given in [6] provided good results, as depicted in Fig. 3.5. The second issue is the generation of interior data points and their auxiliary points. When the input data only includes boundary points, we must

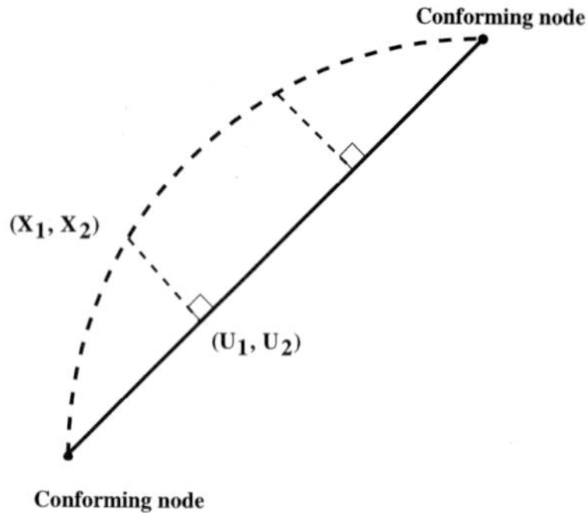

Figure 3.4. Two dimensional projection.

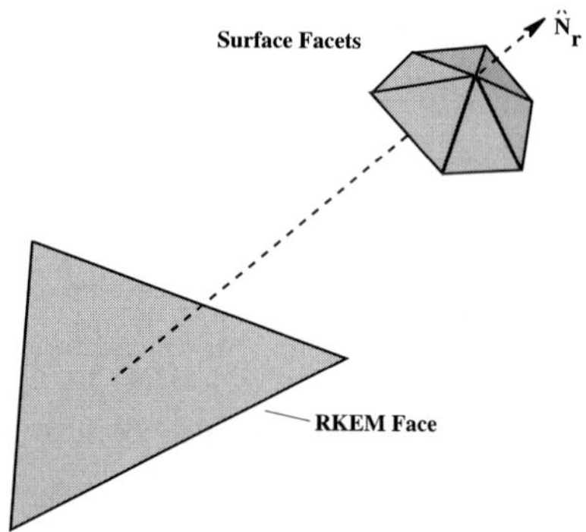

Figure 3.5. Three dimensional projection.

generate the interior points. Our procedure is to take the RKEM mesh points from the boundary auxiliary point list $\mathbf{u}_i$, and use a mesh generator to generate the interior. This gives us a set of $\mathbf{u}_i$ for the entire body. At this point, we only know the associated body points, $\mathbf{x}_i$, for the boundary. To determine the body points for the interior, we use a linear finite element solution on the volume mesh using all essential boundary conditions on the known boundary auxiliary points. This generates a displacement field for the interior points that we use to populate the interior auxiliary point pairs. Given the auxiliary points, we now solve the set of linear algebraic equations, Eqn. 3.6 for the nodal data. Since it may be impractical to require an exact number of points to satisfy the number of unknowns, we solve the equation in the least squares sense using an SVD factorization taking care to insure that the condition number R is not too big, in our examples $R < 10^{12}$. Our work has focused on boundary point sets. Volume point sets, if available may provide an easier procedure, but that is future research.

4 Framework

In this section we suggest a theoretical framework for both understanding and judging the quality of RKEM geometry representation. This framework follows naturally from two observations. First, as already discussed, the geometry representation is analogous to the deformation map in Lagrangian mechanics where the RKEM mesh is viewed as the reference frame, and the interpolated geometry as the current frame ([1]). Second, the input geometry is only known as a discrete point set, hence, a mesh of the point set is a valid definition of the entire body. Thus, the mesh of all the data points could be used as an RKEM mesh for geometry representation. In this case, the interpolation operator is the identity, the RKEM mesh, the actual geometry, and the interpolated geometry are all three identical. These observations lead us to the following conjectures:

Conjecture 1. For a given body $\mathcal{B}$ defined by a discrete point set that can be meshed with a quasi-uniform mesh, there exists a sequence of RKEM meshes, $\{\mathcal{M}_n\}$, and associated RKEM domains, $\{\Omega_n\}$, such that

$$\lim_n \mathcal{I}(\mathcal{M}_n, \Omega_n) \to \mathbb{I}$$

where $\mathbb{I}$ is the identity operator acting on points in the RKEM mesh, $\mathbf{u}$.

Note that each conforming RKEM mesh may define a different RKEM domain, hence the need for associating a sequence of domains with the meshes. Also note that any given refinement may not lead to an improvement. The conjecture is that a sequence of refinements exists, not that all refinements lead to the identity. A technical point arises in that RKEM meshes must meet a quasi-uniform condition, [3], and the mesh of the input point set may not meet that criteria.

Conjecture 2. Let the sequence of sets $\{\mathbf{u}(\mathbf{x})\}_n$ be the preimages corresponding to the sequence of meshes in Conjecture 1. Then

$$\lim_n (\mathbf{u}_i)_n \to \mathbf{x}_i.$$

This conjecture means that, as the RKEM mesh approaches the actual geometry, the preimage points become the geometry points themselves. If true, this justifies the simple procedure we use for finding the auxiliary points. While for any given mesh, the projection technique for finding the auxiliary points may not be optimal for that mesh, we can be sure that, under refinement, it is sufficient.

4.1 Error Indicator

Given an RKEM representation of a geometry, it is desireable to be able to compute a measure that can be regarded as measuring the quality of that representation. Since the representation is a function of the geometry points as a function of the RKEM domain points,

$$\tilde{\mathbf{x}}_i(\mathbf{u}_i)$$

then the determinant of the Jacobian of the transformation

$$J := \det \frac{\partial \tilde{\mathbf{x}}}{\partial \mathbf{u}} \tag{4.7}$$

geometrically represents how much deformation each differential element experiences. This, then, can be used as an error indicator. Assuming the conjectures are true, then the determinant should approach unity under refinement. Further, it can be used to locate regions where the reperesentation is not yet sufficiently refined. The RKEM shape functions are globally smooth, and the necessary derivatives for computing the Jacobian are easily computed, so this is a particularly attractive measure. As a measure of the overall representation, we use the L_2-norm defined by

$$e_{L_2} := \frac{\left[\int_\Omega |J-1|^2 \, d\Omega\right]^{\frac{1}{2}}}{\int_\Omega 1 \, d\Omega} \tag{4.8}$$

which represents deviation from unity of J per unit volume.

4.2 Example

We provide an example using the framework to study the performance of representations. The example consists of data from one slice of a CT scan of a tooth from a bullshark, the input data used by the RKEM representation scheme is shown in Fig. 4.6. Applying the framework, we computed a series

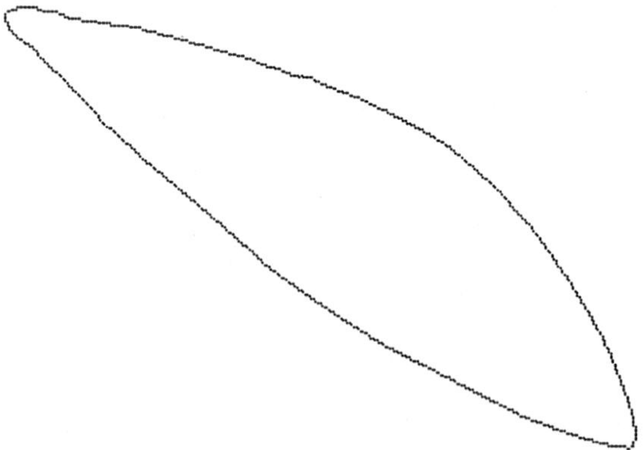

Figure 4.6. Bullshark point set.

of representations for a sequence of refined meshes. Figure 4.7 shows the plot of L_2-norm versus number of nodes in the RKEM mesh. We can see that the error measure decreases nicely under refinement. The point-wise jacobian determinant can also be used to locate regions where refinement is required. The sequence of globally refined (h-refinement) meshes and point-wise jacobian determinants shown in Figs. 4.8–4.11 demonstrate this process. Note that near the tips J shows deviation from one in the coarser mesh, and improves under global refinement. It should be possible to locally refine in the regions of interest to achieve the same result. The following result demonstrates that a graded, rather than uniform, mesh can be used successfully. It is the grading property that should allow local h-refinement. Finally, for this example and input data, we were able to trivially generate a graded quasi-uniform RKEM

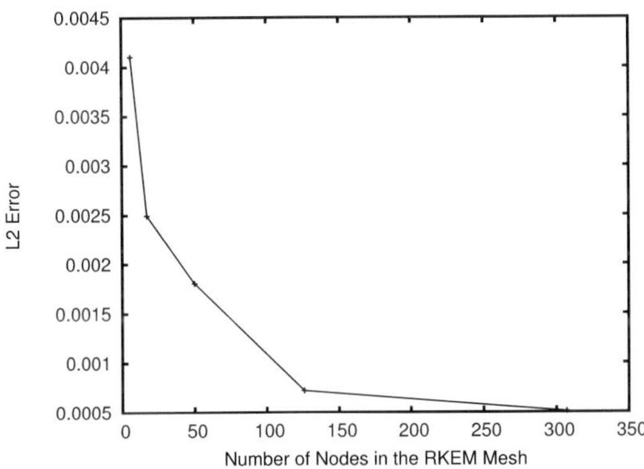

Figure 4.7. L_2-norm convergence of representation for the bullshark tooth.

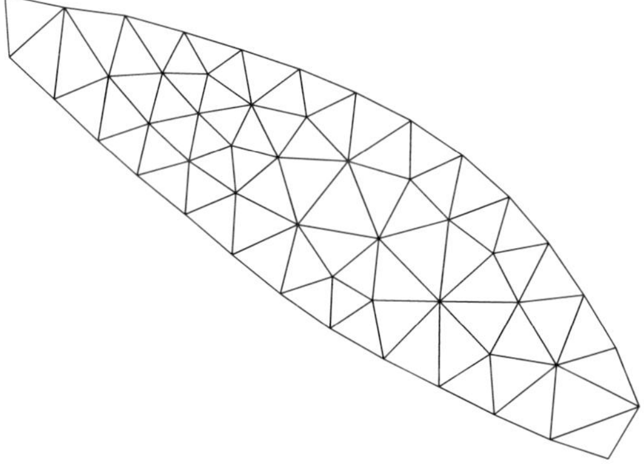

Figure 4.8. 27 node RKEM mesh.

Figure 4.9. Jacobian determinant for 27 node RKEM mesh. (*See also* Color Plate on page 395)

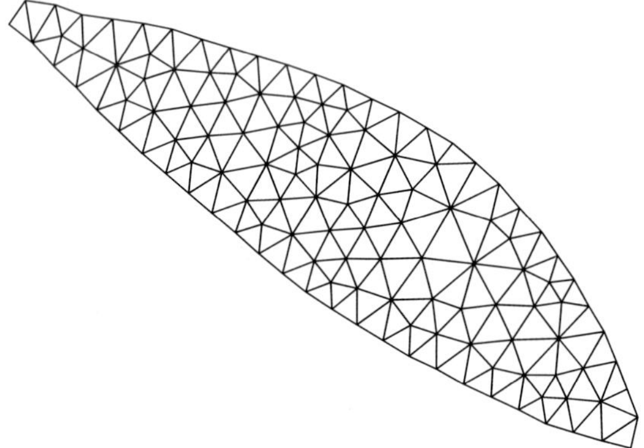

Figure 4.10. 53 node RKEM mesh.

Figure 4.11. Jacobian determinant for 53 node RKEM mesh. (*See also* Color Plate on page 396)

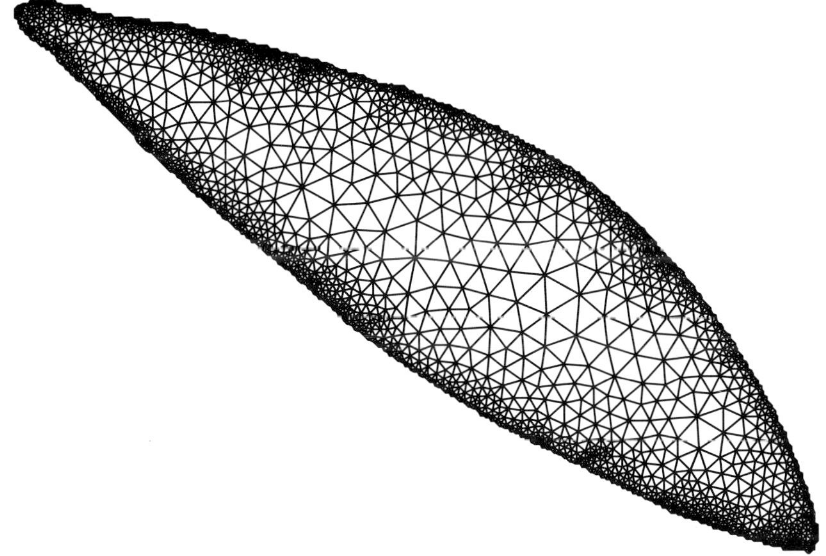

Figure 4.12. Quasi-uniform RKEM mesh exactly representing the input data.

mesh completely resolving the input data set, yielding an RKEM identity operator on this mesh.

5 Conclusion

In this paper we introduced a framework from which we can study the performance of geometry representation using RKEM. We posed some conjectures that apply to discrete points sets to argue the plausibility of both the representation scheme and the error metric used. We provided an example to indicate that this works and is the basis for future work.

References

1. G. Holzapfel, *Nonlinear Solid Mechanics*, John Wiley, New York, 2000.
2. T. Hughes, J. Cottrell, and Y. Bazilevs, *Isogeometric analysis: CAD, finite elements, NURBS, exact geometry and mesh refinement*, Computer Methods in Applied Mechanics and Engineering, 194 (2005), pp. 4135–4195.
3. S. Li, H. Lu, W. Han, W. K. Liu, and D. C. Simkins, Jr., *Reproducing kernel element method, Part II. Global conforming I^m/C^n hierarchy*, Computer Methods in Applied Mechanics and Engineering, 193 (2004), pp. 953–987.
4. W. K. Liu, W. Han, H. Lu, S. Li, and J. Cao, *Reproducing kernel element method: Part I. Theoretical formulation*, Computer Methods in Applied Mechanics and Engineering, 193 (2004), pp. 933–951.

5. H. Lu, S. Li, D. C. Simkins, Jr., W. K. Liu, and J. Cao, *Reproducing kernel element method Part III. Generalized enrichment and applications*, Computer Methods in Applied Mechanics and Engineering, 193 (2004), pp. 989–1011.
6. S. Owen, D. White, and T. Tautges, *Facet-based surfaces for 3d mesh generation*, in Proceedings of 11th International Meshing Roundtable, 2002, pp. 297–312.
7. D. C. Simkins, Jr., *General Reproducing Kernel Element Hierarchies*, PhD thesis, University of California, Berkeley, CA, May 2004.
8. D. C. Simkins, Jr., A. Kumar, N. Collier, and L. Whitenack, *Geometry representation, modification and iterative design using RKEM*, Computer Methods in Applied Mechanics and Engineering, 196 (2007), pp. 4304–4320.
9. D. C. Simkins, Jr., S. Li, H. Lu, , and W. K. Liu, *Reproducing kernel element method Part IV. Globally compatible $C^n(n \geq 1)$ triangular hierarchy*, Computer Methods in Applied Mechanics and Engineering, 193 (2004), pp. 1013–1034.
10. J. Stoer and R. Bulirsch, *Introduction to Numerical Analysis*, Springer-Verlag, New York, 2 ed., 1993.

Coupling of the CFD and the Droplet Population Balance Equation with the Finite Pointset Method

Sudarshan Tiwari[1], Christian Drumm[2], Menwer Attarakih[3], Jörg Kuhnert[1], and Hans-Jörg Bart[2]

[1] Fraunhofer-Institut für Techno- und Wirtschaftmathematik (ITWM) Fraunhofer-Platz 1, D-67663 Kaiserslautern, Germany
tiwari@itwm.fraunhofer.de, kuhnert@itwm.fhg.de

[2] Lehrstuhl f. Thermische Verfahrenstechnik, TU Kaiserslautern, Gottlieb-Daimler-Strasse, 67663 Kaiserslautern, Germany
cdrumm@mv.uni-kl.de, bart@mv.uni-kl.de

[3] Al-Balqa Applied University, Faculty of Eng. Tech., Chemical Eng. Department, 11134 Amman, Jordan
attarakih@yahoo.com

Summary. In this paper we present the liquid-liquid two-phase flow simulations of a stirred extraction column with the help of our own developed meshfree method called the Finite Pointset Method (FPM). The primary (continuous) phase is modeled by the incompressible Navier-Stokes equations. The motion of the secondary (dispersed) phase is simulated by solving the equation of motion in which inertia, drag and buoyancy forces are taken into account. The size of the droplets is obtained by solving the droplet population balance equation (DPBE). The DPBE is solved by the Sectional Quadrature Method of Moments (SQMOM). The coupling between both phases is performed by considering the momentum transfer from each phase. In this work, some simulations in two and three dimensional cases with constant breakage and aggregation kernels are presented.

Key words: Multiphase Flows, Population Balance Equations, Liquid-Liquid Extraction

1 Introduction

Liquid-liquid extraction is a separation process, which is based on the difference of the distribution of the components to be separated between two liquid phases. Liquid-liquid processes are widely applied in chemical and biochemical industries. In the simulation and design of liquid-liquid extraction columns the dispersed phase is classically assumed as a pseudo homogeneous phase, where one parameter accounts for all deviations from the ideal plug flow behavior. Nowadays, the population balance equation (PBE) forms the cornerstone for modeling polydispersed (discrete) systems arising in many engineering applications such as liquid-liquid extraction. To account for this polydispersed nature, the dispersed phase is represented in terms of a density function. In contrast to the classical approach, the evolution of this density function is dictated by different active mechanisms such as breakage and coalescence. Furthermore, the accurate prediction of the dispersed phase evolution depends strongly on the proper modeling of the continuous flow fields and hydrodynamics in which the droplets are dispersed. Up to now the design of an extraction columns without experimental pilot plant experiments has not been feasible. In current droplet population balance model (DPBM) codes [1], the dispersion model is still used for the description of the hydrodynamics and results from small scale devices (single droplet experiments) are needed to predict Sauter mean diameters and hold-up profiles [2, 18]. On the other hand, Computational Fluid Dynamics (CFD), based on a mono-disperse assumption, can predict the flow fields and hydrodynamics in a stirred RDC extraction column and deliver all necessary information for the DPBM [7, 8]. Finally, in order to properly model drop size distributions and flow fields in an extraction column without the need for pilot plant or small scale device experiments, the population balance must be coupled with turbulent CFD modeling. Only a few researchers investigated combined CFD-DPBM models in the field of liquid-liquid extraction so far but achieved encouraging results for the combined models [6, 23]. Hence, this contribution focuses on a combined CFD and DPBM to advance in this promising field of research. The coupling is done with the help of the Finite Pointset Method (FPM), which is a meshfree, Lagrangian, particle method. FPM is an in-house development of the Fraunhofer Institute. Our main goal is to incorporate the DPBM into the FPM code. In comparison to a commercial code the FPM source code is fully modifiable, which eases the implementation of the DPBM. The DPBM equations are solved by the Sectional Quadrature Method of Moments (SQMOM) [3]. This method is found to track accurately any set of low-order moments with the ability to reconstruct the shape of the distribution, so it unites the advantages of the classes method (CM) [10] and the Quadrature Method of Moments (QMOM) [15], while it minimizes the drawbacks. The SQMOM is based on the concept of primary and secondary particles, where the primary particles are responsible for the distribution reconstruction (CM), while the secondary ones are responsible for breakage and coalescence events and carry information about the distribution (QMOM). As a first attempt in the coupling procedure, the normal limitation that only one set of Navier-Stokes equations is solved for all droplets [13] have been retained. Accordingly, only one primary particle is used in the SQMOM which is equal the QMOM using 4 moments for this special case, however with efficient numerical implementation. For validation of the coupled algorithm, a five compartment section of a Rotating Disc Contactor (RDC) was modeled in 2D and 3D. The simple 2D model of the column was used to estimate the parameters for breakage and coalescence, which is not feasible in a complex and CPU-time consuming 3D model. In this paper constant breakage and aggregation kernels are assumed and simulations for the system butylacetate-water were carried out for the same operating conditions as in experiments of Simon [19].

The paper is organized as follows: in section 2, the governing equations are presented, while in section 3, the strategy to incorporate the DPBM in FPM code and numerical schemes for these equations. Section 4 contains results and comparisons of the two phase flows and some conclusions are presented in section 5.

2 Governing equations

2.1 Multiphase flows

We consider the two-fluid model for multiphase flows, where both phases are liquids. We call them as primary phase (aqueous continuous phase) and secondary phase (organic dispersed phase). All the quantities with the index c denotes the continuous phase and the index d denotes the dispersed phase. In this paper we consider all equations in the Lagrangian form. The continuity equations for the continuous and dispersed phases are

$$\frac{d\alpha_c}{dt} = -\alpha_c(\nabla \cdot \boldsymbol{v}_c) \tag{2.1}$$

$$\frac{d\alpha_d}{dt} = -\alpha_d(\nabla \cdot \boldsymbol{v}_d), \tag{2.2}$$

where α_c and α_d are the volume fractions, $\boldsymbol{v}_c$ and $\boldsymbol{v}_d$ are the velocity vectors and $\frac{d}{dt}$ is the material derivative.

In addition to (2.1) and (2.2) the volume fractions must satisfy the following constraint

$$\alpha_c + \alpha_d = 1. \tag{2.3}$$

The conservation of momentum for the continuous phase is given by

$$\frac{d\boldsymbol{v}_c}{dt} = -\frac{\nabla p}{\rho_c} + \frac{1}{\alpha_c \rho_c} \nabla \cdot S_c + \boldsymbol{g} + \frac{1}{\alpha_c \rho_c} \boldsymbol{F}_{drag} \tag{2.4}$$

and the secondary phase is given by

$$\frac{d\boldsymbol{v}_d}{dt} = -\frac{\nabla p}{\rho_d} + \frac{1}{\alpha_d \rho_d} \nabla \cdot S_d + (1 - \frac{\rho_c}{\rho_d})\boldsymbol{g} - \frac{1}{\alpha_d \rho_d} \boldsymbol{F}_{drag}, \tag{2.5}$$

where ρ_c and ρ_d are the densities, p is the pressure shared by both phases, $\boldsymbol{g}$ is the gravitational force and $\boldsymbol{F}_{drag}$ represents the interfacial forces and S is the stress tensor, for example, for the continuous phase it is given by

$$S_c = \alpha_c \mu_c [\nabla \boldsymbol{v}_c + (\nabla \boldsymbol{v}_c)^T - \frac{1}{3}(\nabla \cdot \boldsymbol{v}_c) I], \tag{2.6}$$

where μ_c is the dynamic viscosity of the continuous phase. The inter-phase interaction term consists of different momentum exchange mechanisms. Only the drag force was taken into account, while the virtual mass force and the lift force can be neglected for a liquid-liquid interaction as shown by Wang and Mao [24] in a stirred tank. The interfacial momentum transfer (drag force) between two phases is given by

$$\boldsymbol{F}_{drag} = \frac{3}{4} \alpha_d \rho_c \frac{C_D}{d_{32}} |\boldsymbol{v}_d - \boldsymbol{v}_c| (\boldsymbol{v}_d - \boldsymbol{v}_c), \tag{2.7}$$

where d_{32} is the diameter of the droplets of the dispersed liquid phase and the drag force coefficient C_D is given by Schiller and Naumann [17]

$$C_D = \begin{cases} \frac{24}{Re}(1 + 0.15 Re^{0.687}) & \text{if} \quad Re \leq 1000 \\ 0.44 & \text{if} \quad Re > 1000 \end{cases}$$

and Re is the relative Reynolds number defined as

$$Re = \frac{\rho_c |\boldsymbol{v}_d - \boldsymbol{v}_c| d_{32}}{\mu_c}. \tag{2.8}$$

If the breakage and aggregation between droplets are not taking place, like in the mono-dispersed phase, the diameter of the droplet d_{32} is considered to be constant. In general, this is not constant and droplets are assumed to have a spectrum of sizes. This spectrum can be computed with the help of the population balance equation. One can construct the wide range of classes of droplets based on their sizes, however, the considerations of individual classes may not be feasible in the computer simulations since momentum equations have to be solved for each class. Currently, the applied CFD-PBM models are based on the two-fluid Multiple Size Group (MUSIG) Model [13], where all droplets in the PBM share the same velocity field and only one momentum equation is solved for all droplet classes based on the area averaged droplet size, also called as Sauter mean diameter denoted by d_{32}. This quantity can be computed with the help of the solution of the population balance equation, described in the following subsection. A new strategy, the inhomogeneous MUSIG model tries to divide the dispersed phase into a number N so-called velocity groups, where each of the velocity groups is characterized by its own velocity field, to get rid of the common simplifications. Typically 2-3 velocity groups may be sufficient to capture the fluid dynamics.

The population balance is solved as M subdivisions (classes) of each velocity group, resulting in a multi-fluid approach in which $N \times M$ population balance equations are solve [11]. As a drawback, this multi-fluid approach is based on the CPU-time consuming classes method, where 20-30 or even more classes (scalars) are required to capture the shape of the distribution. In this connection, the SQMOM seems to be another interesting alternative, since each primary particle could move with its own velocity group and hence having its own momentum equation resulting also in a multi-fluid model depending on the number of primary particles. As a benefit, in comparison to the inhomogeneous MUSIG model, SQMOM is less CPU-time consuming, since it is based on the computationally less expensive quadrature method of moments. For example, 6 additional scalars for the moments are required when two velocity groups are applied (3 for each primary particle, the 3rd moment can replace the continuity equation) compared to 20-30 or even more additional scalars in the classes method. The SQMOM and the concept of primary and secondary particles are described in section 3.4.

2.2 Droplet population balance equation (DPBE)

The superstructure of the DPBE and the general derivation based on the Reynolds transport theorem is given in [16]. In the present case this equation can be written as

$$\frac{\partial f(V,\boldsymbol{x},t)}{\partial t} + \nabla \cdot (\boldsymbol{v}_d f(V,\boldsymbol{x},t) = S(f(V,\boldsymbol{x},t),V,\boldsymbol{x},t), \tag{2.9}$$

where

$$\begin{aligned} S = -\Gamma(V)f(V,\boldsymbol{x},t) + \int_V^{V_{max}} & \Gamma(V',\boldsymbol{x},t)\ \beta(V|V')\ f(V',\boldsymbol{x},t)\ dV' \\ & -f(V,\boldsymbol{x},t) \int_{V_{min}}^{V_{max}} \omega(V,V')\ f(V',\boldsymbol{x},t)\ dV' \\ & +\frac{1}{2}\int_{V_{min}}^{V} \omega(V-V',V')f(V-V',\boldsymbol{x},t)f(V',\boldsymbol{x},t)dV' \end{aligned}$$

and $f(V,\boldsymbol{x},t)$ is the number density function with particle size V as an independent variable. The vector $\boldsymbol{v}_d$ is the same particle velocity introduced in previous subsection. The source term S in (2.9) consists of loss term (preceded by minus sign) and gain term (preceded by plus sign) due to breakage and aggregation collisions of droplets. The breakage and aggregation of droplets are governed by breakage and aggregation frequencies, respectively. The breakage frequency $\Gamma(V)$ represents the fraction of droplets breaking per unit time, while the aggregation frequency $\omega(V,V')$ accounts for the probability of successful collisions between a pair of droplets. The splitting of mother droplet of size V' to daughter droplets having a spectrum of sizes is given by the daughter particle distribution $\beta(V|V')$.

The above equation (2.9) is an integro-partial differential equation and has no general solution. Hence, one has to solve the equation by numerical techniques. Several numerical schemes have been reported and many of them are problem specific (see Attarakih et al [2]) for the review of the numerical schemes. We are looking for the one which is feasible to couple with the flow solver. The Quadrature Method of Moments (QMOM) is one of the suitable methods to incorporate easily into flow solvers, however, it may suffer from the ill-conditioned eigenvalue problem in the product-difference algorithm, which is pointed out by Attarakih et al [3]. We present some simulations in two and three dimensional case with constant breakage and aggregation kernels. They have shown that Sectional QMOM (SQMOM) is more stable and compared the numerical solutions with the available analytical solutions of the DPBE. In this paper we incorporate the SQMOM for the DPBE into the FPM code. In the next section we give brief description of the SQMOM, see [3] for details.

3 Numerical methods

3.1 FPM for solving coupled equations

The basis of the computations in FPM is a point cloud, which represents the flow field. The points of the cloud are referred to as particles or numerical grids. They are carriers of all relevant physical informations. The particles have to cover completely the whole flow domain, i.e. the point cloud has to fulfill certain quality criteria (particles are not allowed to form "holes" which means particles have to find sufficiently many neighbors; also, particles are not allowed to cluster; etc.). The point cloud is a geometrical basis, which allows for a numerical formulation making FPM a general finite difference idea applied to continuum mechanics. As a special case, if the point cloud is reduced to a regular cubic point grid, then the FPM would reduce to a classical finite difference method. The idea of general finite differences also means that FPM is not based on a weak formulation like Galerkin's approach. Rather, FPM is a strong formulation which models differential equations by direct approximation of the occurring differential operators. The method used is a moving least squares idea which was especially developed for FPM. In the earlier publications we have reported about the FPM in details, see [12,20–22]. Due to the restriction of the space, we do not repeat the same details in this paper.

To simulate the above presented equations for two-phase flows we establish the separate cluster of points for each phase. Each of these separate cluster of points will act as a numerical grid to approximate the governing differential equations for each phase. These point clouds are decoupled from each other, however they are able to exchange any kind of information among the clusters. In this situation we have to interpolate the quantities at an arbitrary particle in one phase from the surrounding cluster of particles from other phase. This can be easily achieved with the help of least squares method, described in [12,20–22].

3.2 Numerical Scheme for primary liquid phase

The primary liquid phase is solved using the FPM in combination of Chorin's pressure projection method [4]. In this work we decompose the pressure into hydrodynamic and dynamic pressures, i. e. $p = p_{hyd} + p_{dyn}$. The scheme consists of two fractional steps and is of first order accuracy in time. In the first step we compute the intermediate velocities $\boldsymbol{v}^*$ and in the second step we correct the velocity with the constraint that velocity fulfills the continuity equation. In our numerical scheme, we substitute the value of $\nabla \cdot \boldsymbol{v}_c$ which appeared in the stress strain tensor S_c from the equation (2.1). Moreover, after some manipulations, we obtain

$$\nabla \cdot S_c = \nabla \cdot (\alpha_c \mu_c \ \nabla) \boldsymbol{v}_c + \Theta(\boldsymbol{v}_c, \alpha_c \mu_c, \nabla) = \nabla \cdot (\kappa \Delta) \boldsymbol{v}_c + \Theta(\boldsymbol{v}_c, \kappa, \nabla) \quad (3.10)$$

where Θ is some straight forward, but lengthy term and $\kappa = \alpha_c \mu_c$. Moreover, the operator $\nabla \cdot (\kappa \nabla \psi)$ can be re-expressed by

$$\nabla \cdot (\kappa \nabla \boldsymbol{v_c}) = \frac{1}{2} \left[\Delta(\kappa \boldsymbol{v_c}) + \kappa \Delta \boldsymbol{v_c} - \boldsymbol{v_c} \Delta \kappa \right]. \quad (3.11)$$

This means, once we construct the shape functions for the Laplace operator Δ, then the differential operator $\nabla \cdot (\kappa \nabla \boldsymbol{v_c})$ can be approximated by combination of Δ applied to different functions. Hence the momentum equation (2.4) can be re-expressed in the simple form

$$\begin{aligned} \frac{d\boldsymbol{v}_c}{dt} &= -\frac{\nabla p}{\rho_c} + \frac{1}{\alpha_c \rho_c} \left[\nabla \cdot (\kappa \Delta) \boldsymbol{v}_c + \Theta(\boldsymbol{v}_c, \kappa, \nabla) \right] \\ &\quad + \frac{3}{4} \frac{\alpha_d}{\alpha_c} \frac{C_D}{d_{32}} |\boldsymbol{v}_d - \boldsymbol{v}_c| (\boldsymbol{v}_d - \boldsymbol{v}_c) + \boldsymbol{g} \\ &= -\frac{\nabla p}{\rho_c} + \frac{1}{\alpha_c \rho_c} \nabla \cdot (\kappa \Delta) \boldsymbol{v}_c - G_c \boldsymbol{v}_c + \boldsymbol{H}, \end{aligned} \quad (3.12)$$

where $G_c = \frac{C_D}{d_{32}} |\boldsymbol{v}_d - \boldsymbol{v}_c|$ and the vector $\boldsymbol{H}$ consists of all the forces as source term.

In the following we describe the projection scheme for continuous liquid flow in the FPM framework.

(i) Initialize :

$$p_{dyn}^{n+1} := p_{dyn}^n \quad (3.13)$$

(ii) Compute α_c^{n+1} implicitly by

$$\alpha_c^{n+1} = \frac{\alpha_c^n}{1 + \Delta t \ (\nabla \cdot \boldsymbol{v}_c^n)} \quad (3.14)$$

(iii) Compute the actual hydrostatic pressure p_{hyd}^{n+1} from

$$\nabla \cdot \left(\frac{1}{\rho_c} \nabla p_{hyd}^{n+1} \right) = \nabla \cdot \boldsymbol{g} \quad (3.15)$$

with boundary condition

$$\frac{\partial p_{hyd}^{n+1}}{\partial \boldsymbol{n}} = \boldsymbol{g} \cdot \boldsymbol{n} \text{ on } \Gamma_{wall} \text{ and } \Gamma_{inflow}, \; p_{hyd}^{n+1} = p_{hyd}^{0} \text{ on } \Gamma_{outflow} \tag{3.16}$$

(iv) Establish preliminary pressure

$$\tilde{p} = p_{hyd}^{n+1} + p_{dyn}^{n+1} \tag{3.17}$$

(v) Compute implicitly the intermediate velocity $\boldsymbol{v}_c^*$ from

$$\left[(1 + \Delta t G_c)I - \frac{\Delta t}{\alpha_c^{n+1}\rho_c}\nabla \cdot (\kappa \nabla)\right] \boldsymbol{v}_c^* = \boldsymbol{v}_c^n - \frac{\Delta t}{\rho_c}\nabla \tilde{p} + \Delta t \; \boldsymbol{H}^n \tag{3.18}$$

(vi) Correct the velocity

$$\boldsymbol{v}_c^{n+1} = \boldsymbol{v}_c^* - \frac{\Delta t}{\rho_c}\nabla \epsilon^{n+1} \tag{3.19}$$

with the constraint (continuity equation)

$$\nabla \cdot \boldsymbol{v}_c^{n+1} = -\frac{1}{\alpha_c^n}\frac{d\alpha_c}{dt} = -\frac{1}{\alpha_c^n}\frac{\alpha_c^{n+1} - \alpha_c^n}{\Delta t} \tag{3.20}$$

(vii) Updating the dynamic pressure

$$p_{dyn}^{n+1} = p_{dyn}^{n} + \epsilon^{n+1} \tag{3.21}$$

where ϵ^{n+1} is obtained from (this is obtained taking divergence on (3.19))

$$\nabla \cdot (\frac{1}{\rho_c}\nabla \epsilon^{n+1}) = \frac{1}{\Delta t}\left[\frac{1}{\alpha_c^n}\frac{\alpha_c^{n+1} - \alpha_c^n}{\Delta t} + \nabla \cdot \boldsymbol{v}_c^*\right] \tag{3.22}$$

with boundary conditions similar to p_{hyd} given in (3.16).
(viii) Move the particles

$$\boldsymbol{x}_c^{n+1} = \boldsymbol{x}_c^n + \Delta t \; \boldsymbol{v}_c^{n+1} \tag{3.23}$$

We note that in the left hand side of equation (3.18) the operator $(1+\Delta t G_c)I - \frac{\Delta t}{\alpha_c \rho_c}\nabla \cdot (\kappa \; \nabla)$ and the left hand side of equation (3.22) give rise to the construction of a large sparse matrices, each line of which containing the local, discrete approximation of the operators. The right hand sides appear as a load vector. Hence equations (3.15), (3.18) and (3.22) represent large sparse linear systems, which we solve in the meshfree framework with the help of the FPM. In addition to that we approximate the gradient vectors with the help of the FPM.

3.3 Numerical Scheme for secondary phase

Once the velocity, pressure and volume fraction of the continuous phase is known, we interpolate them into the dispersed phase particles. Then, the following steps are followed to compute the quantities on secondary phase.

(i) Compute α_d^{n+1} implicitly by

$$\alpha_d^{n+1} = \frac{\alpha_d^n}{1 + \Delta t \left(\nabla \cdot \boldsymbol{v}_d^n\right)} \tag{3.24}$$

and normalize the volume fractions in order to fulfill the consistency condition (2.3) as follows

$$\alpha_d^{n+1} = \frac{\alpha_d^{n+1}}{\alpha_d^{n+1} + \alpha_c^{n+1}} \quad \text{and} \quad \alpha_c^{n+1} = \frac{\alpha_c^{n+1}}{\alpha_d^{n+1} + \alpha_c^{n+1}} \tag{3.25}$$

(ii) compute the velocity $\boldsymbol{v}_d^{n+1}$ implicitly as

$$\begin{aligned} &\left[(1 + \Delta t G_d) I - \frac{\Delta t}{\alpha_d^{n+1} \rho_d} \nabla \cdot (\kappa \nabla)\right] \boldsymbol{v}_d^{n+1} = \\ &\boldsymbol{v}_d^n - \Delta t \left[\frac{\nabla p^{n+1}}{\rho_d} + (1 - \frac{\rho_c}{\rho_d}) \boldsymbol{g} + G_d \, \boldsymbol{v}_c^{n+1}\right] \end{aligned} \tag{3.26}$$

where $G_d = \frac{3}{4} \frac{\rho_c}{\rho_d} \frac{C_D}{d_{32}} \|\boldsymbol{v}_c^n - \boldsymbol{v}_d^n\|$, $\kappa = \alpha_d^{n+1} \mu_d$.

(iii) Move particles

$$\boldsymbol{x}_d^{n+1} = \boldsymbol{x}_d^n + \Delta t \, \boldsymbol{v}_d^{n+1} \tag{3.27}$$

3.4 SQMOM for the DPBE

There exist many numerical methods in the literature as attempts to solve certain type of the PBEs. In the sectional methods (e.g. [10]) the particle size (here it is denoted by the particle diameter D) is discretized into finite number of sections. One limitation of the finite difference schemes is their inability to predict accurately integral quantities (low-order moments as a especial case) associated with populations of sharp shapes [2, 16]. A Large number of primary particles in the classical sectional methods is required, not only to reconstruct the shape of the distribution, but also to estimate the desired integral quantities associated with the distribution. The quadrature method of moments (QMOM) as first introduced by McGraw [15] is found very efficient from accuracy and computational cost point of view. Unlike the sectional methods, the QMOM has a drawback of destroying the shape of the distribution and the information about the distribution is only stored in its moments. On the other hand, the QMOM tracks the population moments and hence it conserves the integral quantities.

The idea behind the SQMOM is to divide the population density function into sections followed by the application of the QMOM to each section. In the SQMOM framework of discretization, the single particle from the sectional methods will be called the primary particle N_{pp} and it will be responsible for the reconstruction of the distribution. To overcome the fundamental problem of the sectional methods, N_{sp} secondary particles are generated in each section with positions (abscissas) and weights that are given by: $D_j^{<i>}, w_j^{<i>};\ j = 1, \ldots, N_{sp},\ i = 1, N_{pp}$, respectively. Note that, the secondary particles are exactly equivalent to the number of quadrature points in Gauss-like quadrature or the QMOM [9, 14, 15]. Accordingly, each secondary particle could conserve or reproduce two low-order moments and in general moments in each section; and hence the method is given the name Sectional QMOM (SQMOM). The way in which the SQMOM works is started by dividing the particle size into (N_{pp} primary particles) contiguous sections. Each section is then seeded by the desired number of secondary particles which carry detailed information about the distribution. In this framework, the active particle mechanisms such as splitting and aggregation occur through interactions between the secondary particles. Therefore, $N_{sp} \times N_{pp}$ particles are contributing in the splitting and aggregation events. The distribution could be reconstructed from the secondary particles by averaging the total weights of the secondary particles with respect to the section width (ΔD_i). These operations of averaging are carried out for all primary particles as follows:

$$\bar{w}_i(\boldsymbol{x}, t) = \frac{1}{\Delta D_i} \sum_{j=1}^{N_{sp}} w_j^{<i>}, \bar{D}_i(\boldsymbol{x}, t) = \frac{\sum_{j=1}^{N_{sp}} w_j^{<i>} D_j^{<i>}}{\sum_{j=1}^{N_{sp}} w_j^{<i>}}, i = 1, \ldots, N_{pp} \quad (3.28)$$

In pure mathematical sense, the above presentation is equivalent to applying the QMOM to each section of an arbitrary width $[D_{i-1/2}, D_{i+1/2}], i = 1, \ldots, N_{pp}$ resulting in a set of sectional moments that could be written as

$$m_r^{<i>}(\boldsymbol{x}, t) = \int_{D_{i-1/2}}^{D_{i+1/2}} D^r f(D, \boldsymbol{x}, t)\, dD, \qquad r = 0, 1, \ldots, 2N_{sp} - 1. \quad (3.29)$$

To this end it remains how to relate the positions and weights (appearing in (3.28) of the secondary particles in the ith section to the sectional moments of the unknown function $f(D, \boldsymbol{x}, t)$ given by (3.29). The sectional moments $m_r^{<i>}$ can be approximated by the weights and abscissas of the secondary particles

$$m_r^{<i>}(\boldsymbol{x}, t) = \sum_{j=1}^{N_{sp}} w_j^{<i>} \left(D_j^{<i>}\right)^r, \quad r = 0, 1, \ldots, 2N_{sp} - 1. \quad (3.30)$$

Now, the problem of sectional moments could be stated as: Given a set of sectional moments $m_r^{<i>}$ for $r = 0, 1, \ldots, 2N_{sp} - 1$ and $i = 1, 2, \ldots, N_{pp}$, find a set of weights $w_j^{<i>}$ and abscissas $D_j^{<i>}$ of the secondary particles for

$j = 1, 2, \ldots, N_{sp}$ and $i = 1, 2, \ldots, N_{pp}$. Although, the set of weights and positions of the secondary particles is unique for a specified number of sectional moments, there are several ways for finding them. The product-difference algorithm of Gordon and the direct tracking of the weights and positions are examples of these methods [9, 14]. It is worthwhile to mention here that these methods were only applied to population densities with a whole section ranging from zero to infinity. These methods suffer from ill-conditioning if the number of secondary particles is large (usually if greater than four) [2]. This increase in the number of secondary particles in a one-population section could only increase the accuracy of the integration quadrature at the expense of solving a large eigenvalue problem. To overcome this difficulty, the number of secondary particles is fixed to one or two in each section in the SQMOM and the accuracy of the integration quadrature is achieved by controlling the width of each section (like adaptive integration). In this way, analytical solutions are derived for the weights and positions of secondary particles using equal-weight quadratures. In order to obtain the abscissas and weights of the secondary particles analytically three low order moments are considered [6]. To conserve the total droplet number, length and mass in each section, $m_r^{<i>}, r = 0, 1, 3,\ i = 1, 2, \ldots, N_{pp}$ are chosen, where Eq.(3.30) is reduced to three nonlinear algebraic equations which are analytically solved for the secondary particle weights and abscissas

$$w_{1,2}^{<i>}(\boldsymbol{x}, t) = \frac{1}{2}\hat{m}_0^{<i>}, \quad D_{1,2}^{<i>}(\boldsymbol{x}, t) = \hat{m}_1^{<i>} \mp \frac{1}{3}\sqrt{\frac{\hat{m}_3^{<i>}}{\hat{m}_1^{<i>}} - (\hat{m}_1^{<i>})^2}\,, \quad (3.31)$$

where $\hat{m}_r = m_r/m_0$.

Finally, the continuity equation for the rth sectional moment could be written in Lagrangian form as

$$\frac{dm_r^{<i>}(\boldsymbol{x}, t)}{dt} + (\nabla \cdot \boldsymbol{v}_d)m_r^{<i>}(\boldsymbol{x}, t) =$$
$$-D_r^{<i>}[\Gamma^{<i>} \bullet w^{<i>}]^T + \sum_{m=1}^{N_{pp}} C_r^{<i,m>}[\Gamma^{<i>} \bullet w^{<i>}]^T +$$
$$\sum_{k=1}^{i \times N_{sp}} \left[\sum_{j=k}^{i \times N_{sp}} \Psi_{k,j,r}^{<i>} \omega_{j,k} w_j' w_k' - \eta_k \sum_{n=1}^{N_{pp} \times N_{sp}} (d_k')^r \omega_{k,n} w_k' w_n' \right], \quad (3.32)$$

where $r = 0, 1, \ldots, 2N_{sp} - 1,\ i = 1, 2, \ldots, N_{pp}$ and the symbols T and $\bullet$ are used for matrix transpose and element by element multiplication, respectively. The matrices $D^{<i>}$ and $w^{<i>}$ contain the secondary abscissas and weights in the ith section. The matrix $\Gamma^{<i>}$ contains the breakage frequencies and $C^{<i,m>}$ consists of integrals that preserve the low-order moments of the newly birthed particles in the ith section due to splitting of a mother particle of size $d_j^{<m>}$. The function η_k is used to select the secondary particles disappearing due to aggregation in the ith section, while $\Psi_{r,k,j}^{<i>}$ is an aggregation matrix whose nonzero elements represent the successful aggregation events between any pair of secondary particle with abscissas and weights that are given by the augmented vectors: D' and w', respectively. This matrix preserves exactly the first $2N_{sp}$ low-order moments of the newly birthed particles by aggregation. The velocity of the dispersed phase $\boldsymbol{v}_d$ is calculated in the momentum equation of the dispersed phase. Once the moments are computed as above, the important parameters like Sauter mean diameter is defined as the ratio between the third and second moments

$$d_{32}^{<i>}(\boldsymbol{x}, t) = m_3^{<i>}/m_2^{<i>} \tag{3.33}$$

and the volume fraction can be computed from the third moment

$$\alpha_d^{<i>}(\boldsymbol{x}, t) = S_V m_3^{<i>}(\boldsymbol{x}, t), \tag{3.34}$$

where S_V is a shape factor, for example, if the droplets are spherical, $S_V = \pi/6$. Note that, in the case of two fluid model, where one considers only one primary particle, $d_{32}^{<i>} = d_{32}$ and $\alpha_d^{<i>} = \alpha_d$. If one considers the primary particle more than one, we obtain the multi-fluid model, where one has to establish the same number of primary particle of droplets and the volume fraction is equal to the sum of all volume fractions of primary particle. In this paper, we have considered only one class of primary particles. Hence, together with the continuity and momentum equations for the two-phase flows equations, we have to solve for each primary particle the additional $2N_{sp}$ transport equations for the moments (3.32).

4 Numerical examples

Our main goal is to incorporate the SQMOM in the FPM to simulate the 3d extraction column as shown in Fig. 4.1, which consist of rotors and stators. The rotor rotates with 150 rpm. The volume flow of the aqueous (water) and organic (butyl-acetate) phases are 100 l/h which corresponds to the inflow velocities equal to 0.0342 m/s. On the top plane the small central ring is the inflow boundary for the aqueous phase and the rest is the outflow boundary for organic phase. Similarly, on the bottom the small central ring is the inflow boundary for the organic phase and the rest is the outflow boundary for the aqueous phase. The rest of the boundaries are the no-slip boundaries for both phases. The densities of the aqueous and organic phases are 1000 kg/m^3 and 880 kg/m^3, respectively. The viscosities for the aqueous and organic phases are 0.001 $kg/(ms)$ and 0.0007 $kg/(ms)$. Moreover, the gravitational force acts in the negative z-direction. The moments in the inlet boundary of the dispersed phase are $m_0 = 1145352.0$, $m_1 = 2863.3805$, $m_2 = 7.15845$, $m_4 = 0.01789$ such that the abscissas are: 1.63 mm and 2.58 mm and both weights equal to 39699.25. These moments are used for the experiments by Simon [19]. The droplets Sauter mean diameter in the experiments of Simon was $d_{32} =$

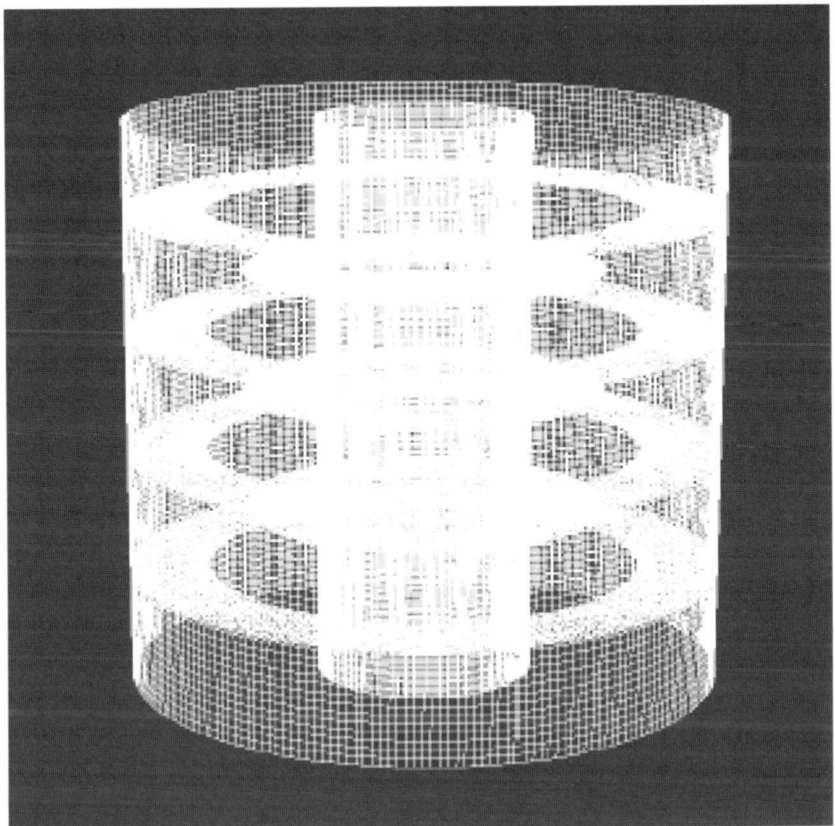

Figure 4.1. 3D extraction column.

2.6 mm at the outlet. In this paper we have considered constant breakage and aggregation kernels. The computations are continued until the flow reached quasi-steady state. In order to find some estimation on the aggregation and breakage kernel, we performed some parameter study in $2D$ geometry where the rotor is fixed. The 2D computational domain is depicted in Fig. 4.2, where 5 compartments (as in the 3D domain) are considered. The size of the domain is 310 $mm \times 48$ mm. On the left boundary $BC = 4mm$ is the inflow boundary for primary phase and the rest is no-slip boundary for the primary phase and outflow boundary for the secondary phase. On the right wall $FG = 4mm$ is inflow boundary for the droplet and the rest EF and GH are the no-slip boundary for the secondary phase and outflow boundary for the primary phase. Rest of the walls are considered as no-slip boundary for both phases. The size of compartments is $ML = 29$ mm and they are separated by $LK = 1$ mm. The length of other walls like $AN = KH = 80$ mm and $DI = JE = 94$ mm. The dimensions are the same in the 3D domain. The approach of the

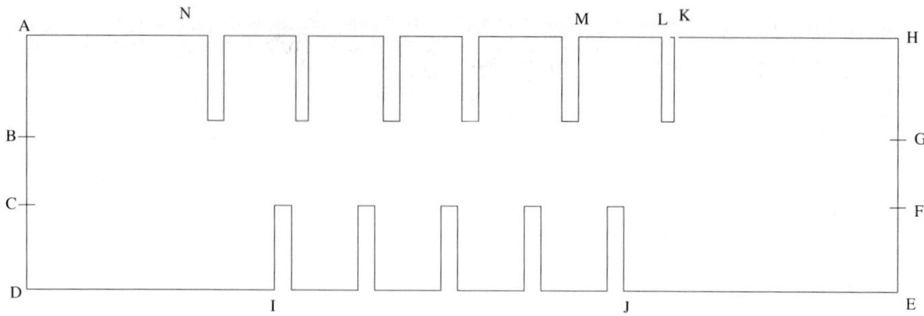

Figure 4.2. 2d computational domain.

2D domain for parameter studies was chosen because parameter studies are not feasible (large CPU time) in a complex 3D domain.

The following approach is followed: we have considered three cases: only breakage, only aggregation and breakage and aggregation both. In Fig. 4.3 we have plotted the Sauter mean diameter for all three cases. Best results were achieved for the constant breakage kernel of magnitude equals 1.0 and a constant aggregation kernel of magnitude equals $1e-10$. On the top the results are only with the breakage. Here we see the breakage take place mainly in the middle of the domain. The Sauter mean diameter is decreasing from 2.5 mm to below 2 mm at the outlet. In the middle we have plotted the Sauter diameter for aggregation only. The diameter is increasing to around 4.5 mm. In the bottom, we see the Sauter diameter for both breakage and aggregation. As in the experiment, we have obtained the Sauter diameter approximately equal to 2.6 mm. In the next step, the same breakage and coalescence kernels were applied in the 3D domain. The results are depicted in Figs. 4.4 and 4.5. As in the 2D domain, the Sauter diameter is equal to 2.6 mm and can match the experiments. Now it is obvious that the same parameters, which were achieved in the CPU-time saving 2D domain, are also suitable for the simulation of the real column. This is possible, since only constant kernels were applied. Real models for coalescence and breakage (e.g. [5]) are strong functions of the energy input (turbulent energy dissipation). Since the main purpose of this work is to couple the FPM and the SQMOM (CFD-PBM) this simplification seems to be a good starting point to focus on the basic principles without considering complex models. The current approach allows for a good description of the real column behavior but it is not adequate for a prediction of the Sauter mean diameter. That's why real models should and will be applied in future work.

The results of the velocity magnitudes and vectors are shown in Fig. 4.5. Two large vortices can be estimated in the compartment; one between the stators and the other between the stirrer. Because of the rising droplets the vortices shift or turn around and are directed to the top of the column, so that the vortex between the stirrer moves to the position above the stirrer and vice

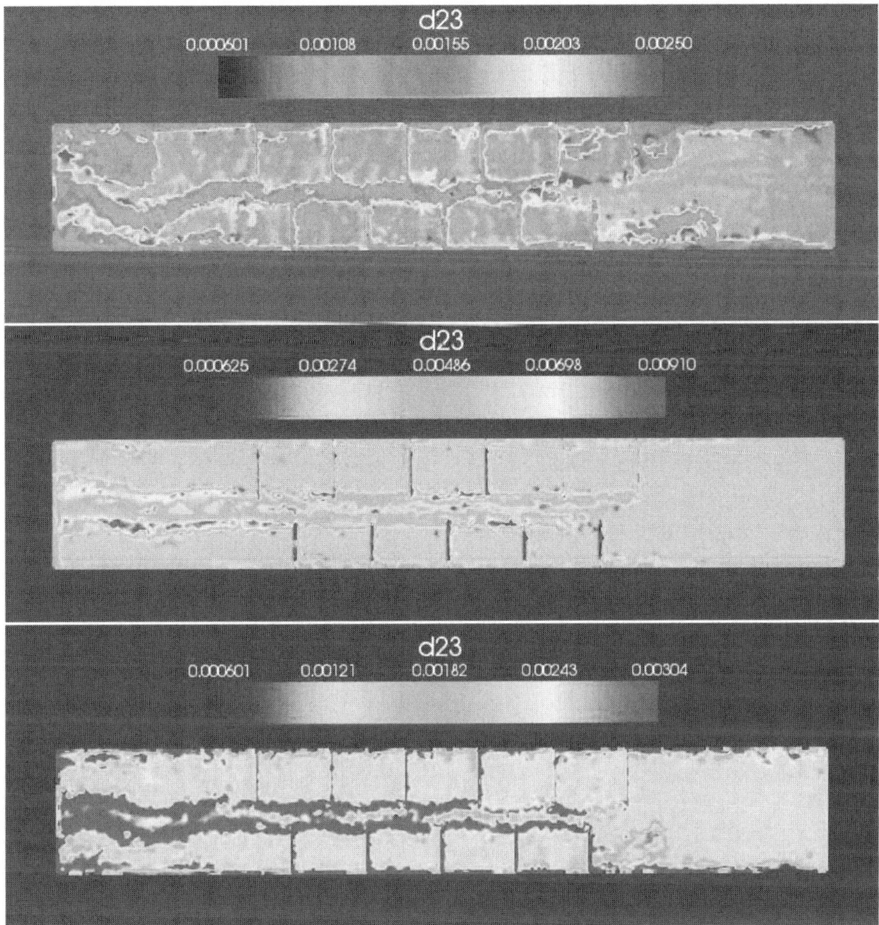

Figure 4.3. Sauter mean diameter with breakage only (top), aggregation only (middle) and breakage and aggregation (bottom) (*See also* Color Plate on page 397)

versa. As expected, the highest velocities are near the stirrer decreasing to zero near the wall. It could already be shown that FPM can predict flow fields and velocities in the stirred extraction column, whereas all flow phenomena such as vortices can be described. In addition, the FPM results match the overall experimental data and the predicted velocities are in good agreement with the experimental ones [8]. The results of the holdup of the dispersed phase are shown in the right hand side of Fig. 4.4. It can be seen that the organic droplets accumulate under stirrer and stators and move mainly through the middle of the column without penetrating the outer regions of the column. This agree well with the real column behavior, since the same observations were also seen in experiments [7].

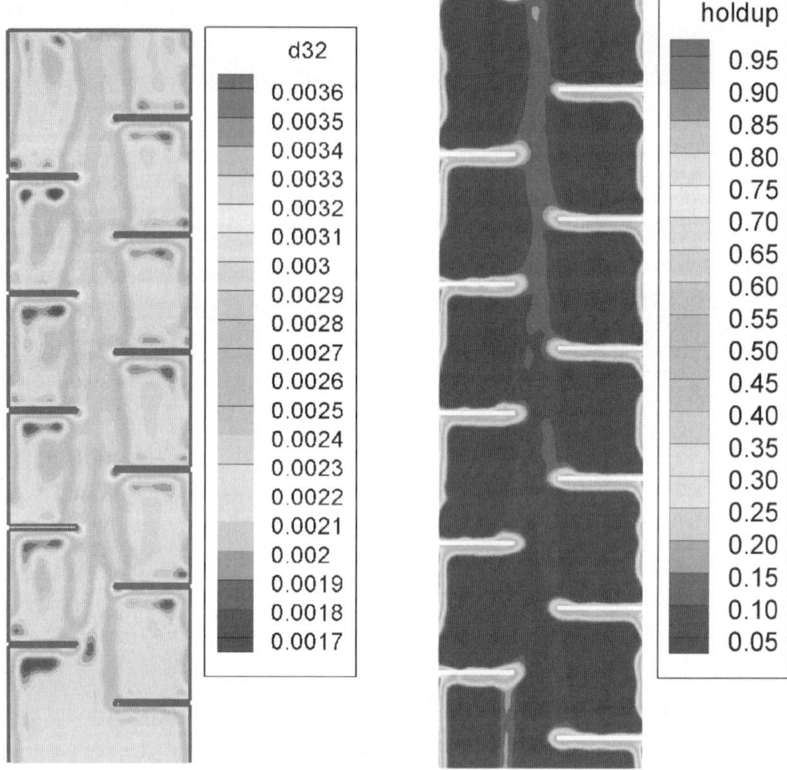

Figure 4.4. Sauter diameter in m (left) and Holdup of dispersed phase (right) (*See also* Color Plate on page 398)

Convergence study

For the purpose of convergence study, we have considered a simple 2D rectangular geometry of size $[1, 1.5]\ m[0, 1.5]\ m$. For the liquid phase it is considered as a closed tank without in- and outflow boundaries. The gravity acts in the negative y-direction. For the droplet, we considered the top boundary as the outflow and the interval $[1.145, 1.16]\ m$ on the bottom as the inflow boundary with the velocity $v = (0, 0.01)\ m/s$. Both effects of aggregation and breakage have been considered with constant kernels. The rest of the parameters are same as in the above mentioned numerical examples. We have simulated upto 10 seconds. In this time state Table 4.1 shows the size of h as well as the corresponding total number of particles in both phases and maximum velocities of both phases. Here, we see that the scheme converges with the order of one. In Fig. 4.6, we have plotted the velocity field for the continuous phase with $h = 0.02\ m$ at the time $t = 8.86\ s$.

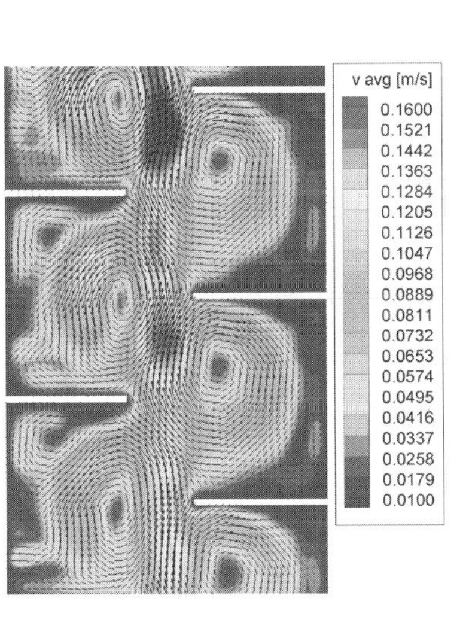

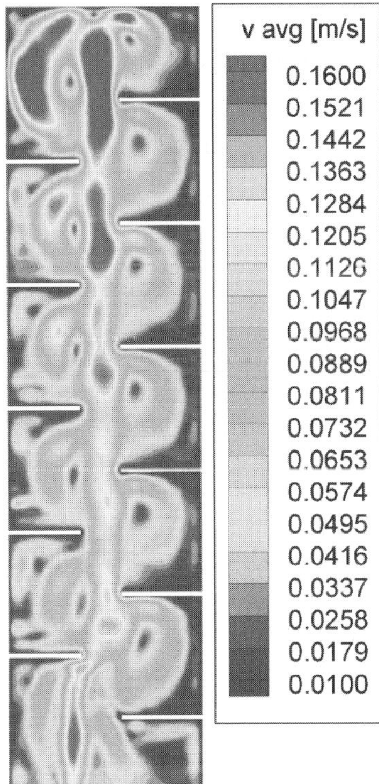

Figure 4.5. velocity contour of continuous phase (*See also* Color Plate on page 399)

Table 4.1.

h	N	$\|\boldsymbol{v}_c\|_{max}$	$\|\boldsymbol{v}_d\|_{max}$
0.08	1548	0.8031	1.0148
0.04	5628	0.6220	0.8058
0.02	21451	0.5512	0.7538
0.01	83133	0.5333	0.7216
0.0075	145840	0.5304	0.7022

5 Conclusion

In this paper we have considered the coupling of CFD with DPBM. The PBE is solved using the SQMOM with one primary particle. We have considered constant breakage and aggregation kernels and they have been fitted to experimental data in a 2D domain in which the numerical results are close to experimental ones. The same parameters were applied in the 3D domain and produce suitable results for the Sauter mean diameter, holdup and aqueous

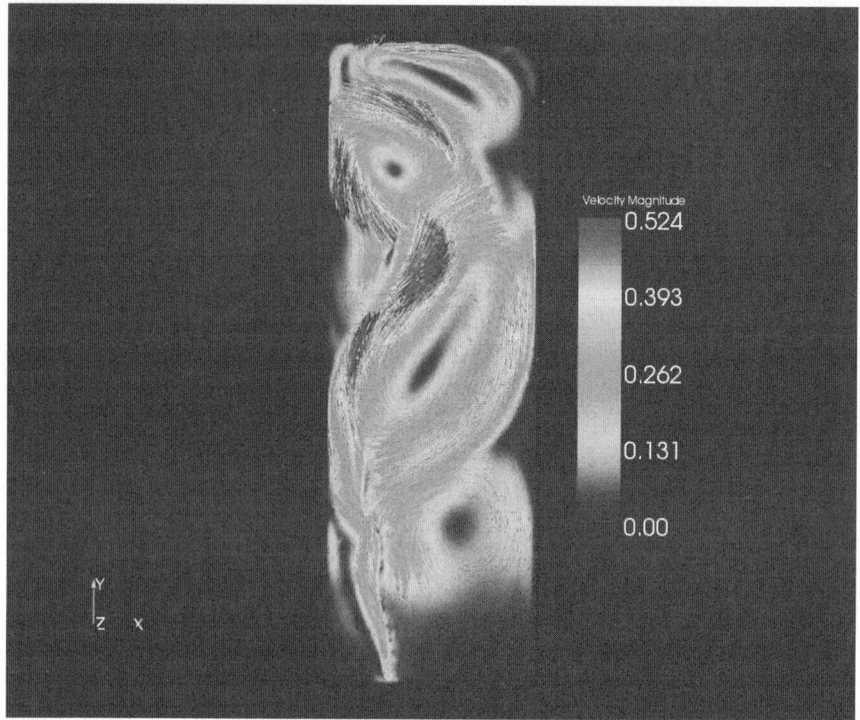

Figure 4.6. Velocity contour for continuous phase (*See also* Color Plate on page 400)

phase velocities. In summary, the coupling of FPM and SQMOM is accomplished using one primary particle resulting is a two-fluid model and constant kernels for breakage and coalescence. In future work, more primary particles will be considered to achieve a multi-fluid CFD-DPBM model. Furthermore, instead of the constant breakage and aggregation kernels real coalescence and breakage models will be used in the DPBE to permit a prediction of the column behavior. Such kernels in general depend on the turbulent energy, therefore, the turbulence model will be included in the FPM code.

Acknowledgment: We would like to thank the "Deutsche Forschungsgemeinschaft (DFG) " for the financial support of our research.

References

1. M. M. Attarakih, H.-J. Bart, L.G. Lagar and N.M. Faqir, *LLECMOD: A Windows-based program for hydrodynamics simulation of liquid-liquid extraction columns*, Chem. Eng. Proc., **45**(2), (2006), 113–123.

2. M. M. Attarakih, H.-J. Bart and N.M Faqir, *Numerical solution of the bivariate population balance equation for the interacting hydrodynamics and mass transfer in liquid-liquid extraction column*, Chem. Eng. Sci., **61**(1), (2006), 113–123.
3. M. M. Attarakih, C. Drumm and H.-J. Bart, *Solution of the population balance equation using the sectional quadrature method of moments*, submitted in Chem. Eng. Sci., special issue 3rd international conference on population balance modelling, Quebec, Canada, 19–21 Sept. 2007.
4. A. Chorin, *Numerical solution of the Navier-Stokes equations*, J. Math. Comput., **22** (1968), 745 762.
5. C. A. Coulaloglou and L. L. Tavlarides, *Description of interaction processes in agitated liquid-liquid dispersions*, Chem. Eng. Sci., **32**, (1977), 1289-1297.
6. C. Drumm, M. M. Attarakih and H.-J. Bart, *Coupling of CFD with DPBM for a RDC extractor*, submitted in Chem. Eng. Sci., special issue 3rd international conference on population balance modelling, Quebec, Canada, 19–21 Sept. 2007.
7. C. Drumm and H.-J. Bart, *Hydrodynamics in a RDC extractor: single and two-phase PIV measurements and CFD simulations*, Chem. Eng. Techn., **29**(11), (2006), 1–8.
8. C. Drumm, S. Tiwari, J. Kuhnert and H.–J. Bart, *Finite pointset method for simulation of the liquid-liquid flow field in an extractor*, (2007), submitted Comp. Chem. Eng.
9. R. G. Gordon, *Error bounds in equilibrium statistical mechanics*, J. Math. Phys. **9**, (1968), 655–663.
10. M.J. Hounslow, R.L. Ryall and V.R. Marshall, *A discretized population balance for nucleation, growth and aggregation*, AIChE Journal, **34**(11), (1988), 1821–1832.
11. E. Krepper, T. Frank, D. Lucas, H.-M. Prasser and P.J. Zwart, *Inhomogeneous MUSIG model - a population balance approach for polydispersed bubbly flows*, Proceedings of the ICMF-2007, M. Sommerfeld (Ed.), 6th International Conference on Multiphase Flow, Leipzig, (2007).
12. J. Kuhnert, *An upwind finite pointset method (FPM) for compressible Euler and Navier-Stokes equations*, (M. Griebel and M. A. Schweitzer, eds.), Lecture Notes in Computational Science and Engineering, vol. 26, Springer, 2002, pp. 239–249.
13. S. Lo, *Application of population balance to CFD modelling of gas-liquid reactors*, Proc. "Trends in numerical and physical modelling for industrial multiphase flows", Corsica, France, 2002.
14. L.D. Marchisio and R.O. Fox, *Solution of the population balance equations using the direct quadrature method of moments*, J. Aerosol Sci. **36**, (2005), 43–73.
15. R. McGraw, *Description of aerosol dynamics by the quadrature method of moments*, Aerosol Sci. & Tech., **27**(2), (1997), 255–265.
16. D. Ramkrishna, *Population Balances*, Academic Press, San Diego, 2000.
17. L. Schiller and Z. Naumann, *A drag coefficient correlation*, Z. Ver. Deutsch. Ing., **77**, (1935), 318.
18. S. A. Schmidt, M. Simon, M. M. Attarakih, L. Lager and H.-J. Bart, *Droplet population balance modelling - hydrodynamics and mass transfer*, Chem. Eng. Sci., **61**(1), (2006), 246–256.
19. M. Simon, *Koaleszenz von Tropfen und Tropfenschwärmen*, Dr. Ing. Thesis, TU Kaiserslautern, Germany, 2002.
20. S. Tiwari and J. Kuhnert, *Finite pointset method based on the projection method for simulations of the incompressible Navier-Stokes equations*, (M. Griebel and

M. A. Schweitzer, eds.), Lecture Notes in Computational Science and Engineering, vol. 26, Springer, 2002, pp. 373–387.

21. S. Tiwari and J. Kuhnert, *A numerical scheme for solving incompressible and low Mach number flows by Finite Pointset Method*, (M. Griebel and M. A. Schweitzer, eds.), Lecture Notes in Computational Science and Engineering, vol. 43, Springer, 2005, pp. 191–206.
22. S. Tiwari, S. Antonov, D. Hietel, J. Kuhnert and R. Wegener, *A Meshfree Method for Simulations of Interactions between Fluids and Flexible Structures*, (M. Griebel and M. A. Schweitzer, eds.), Lecture Notes in Computational Science and Engineering, vol. 57, Springer, 2006, pp. 249–264.
23. A. Vikhansky, M. Kraft, M. Simon, S. Schmidt and H.-J. Bart, *Droplets Population balance in a rotating disc contactor: an inverse problem approach*, AIChE Journal, **52**(4), (2006), 1441–1450.
24. F. Wang and Z.-S. Mao, *Numerical and experimental investigation of liquid-liquid two-phase flow in stirred tanks* Ind. Eng.Chem. Res., **44**, (2005), 5776.

Hybrid Methods for Fluid-Structure-Interaction Problems in Aeroelasticity

Holger Wendland[1]

Department of Mathematics, University of Sussex, Brighton, BN1 9RF, England
H.Wendland@sussex.ac.uk

Summary. In this paper, we describe a hybrid method for problems coming from computational aeroelasticity. The method is hybrid in the sense that it comprises different discretisation techniques such as finite elements, finite volumes and meshfree approximation. After a thorough introduction into the field of aeroelasticity and the involved mathematical models from elasticity and aerodynamics, the paper concentrates on the coupling process, which is realised by meshfree, kernel-based interpolation techniques, taking new developments like partition of unity and matrix-valued kernels into account. The paper ends with examples demonstrating the feasibility of the proposed method.

Key words: Aeroelasticity, Staggered Algorithm, Meshfree Interpolation.

1 Introduction

Fluid-structure interaction (FSI) investigates the interplay between fluid dynamics, structural dynamics and structural elasticity.

This interaction is evident in many of the problems encountered in aircraft design. Examples of such problems are flutter, buffeting, gust response, acoustic resonances and flexibility-induced effects on stability and control. But it also has many non aircraft related applications such as the modelling of cardiovascular flow or the modelling of the impact of winds on suspension bridges.

In this paper we study the mutual interaction between aerodynamical and elastic forces for an aerospace vehicle. An airborne aircraft structure is subjected to surface pressures induced by its surrounding flow. For an introduction into the field of *aeroelasticity* we refer the reader for example to [3, 5, 9, 12].

In the development process of future air vehicles, multidisciplinary simulation has become a key technology. For the design and clearance of modern aircrafts, such multi-disciplinary simulations are used to predict and to analyse the behaviour of the elastic aircraft in flight and during manoeuvre [8, 13].

This is particularly crucial for highly critical flight conditions representing specific parts of the flight envelope. The behaviour of the elastic aircraft in flight is influenced by the interaction between deformations of the elastic structure caused by fluid flow and the impact of the aerodynamic forces on the structure. Hence, the interaction between the fluid flow and the elastic structure has to be studied in depth [4,6,10,11].

In principle, there are the following approaches to the numerical solution of the FSI problem. In a *monolithic* approach, one tries to model the fluid, the structure, and the interaction in one *single* model. While this has, from the mathematical point of view certain advantages, its drawbacks in practical applications come in particular from the fact that the new mathematical model forces a completely new programming of the solver. In particular, for complex aeroelastic problems, the fluid and the structural domain show different mathematical and numerical properties, requiring highly tuned and adapted solvers. Hence, the simultaneous solution by a monolithic scheme is in general computationally challenging, mathematically and economically suboptimal, and software-wise almost unmanageable [7].

A more practical point of view is taken in the *loose coupling* approach. Here, for each time step, the fluid and the structural problem are solved independently, and the influence of the other problem is restricted to the exchange of boundary conditions. To be more precise, a loose coupling scheme consists mainly of the iteration of the following four steps:

1. Compute the fluid solution and the resulting forces on fluid side.
2. Transfer the forces to the structure.
3. Compute the solution of the structure problem resulting in displacements on the structural side.
4. Transfer the deformations to the fluid model.

This has the important advantage that problem-specific models and existing solvers for both the fluid and the structure problem can still be used in FSI. However, since now we cannot expect the underlying discretisations of the single problems to match, special methods have to be developed to exchange these boundary conditions. This explicit approach usually works fine in aeroelastic problems or, more generally, in problems dealing with compressible fluids, though it is not unconditionally stable.

There is also another approach, which is both iterative and strong. In this approach, also existing solvers for the different problems can be used, see [18].

2 The Three Field Formulation

2.1 Mathematical Formulation - Fluid

From a mathematical point of view fluid-structure interaction comprises a system of nonlinear partial differential equations, describing the fluid flow

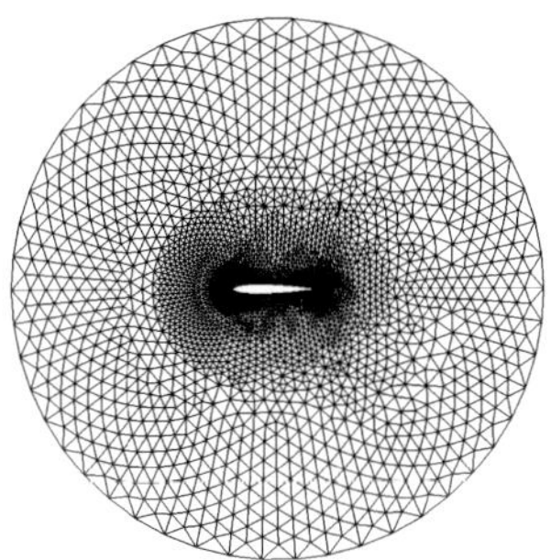

Figure 2.1. Left: the two-dimensional Naca-012 profile with a circular far-field boundary.

around the aircraft, the structural behaviour of the aircraft under the influence of external forces and the coupling of both models.

To this end, the domain of interest comprising the flow field as well as the structure is subdivided into

$$\mathbb{R}^3 = \Omega_F(t) \cup \Omega_S(t),$$

with the joined boundary $\Gamma(t) = \overline{\Omega_F(t)} \cap \overline{\Omega_S(t)}$. For computational reasons, the fluid domain $\Omega_F(t)$, which is roughly the exterior of the structural domain $\Omega_S(t)$ is further restricted by an outer ball or box, which introduces an additional far-field boundary (see Figure 2.1).

For the design of civil aircrafts with a usual cruising altitude of about 9,000 metres and a cruising speed of about $v_c = 965 km/h$ we have transonic flight conditions. Thus, compressibility is an issue and hence the flow has to be modelled using equations for compressible fluids. These are usually the Navier-Stokes equations, though, particular for steady state solutions, often the Euler equations are sufficient. Both equations can be written on the time-dependent flow domain $\Omega_F(t)$ in conservative form as follows:

$$\partial_t \rho_F + \nabla \cdot (\rho_F \mathbf{u}) = 0 \qquad \text{(conservation of mass)} \tag{2.1}$$

$$\partial_t (\rho_F \mathbf{u}) + \nabla \cdot (\rho_F \mathbf{u}\mathbf{u}^T - \sigma_F) = 0 \qquad \text{(conservation of momentum)} \tag{2.2}$$

$$\partial_t (\rho_F E) + \nabla \cdot ((\rho_F E - \sigma_F)\mathbf{u}) = 0 \qquad \text{(conservation of energy)} \tag{2.3}$$

Here, we employed the stress tensor for Newtonian fluids given by

$$\sigma_F = -pI + \mu_F(\nabla \mathbf{u} + \nabla \mathbf{u}^T) + \lambda_F(\nabla \cdot \mathbf{u})I =: -pI + \tau_F,$$

having a divergence

$$\nabla \cdot \sigma_F = -\nabla p + \mu_F \Delta \mathbf{u} + (\lambda_F + \mu_F)\nabla(\nabla \cdot \mathbf{u}).$$

As usual λ_F and μ_F denote constants describing the viscosity of the fluid. The first one refers to the volume viscosity while the latter is the shear viscosity.

The difference between Euler and Navier-Stokes in this form is simply given by the fact that, in the first case, the stress tensor consists only of the pressure term, i.e. $\sigma_F = -pI$.

Physically, the difference between Navier-Stokes and Euler is substantial. However, in industrial aeroelastic computations with the goal of predicting the aeroelastic equilibrium the Euler equations are often completely sufficient.

We have to solve this system of five equations for the six unknowns $\mathbf{u}$ (velocity), ρ_F (density), p (pressure) and $E = e + \|\mathbf{x}\|^2/2$ ((total) energy). Thus, an additional equation is missing, which is, in this context, usually given by the perfect gas law

$$\frac{p}{\rho} = RT[1+\alpha], \qquad e = c_\nu T,$$

which relates pressure, density and temperature. Here, $\alpha = \alpha(p,T)$ denotes the effective mass fraction of diatomic molecules in the dissociated condition, while R is the "undissociated" gas constant.

2.2 Mathematical Formulation - Structure

The structural behaviour is usually described by standard equations from elasticity. While a flutter analysis might require non-linear elasticity, it often suffices for the computation of an aeroelastic equilibrium to employ equations from linear elasticity. Here, usually the reference domain $\Omega_S(0)$ and the time-dependent domain $\Omega_S(t)$ are identified and the deformations or displacements $\mathbf{s}$ are modelled over the reference domain. They have to satisfy the equation

$$\rho_S \ddot{\mathbf{s}} = \nabla \cdot \sigma_S + \mathbf{b}, \tag{2.4}$$

where, $\mathbf{b}$ denotes volume forces like gravity, while ρ_S denotes the structural density, which is supposed to be constant here. Furthermore, we have the stress tensor

$$\begin{aligned} \sigma_S = \tau_S &= \mu_S(\nabla \mathbf{s} + \nabla \mathbf{s}^T) + \lambda_S(\nabla \cdot \mathbf{s})I \\ &= 2\mu_S \nabla \cdot \epsilon(\mathbf{s}) + \lambda_S \nabla(\nabla \cdot \mathbf{s}) + \mathbf{b}, \end{aligned} \tag{2.5}$$

which involves the Lamé constants μ_S and λ_S and the strain tensor

$$\epsilon(\mathbf{s}) = \frac{1}{2}\left(\nabla \mathbf{s} + \nabla \mathbf{s}^T\right).$$

Inserting (2.5) into (2.4) gives the well-known Lamé equation

$$\rho_S \ddot{\mathbf{s}} = 2\mu_S \nabla \cdot \epsilon(\mathbf{s}) + \lambda_S \nabla(\nabla \cdot \mathbf{s}) + \mathbf{b}. \tag{2.6}$$

2.3 Boundary conditions

The coupled system consisting of the Navier-Stokes/Euler equations (2.1)–(2.3), and the elasticity equation (2.6) requires additional initial and boundary conditions.

The boundary conditions are, in the context of aeroelasticity, usually the following ones. For the fluid, there is a prescribed velocity $\mathbf{u} = \mathbf{u}_\infty$ at the far field boundary. For the rigid part of the aircraft, there are either the conditions $\mathbf{u} = \mathbf{0}$ if the Navier-Stokes equations are employed or $\mathbf{u} \cdot \mathbf{n} = 0$ if the Euler equations are used. In the latter case, $\mathbf{n}$ denotes the outer normal of the boundary.

On fixed parts of the structure, we usually simply have $\mathbf{s} = \mathbf{0}$. In both cases, for the fluid and the structural problem, these types of boundary conditions are standard. For the coupled problem, however, are the following boundary conditions on the joined interface $\Gamma(t)$ more important:

- The equilibrium of traction, which is given by $\sigma_S \mathbf{n} = \sigma_F \mathbf{n} = -p\mathbf{n} + \tau_F \mathbf{n}$.
- The equilibrium of velocity fields, which is either given by $\mathbf{u} = \partial_t \mathbf{s}$ or by $\mathbf{u} \cdot \mathbf{n} = (\partial_t \mathbf{s}) \cdot \mathbf{n}$ for Navier-Stokes and Euler, respectively.

3 Typical Discretisations

As mentioned before there are different ways of discretising the coupled problem. One major drawback is the fact that the fluid equations are usually discretised using an Eulerian description while the structural equations are usually formulated and discretised using a Lagrangian approach. For a monolithic description it is therefore necessary to either convert the Eulerian description of the fluid problem to a Lagrangian one or to convert the Lagrangian description of the structural problem to an Eulerian one. Both options have severe disadvantages and the usual strategy is to employ an ALE formulation.

However, as outlined in the introduction, such an approach requires a significant adaption or rewriting of the involved software. Hence, in particular industry partners are keen on employing strategies which allow them to reuse their sophisticated software.

Thus, in this section we will discuss the discretisation techniques dominant in industrial applications.

3.1 Fluid - Finite Volumes

The fluid equations (2.1)–(2.3) are given in conservative form, i.e. they are of the form

$$\partial_t \mathbf{v} + \nabla \cdot \mathbf{F}(\mathbf{v}) = \mathbf{0}, \tag{3.7}$$

where the vector of unknowns $\mathbf{v}$ is given by

$$\mathbf{v} = \begin{pmatrix} \rho_F \\ \rho_F \mathbf{u} \\ \rho_F E \end{pmatrix},$$

and the flux vector F is determined by (2.1)–(2.3). The nonlinear equation (3.7) has to be satisfied on $\Omega_F(t)$ and is equipped with appropriate initial and boundary conditions.

An Eulerian discretisation of (3.7) on $\Omega_F(t)$ using finite volumes decomposes the region $\Omega_F(t)$ into a set of geometric primitives $\{V_j\}_j$, which are usually tetrahedra. Then, (3.7) is integrated over each test volume V_j. Integration by parts leads to

$$\frac{d}{dt}\int_{V_j} \mathbf{v} d\mathbf{x} = -\int_{V_j} \nabla \cdot \mathbf{F}(\mathbf{v}) d\mathbf{x} = -\int_{\partial V_j} \mathbf{F}(\mathbf{v}) \mathbf{n} dS.$$

Discretising the spatial boundary integrals and introducing a numerical flux function $\mathbf{H}$ gives

$$\frac{d}{dt}\int_{V_j} \mathbf{v} d\mathbf{x} = \sum_{V \in \mathcal{N}_j} \sum_{k=1}^{n_Q} w_k^{V\cap V_j} \mathbf{H}(\mathbf{v}(\mathbf{x}_k^{V\cap V_j}), \mathbf{v}(\mathbf{x}_k^{V\cap V_j}), \mathbf{n}^{V\cap V_j}), \tag{3.8}$$

where we have suppressed the time variable. Here, $\mathcal{N}_j$ consists of all geometric primitives having a joint face with V_j. The outer normal to that face is denoted by $\mathbf{n}^{V\cap V_j}$. Moreover, the integration points on $V \cap V_j$ and weights are denoted by $\mathbf{x}_k^{V\cap V_j}$ and $w_k^{V\cap V_j}$, respectively.

The discretisation is usually completed by an explicit time discretisation. More details on the finite volume method, particularly on how to choose the numerical flux function, can be found in any book an finite volumes.

3.2 Finite Volume Reconstructions and Optimal Recovery

An explicit time discretisation of (3.8) leads to a specific spatial reconstruction problem. We are given the cell averages $\int_{V_j} \mathbf{v} d\mathbf{x}$ of the previous time step and require the point-wise information of $\mathbf{v}$ at the integration points $\mathbf{x}_k^{V\cap V_j}$. We will discuss this problem now, though it is worth pointing out that we only require point-wise information because of the employed spatial quadrature rule. It would be more natural to try to recover the boundary integrals from the spatial integrals.

While classical finite volume methods use polynomial reconstructions to recover the unknown function point-wise from given cell average information, it is possible to express the reconstruction problem in a more general setting.

Assuming that we can reconstruct the vector-valued function $\mathbf{v}$ component-wise, we can work within the following framework:

- Our true solution v belongs to a Hilbert space $\mathcal{H}$ of functions defined on some region $\Omega \subseteq \mathbb{R}^d$.

- We are given information $v_1, \dots, v_N$ of v and we know that this information is of the form $v_j = \lambda_j(v)$, where $\lambda_1, \dots, \lambda_N \in \mathcal{H}^*$ are linearly independent.

Then, it would be natural to look for on *optimal* solution in the following sense.

Definition 1. *The norm-minimal quasi-interpolant to an unknown function $u \in \mathcal{H}$, based on the information generated by $\lambda_1, \dots, \lambda_N \in \mathcal{H}^*$ is the unique solution s^* of*

$$\min \{ \|s\|_{\mathcal{H}} : \lambda_j(s) = \lambda_j(v), 1 \le j \le N \} .$$

Fortunately, the norm-minimal interpolant can be computed directly.

Theorem 1 ([27]). *The norm-minimal quasi-interpolant to an unknown function $v \in \mathcal{H}$ is given by*

$$s^* = \sum_{j=1}^{N} \alpha_j v_j,$$

where v_j is the Riesz representer of λ_j. The coefficients α_j solve the linear system

$$(\lambda_i(v_j))\alpha = (\lambda_i(v)).$$

Since we want to reconstruct the function point-wise, it is natural to assume that $\mathcal{H} \subseteq C(\Omega)$. However, in this particular situation, it is well-known again, that the Hilbert space $\mathcal{H}$ has a unique reproducing kernel, i.e. a function $\Phi : \Omega \times \Omega \to \mathbb{R}$, satisfying

- $\Phi(\cdot, \mathbf{x}) \in \mathcal{H}$, for all $\mathbf{x} \in \Omega$,
- $v(\mathbf{x}) = (v, \Phi(\cdot, \mathbf{x}))_{\mathcal{H}}$, for all $\mathbf{x} \in \Omega$ and all $v \in \mathcal{H}$.

Moreover, the Riesz representer of a functional $\lambda \in \mathcal{H}^*$ is then simply given by $\lambda^{\mathbf{y}} \Phi(\cdot, \mathbf{y})$, where the upper index indicates that λ acts with respect to $\mathbf{y}$. Thus, in our particular situation where $\lambda_j(v) = \int_{V_j} v d\mathbf{x}$, the reconstruction is given by

$$s^* = \sum_{j=1}^{N} \alpha_j \int_{V_j} \Phi(\cdot, \mathbf{y}) d\mathbf{y},$$

and the coefficients are determined by the interpolation conditions

$$\sum_{j=1}^{N} \alpha_j \int_{V_i} \int_{V_j} \Phi(\mathbf{x}, \mathbf{y}) d\mathbf{x} d\mathbf{y}, \qquad 1 \le i \le N. \tag{3.9}$$

Since the V_j are tetrahedra, for many different choices of Φ, the integrals in (3.9) can be computed explicitly.

While the general approach for solving these kind of generalised interpolation problems has been known for quite a while (see [19, 24, 27, 29] for the general framework and [20–22] for particular applications in the context of finite volumes), only recently convergence results were provided.

Theorem 2 ([26]). *If $\mathcal{H} = W_2^\tau(\Omega)$ with $\tau > d/2$ then the error between the true solution and the optimal reconstruction can be bounded by*

$$\|v - s^*\|_{L_\infty(\Omega)} \leq Ch^{\tau - d/2} \|v\|_{W_2^\tau(\Omega)},$$

where h is a fixed multiple of the mesh norm.

It is important to know that Theorem 2 can be generalised in the following two ways. As a matter of fact the proof of Theorem 2 in [26] reflects both generalisations. First, the reconstruction process can be limited to a local reconstruction process using only cell information from the neighbouring cells. The actual number of necessary cells depends on the smoothness of the target function. Second, to derive the stated error estimate it is sufficient to have cell information from one "direction". To be more precise, it is possible to define a cone, which is centred in the target cell, and only neighbouring cells inside that cone have to be used for the reconstruction process. This opens the road for deriving error estimates for upwind and (w)eno schemes.

3.3 Structure - Finite Elements

The structural problem consists of the elliptic equation

$$\begin{aligned} \rho_S \ddot{\mathbf{s}} &= 2\mu_S \nabla \cdot \epsilon(\mathbf{s}) + \lambda_S \nabla(\nabla \cdot \mathbf{s}) + \mathbf{b} \qquad \text{in } \Omega_S(t) \\ \sigma_S \mathbf{n} &= \mathbf{g} \qquad \text{on } \Gamma(t), \end{aligned} \tag{3.10}$$

with possibly additional boundary conditions on fixed parts of the structure. Here, the vector $\mathbf{g}$ ideally comes from the solution of the fluid problem, practically, it has to be constructed from that solution. For the time being, we assume it to be known.

The predominant discretisation method for these kind of problems is the finite element method, which requires to rewrite (3.10) in its weak form. To this end, (3.10) is multiplied by a vector-valued test function $\boldsymbol{\varphi} \in C_0(\Omega)^3$. Integration by parts then yields

$$\int_\Omega [\rho_S \ddot{\mathbf{s}} \cdot \boldsymbol{\varphi} + 2\mu_S(\epsilon(\mathbf{s}) : \epsilon(\boldsymbol{\varphi})) + \lambda_S(\nabla \cdot \mathbf{s})(\nabla \cdot \boldsymbol{\varphi})]\, d\mathbf{x} = \int_\Omega \mathbf{b} \cdot \boldsymbol{\varphi} d\mathbf{x} + \int_{\partial\Omega} \mathbf{g} \cdot \boldsymbol{\varphi} dS,$$

where we used the usual notation $A : B = \text{trace}(A^T B)$. This equation is then discretised in space by using a standard finite element space, like piecewise linear elements over a tetrahedral decomposition.

In particular in aircraft design, the *Modes of Vibration* play an important role. They can be computed by splitting the time and space variable for discretising the weak form of the elasticity equation, i.e. by making the Ansatz

$$\mathbf{s} = \sum_n q_n(t) \boldsymbol{\psi}_n(\mathbf{x}).$$

A straight-forward calculation then leads to

$$M\ddot{\mathbf{q}} + K\mathbf{q} = \mathbf{e},$$

with mass matrix M and stiffness matrix K given by

$$M_{kn} = \int \rho_S \boldsymbol{\psi}_n \cdot \boldsymbol{\psi}_k d\mathbf{x},$$

$$K_{kn} = 2\mu_S \int \epsilon(\boldsymbol{\psi}_n) : \epsilon(\boldsymbol{\psi}_k) d\mathbf{x} + \lambda_S \int (\nabla \cdot \boldsymbol{\psi}_n)(\nabla \cdot \boldsymbol{\psi}_k) d\mathbf{x}.$$

The natural modes of vibration can then be determined by computing the eigenvalues and functions of

$$K\mathbf{z} = \lambda M\mathbf{z}.$$

4 The Coupling Procedure

4.1 Strategy for Coupling

As pointed out before, the predominant strategy for solving the coupled fluid-structure problem in aeroelasticity is given by a loose coupling approach. There are plenty of advantages for such an approach, such as:

- Each problem can be discretised separately.
- Single discretisations can reflect the physical needs of each problem.
- Distinguished solvers for each problem can be used.
- Exchange of information is restricted to boundary information.

However, there are also some drawbacks involved with this process. For example, we will usually encounter non-matching discretisations, which means that, on the one hand, the motion of the structure has to be followed by the motion of the fluid, and, on the other hand, loads from the fluid side of the fluid-structure-interface have to be transferred to the structural side.

Another) drawback is that there are no convergence results for such a loose coupling procedure and that the explicit time discretisation is not unconditionally stable. Finally, if we allow the fluid and the structural solver to be rather arbitrary, we need a rather universal coupling scheme, which can handle a large variety of different discretisations. To this end, we will restrict ourselves mainly to point-based information.

Typical requirements for a coupling process comprise the following. For the subprocess being responsible for transferring the motion of the structure to the fluid model, it is important to have an exact reproduction on the structural nodes. Furthermore, rigid body motions should be exactly recovered. Finally, the reconstruction process should be linear such that it can be realised by a coupling matrix.

The whole coupling procedure must have the ability of dealing with rather arbitrary models, including large-scale models but also flat models.

Finally and most importantly, the reconstruction process has to conserve virtual work and energy.

4.2 Scattered Data Interpolation

To guarantee the universal character of our coupling scheme, we will restrict ourselves to point-based information. Furthermore, we will assume that the displacement of the structure is governed by a smooth deformation field such that interpolation at the structure points is a feasible tool. Again, we will consider the vector-valued deformation information component-wise, though we will later on also discuss a real vector-valued approach. Hence, we now shortly discuss the ideas behind scattered data interpolation, which are based on the same principles as the reconstruction process described in Section 3.2. This time our functionals are simply given by point evaluations, i.e. we have $\lambda_j(v) = \delta_{\mathbf{x}_j}(v) = v(\mathbf{x}_j)$. Furthermore, we relax the condition on the Hilbert space and allow also a semi-Hilbert space having a semi-norm generated by a positive semi-definite bilinear form with a finite dimensional kernel. This kernel consists usually of polynomials of low degree.

Hence, the general setup is now that we are given

- Structure Nodes $X = \{\mathbf{x}_1, \dots, \mathbf{x}_N\} \subseteq \mathbb{R}^3$,
- deformations in one coordinate direction $v_1, \dots, v_N$.

We form the interpolant to the deformation field in this coordinate direction as

$$s(\mathbf{x}) = \sum_{j=1}^{N} \alpha_j \Phi(\mathbf{x}, \mathbf{x}_j) + \sum_{j=1}^{Q} \beta_j p_j(\mathbf{x}) \tag{4.11}$$

with a (conditionally) positive definite kernel Φ of order, say, m and a basis $p_1, \dots, p_Q$ of multivariate polynomials of degree at most $m-1$. The coefficients α_j, β_j are determined by by solving the interpolation equations

$$s(x_j) = v_j, \qquad 1 \le j \le N,$$

together with the *discrete moment* conditions

$$\sum_{j=1}^{N} \alpha_j p_k(x_j) = 0, \qquad 1 \le k \le Q.$$

The corresponding linear system is uniquely solvable if the structural points form a uni-solvent system for the polynomials, i.e. that the only polynomial from span$\{p_1, \dots, p_Q\}$ which vanishes on all $\mathbf{x}_j$ is the zero polynomial. It is important to realise that a conditionally positive definite kernel of order m is also conditionally positive definite of order $\ell > m$. Hence, adding higher order polynomials to the interpolant (4.11) is always possible and in aeroelastic applications often even essential.

In our applications, using for example linear polynomials, this means that there are at least four structural points, which are not in a plane. Unfortunately, it sometimes happens that structural models are designed such that all points are in a plane or even on a one dimensional rod. In these cases it is still possible to solve the linear system but the solution is not unique any more (see [28]).

For recovering the structural deformations in fluid-structure interaction problems it is essential to include linear polynomials even if the kernel is positive definite, i.e. conditionally positive definite of order $m = 0$. This ensures that rigid body transformations are recovered exactly since they can be described by a linear polynomial.

4.3 Typical Kernels

Usually, the kernels employed are defined on all of $\mathbb{R}^3$ and are both translation and rotational invariant, meaning that $\Phi(x, y) = \phi(\|x-y\|_2)$ with a univariate function $\phi : [0, \infty) \to \mathbb{R}$.

There are a variety of kernels used in aeroelasticity, some of them are listed in Table 4.1 together with their degree of positive definiteness.

The table in particular lists the thin-plate splines originating in the fundamental paper [15] as well as the compactly supported functions introduced in [23] and used in the context in aeroelasticity in [2] for the first time.

4.4 Coupling Matrix *H*

Software packages as MSC.Nastran [17] usually require that the coupling process is described by a coupling matrix. This means that we have to define a matrix or linear mapping $H : \mathbb{R}^N \to \mathbb{R}^M$ mapping the deformations (in one coordinate direction) on structural side to the deformations (in the same

Table 4.1. (Conditionally) positive definite functions

$\Phi(\mathbf{x}) = \phi(r)$ with $r = \|\mathbf{x}\|_2$	Name	CPD
r	VS	$m \geq 1$
$r^2 \log r$	TPS	$m \geq 2$
$\sqrt{r^2+1}$	MQ	$m \geq 1$
$1/\sqrt{r^2+1}$	IMQ	$m \geq 0$
$\exp(-r^2)$	G	$m \geq 0$
$(1-r)_+^2$	W1	$m \geq 0$
$(1-r)_+^4(1+4r)$	W2	$m \geq 0$
$(1-r)_+^6(35r^2+18r+3)$	W3	$m \geq 0$
$\frac{\pi}{12}(r+2)(1-r)_+^2$	EH	$m \geq 0.$

direction) on fluid side. In our context, this can be achieved by introducing the matrices

$$\begin{aligned} A_{\Phi,X} &= (\Phi(\mathbf{x}_i,\mathbf{x}_j)) \in \mathbb{R}^{N\times N}, & P_X &= p_k(\mathbf{x}_i) \in \mathbb{R}^{N\times Q}, \\ A_{\Phi,X,Y} &= (\Phi(\mathbf{y}_i,\mathbf{x}_j)) \in \mathbb{R}^{M\times N}, & P_Y &= (p_j(\mathbf{y}_i)) \in \mathbb{R}^{M\times Q}, \end{aligned}$$

employing the evaluation points $Y = \{\mathbf{y}_1, \ldots, \mathbf{y}_M\}$ on fluid side.

Writing the evaluation of the interpolant at the evaluation points in matrix form

$$s|Y = \begin{pmatrix} A_{\Phi,X,Y} & P_Y \end{pmatrix} \begin{pmatrix} A_{\Phi,X} & P_X \\ P_X^T & 0 \end{pmatrix}^{-1} \begin{pmatrix} \mathbf{v} \\ 0 \end{pmatrix} =: \begin{pmatrix} H & \widetilde{H} \end{pmatrix} \begin{pmatrix} \mathbf{v} \\ 0 \end{pmatrix}$$

determines the coupling matrix $H : \mathbb{R}^N \to \mathbb{R}^M$, which satisfies not only $H\mathbf{v} = s|Y$ but also

$$H\mathbf{1} = \mathbf{1},$$

where $\mathbf{1}$ denotes the vector from $\mathbb{R}^N$ or $\mathbb{R}^M$, respectively, having all entries equal to one.

4.5 Partition of Unity

Recent development in model sizes lead to a problem of the just described direct approach. Its computational complexity becomes sometimes unmanageable. This refers to both resources: time and space. To be more precise, we need

- $\mathcal{O}(N^3 + N^2M)$ time for building the coupling matrix,
- $\mathcal{O}(N \cdot M)$ memory for storing the coupling matrix.

Obviously, the memory requirement could be resolved by not explicitly storing a coupling matrix, which, however, would then require to rewrite large parts of existing software packages. The use of compactly supported basis functions together with a sparse matrix representation also leads to some improvement.

However, a recently successfully rediscovered strategy is to combine the radial basis function approach with a localisation technique called *partition of unity*. The idea behind this is easily described. Instead of solving one large linear system, the problem is broken up into a large number of small systems, which are solved individually and the local solutions are then combined to a global one by blending.

To be more precise, let Ω be a superset of the bounding box of $X \cup Y$. Then, the partition of unity approach can be described in the following four steps.

1. Cover Ω by overlapping patches $\{\Omega_j\}$, i.e. $\Omega \subseteq \bigcup \Omega_j$.
2. Solve the local problems. For each patch Ω_j, collect the structural points $X_j = X \cap \Omega_j$. Compute the local interpolant s_j satisfying $s_j(x_i) = v_i$ for all $x_i \in X_i$.
3. Choose a partition of unity corresponding to the patches Ω_j, i.e. a family of functions $w_j \in C_0(\Omega_j)$ with

$$\sum w_j(\mathbf{x}) = 1, \qquad w_j(\mathbf{x}) \geq 0, \qquad \mathbf{x} \in \Omega.$$

4. Form a global interpolant by summing up the weighted local interpolants

$$s(\mathbf{x}) := \sum w_j(\mathbf{x}) s_j(\mathbf{x}), \qquad \mathbf{x} \in \Omega.$$

It is easy to see that the global function s is indeed an interpolant in all of the points. Furthermore, if all the local interpolants recover polynomials of a certain degree exactly so does the global function.

To improve the computational complexity of the interpolation process several geometric issues have to be satisfied by the construction of the patches.

Each patch should contain about the same small number of points. This number should be considered to be constant compared to the total number of points. As a consequence, the number of patches is linear in the number of points. Furthermore, each local problem can be solved in constant time, such that all local problems can be solved in linear time.

For each evaluation point $\mathbf{x} \in \Omega$ the number $n(\mathbf{x})$ of patches Ω_j with $\mathbf{x} \in \Omega_j$ should be constant compared to N. Furthermore, it must be possible to locate these patches easily meaning in constant or at most logarithmic time.

This together ensures efficient solving and evaluating but it requires an additional data structure for building the patches. Typically ([14, 25]), such data structures are based on tree-like constructions. Most popular examples comprises kd-trees, Range-trees, and Oct-trees.

It is possible to show that such an "intelligent" data structure requires about $\mathcal{O}(N \log N)$ time to be built, while the memory overhead is linear. The time necessary for evaluation is logarithmic.

4.6 Transferring Forces

So far, we have only dealt with one part of the coupling procedure: the transfer of deformations from the structural field to the aerodynamic field. It remains to discuss the second part, the transfer of forces from the aerodynamic grid to the structural grid. Unfortunately, we cannot employ interpolation to do this since it contradicts elementary physical requirements like the conservation of the total virtual work and the total forces. However, the principle of conservation of virtual work performed by the external aerodynamic loads and the internal structural forces,

$$\delta W = \delta \mathbf{v}_s^T \mathbf{f}_s = \delta \mathbf{v}_f^T \mathbf{f}_f,$$

where the indices s and f refer to the structural and fluid model, and $\mathbf{v}$ and $\mathbf{f}$ denote the displacement and force vectors, respectively, together with the fact that we assume a linear context for the transfer of displacements

$$\mathbf{v}_f = H\mathbf{v}_s$$

shows that, at least on an infinitesimal scale, the forces should be transferred by

$$\mathbf{f}_s = H^T\mathbf{f}_f.$$

Hence, in all our examples we will use the transposed of the coupling matrix for transferring forces.

In our discrete setting with structural nodes $X = \{\mathbf{x}_1, \dots, \mathbf{x}_N\}$ and fluid nodes $Y = \{\mathbf{y}_1, \dots, \mathbf{y}_N\}$ let us denote the forces on structural side by $\mathbf{f}$ and the forces on fluid side with $\mathbf{F}$. Then, we have $\mathbf{f} = H^T\mathbf{F}$ and conservation of virtual work means

$$\sum_{i=1}^{N} f_j(\mathbf{x}_i)v_j(\mathbf{x}_i) = \sum_{i=1}^{M} F_j(\mathbf{y}_i)s_j(\mathbf{y}_i), \qquad 1 \le j \le 3,$$

where $\mathbf{s}(\mathbf{y}_j) \in \mathbb{R}^3$ and $\mathbf{v}(\mathbf{x}_j) \in \mathbb{R}^3$ are the deformations on fluid and structure side, respectively.

Since our coupling matrix H also satisfies $H\mathbf{1} = \mathbf{1}$, we have in addition the conservation of forces in the form

$$\sum_{i=1}^{N} \mathbf{f}(\mathbf{x}_i) = \sum_{i=1}^{M} \mathbf{F}(\mathbf{y}_i).$$

However, in some applications the use of the transposed coupling matrix results to an unnatural and unphysical distribution of the forces on the structural grid. We will see that the partition of unity approach leads also to a better and more physical distribution. Theoretically, this is corroborated by the fact that we have now not only conservation of virtual work and forces in the global context but also on each patch.

5 Recovery of Rotations

In contrast to high-resolution models there are also very reduced structural models in use. The best known case is the structural grid of the AMP test wing ([16, 30]), which is sometimes reduced to a simple rod, see Figure 5.2.

For such reduced models, often not only displacement vectors at the nodes are known but also rotational information is given. In our terminology, this means that for each $\mathbf{x}_j$ we are given not only the displacements $\mathbf{v}(\mathbf{x}_j)$ but also angles $\boldsymbol{\theta}_j \in \mathbb{R}^3$. Since (small) angles can be related to derivatives of the displacement field via

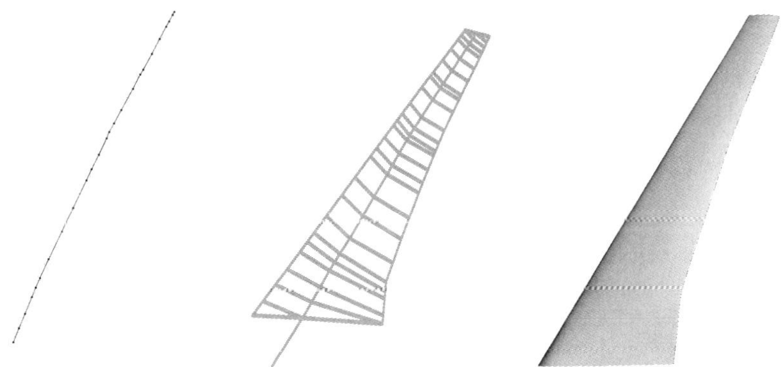

Figure 5.2. Two structural models (left and middle) and the aerodynamical model (right) of the AMP wing.

$$\begin{pmatrix} \theta_x \\ \theta_y \\ \theta_z \end{pmatrix} = \frac{1}{2} \begin{pmatrix} 0 & -\frac{\partial}{\partial z} & \frac{\partial}{\partial y} \\ \frac{\partial}{\partial z} & 0 & -\frac{\partial}{\partial x} \\ -\frac{\partial}{\partial y} & \frac{\partial}{\partial x} & 0 \end{pmatrix} \begin{pmatrix} v_1 \\ v_2 \\ v_3 \end{pmatrix} =: \frac{1}{2} \nabla \times \mathbf{v},$$

we are now looking for an interpolant $\mathbf{s} : \mathbb{R}^3 \to \mathbb{R}^3$ satisfying

$$\mathbf{s}(\mathbf{x}_i) = \mathbf{v}(\mathbf{x}_i), \qquad \nabla \times \mathbf{s}(\mathbf{x}_i) = 2\boldsymbol{\theta}(\mathbf{x}_i), \qquad 1 \leq i \leq N.$$

This is a generalised interpolation problem for vector-valued functions, where the components cannot be treated independently anymore. A generalisation of the classical approach has been suggested in [19], which can be summarised as follows. First of all, we define the action of a *vector-valued* functional $\boldsymbol{\lambda}$ to a vector-valued function $\mathbf{f}$ by

$$\boldsymbol{\lambda}(\mathbf{f}) := \lambda_1(f_1) + \lambda_2(f_2) + \lambda_3(f_3). \tag{5.12}$$

Given K vector-valued functionals $\boldsymbol{\lambda}_1, \ldots, \boldsymbol{\lambda}_K$ we then can form a vector-valued interpolant using a *matrix-valued* kernel $\Phi : \mathbb{R}^3 \times \mathbb{R}^3 \to \mathbb{R}^{3\times3}$ by

$$\mathbf{s}(\mathbf{x}) = \sum_{j=1}^{K} \alpha_j \boldsymbol{\lambda}_j^{\mathbf{y}} \Phi(\mathbf{x}, \mathbf{y}) + \mathbf{p}(\mathbf{x}), \tag{5.13}$$

where $\mathbf{p}$ is a vector-valued low order degree polynomial, i.e. $\mathbf{p} \in \pi_{m-1}(\mathbb{R}^3)^3$. The expression $\boldsymbol{\lambda}_j^{\mathbf{y}} \Phi(\mathbf{x}, \mathbf{y})$ means that the functional $\boldsymbol{\lambda}$ acts with respect to the variable $\mathbf{y}$. Since Φ is matrix-valued, the action of $\boldsymbol{\lambda}$ is on the rows of Φ in the sense of (5.12). This means in particular for any two functionals $\boldsymbol{\lambda}$ and $\boldsymbol{\mu}$ that $\boldsymbol{\lambda}^{\mathbf{y}} \Phi(\cdot, \mathbf{y})$ is a vector valued function and $\boldsymbol{\mu}^{\mathbf{x}} \boldsymbol{\lambda}^{\mathbf{y}} \Phi(\mathbf{x}, \mathbf{y})$ is a scalar.

As in the scalar-valued case the coefficients α_j and the polynomial $\mathbf{p}$ are determined by the conditions

$$\boldsymbol{\lambda}_i(\mathbf{s}) = \boldsymbol{\lambda}_i(\mathbf{f}), \qquad 1 \le i \le K,$$
$$\sum_{j=1}^{K} \alpha_j \boldsymbol{\lambda}_j(\mathbf{q}) = 0, \qquad \mathbf{q} \in \pi_{m-1}(\mathbb{R}^3)^3.$$

The resulting interpolation matrix is invertible if the kernel Φ is (conditionally) positive definite.

Definition 1. *A function* $\Phi \in C^{2\ell}(\Omega \times \Omega; \mathbb{R}^{d\times d})$ *is called* positive definite *if for all* $\boldsymbol{\lambda} \in (C(\Omega)^*)^d$,

$$\boldsymbol{\lambda}^{\mathbf{x}} \boldsymbol{\lambda}^{\mathbf{y}} \Phi(\mathbf{x}, \mathbf{y}) > 0.$$

In our specific situation, we have $K = 6N$ functionals, $3N$ functionals representing point evaluation at $\mathbf{x}_j$ and $3N$ functionals representing the rotational information via the curl operator. Hence, we could define

$$\boldsymbol{\lambda}_{3(N+j)-2} = \begin{pmatrix} \delta_{\mathbf{x}_j} \\ 0 \\ 0 \end{pmatrix}, \boldsymbol{\lambda}_{3(N+j)-1} = \begin{pmatrix} 0 \\ \delta_{\mathbf{x}_j} \\ 0 \end{pmatrix}, \boldsymbol{\lambda}_{3(N+j)} = \begin{pmatrix} 0 \\ 0 \\ \delta_{\mathbf{x}_j} \end{pmatrix},$$

$$\boldsymbol{\lambda}_{3j-2} = \begin{pmatrix} 0 \\ -\delta_{\mathbf{x}_j} \circ \partial_z \\ \delta_{\mathbf{x}_j} \circ \partial_y \end{pmatrix}, \boldsymbol{\lambda}_{3j-1} = \begin{pmatrix} \delta_{\mathbf{x}_j} \circ \partial_z \\ 0 \\ -\delta_{\mathbf{x}_j} \circ \partial_x \end{pmatrix}, \boldsymbol{\lambda}_{3j} = \begin{pmatrix} -\delta_{\mathbf{x}_j} \circ \partial_y \\ \delta_{\mathbf{x}_j} \circ \partial_x \\ 0 \end{pmatrix}$$

for $1 \le j \le N$.

In [1], this modification of the general interpolation has been proposed and analysed to meet these generalised interpolation conditions. From that paper it follows that the generalised interpolant (5.13) can also be written in the form

$$\mathbf{s}(\mathbf{x}) := \sum_{j=1}^{N} \Phi(\mathbf{x}, \mathbf{x}_j)\boldsymbol{\alpha}_j + \sum_{j=1}^{N} (\delta_{\mathbf{x}_j} \circ (\mathrm{curl}))^{\mathbf{y}} \Phi(\mathbf{x}, \mathbf{y}) \boldsymbol{\alpha}_{N+j} + \mathbf{p}(\mathbf{x}), \qquad (5.14)$$

where this time $\boldsymbol{\alpha}_j \in \mathbb{R}^3$ is a coefficient vector for each functional.

6 An Aeroelastic Analysis

In this section, we present some numerical results for the coupling process and the aeroelastic problem.

We applied the partition of unity technique as well as the global one based only on radial basis functions to Alenia's SMJ model, which has been developed and provided as a test case for the TAURUS project by Alenia Aeronautica. It describes a reference for a 70 seats passenger aircraft.

6.1 Structural Model

The structural model consists of a fuselage, fixed in its plane of symmetry, a wing structure, and a horizontal and a vertical tail (see the left part of Figure 6.3). The structure is assumed to be made of ideal aluminium with a mass density of 2700 $\frac{Kg}{m^3}$. Furthermore, Young's modulus is 71 GPa and Poisson's ratio is 0.33.

For the numerical computation of the static aeroelastic equilibrium of the aircraft, symmetric boundary conditions have been imposed.

The discretised model consists of 5630 nodes and 12,388 finite elements of different types. The structural finite element equations are solved by an analysis code using the formulation of the structural model in a generalised form. The related generalised coordinates, stiffness and mass matrices result from a normal-modes solution, which has been computed using the finite element analysis code MSC.Nastran [17].

A series of tests has been performed to estimate the number of Lagrangian coordinates necessary to minimise any inaccuracy introduced by the reduced model. The results of several static load cases for a modal based structure having different numbers of eigenmodes have been compared to those obtained by a direct solution of the complete FE matrices using MSC.Nastran.

Particularly, the test cases comprised a concentrated load at the tip of the wing and the horizontal tail-plane to produce a bending deformation of the lifting surface. Another test case comprised a distribution of torque along the wing and horizontal tail-plane span to produce a torsional deformation. Table 6.2 contains the details, and Figures 6.4 and 6.5 show the deflection and rotations of the wing along the elastic axis. It follows that about 100 eigenmodes are sufficient to represent the structural behaviour. It is worth noting that in particular the correct representation of wing torsion requires high and hence local frequency modes.

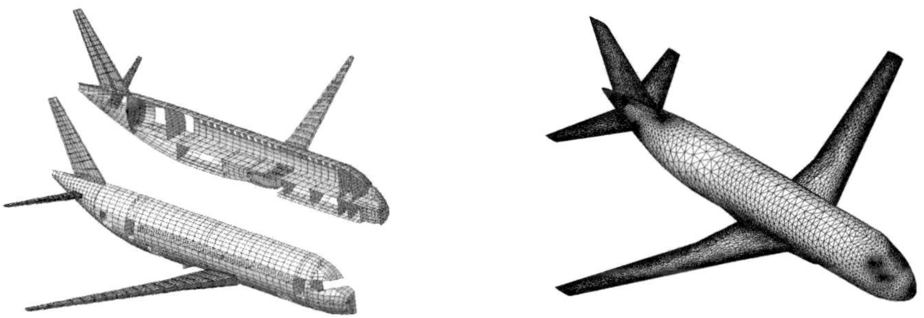

Figure 6.3. The structural (left) and aerodynamical (right) model.

Table 6.2. Test cases for the structural model

Applied load	Magnitude	Applied load	Magnitude
Wing bending load	20000 N	Wing torsion load	6×2500 Nm
HTP bending load	10000 N	HTP torsion load	1×5000 Nm

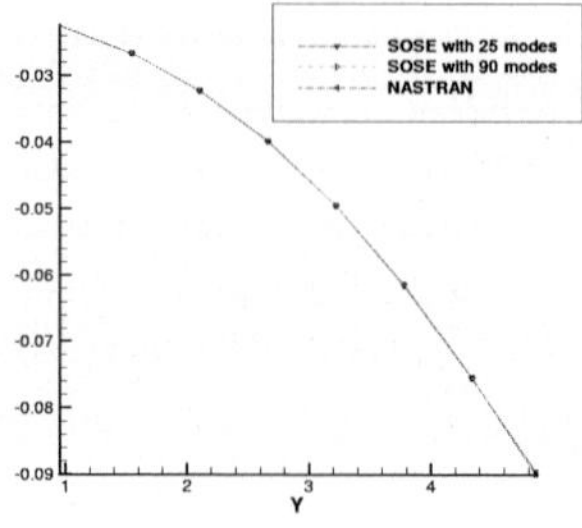

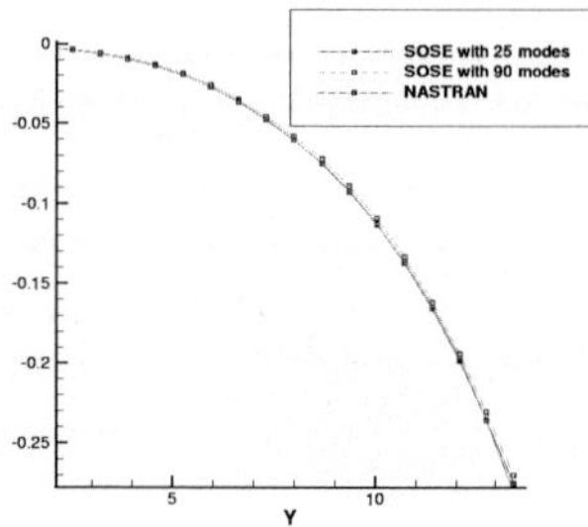

Figure 6.4. Vertical displacements along the elastic axis of the tail-plane (left) and wing (right) for different structure models.

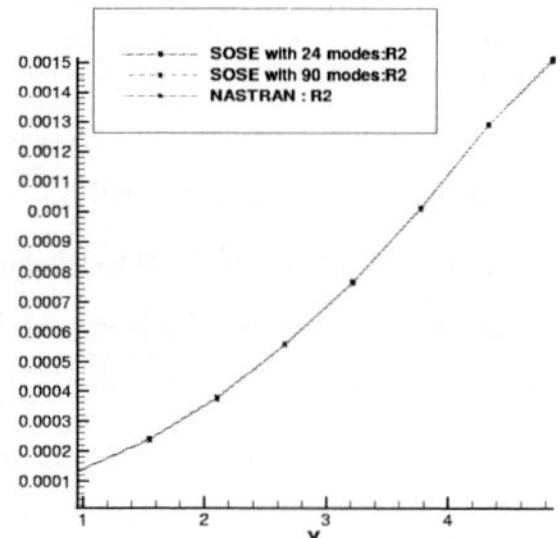

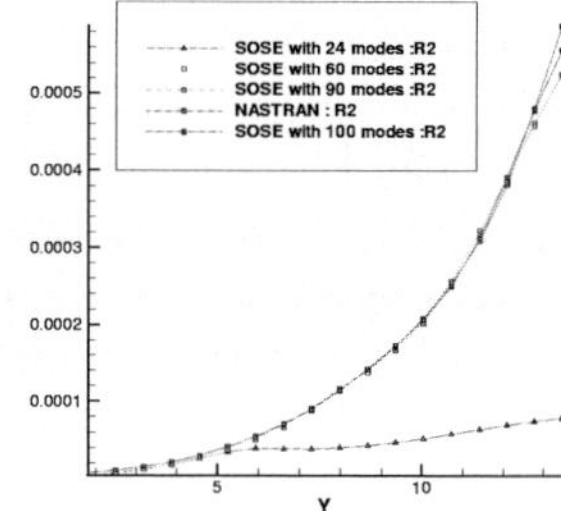

Figure 6.5. Rotations about the elastic axis of the wing for different structure models.

Finally, just for illustrational purposes, the first four eigenmodes corresponding to 2.00Hz, 2.81Hz, 3.71Hz and 5.28Hz are depicted in Figure 6.6.

6.2 Aerodynamic Model

For the aerodynamic model of the transport aircraft, the fluid flow is described by the nonlinear three-dimensional Euler equations. which are solved by a specific upwind-scheme based on finite volumes. In our situation, spatial discretisation is given by an unstructured finite volume mesh consisting of four million tetrahedrons. The surface mesh of the complete aircraft (see the right part of Figure 6.3) is based on about 300 000 triangles.

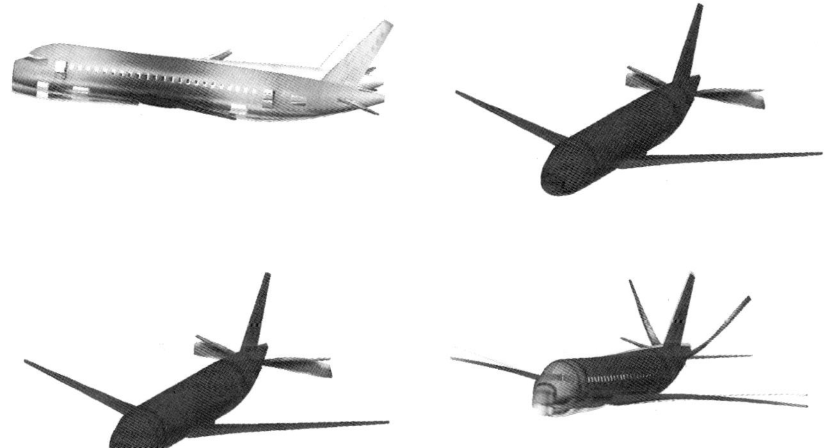

Figure 6.6. The first four eigenmodes of the structural model. (*See also* Color Plate on page 400)

6.3 Aeroelastic Computation

All aeroelastic computations were performed using the following flight and flow conditions (at infinity). We assumed a Mach number of 0.8, a temperature of 216.65 K, a static pressure of 1662.68 Pa and a dynamic pressure of 9739.54 Pa. The crusing altitude was suppose to be 10498.2 meters at a velocity of 236.457 m/s and an angle of attack of $\alpha = -0.0875°$.

All computations were done on a small Linux workstation cluster with 8 Intel Xeon 2.66 GHz CPUs, where the CFD solver occupied permanently 6 CPUs and both the FE solver and the coupling program had one CPU each.

The deformation of the structure is shown in the left part of Figure 6.7. It leads to a reduction of the total lift C_l from 0.508 to 0.431 which corresponds to a loss of about 15%. But we also have the expected reduction of drag C_d from 0.0368 to 0.0299, which is about 15%.

Besides the reduced lift also a different behaviour of the span-wise C_p distribution at the wing tip is given, see the right hand side of Figure 6.7. This phenomenon can be explained by a vortex at the trailing edge of the wing tip. In this region, the fluid particles undergo a rapid acceleration when passing the tip and produce a local vortex that increases the Mach number to a maximum value of 1.4.

Next, we want to compare the performance of the global coupling method based on radial basis function to the performance of the partition of unity method combined with radial basis functions.

The local method based on partition of unity leads to convergence after 5 iterations with a residual of $\varepsilon = 3.44e^{-4}$. Here, the residual is defined to be the relative difference in displacements of two consecutive solution steps. For

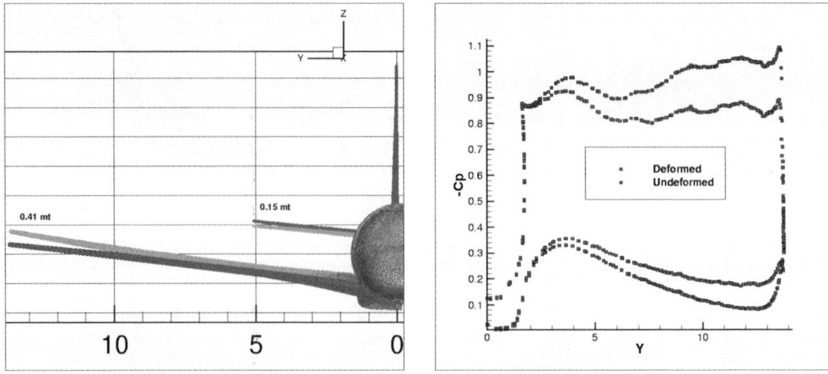

Figure 6.7. Comparison rigid versus elastic: Deflection and Pressure Distribution. (*See also* Color Plate on page 401)

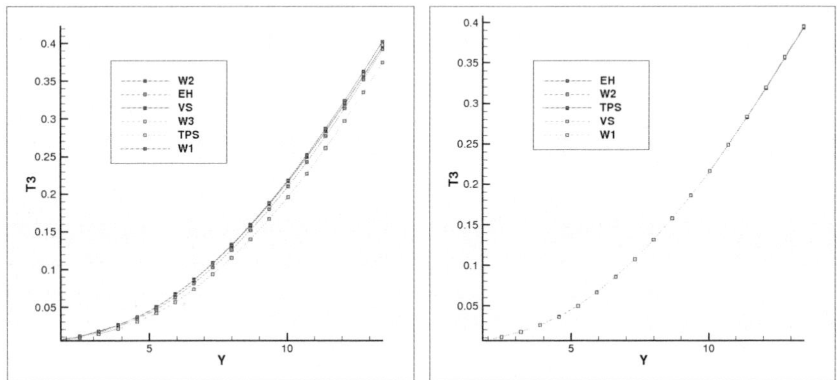

Figure 6.8. Influence of the chosen RBF on the global (left) and local (right) method. (*See also* Color Plate on page 401)

comparison, the classical, global spatial coupling approach needed 6 iterations to stop with an even worse residual of $\varepsilon = 1.16e^{-3}$.

As expected the total time necessary for the coupling process dropped from 5000 seconds to 20 seconds. Interestingly, also the time spend on solving the CFD problem dropped from 7500 seconds to 6700 seconds when comparing the global to the local method. The reason for this is that the local method produces more reliable structural deformations and force distributions resulting in better meshes for the CFD solver.

To study the influence of the chosen basis function, different basis functions were tested in the global and the partition of unity setting. Figure 6.8 shows the deflections along the elastic axis of the wing for a variety of basis functions. The maximum difference between the basis functions is 7.5% in the global and 0.5% in the local setting. Hence, the local method produces more reliable results being almost independent of the chosen radial basis function.

Even though the conservation of forces and work is guaranteed for each method, an advantage of the local algorithm comes from the local way in which it couples the force fields.

In most cases, global methods result in non-sparse coupling matrices so that using the transposed matrix for exchanging forces will eventually destroy any local character of the force distribution. This leads, in certain cases, to a non-physical distribution of loads. Local methods instead, result in a distribution which is similar to the "physical" one.

As shown in Figures 6.9 and 6.10, the global coupling algorithm returns a certain number of concentrated high modulus forces, where most of the lift is produced. On the other hand, with the local method high peaks of forces are more uniformly distributed along the leading and trailing edges of the lifting surfaces which is physically more reasonable.

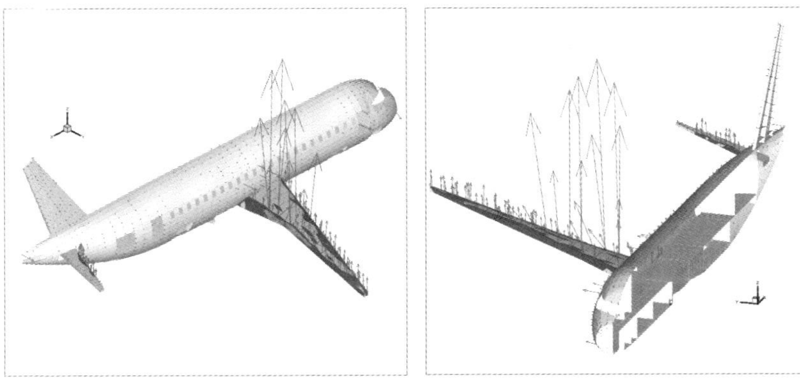

Figure 6.9. Transformed forces on the FE model, global method. (*See also* Color Plate on page 402)

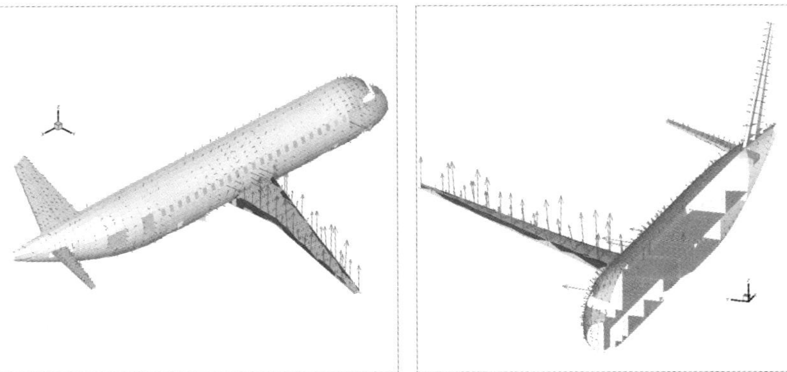

Figure 6.10. Transformed forces on the FE model, local method. (*See also* Color Plate on page 403)

7 Reconstruction of Rotations

We are finally coming back to the problem of recovering rotations. To demonstrate the matrix- and vector-valued approach, we use the AMP test wing, which is shown in Figure 5.2. In particular, we employ the rod-like structural model consisting of 25 nodes and 24 elements, while the CFD surface model consists of 54,653 nodes and 109,216 elements. For testing purposes, we applied the following simple, analytic displacements to the structural grid:

$$\mathbf{g}(\mathbf{x}) = \left(0, 0, 0.1y^2 + 0.1x\right)^T, \tag{7.15}$$

which induces the rotations

$$2\boldsymbol{\theta}(\mathbf{x}) = \nabla \times \mathbf{g}(\mathbf{x}) = (0.2y, -0.1, 0)^T. \tag{7.16}$$

Figure 7.11 shows the analytical result of this deformation when applied to the CFD mesh. For reasons of comparison, the figure contains both the undeformed and the deformed CFD surface. Obviously, the maximum deflection is at the wing tip and it is of size 0.186021.

Figure 7.12 shows the errors for both methods on the CFD mesh, i.e. the differences between the analytic deformation and the one computed. Clearly,

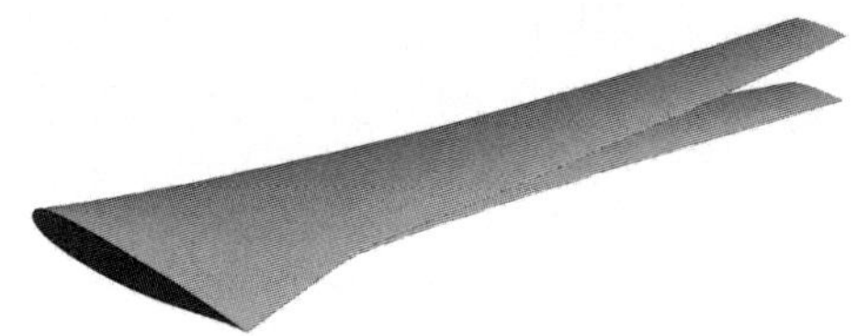

Figure 7.11. The results of the analytical deformation applied to the CFD surface.

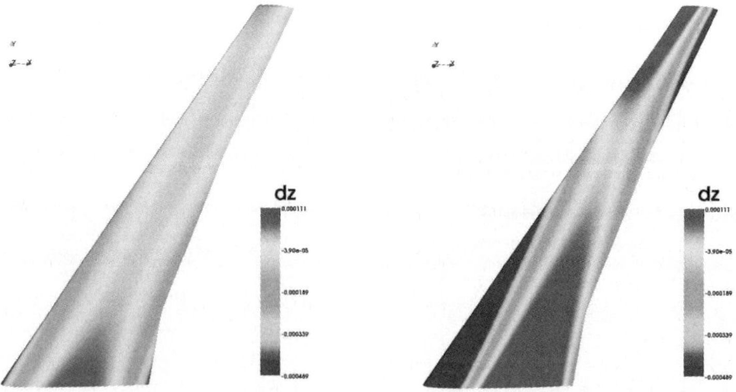

Figure 7.12. Results for the AMP test wing with (left) and without (right) the reconstruction of rotations. (*See also* Color Plate on page 403)

in both cases the error is the largest close to the root of the wing and at the leading and trailing edge, where structural and aerodynamic meshes differ most. But it is also apparent that the new method is much more capable to recover the rotational caused deformations at the leading and trailing edge of the aerodynamic model.

References

1. R. Ahrem, A. Beckert, and H. Wendland, *Recovering rotations in spatial coupling for fluid-structure-interaction problems*, Journal of Fluid and Structures, 23 (2007), pp. 874–884.
2. A. Beckert and H. Wendland, *Multivariate interpolation for fluid-structure-interaction problems using radial basis functions*, Aerosp. Sci. Technol., 5 (2001), pp. 125–134.
3. R. L. Bisplinghoff and H. Ashley, *Principles of Aeroelasticity*, John Wiley & Sons, Inc., New York, 1962.
4. G. T. S. Done, *Past and future progress in fixed and rotary wing aeroelasticity*, Aeronautic Journal, 100 (1996), pp. 260–279.
5. E. H. Dowell, E. F. Crawley, H. C. Jr., D. A. Peter, R. H. Scanlan, and F. Sisto, *A Modern Course in Aeroelasticity*, Kluwer Acad. Pub., Dordrecht/Boston, 1995.
6. J. W. Edward, *Computational aeroelasticity*, in Flight-Vehicles, Materials, Structures and Dynamics – Assessment and Future Directions, vol. 5, The American Society of Mechanical Engineers, New York, 1993.
7. C. Farhat and M. Lesoinne, *Higher-order staggered and subiteration free algorithms for coupled dynamic aeroelasticity problems*, in 36th Aerospace Sciences Meeting and Exhibit,AIAA 98-0516, Reno/NV, 1998.
8. L. Fornasier, H. Rieger, U. Tremel, and E. V. der Weide, *Time-dependent aeroelastic simulation of rapid manoeuvring aircraft*, in In 40th AIAA Aerospace Sciences Meeting & Exhibit, 14–17 January 2002, Reno Nevada, AIAA, 2002.
9. H. Försching, *Grundlagen der Aeroelastik*, Springer, Heidelberg-Berlin-New York, 1974.
10. ——, *New ultra high capacity aircraft (uhca) - challenges and problems from an aeroelastic point of view*, ZFW, 18 (1994), pp. 219–231.
11. ——, *Challenges and perspectives in computational aeroelasticity*, in Proc. of the International Forum on Aeroelasticity and Structural Dynamics, vol. 1, Manchester (UK), 1995, pp. 1.1–1.9.
12. Y. C. Fung, *An introduction to the theory of aeroelasticity*, John Wiley, New York, 1955.
13. P. Geuzaine, G. Brown, C. Harris, and C. Farhat, *Aeroelastic dynamic analysis of a full f-16 configuration for various flight conditions*, AIAA Journal, 41 (2003), pp. 363–371.
14. M. Griebel and M. A. Schweitzer, *A particle-partition of unity method - part II: Efficient cover construction and reliable integration*, SIAM J. Sci. Comput., 23 (2002), pp. 1655–1682.
15. R. L. Harder and R. N. Desmarais, *Interpolation using surface splines*, Journal of Aircraft, 9 (1972), pp. 189–197.

16. H. Hönlinger, R. Manser, A. Gravelle, and P. Viguier, *Measurement of wind tunnel model deformation under airload*, Tech. Rep. 91-069, DGLR-Bericht, Aachen, 1991. European Forum on Aeroelasticity and Structural Dynamics.
17. MacNeal-Schwendler Corp., *MSC.Nastran - Handbook for Aeroelastic Analysis I/II*, 1987. ISSN 0741-8043.
18. H. G. Matthies and J. Steindorf, *Partitioned strong coupling algorithms for fluid-structure interaction*, Computers & Structures, 81 (2003), pp. 805–812.
19. F. J. Narcowich and J. D. Ward, *Generalized Hermite interpolation via matrix-valued conditionally positive definite functions*, Math. Comput., 63 (1994), pp. 661–687.
20. T. Sonar, *Optimal recovery using thin plate splines in finite volume methods for the numerical solution of hyperbolic conservation laws*, IMA J. Numer. Anal., 16 (1996), pp. 549–581.
21. ——, *On the construction of essentially non-oscillatory finite volume approximations to hyperbolic conservation laws on general triangulations: Polynomial recovery, accuracy, and stencil selection*, Comput. Methods in Appl. Mechanics and Engineering, 140 (1997), pp. 157–181.
22. ——, *On families of pointwise optimal finite volume ENO approximations*, SIAM J. Numer. Anal., 35 (1998), pp. 2350–2369.
23. H. Wendland, *Piecewise polynomial, positive definite and compactly supported radial functions of minimal degree*, Adv. Comput. Math., 4 (1995), pp. 389–396.
24. ——, *Numerical solutions of variational problems by radial basis functions*, in Approximation Theory IX, Vol. 2: Computational Aspects, C. K. Chui and L. L. Schumaker, eds., Nashville, 1998, Vanderbilt University Press, pp. 361–368.
25. ——, *Fast evaluation of radial basis functions: Methods based on partition of unity*, in Approximation Theory X: Wavelets, Splines, and Applications, C. K. Chui, L. L. Schumaker, and J. Stöckler, eds., Nashville, 2002, Vanderbilt University Press, pp. 473–483.
26. ——, *On the convergence of a general class of finite volume methods*, SIAM J. Numer. Anal., 43 (2005), pp. 987–1002.
27. ——, *Scattered Data Approximation*, Cambridge Monographs on Applied and Computational Mathematics, Cambridge University Press, Cambridge, UK, 2005.
28. ——, *Spatial coupling in aeroelasticity by meshless kernel-based methods*, in Proceedings to ECCOMAS CFD 2006, P. Wesseling, E. Onate, and J. Periaux, eds., Egmond aan Zee, The Netherlands, 2006.
29. Z. Wu, *Hermite-Birkhoff interpolation of scattered data by radial basis functions*, Approximation Theory Appl., 8 (1992), pp. 1–10.
30. H. Zingel, U. Jajes, and S. Vogel, *Aeroelastisches Modellprogramm I, Stationäre und instationäre Luftkräfte*, Tech. Rep. DA/BRE/91-111, Deutsche Airbus, 1991.

Color Plates

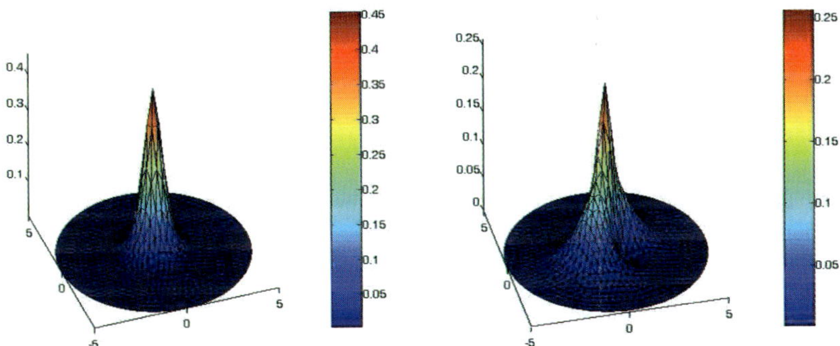

Figure 4.2. Evolution of the electronic distribution of a system composed of a single nucleus ($Z = 3$) and an increasing number of electrons (two-dimensional physical space). (*See also* on page 15)

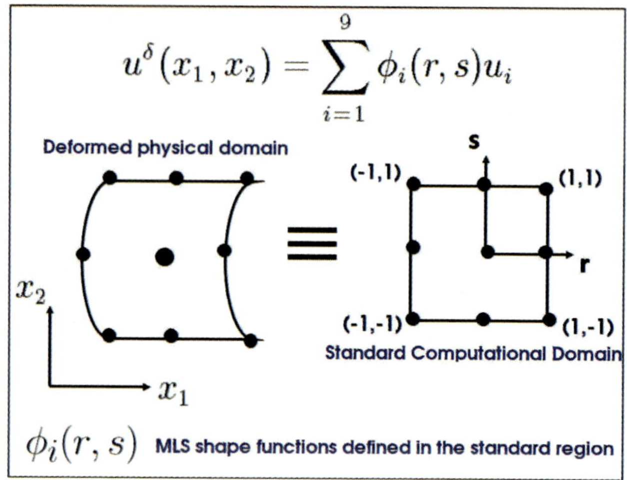

Figure 3.1. Deformed physical domain mapped to a local standard computationl domain. The MLS shape functions are digitally generated within the local domain (r,s) by using 9 neighbours located in the near vicinity of each particle. (*See also* on page 25)

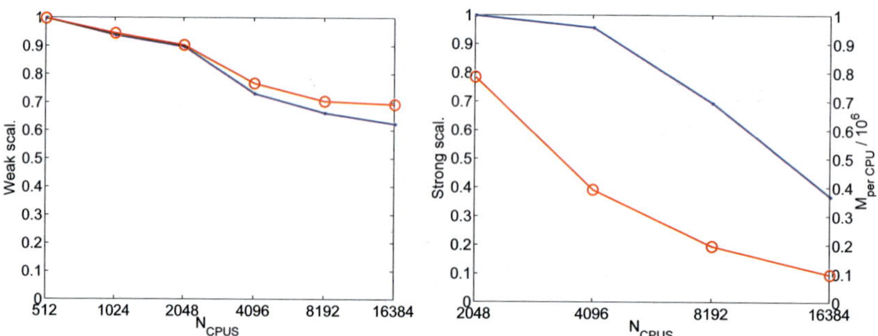

(a) Weak scalability for a per-processor problem size $= 4 \cdot 10^5$;full problem (solid dots) and excluding the Poisson solver (circles)

(b) Strong scalability (solid dots) and per-processor size (circles)

Figure 3.3. Medium-wavelength instability of counter-rotating vortices: parallel efficiencies on IBM BlueGene/L. (*See also* on page 42)

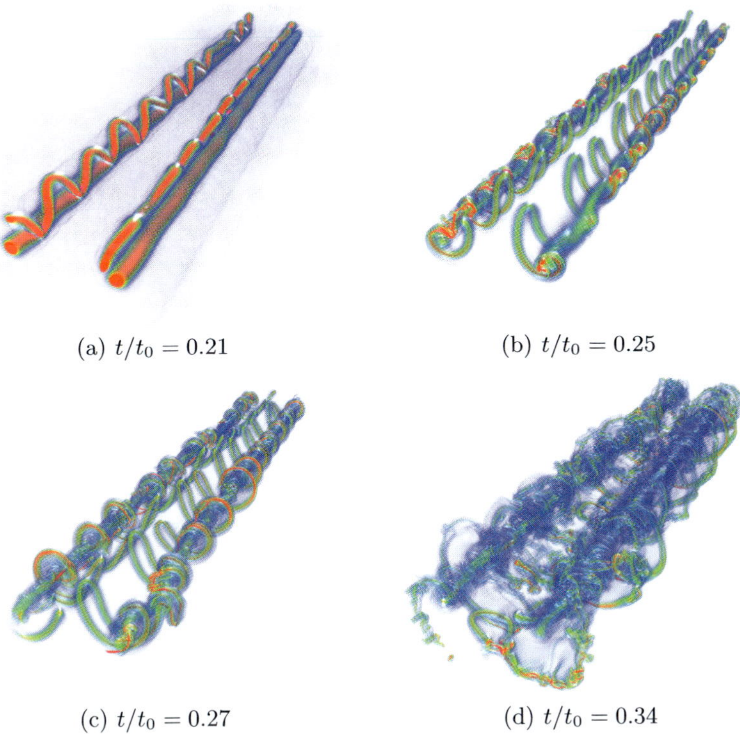

(a) $t/t_0 = 0.21$ (b) $t/t_0 = 0.25$

(c) $t/t_0 = 0.27$ (d) $t/t_0 = 0.34$

Figure 3.4. Trailing vortices, instability initiation by ambient noise: visualization of the vorticity structures by volume rendering. High vorticity norm regions correspond to red and opaque; low vorticity ones are blue and transparent. (*See also* on page 43)

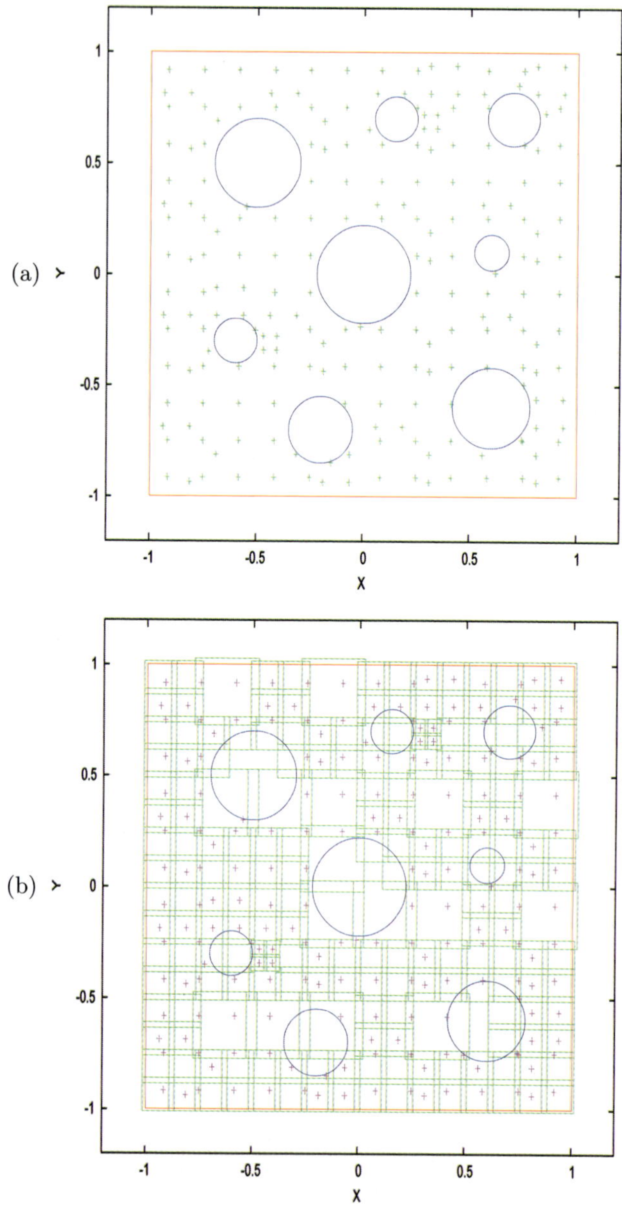

Figure 2.2. (a) View of a domain containing 208 points and eight holes of varying sizes, and (b) domain with cover, with a cover factor of 1.2. (*See also* on page 78)

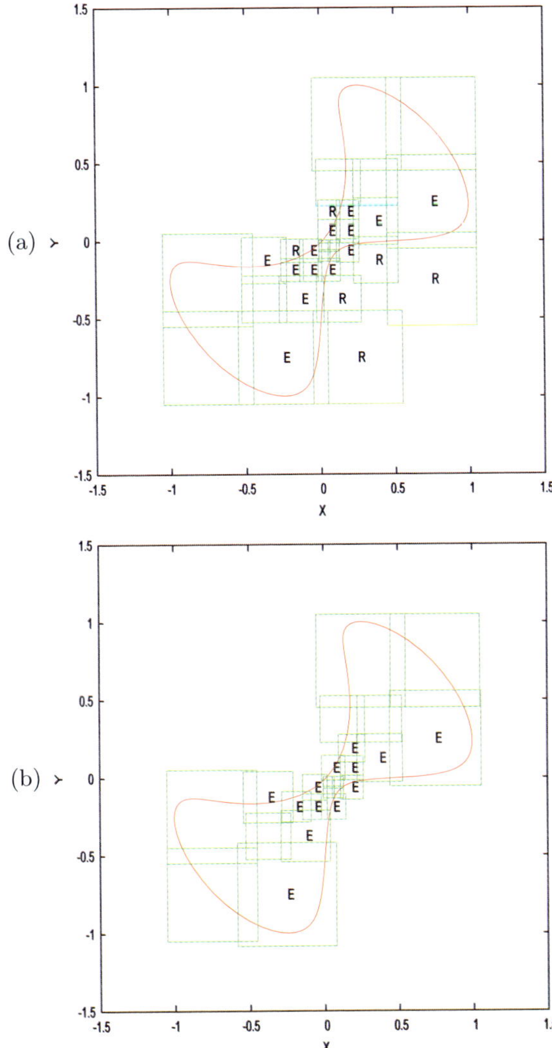

Figure 2.3. (a) A cover created by Algorithm 1, with patches to be removed labelled 'R' and patches which will be extended to maintain a complete cover labelled 'E'. (b) The cover generated at the end of Algorithm 2, with six patches from the original cover removed and thirteen patches (labelled 'E') extended. (*See also* on page 79)

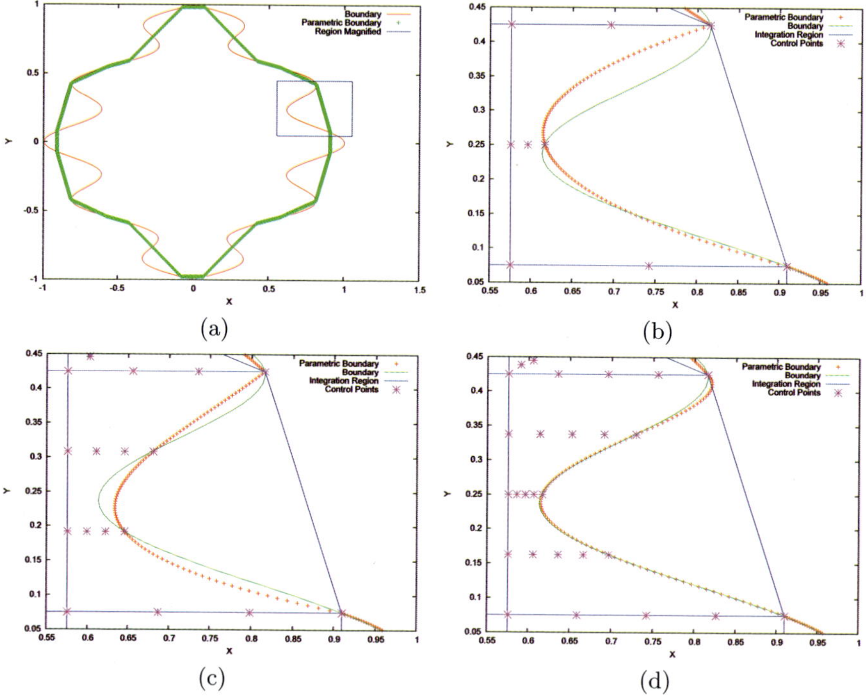

Figure 2.4. (a) A domain with a linear approximation to its boundary. The region enclosed in the small box is shown magnified in the remaining frames. (b) Close-up of a quadratic approximation to the boundary. (c) Close-up of a cubic approximation to the boundary. (d) Close-up of a quartic approximation to the boundary. In the last three frames, we also plot the control points [24] used to create the parametric maps. (*See also* on page 81)

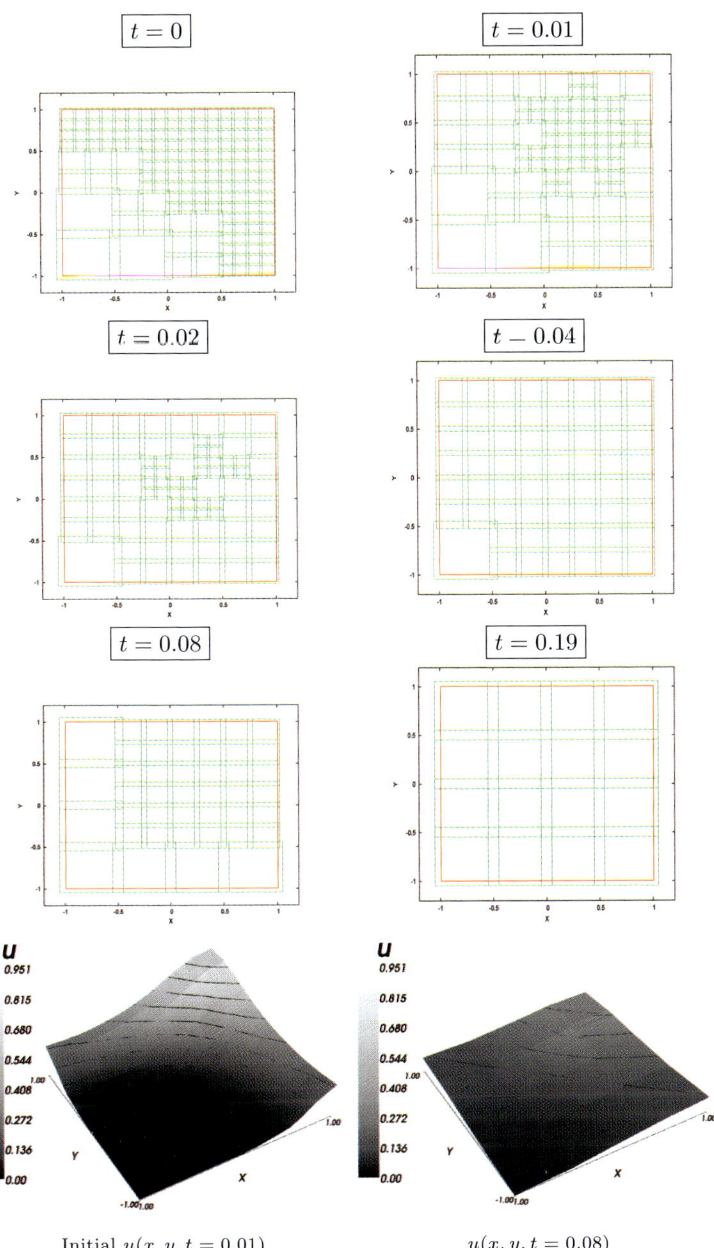

Initial $u(x, y, t = 0.01)$ $\quad$ $u(x, y, t = 0.08)$

Figure 3.6. Successive refinement and coarsening of the cover in an adaptive parabolic problem at selected timesteps (TS). The initial and final (see TS = 19) cover consists of 16 equally-sized patches. The solution $u(x, y, t)$ after an initial time step and at time step 8 of the simulation are shown in the last two frames. (*See also* on page 84)

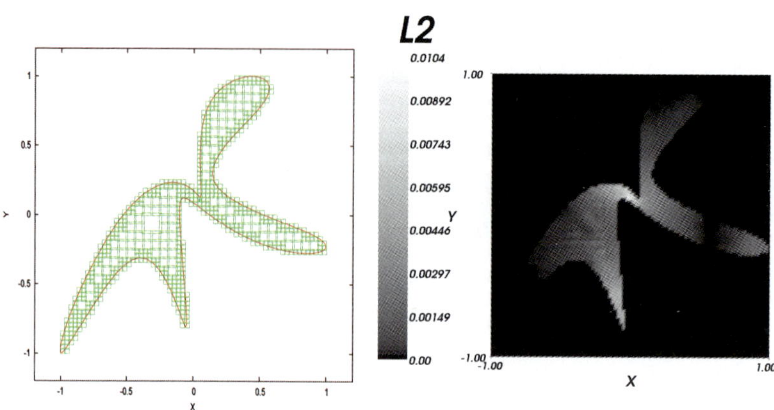

Figure 3.7. (a) The domain of Example 3.2 with a cover of 642 patches with cover factor 1.2. (b) The point-wise L^2 error. (*See also* on page 86)

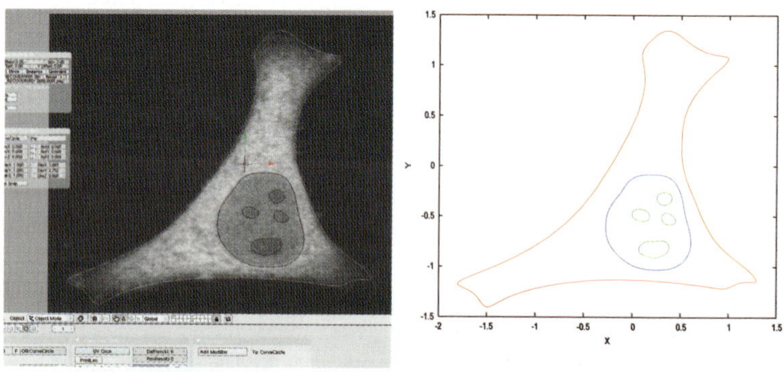

Figure 3.8. Domain with nested subdomains. Left the original fibroplast microscopy image inside a modelling programme to retrieve the (sub-)domain shapes, courtesy of Gimmi Ratto, see also [6]. Right: The main domain represents the cell body, the main sub-domain represents the nucleus, and the four inner sub-subdomains with interfaces $\partial\Omega_{1...4}$ (green) represent nucleoli. (*See also* on page 88)

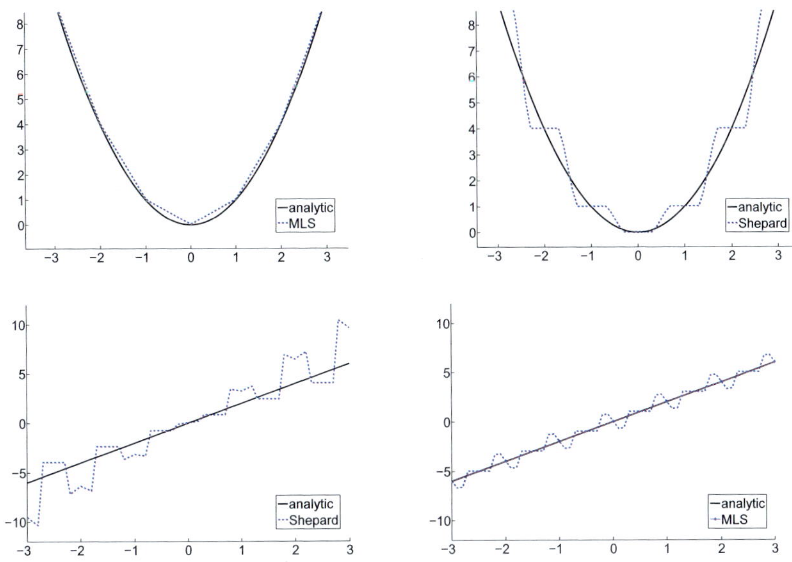

Figure 3.1. Approximation (top row) by Shepard's Method (left) and MLS (right) and the respective derivatives (bottom row). (*See also* on page 117)

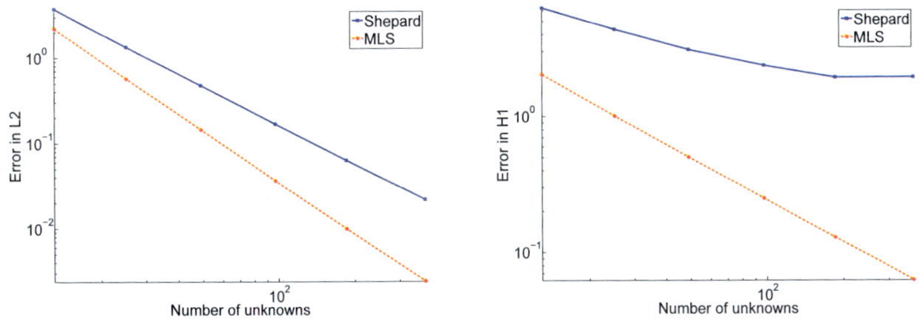

Figure 3.2. Error in the L^2-norm (left) and H^1-norm (right) of Shepard's method (solid) and the MLS (dashed). (*See also* on page 118)

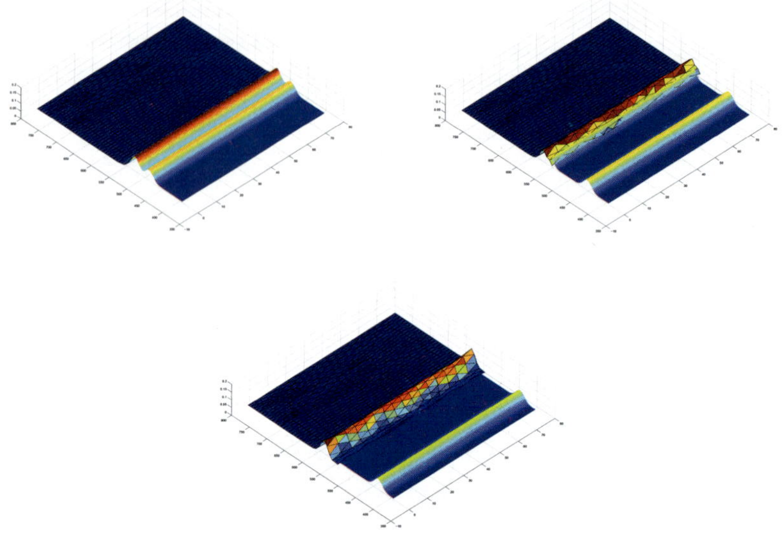

Figure 3.3. Weak scale transfer ($2d$) based on (1.4) using Shepard's approach ($m = 0$). (*See also* on page 118)

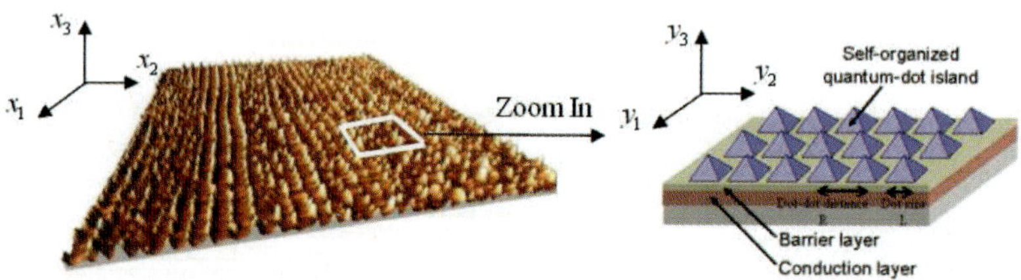

Figure 2.1. A two-scale model of QDA with features in different length scales. (*See also* on page 123)

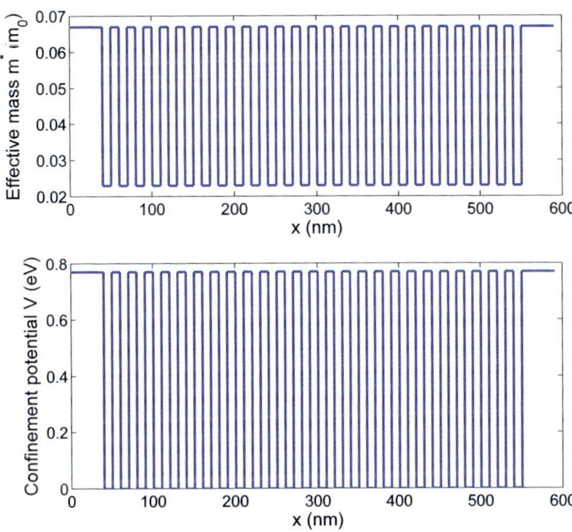

Figure 4.5. 1-D QDA model with uniform oscillating effective mass and confinement potential. (*See also* on page 130)

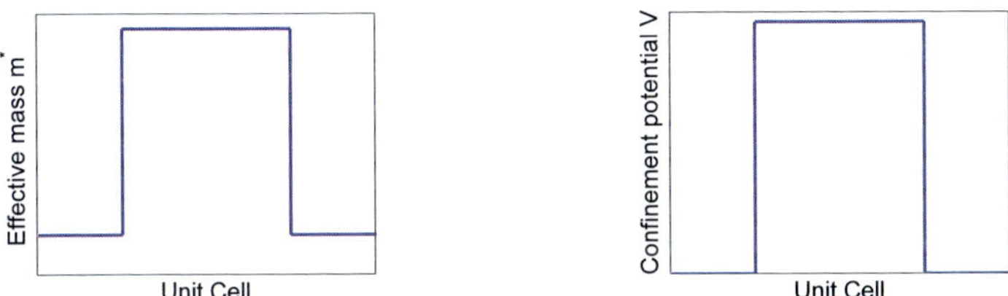

Figure 4.6. Effective mass and confinement potential in the fine scale unit cell. (*See also* on page 130)

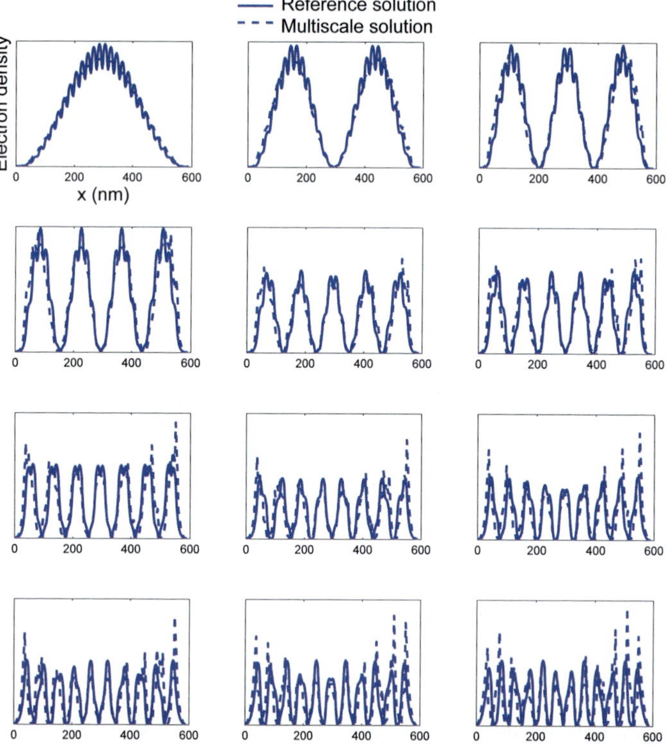

Figure 4.8. Comparison of the electron density distribution associated with the first 12 energy levels between the reference solution and the multiscale solution. (*See also* on page 132)

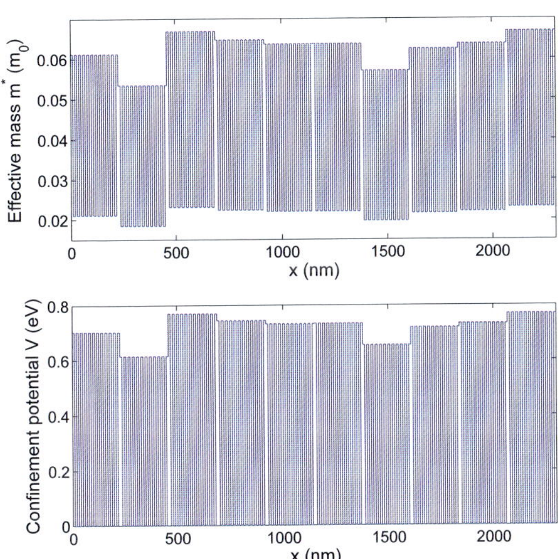

Figure 4.9. 1-D model of a series of QDA with slightly nonuniform oscillating effective mass and confinement potential. (*See also* on page 133)

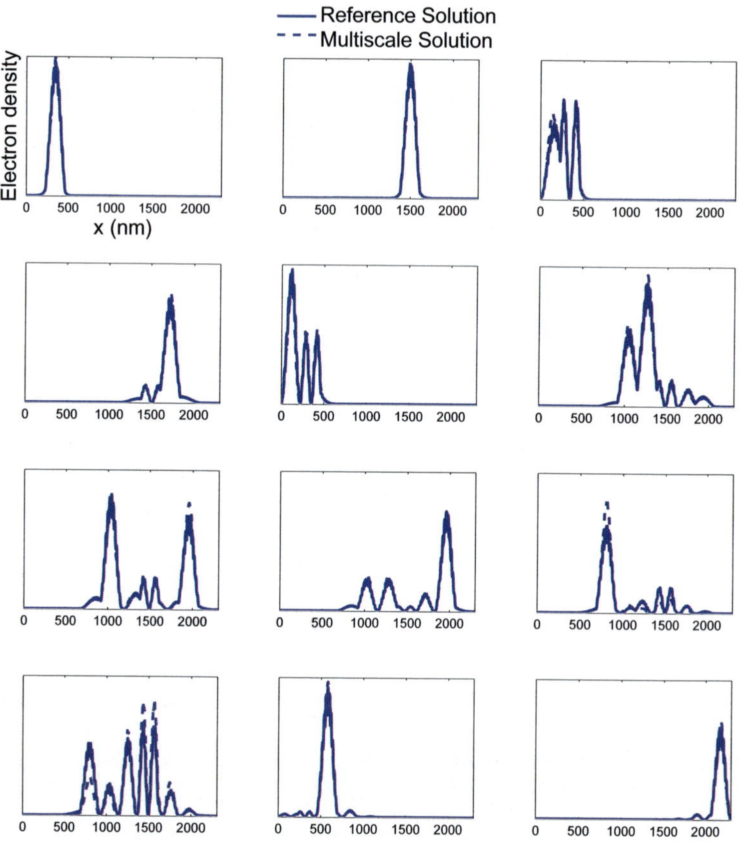

Figure 4.11. Comparison of the electron density distribution associated with the first 12 energy levels between the reference solution and the multiscale solution. (*See also* on page 134)

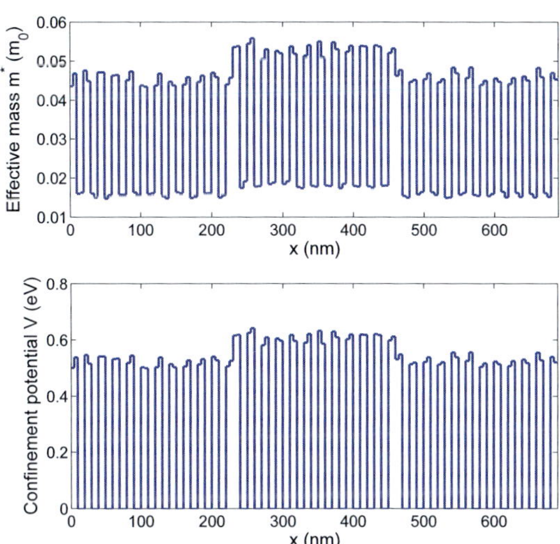

Figure 4.12. 1-D model of a series of QDA with randomly oscillating effective mass and confinement potential. (*See also* on page 135)

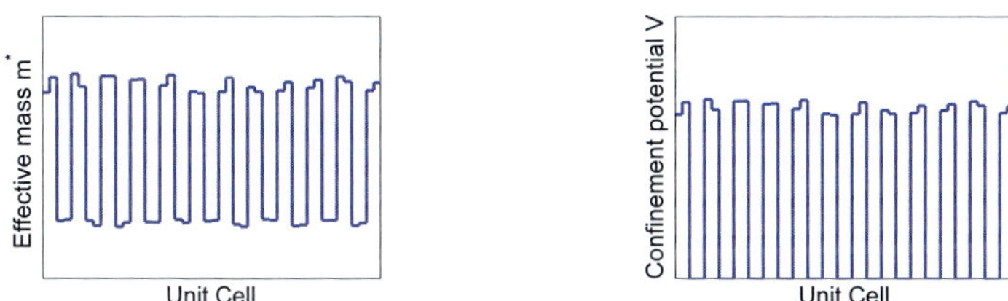

Figure 4.13. Effective mass and confinement potential in the fine scale unit cell. (*See also* on page 135)

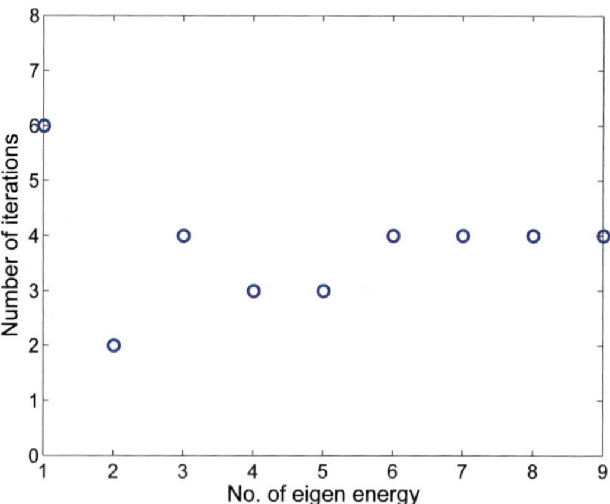

Figure 4.14. Number of iteration steps for the first 9 eigen pairs by using the iterative multiscale method. (*See also* on page 135)

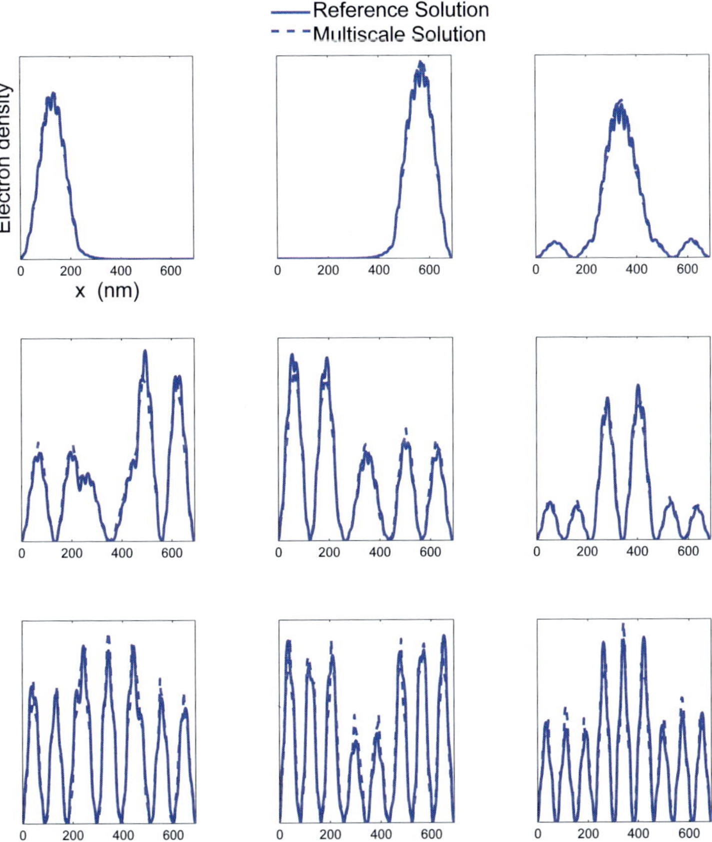

Figure 4.16. Comparison of the electron density distribution associated with the first 9 energy levels between the reference solution and the multiscale solution. (*See also* on page 136)

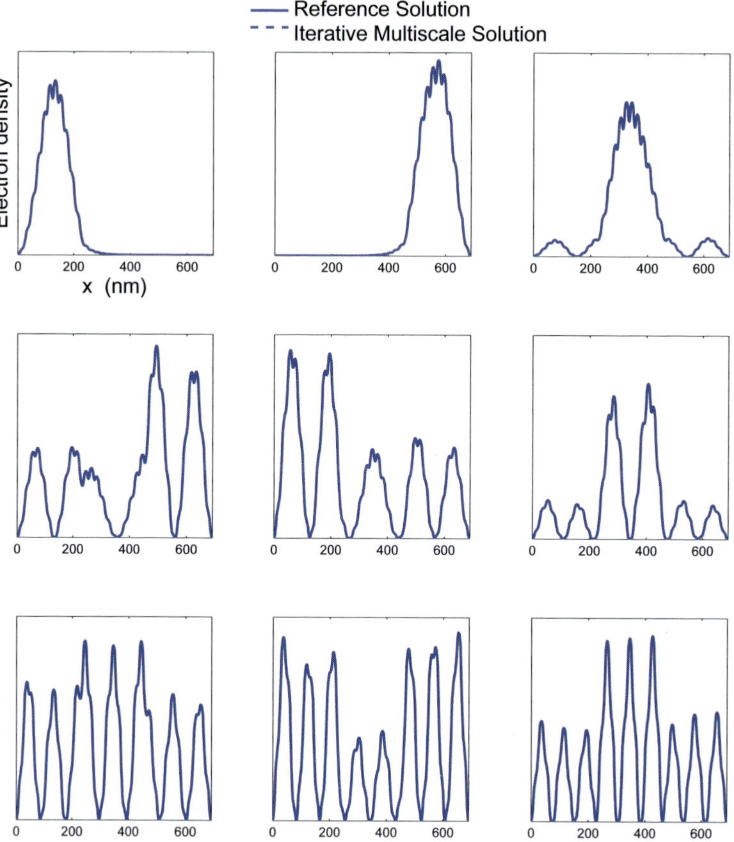

Figure 4.17. Comparison of the electron density distribution associated with the first 9 energy levels between the reference solution and the iterative multiscale solution. (*See also* on page 137)

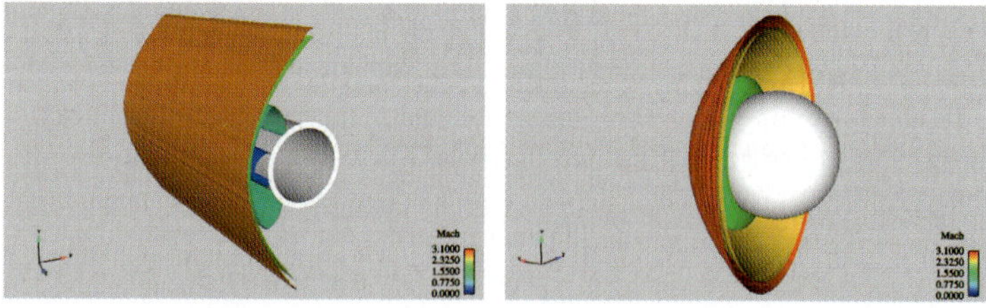

Figure 5.6. Mach isosurfaces for flow past a cylinder, third order reconstruction (left) and past a sphere, second order reconstruction (right). (*See also* on page 170)

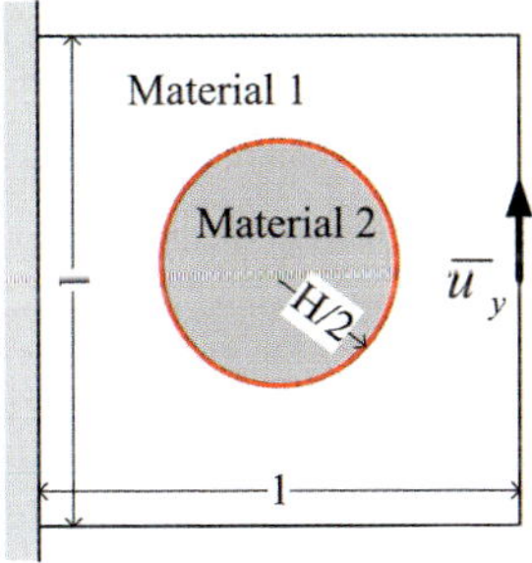

Figure 5.9. Problem statement and initial design (Example 3) (*See also* on page 185)

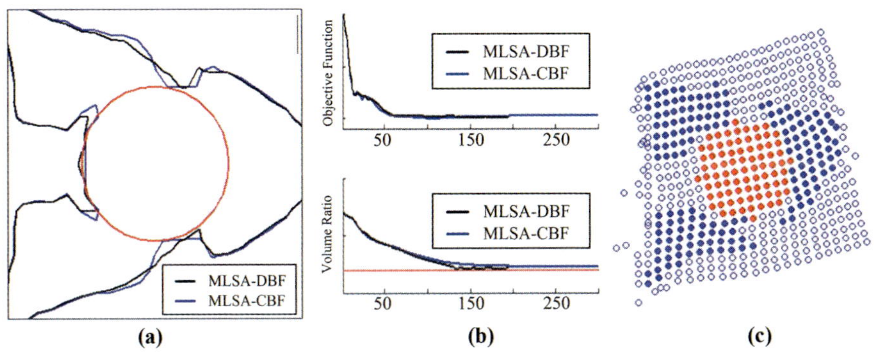

Figure 5.10. Solutions of Example 3: (**a**) Optimal design; (**b**) convergence histories; (**c**) deformation (*See also* on page 185)

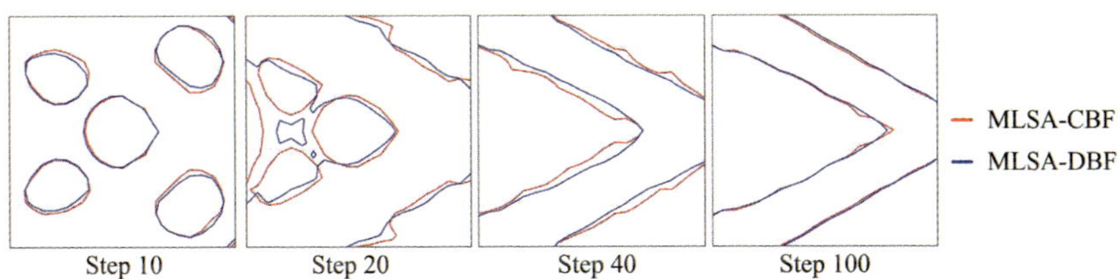

Figure 5.12. Material interfaces at different steps of topology evolution (*See also* on page 186)

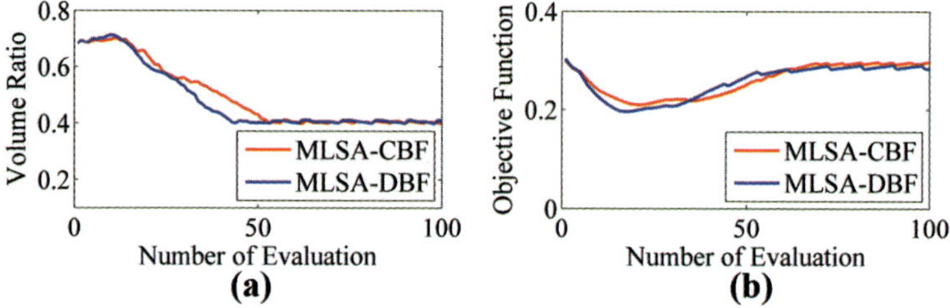

Figure 5.13. Convergence histories: (**a**) objective function; (**b**) volume ratio (*See also* on page 186)

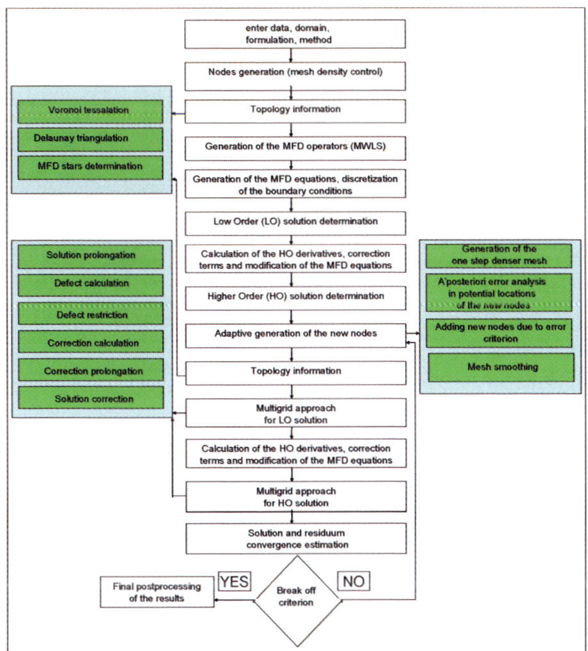

Figure 6.2. HOA multigrid adaptive solution approach - flow chart (*See also* on page 204)

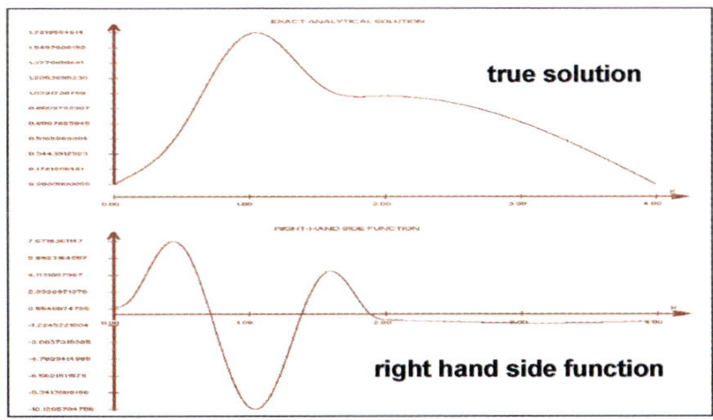

Figure 6.3. Analytical solution and right hand side of 1D test (*See also* on page 205)

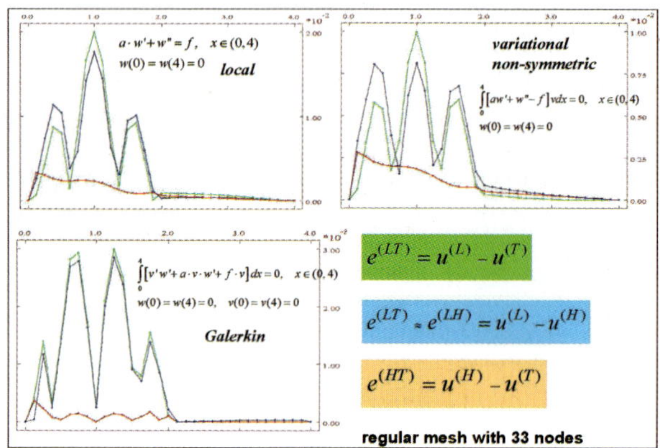

Figure 6.4. Local solution error estimation (*See also* on page 205)

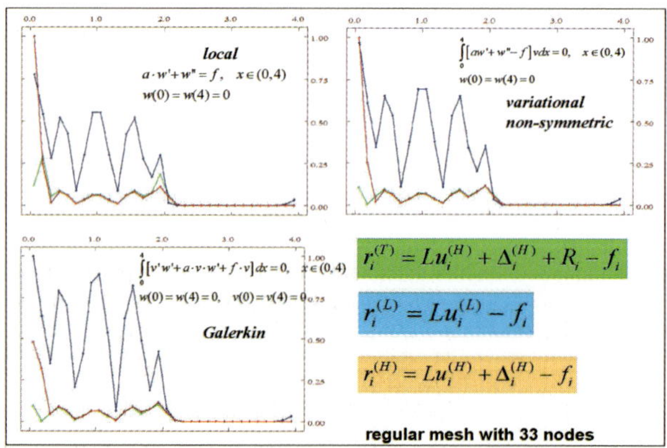

Figure 6.5. Local residual error estimation (*See also* on page 206)

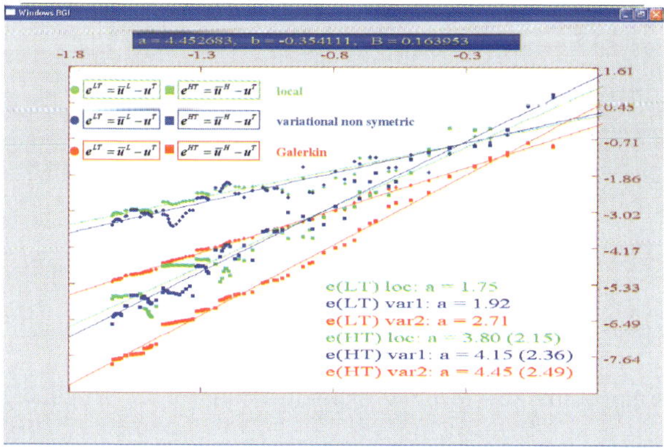

Figure 6.6. Solution convergence rates on the set of adaptive irregular meshes (*See also* on page 206)

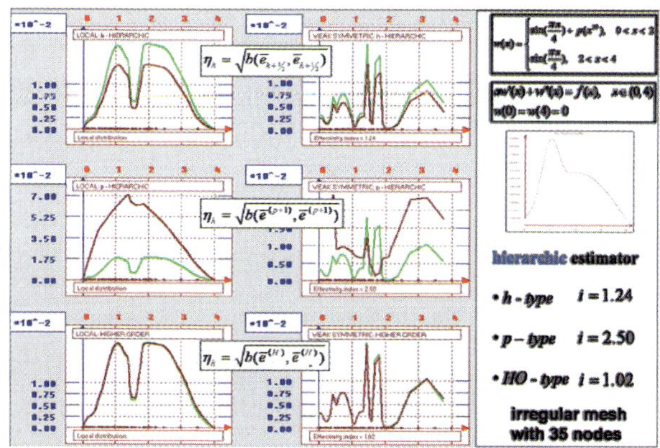

Figure 6.7. Hierarchic estimators on irregular mesh (*See also* on page 206)

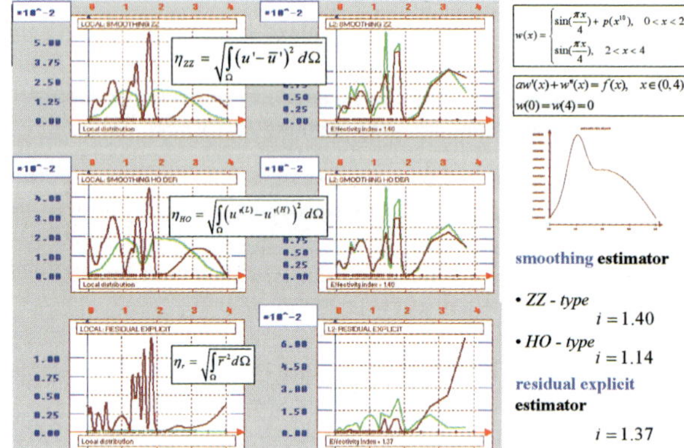

Figure 6.8. Smoothing and residual explicit estimator on irregular mesh (*See also* on page 207)

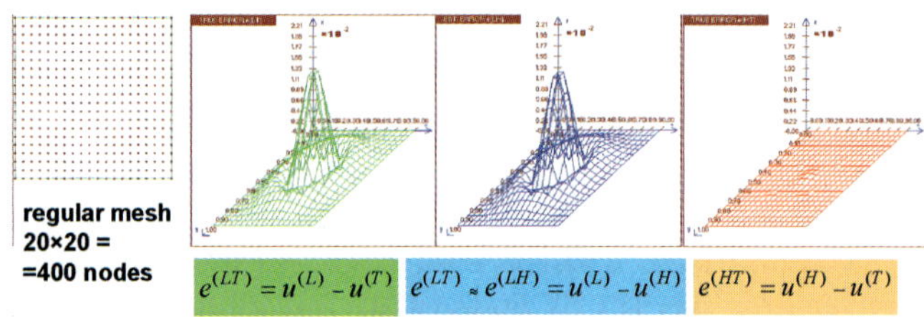

Figure 6.9. Solution error estimation (*See also* on page 208)

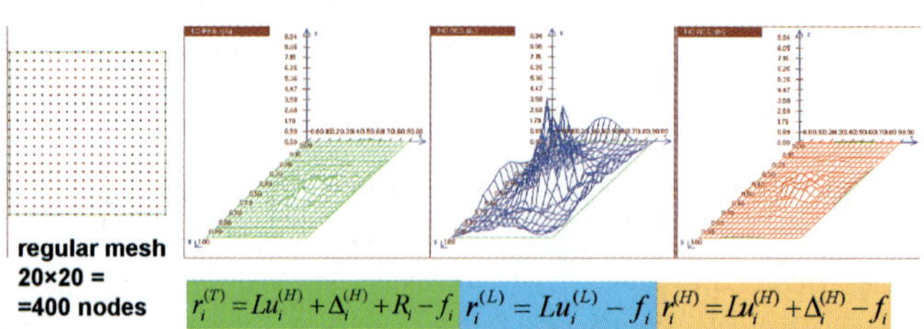

Figure 6.10. Residual error estimation (*See also* on page 208)

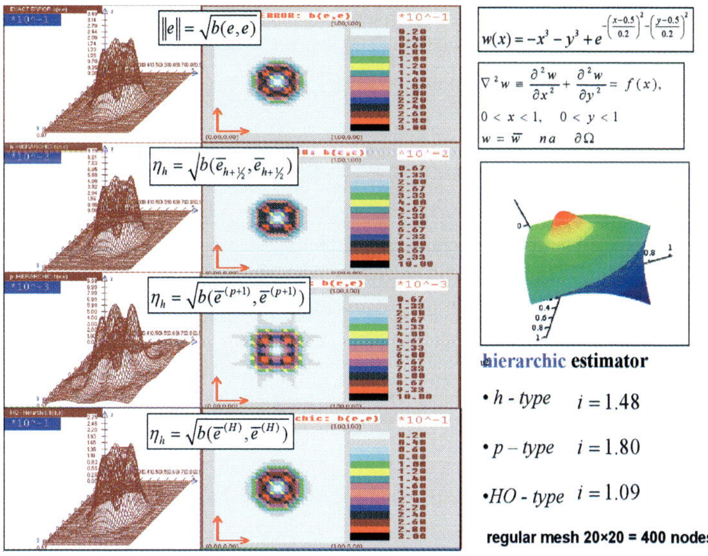

Figure 6.11. Hierarchic estimators (*See also* on page 208)

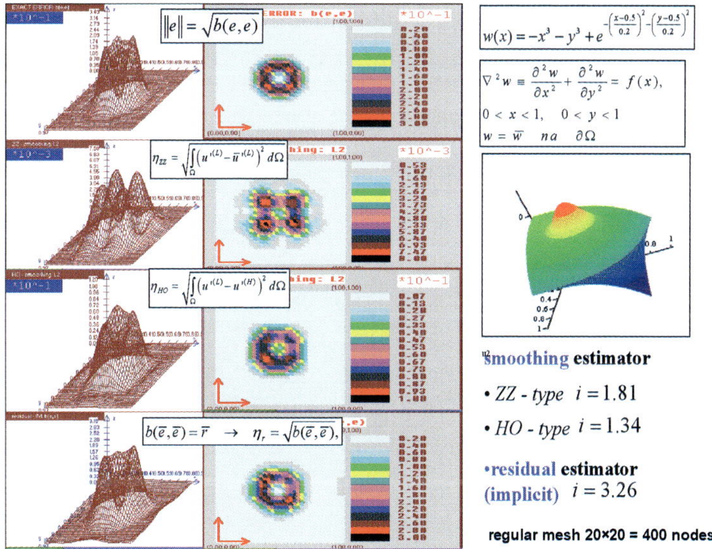

Figure 6.12. Smoothing and residual estimators (*See also* on page 209)

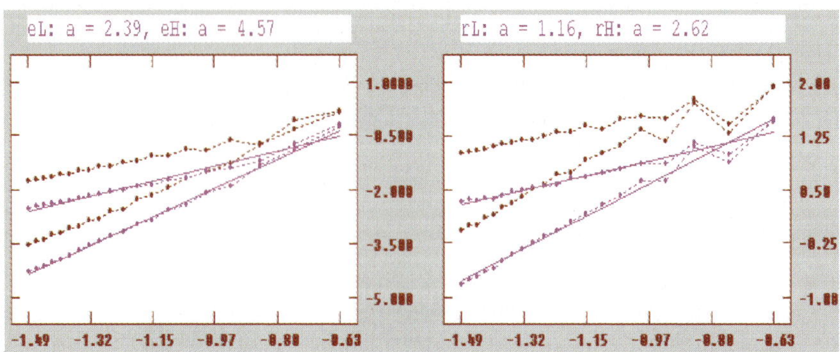

Figure 6.13. Solution and residual convergence on the set of regular meshes (*See also* on page 209)

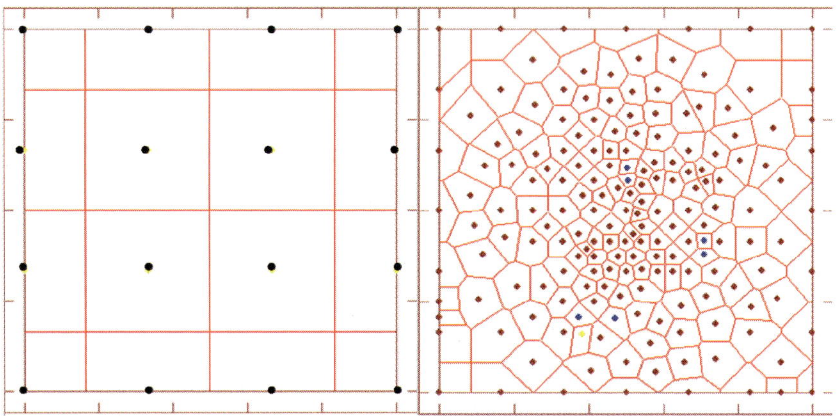

Figure 6.14. Adaptation process, the first regular and last irregular mesh (*See also* on page 210)

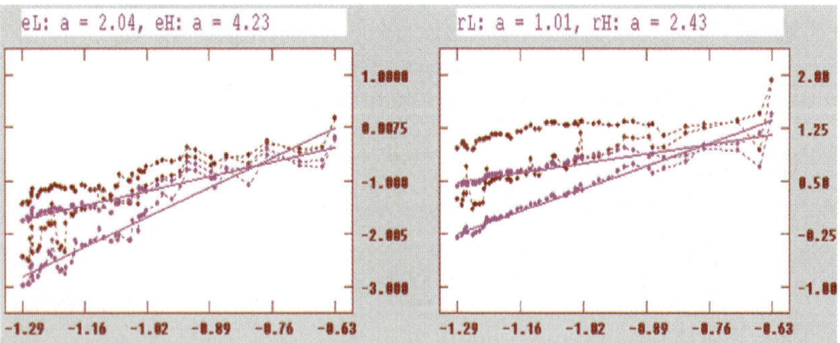

Figure 6.15. Solution and residual convergence on the set of adaptive irregular meshes (*See also* on page 210)

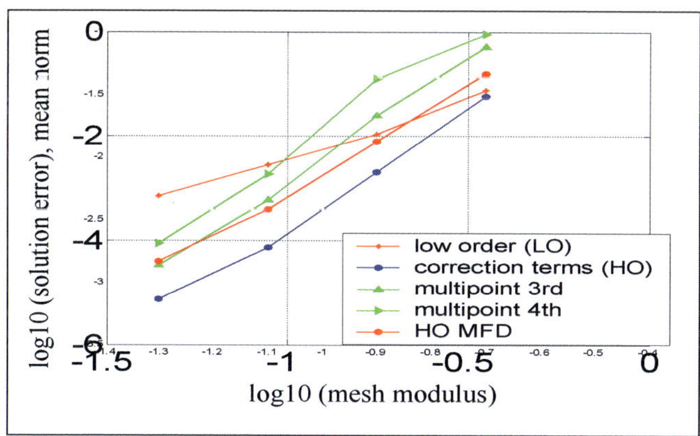

Figure 6.16. Comparison with the other HO techniques (*See also* on page 211)

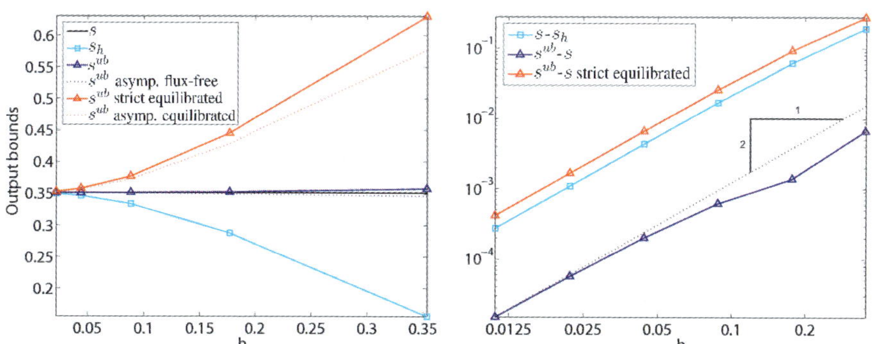

Figure 6.1. Example 1: series of uniformly h-refined linear triangular meshes. Bounds (left) and their convergence (right). (*See also* on page 226)

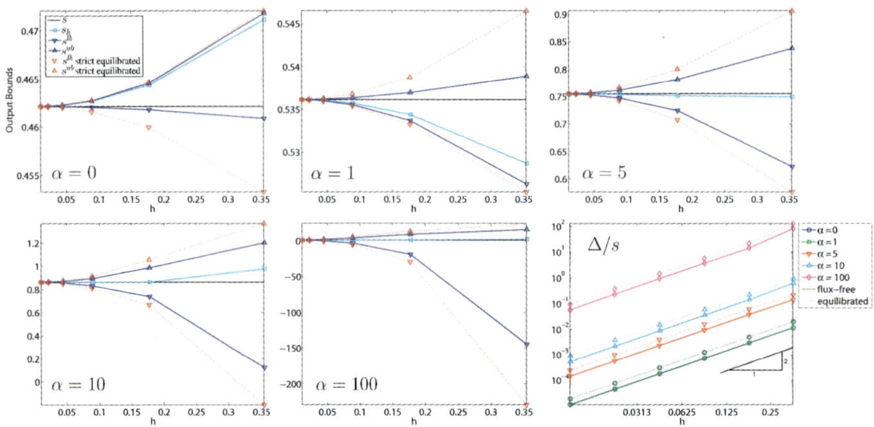

Figure 6.2. Example 2: bounds and convergence of the bound gap for different values of the convection parameter, $\alpha = 0, 1, 5, 10$ and 100 for $\sigma = 1$. (*See also* on page 228)

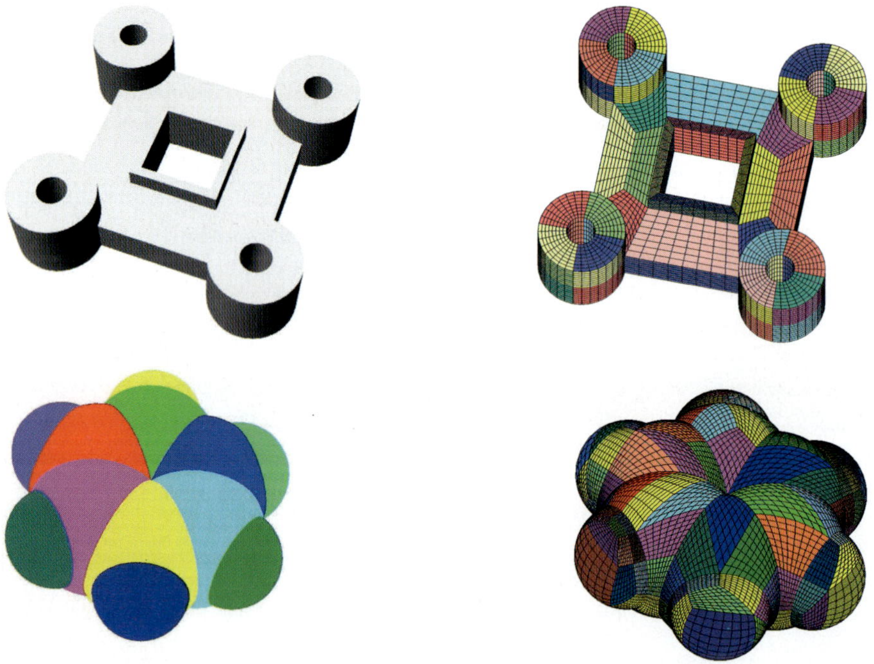

Figure 1.1. CAD and molecular surfaces: before and after decomposition (*See also* on page 232)

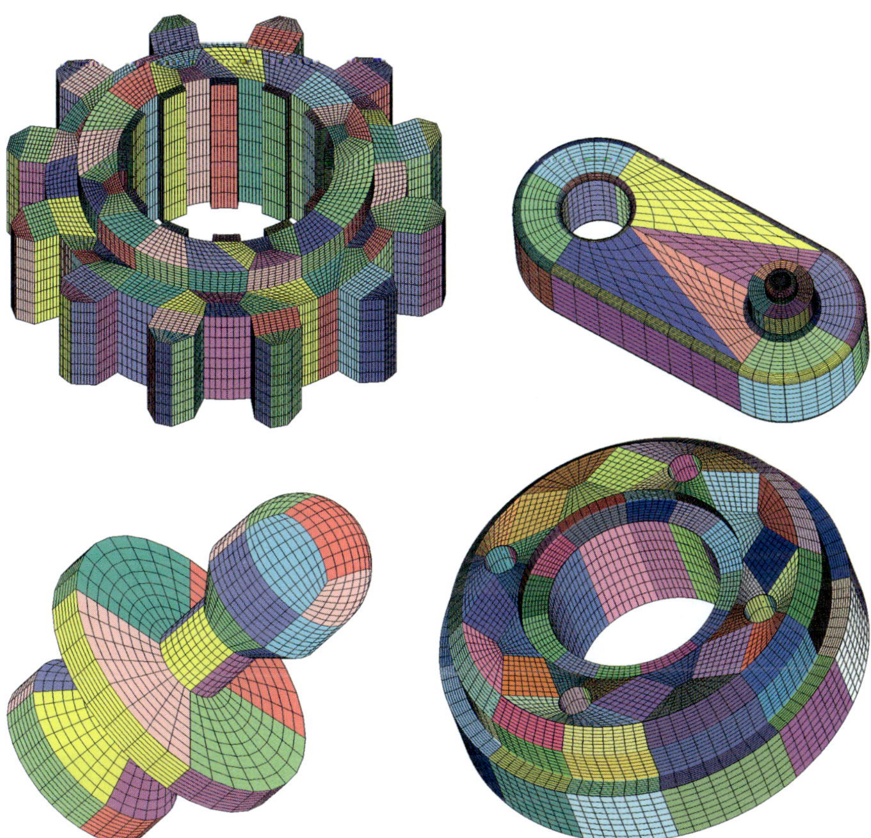

Figure 5.2. Decomposition of simple CAD models (*See also* on page 240)

Figure 5.3. Decomposition of realistic CAD models (*See also* on page 241)

Figure 5.4. Molecules: pentane, fullerene, propane, ice. (*See also* on page 243)

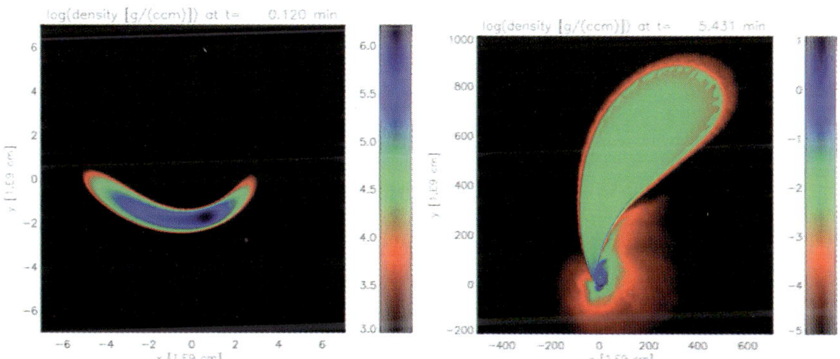

Figure 1.1. Two snapshots from the simulation of the tidal disruption of a white dwarf by a black hole. (*See also* on page 248)

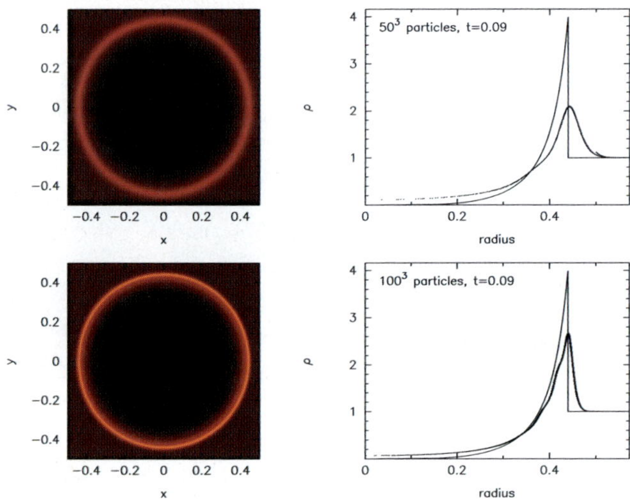

Figure 3.4. Results of the hydrodynamic Sedov blast wave test in 3D at $t = 0.09$ at resolutions of 125,000 (top) and 1 million (bottom) particles respectively. The density and radial position of each SPH particle are shown in each case, which may be compared to the exact solution given by the solid line. (*See also* on page 264)

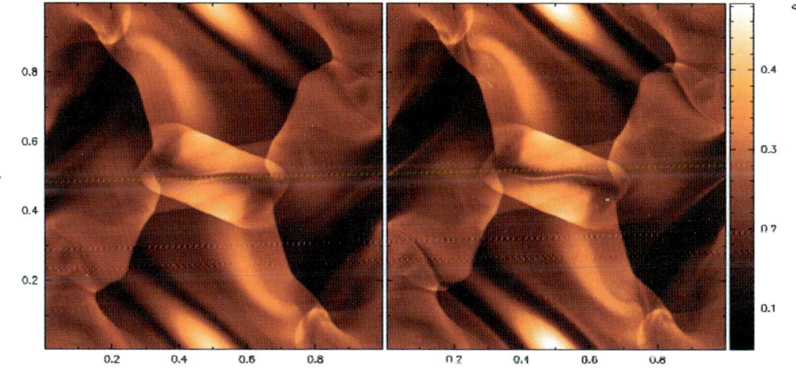

Figure 3.7. Density distribution in the two dimensional Orzsag-Tang vortex problem at $t = 0.5$. The initial vortices in the velocity field combined with a doubly periodic field geometry lead to a complex interaction between propagating shocks and the magnetic field. Results are shown using 512×590 particles using a SPMHD formalism of [48] (left) and using the Euler potentials (right). The reduced artificial resistivity required in the Euler potential formalism leads to a much improved effective resolution. (*See also* on page 269)

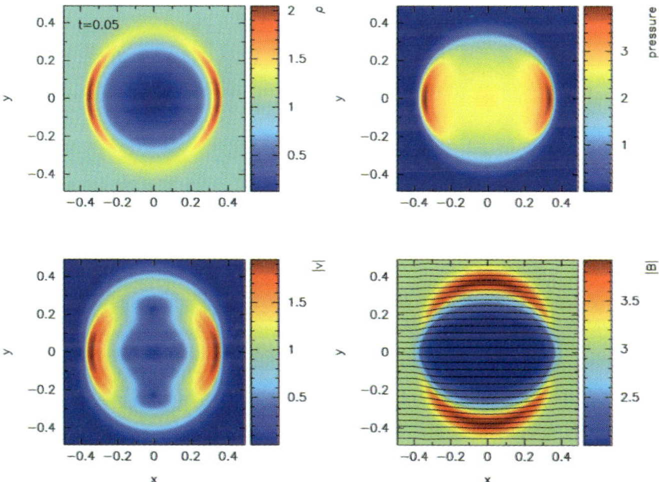

Figure 3.8. Results of the 3D MHD blast wave test at $t = 0.05$ at a resolution of 1 million (100^3) particles. Plots show (left to right, top to bottom) density, pressure, magnitude of velocity and magnetic field strength (with overlaid field lines), plotted in a cross-section slice through the $z = 0$ plane. (*See also* on page 270)

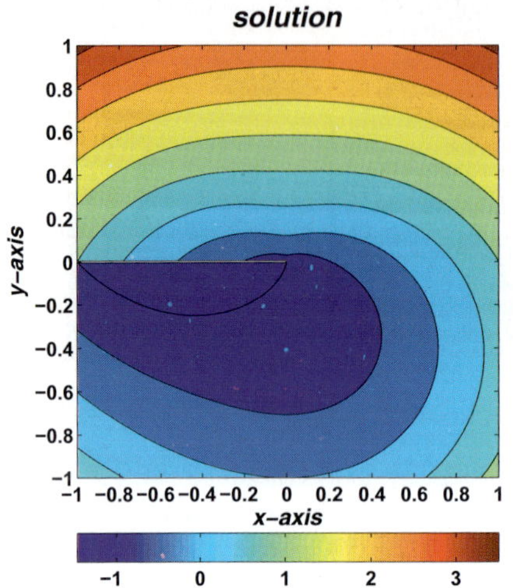

Figure 4.2. Contour plot of the solution (4.23). (*See also* on page 292)

Figure 4.3. Convergence history of the measured relative errors e (4.20) with respect to the complete domain Ω in the L^{∞}-norm, the L^2-norm, the H^1-norm, and the energy-norm on the respective level (denoted by E in the legend). (*See also* on page 292)

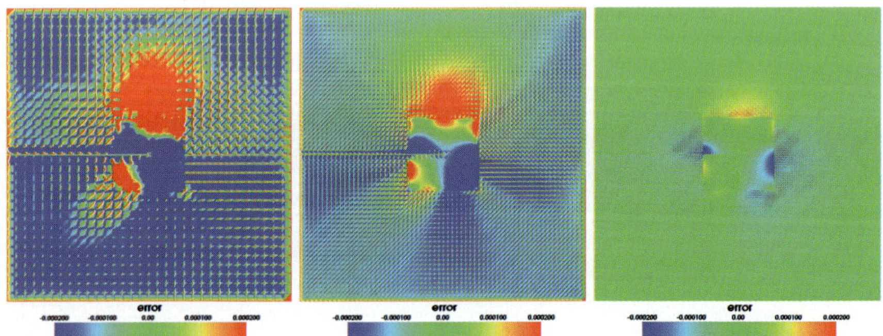

Figure 4.4. Contour plot of the error $u_k^{\mathrm{PU}} - u$ for $k = 5, 6, 7$ (from left to right). (*See also* on page 292)

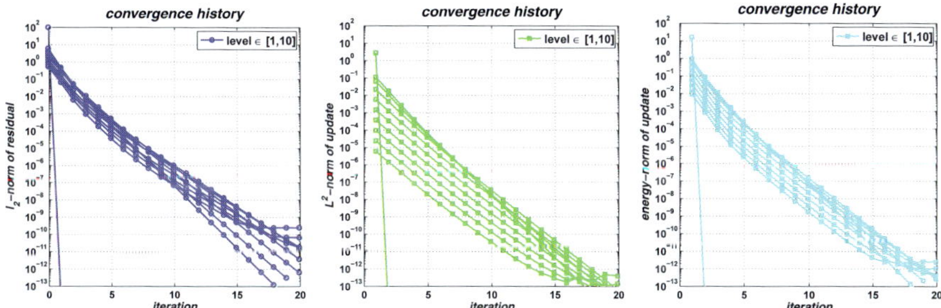

Figure 4.5. Convergence history for a $V(1,1)$-cycle multilevel iteration with block-Gauß–Seidel smoother and nested iteration initial guess (left: convergence of residual vector (4.27) in the l^2-norm, center: convergence of iteration update (4.26) in the L^2-norm, right: convergence of iteration update (4.26) in the energy-norm). (*See also* on page 293)

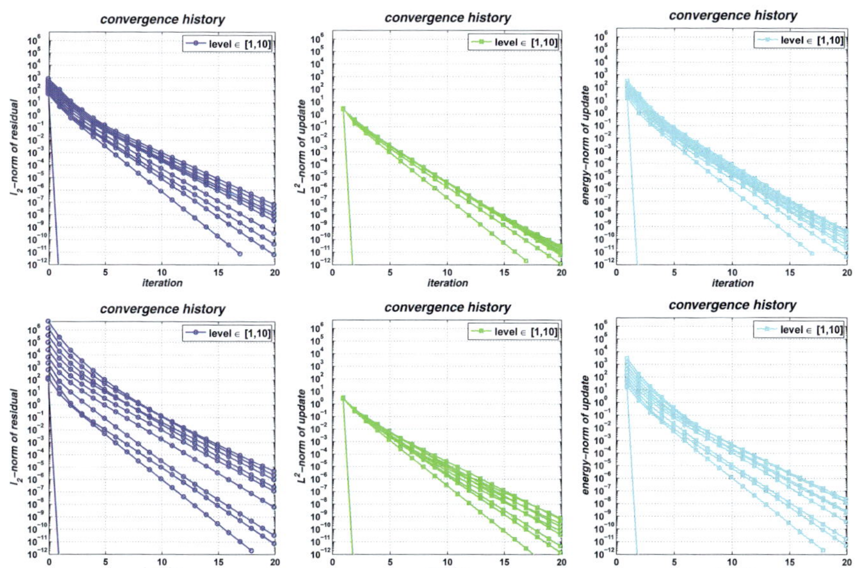

Figure 4.6. Convergence history for a $V(1,1)$-cycle multilevel iteration with block-Gauß–Seidel smoother and zero (upper row) and random (lower row) initial guess (left: convergence of residual vector (4.27) in the l^2-norm, center: convergence of iteration update (4.26) in the L^2-norm, right: convergence of iteration update (4.26) in the energy-norm). (*See also* on page 294)

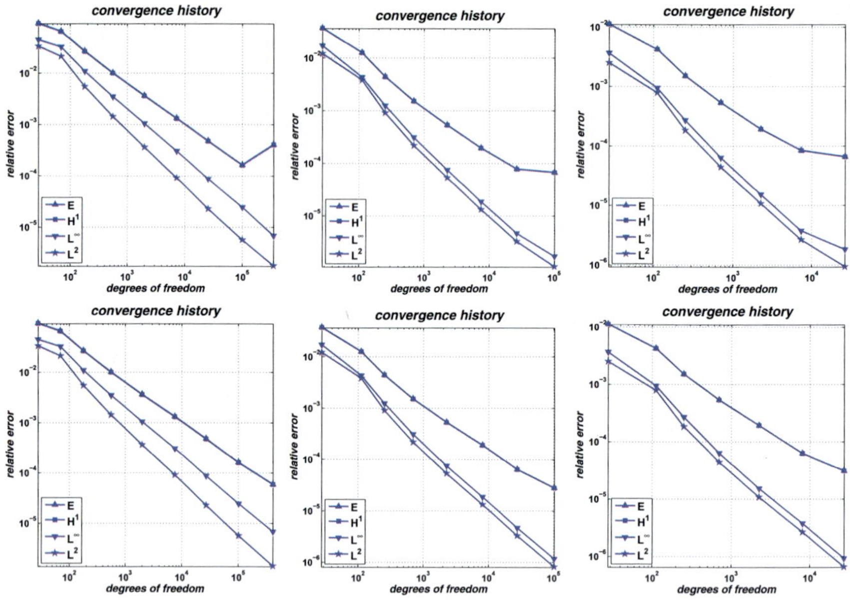

Figure 4.7. Convergence history of the measured relative errors e (4.20) with respect to the subdomains E_1 (left), E_2 (center), and E_3 (right) given in (4.28) with respect to the L^∞-norm, the L^2-norm, the H^1-norm, and the energy-norm on the respective level (denoted by E in the legend). The upper row refers to the enriched PPUM using a preconditioner based on the L^2-norm, i.e., the local mass matrix, the lower row refers to the enriched PPUM using a preconditioner based on the H^1-norm, i.e., the local stiffness matrix. (*See also* on page 296)

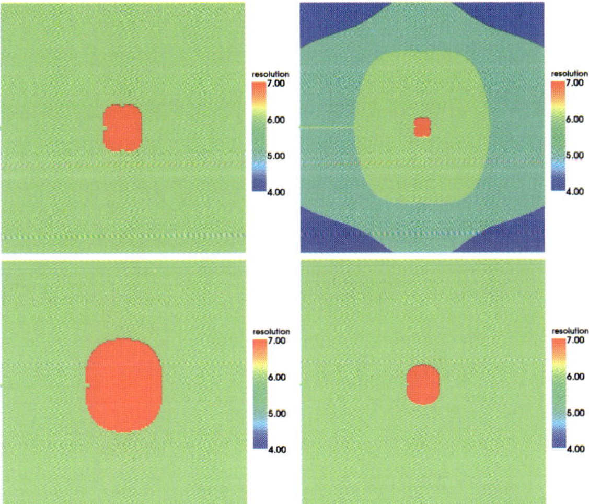

Figure 4.8. Enrichment pattern on levels $k = 9$ (left) and $k = 10$ (right) within the enrichment zone E_{tip}. Color coded is the dimension of the local approximation space $\dim(V_{i,k}) = \text{card}(\{\tilde{\vartheta}^m_{i,k}\})$ (denoted as 'resolution' in the legend). The upper row refers to the enriched PPUM using a preconditioner based on the L^2-norm, i.e., the local mass matrix, the lower row refers to the enriched PPUM using a preconditioner based on the H^1-norm, i.e., the local stiffness matrix. In both cases we used a cut-off parameter of $\epsilon = 10^{-12}$. (*See also* on page 297)

Figure 4.9. Jacobian determinant for 27 node RKEM mesh. (*See also* on page 312)

Figure 4.11. Jacobian determinant for 53 node RKEM mesh. (*See also* on page 312)

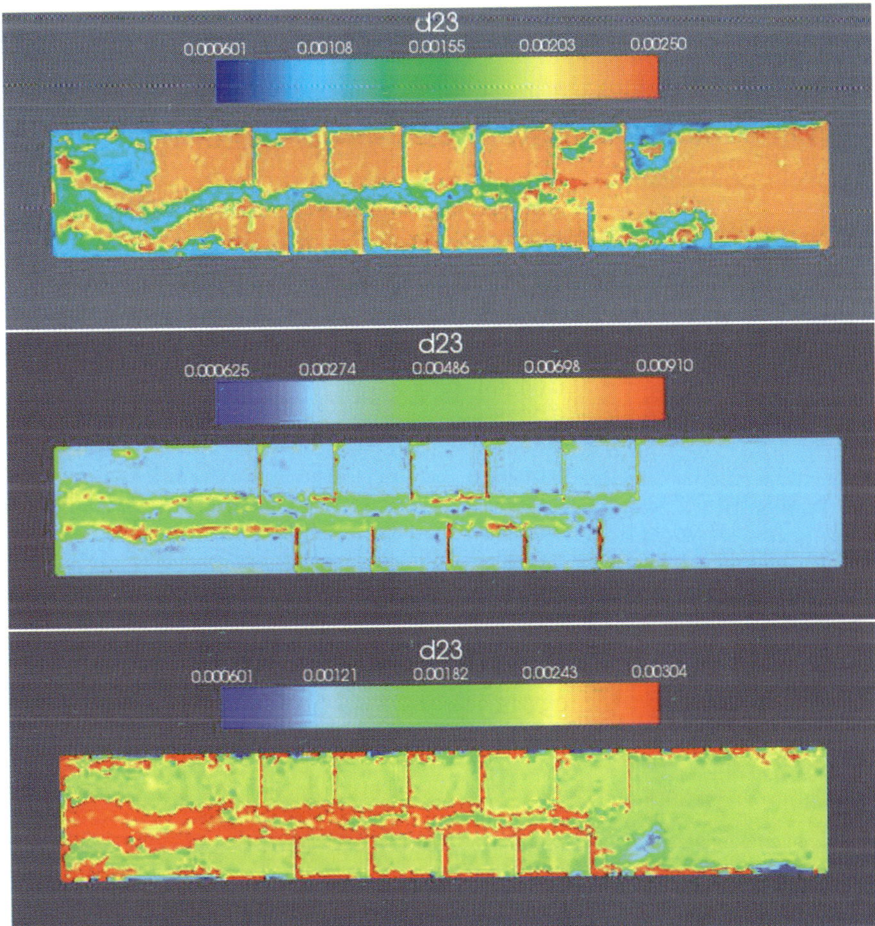

Figure 4.3. Sauter mean diameter with breakage only (top), aggregation only (middle) and breakage and aggregation (bottom) (*See also* on page 329)

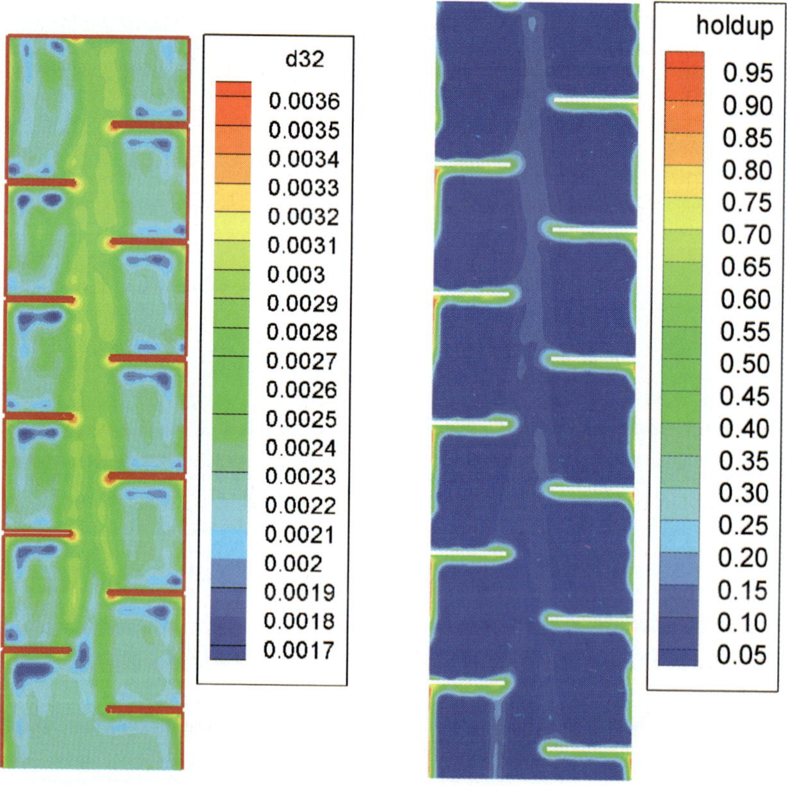

Figure 4.4. Sauter diameter in m (left) and Holdup of dispersed phase (right) (*See also* on page 330)

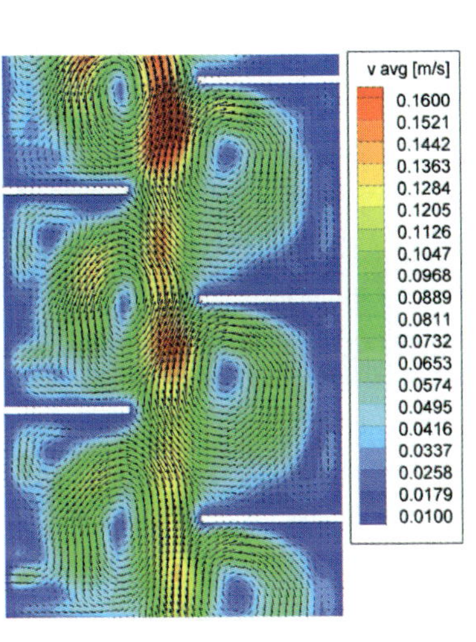

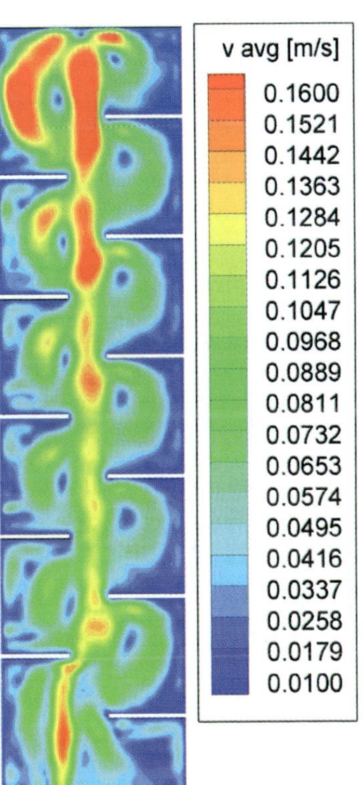

Figure 4.5. Velocity contour of continuous phase (*See also* on page 331)

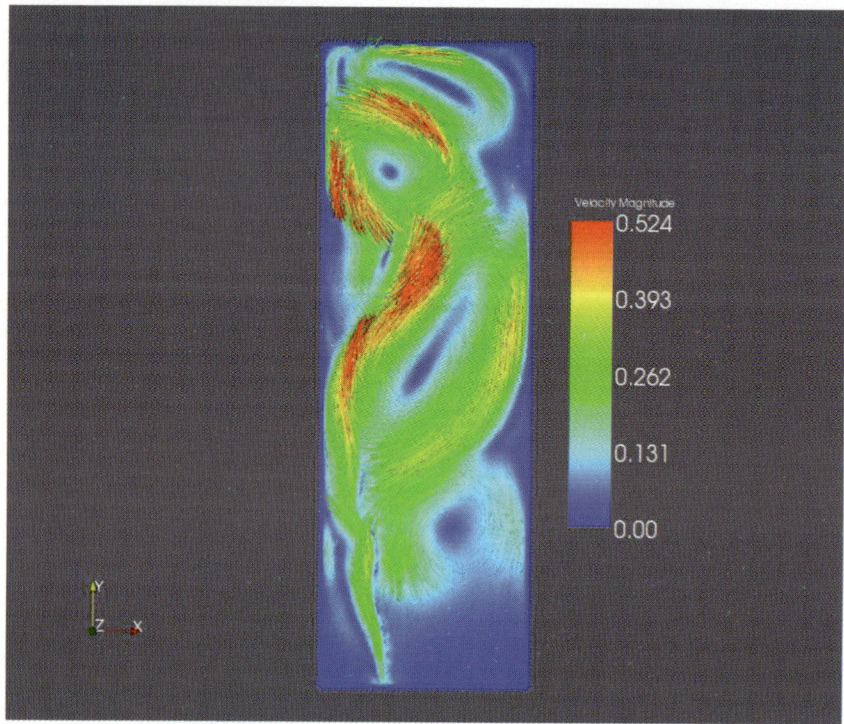

Figure 4.6. Velocity contour for continuous phase (*See also* on page 332)

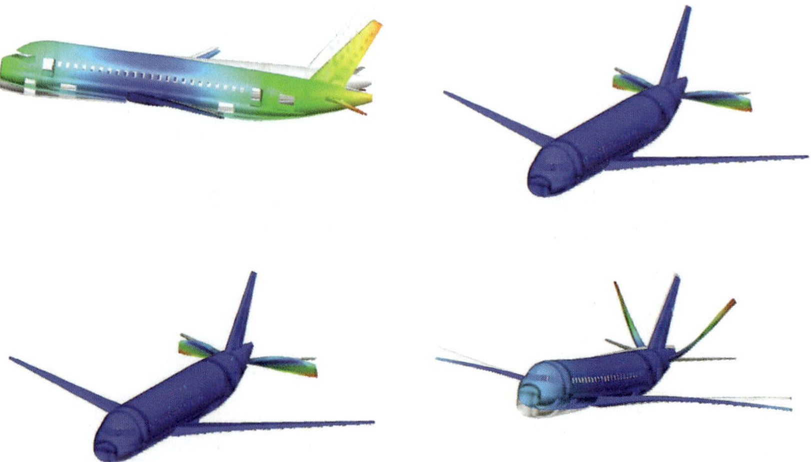

Figure 6.6. The first four eigenmodes of the structural model. (*See also* on page 353)

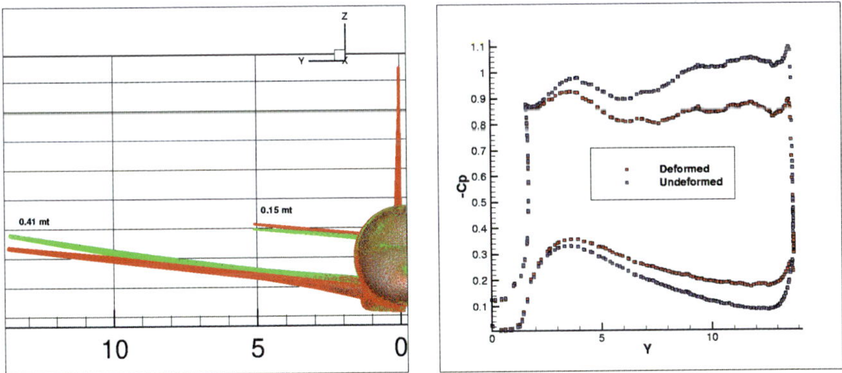

Figure 6.7. Comparison rigid versus elastic: Deflection and Pressure Distribution. (*See also* on page 354)

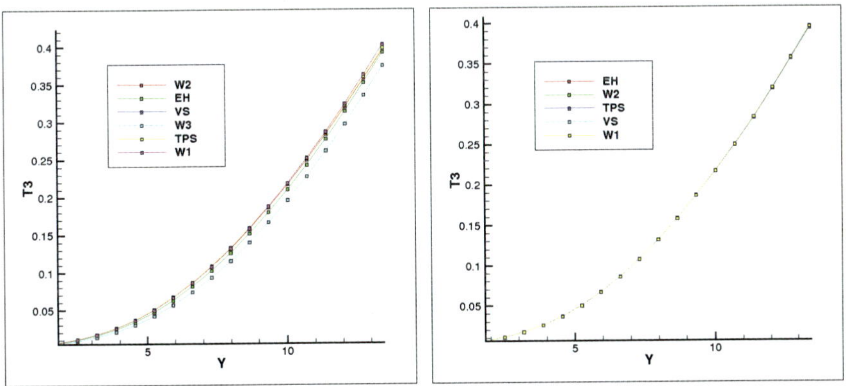

Figure 6.8. Influence of the chosen RBF on the global (left) and local (right) method. (*See also* on page 354)

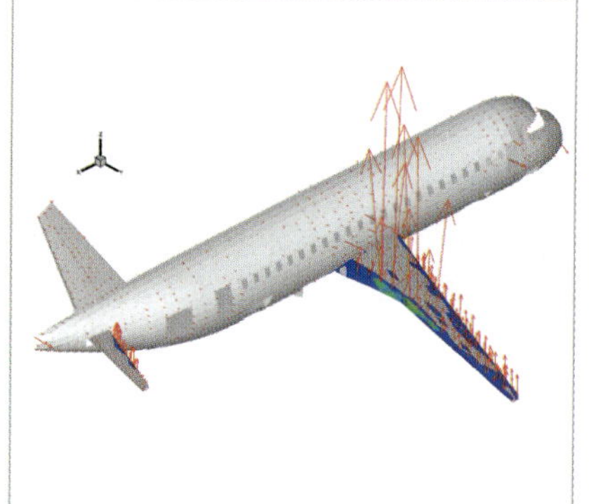
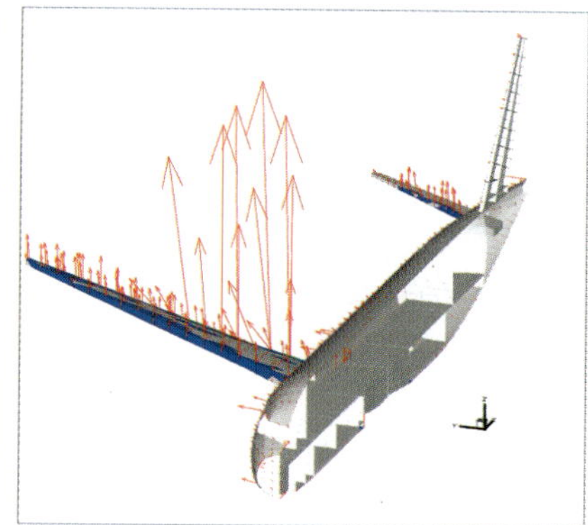

Figure 6.9. Transformed forces on the FE model, global method. (*See also* on page 355)

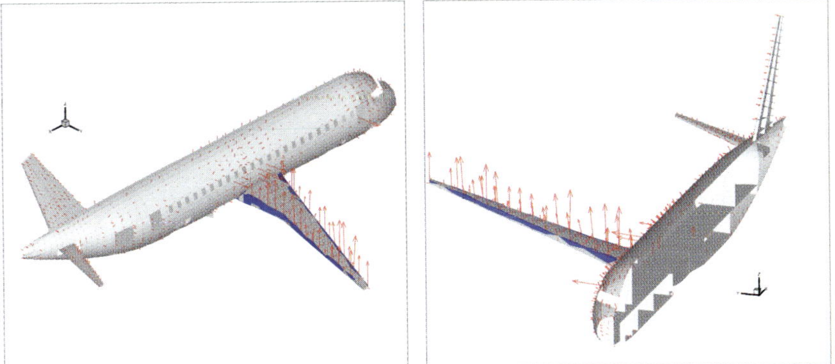

Figure 6.10. Transformed forces on the FE model, local method. (*See also* on page 355)

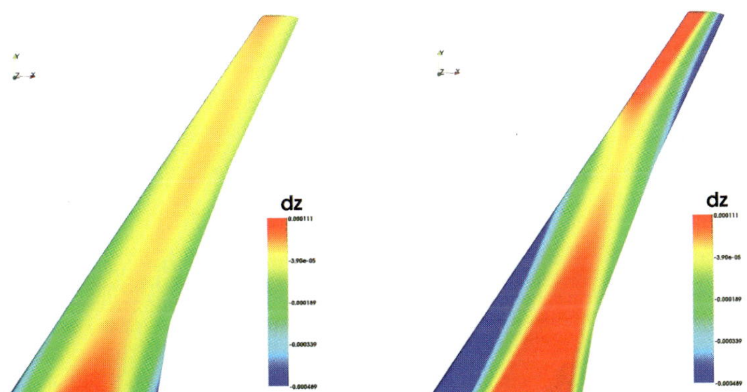

Figure 7.12. Results for the AMP test wing with (left) and without (right) the reconstruction of rotations. (*See also* on page 356)

Editorial Policy

1. Volumes in the following three categories will be published in LNCSE:

i) Research monographs
ii) Lecture and seminar notes
iii) Conference proceedings

Those considering a book which might be suitable for the series are strongly advised to contact the publisher or the series editors at an early stage.

2. Categories i) and ii). These categories will be emphasized by Lecture Notes in Computational Science and Engineering. **Submissions by interdisciplinary teams of authors are encouraged.** The goal is to report new developments – quickly, informally, and in a way that will make them accessible to non-specialists. In the evaluation of submissions timeliness of the work is an important criterion. Texts should be well-rounded, well-written and reasonably self-contained. In most cases the work will contain results of others as well as those of the author(s). In each case the author(s) should provide sufficient motivation, examples, and applications. In this respect, Ph.D. theses will usually be deemed unsuitable for the Lecture Notes series. Proposals for volumes in these categories should be submitted either to one of the series editors or to Springer-Verlag, Heidelberg, and will be refereed. A provisional judgment on the acceptability of a project can be based on partial information about the work: a detailed outline describing the contents of each chapter, the estimated length, a bibliography, and one or two sample chapters – or a first draft. A final decision whether to accept will rest on an evaluation of the completed work which should include

- at least 100 pages of text;
- a table of contents;
- an informative introduction perhaps with some historical remarks which should be accessible to readers unfamiliar with the topic treated;
- a subject index.

3. Category iii). Conference proceedings will be considered for publication provided that they are both of exceptional interest and devoted to a single topic. One (or more) expert participants will act as the scientific editor(s) of the volume. They select the papers which are suitable for inclusion and have them individually refereed as for a journal. Papers not closely related to the central topic are to be excluded. Organizers should contact Lecture Notes in Computational Science and Engineering at the planning stage.

In exceptional cases some other multi-author-volumes may be considered in this category.

4. Format. Only works in English are considered. They should be submitted in camera-ready form according to Springer-Verlag's specifications.
Electronic material can be included if appropriate. Please contact the publisher.
Technical instructions and/or LaTeX macros are available via http://www.springer.com/authors/book+authors?SGWID=0-154102-12-417900-0. The macros can also be sent on request.

General Remarks

Lecture Notes are printed by photo-offset from the master-copy delivered in camera-ready form by the authors. For this purpose Springer-Verlag provides technical instructions for the preparation of manuscripts. See also *Editorial Policy*.

Careful preparation of manuscripts will help keep production time short and ensure a satisfactory appearance of the finished book.

The following terms and conditions hold:

Categories i), ii), and iii):
Authors receive 50 free copies of their book. No royalty is paid. Commitment to publish is made by letter of intent rather than by signing a formal contract. Springer- Verlag secures the copyright for each volume.

For conference proceedings, editors receive a total of 50 free copies of their volume for distribution to the contributing authors.

All categories:
Authors are entitled to purchase further copies of their book and other Springer mathematics books for their personal use, at a discount of 33.3% directly from Springer-Verlag.

Addresses:

Timothy J. Barth
NASA Ames Research Center
NAS Division
Moffett Field, CA 94035, USA
e-mail: barth@nas.nasa.gov

Michael Griebel
Institut für Numerische Simulation
der Universität Bonn
Wegelerstr. 6
53115 Bonn, Germany
e-mail: griebel@ins.uni-bonn.de

David E. Keyes
Department of Applied Physics
and Applied Mathematics
Columbia University
200 S. W. Mudd Building
500 W. 120th Street
New York, NY 10027, USA
e-mail: david.keyes@columbia.edu

Risto M. Nieminen
Laboratory of Physics
Helsinki University of Technology
02150 Espoo, Finland
e-mail: rni@fyslab.hut.fi

Dirk Roose
Department of Computer Science
Katholieke Universiteit Leuven
Celestijnenlaan 200A
3001 Leuven-Heverlee, Belgium
e-mail: dirk.roose@cs.kuleuven.ac.be

Tamar Schlick
Department of Chemistry
Courant Institute of Mathematical
Sciences
New York University
and Howard Hughes Medical Institute
251 Mercer Street
New York, NY 10012, USA
e-mail: schlick@nyu.edu

Mathematics Editor at Springer:
Martin Peters
Springer-Verlag
Mathematics Editorial IV
Tiergartenstrasse 17
D-69121 Heidelberg, Germany
Tel.: *49 (6221) 487-8185
Fax: *49 (6221) 487-8355
e-mail: martin.peters@springer.com

Lecture Notes in Computational Science and Engineering

1. D. Funaro, *Spectral Elements for Transport-Dominated Equations.*
2. H. P. Langtangen, *Computational Partial Differential Equations.* Numerical Methods and Diffpack Programming.
3. W. Hackbusch, G. Wittum (eds.), *Multigrid Methods V.*
4. P. Deuflhard, J. Hermans, B. Leimkuhler, A. E. Mark, S. Reich, R. D. Skeel (eds.), *Computational Molecular Dynamics: Challenges, Methods, Ideas.*
5. D. Kröner, M. Ohlberger, C. Rohde (eds.), *An Introduction to Recent Developments in Theory and Numerics for Conservation Laws.*
6. S. Turek, *Efficient Solvers for Incompressible Flow Problems.* An Algorithmic and Computational Approach.
7. R. von Schwerin, ***M**ulti **B**ody **S**ystem **SIM**ulation.* Numerical Methods, Algorithms, and Software.
8. H.-J. Bungartz, F. Durst, C. Zenger (eds.), *High Performance Scientific and Engineering Computing.*
9. T. J. Barth, H. Deconinck (eds.), *High-Order Methods for Computational Physics.*
10. H. P. Langtangen, A. M. Bruaset, E. Quak (eds.), *Advances in Software Tools for Scientific Computing.*
11. B. Cockburn, G. E. Karniadakis, C.-W. Shu (eds.), *Discontinuous Galerkin Methods.* Theory, Computation and Applications.
12. U. van Rienen, *Numerical Methods in Computational Electrodynamics.* Linear Systems in Practical Applications.
13. B. Engquist, L. Johnsson, M. Hammill, F. Short (eds.), *Simulation and Visualization on the Grid.*
14. E. Dick, K. Riemslagh, J. Vierendeels (eds.), *Multigrid Methods VI.*
15. A. Frommer, T. Lippert, B. Medeke, K. Schilling (eds.), *Numerical Challenges in Lattice Quantum Chromodynamics.*
16. J. Lang, *Adaptive Multilevel Solution of Nonlinear Parabolic PDE Systems.* Theory, Algorithm, and Applications.
17. B. I. Wohlmuth, *Discretization Methods and Iterative Solvers Based on Domain Decomposition.*
18. U. van Rienen, M. Günther, D. Hecht (eds.), *Scientific Computing in Electrical Engineering.*
19. I. Babuška, P. G. Ciarlet, T. Miyoshi (eds.), *Mathematical Modeling and Numerical Simulation in Continuum Mechanics.*
20. T. J. Barth, T. Chan, R. Haimes (eds.), *Multiscale and Multiresolution Methods.* Theory and Applications.
21. M. Breuer, F. Durst, C. Zenger (eds.), *High Performance Scientific and Engineering Computing.*
22. K. Urban, *Wavelets in Numerical Simulation.* Problem Adapted Construction and Applications.

23. L. F. Pavarino, A. Toselli (eds.), *Recent Developments in Domain Decomposition Methods.*

24. T. Schlick, H. H. Gan (eds.), *Computational Methods for Macromolecules: Challenges and Applications.*

25. T. J. Barth, H. Deconinck (eds.), *Error Estimation and Adaptive Discretization Methods in Computational Fluid Dynamics.*

26. M. Griebel, M. A. Schweitzer (eds.), *Meshfree Methods for Partial Differential Equations.*

27. S. Müller, *Adaptive Multiscale Schemes for Conservation Laws.*

28. C. Carstensen, S. Funken, W. Hackbusch, R. H. W. Hoppe, P. Monk (eds.), *Computational Electromagnetics.*

29. M. A. Schweitzer, *A Parallel Multilevel Partition of Unity Method for Elliptic Partial Differential Equations.*

30. T. Biegler, O. Ghattas, M. Heinkenschloss, B. van Bloemen Waanders (eds.), *Large-Scale PDE-Constrained Optimization.*

31. M. Ainsworth, P. Davies, D. Duncan, P. Martin, B. Rynne (eds.), *Topics in Computational Wave Propagation.* Direct and Inverse Problems.

32. H. Emmerich, B. Nestler, M. Schreckenberg (eds.), *Interface and Transport Dynamics.* Computational Modelling.

33. H. P. Langtangen, A. Tveito (eds.), *Advanced Topics in Computational Partial Differential Equations.* Numerical Methods and Diffpack Programming.

34. V. John, *Large Eddy Simulation of Turbulent Incompressible Flows.* Analytical and Numerical Results for a Class of LES Models.

35. E. Bänsch (ed.), *Challenges in Scientific Computing - CISC 2002.*

36. B. N. Khoromskij, G. Wittum, *Numerical Solution of Elliptic Differential Equations by Reduction to the Interface.*

37. A. Iske, *Multiresolution Methods in Scattered Data Modelling.*

38. S.-I. Niculescu, K. Gu (eds.), *Advances in Time-Delay Systems.*

39. S. Attinger, P. Koumoutsakos (eds.), *Multiscale Modelling and Simulation.*

40. R. Kornhuber, R. Hoppe, J. Périaux, O. Pironneau, O. Wildlund, J. Xu (eds.), *Domain Decomposition Methods in Science and Engineering.*

41. T. Plewa, T. Linde, V.G. Weirs (eds.), *Adaptive Mesh Refinement – Theory and Applications.*

42. A. Schmidt, K.G. Siebert, *Design of Adaptive Finite Element Software.* The Finite Element Toolbox ALBERTA.

43. M. Griebel, M.A. Schweitzer (eds.), *Meshfree Methods for Partial Differential Equations II.*

44. B. Engquist, P. Lötstedt, O. Runborg (eds.), *Multiscale Methods in Science and Engineering.*

45. P. Benner, V. Mehrmann, D.C. Sorensen (eds.), *Dimension Reduction of Large-Scale Systems.*

46. D. Kressner, *Numerical Methods for General and Structured Eigenvalue Problems.*

47. A. Boriçi, A. Frommer, B. Joó, A. Kennedy, B. Pendleton (eds.), *QCD and Numerical Analysis III.*

48. F. Graziani (ed.), *Computational Methods in Transport.*

49. B. Leimkuhler, C. Chipot, R. Elber, A. Laaksonen, A. Mark, T. Schlick, C. Schütte, R. Skeel (eds.), *New Algorithms for Macromolecular Simulation.*

50. M. Bücker, G. Corliss, P. Hovland, U. Naumann, B. Norris (eds.), *Automatic Differentiation: Applications, Theory, and Implementations.*

51. A.M. Bruaset, A. Tveito (eds.), *Numerical Solution of Partial Differential Equations on Parallel Computers.*

52. K.H. Hoffmann, A. Meyer (eds.), *Parallel Algorithms and Cluster Computing.*

53. H.-J. Bungartz, M. Schäfer (eds.), *Fluid-Structure Interaction.*

54. J. Behrens, *Adaptive Atmospheric Modeling.*

55. O. Widlund, D. Keyes (eds.), *Domain Decomposition Methods in Science and Engineering XVI.*

56. S. Kassinos, C. Langer, G. Iaccarino, P. Moin (eds.), *Complex Effects in Large Eddy Simulations.*

57. M. Griebel, M.A Schweitzer (eds.), *Meshfree Methods for Partial Differential Equations III.*

58. A.N. Gorban, B. Kégl, D.C. Wunsch, A. Zinovyev (eds.), *Principal Manifolds for Data Visualization and Dimension Reduction.*

59. H. Ammari (ed.), *Modeling and Computations in Electromagnetics: A Volume Dedicated to Jean-Claude Nédélec.*

60. U. Langer, M. Discacciati, D. Keyes, O. Widlund, W. Zulehner (eds.), *Domain Decomposition Methods in Science and Engineering XVII.*

61. T. Mathew, *Domain Decomposition Methods for the Numerical Solution of Partial Differential Equations.*

62. F. Graziani (ed.), *Computational Methods in Transport: Verification and Validation.*

63. M. Bebendorf, *Hierarchical Matrices.* A Means to Efficiently Solve Elliptic Boundary Value Problems.

64. C.H. Bischof, H.M. Bücker, P. Hovland, U. Naumann, J. Utke (eds.), *Advances in Automatic Differentiation.*

65. M. Griebel, M.A. Schweitzer (eds.), *Meshfree Methods for Partial Differential Equations IV.*

For further information on these books please have a look at our mathematics catalogue at the following URL: www.springer.com/series/3527

Monographs in Computational Science and Engineering

1. J. Sundnes, G.T. Lines, X. Cai, B.F. Nielsen, K.-A. Mardal, A. Tveito, *Computing the Electrical Activity in the Heart.*

For further information on this book, please have a look at our mathematics catalogue at the following URL: www.springer.com/series/7417

Texts in Computational Science and Engineering

1. H. P. Langtangen, *Computational Partial Differential Equations.* Numerical Methods and Diffpack Programming. 2nd Edition
2. A. Quarteroni, F. Saleri, *Scientific Computing with MATLAB and Octave.* 2nd Edition
3. H. P. Langtangen, *Python Scripting for Computational Science*. 3rd Edition
4. H. Gardner, G. Manduchi, *Design Patterns for e-Science.*
5. M. Griebel, S. Knapek, G. Zumbusch, *Numerical Simulation in Molecular Dynamics.*

For further information on these books please have a look at our mathematics catalogue at the following URL: www.springer.com/series/5151

Printing: Krips bv, Meppel, The Netherlands
Binding: Stürtz, Würzburg, Germany